AF473891

New Developments in Algebraic Geometry, Integrable Systems and Mirror Symmetry (RIMS, Kyoto, 2008)

ADVANCED STUDIES
IN PURE MATHEMATICS 59

Chief Editor of the Series: Sadayoshi Kojima (Tokyo Institute of Technology)

New Developments in Algebraic Geometry, Integrable Systems and Mirror Symmetry (RIMS, Kyoto, 2008)

Edited by

Masa-Hiko Saito (Kobe University)(Chief)
Shinobu Hosono (The University of Tokyo)
Kōta Yoshioka (Kobe University)

Mathematical Society of Japan

This book was typeset by AMS-TEX and AMS-LATEX, the TEX macro systems of the American Mathematical Society, together with the style files aspm.sty *and* aspmfm.sty *for AMS-TEX written by Dr. Chiaki Tsukamoto and* aspmproc.sty *for AMS-LATEX written by Dr. Akihiro Munemasa and* aspm.cls *for AMS-LATEX provided by Livretech Co., Ltd.*

TEX is a trademark of the American Mathematical Society.

Edited by the Mathematical Society of Japan.

Published by the Mathematical Society of Japan.

Distributed by the Mathematical Society of Japan, the American Mathematical Society and the World Scientific Publishing Co., Ltd.
Distributed exclusively in North America by the American Mathematical Society.

Advanced Studies in Pure Mathematics 59
ISBN 978-4-931469-62-4

PRINTED IN JAPAN
by Livretech Co., Ltd.

2000 Mathematics Subject Classification.
Primary 14N.
Secondary 14D, 14H, 14J, 35Q, 53D, 83B20.

Preface

For the last two decades, many important developments and interactions in algebraic geometry and integrable systems have been arising from the ideas of mirror symmetry. The RIMS International Project Research 2007, "Mirror Symmetry and Topological Field Theory" was held in the Japanese fiscal year April 2007–March 2008 at RIMS, Kyoto University to explore recent further progress.

As a part of the activities, the conference "New Developments in Algebraic Geometry, Integrable Systems and Mirror Symmetry" was held in Kyoto on January 7–11, 2008. Also, prior to the conference, a workshop on "Quantum Cohomology and Mirror Symmetry" was held at Kobe University on January 4–5, 2008. During the conferences, there were research talks on new results in algebraic geometry, integrable systems, Gromov–Witten theory, topological field theory and symplectic geometry. Also, it was helpful for the participants that we had good expository lectures on developing theories related to these subjects.

This volume is the outcome of those conferences and consists of 12 contributed papers from participants. We would like to thank the authors for their contributions and the referees for their invaluable efforts. Special thanks are due as well to the secretaries at Kobe University for editorial work.

March 2010

The Editors
Masa-Hiko Saito (Chief)
Shinobu Hosono
Kōta Yoshioka

All papers in this volume have been refereed and are in final form. No version of any of them will be submitted for publication elsewhere.

New developments in Algebraic Geometry, Integrable Systems and Mirror symmetry

The conference "New developments in Algebraic Geometry, Integrable Systems and Mirror Symmetry" was held at RIMS (Research Institute for Mathematical Sciences), Kyoto University for January 7–11, 2008.

The main purpose of the conference was to explore those recent developments and interactions of various mathematical fields, such as algebraic geometry, integrable systems, Gromov–Witten theory and symplectic geometry, that have come from ideas in mirror symmetry.

There were 97 participants in the conference including 18 from overseas, and there were 19 one-hour lectures by invited speakers, followed by many discussions. We thank all the speakers and participants for their contributions, which made the conference very active and productive.

This conference was held as a part of activities of the RIMS International Project Research 2007, "Mirror Symmetry and Topological Field Theory" (April, 2007–March, 2008). We would like to thank RIMS for their support in organizing the conference. Special thanks are due to the secretaries of RIMS for their kind arrangements for participants. The conference was also supported by the Grant-in-Aids for Scientific Research [1] of the Japan Society for the Promotion of Science. We are grateful for their generous financial support.

March 2010

Organizing Committee
Kenji Fukaya
Shinobu Hosono
Masa-Hiko Saito
Kōta Yoshioka

[1]S-18104001 (K. Fukaya), S-19104002 (M.-H. Saito), B-18340010 (K. Yoshioka), C-18540014 (S. Hosono)

Program

Dates: January 7–11, 2008

Venue: Room 420, Research Institute for Mathematical Sciences, Kyoto University, Kyoto, Japan

January 7 (Monday)

9:00–9:30 Registration

9:30–10:30 **R. Pandharipande** (Princeton Univ.)
Open descendent integrals

10:50–11:50 **J. Li** (Stanford Univ.)
Towards high genus GW-invariants of quintic Calabi–Yau threefold

13:30–14:30 **N. C. Leung** (Chin. Univ. of Hong Kong)
On the SYZ mirror transformation

14:50–15:50 **H. Iritani** (Kyushu Univ.)
Wall-crossing in toric Gromov–Witten theory

16:10–17:10 **M-H. Saito** (Kobe Univ.)
Moduli spaces of linear connections and Riemann–Hilbert correspondences

January 8 (Tuesday)

9:30–10:30 **K. Fukaya** (Kyoto Univ.)
Floer theory of orbits in toric manifolds

10:50–11:50 **M. Gross** (Univ. of California, San Diego)
The tropical vertex

13:30–14:30 **A. Takahashi** (Osaka Univ.)
Mirror symmetry of isolated hypersurface singularities

14:50–15:50 **T. Pantev** (Univ. of Pennsylvania)
Generalized Hodge structures and mirror symmetry

16:10–17:10 **Y. Toda** (Univ. of Tokyo)
Limit stable objects on Calabi–Yau 3-folds

January 9 (Wednesday)

9:30–10:30 **N. Nekrasov** (IHÉS)
Lessons from low dimensional topological strings

10:50–11:50 **R. Donagi** (Univ. of Pennsylvania)
Hitchin systems, mirror symmetry, and geometric Langlands duality

January 10 (Thursday)

9:30–10:30 : **C. Sabbah** (École Polytechnique)
Quantum cohomology of the Grassmannian and alternate Thom–Sebastiani

10:50–11:50 **T. Mochizuki** (Kyoto Univ.)
On wild harmonic bundles

13:30–14:30 **C. Hertling** (Univ. Mannheim)
Hypersurface singularities and tt^* geometry

14:50–15:50 **B. Kim (KIAS)**
A compactification of the space of maps from curves

16:10–17:10 **K. Takasaki** (Kyoto Univ.)
Integrable structure in melting crystal model of 5D gauge theory

January 11 (Friday)

9:30–10:30 **J. Stienstra** (Utrecht Univ.)
Two-variable hypergeometric systems and dessins d'enfants

10:50–11:50 **M. Mulase** (Univ. of California, Davis)
Matrix integral approach to character varieties

List of Participants

(The affiliation at the time of the conference)

Abe, Takeshi (RIMS, Kyoto University)
Agou, Takashi (Tokyo University of Science)
Akahori, Takao (University of Hyogo)
Awata, Hidetoshi (Nagoya University)
Chan, Kwokwai (Chinese University of Hong Kong)
Chekonov, Yuri (RIMS, Kyoto University/Moscow)
Donagi, Ron (University of Pennsylvania)
Eguchi, Tohru (YITP, Kyoto University)
Forbes, Brian (RIMS, Kyoto University)
Fukaya, Kenji (Kyoto University)
Futaki, Masahiro (The University of Tokyo)
Gross, Mark (University of California, San Diego)
Guzzetti, Davide (RIMS, Kyoto University)
Hanamura, Masaki (Tohoku University)
Haraoka, Yoshishige (Kumamoto University)
Hertling, Claus (Universität Mannheim)
Hikami, Shinobu (The University of Tokyo)
Hosono, Shinobu (The University of Tokyo)
Inoue, Rei (The University of Tokyo)
Iritani, Hiroshi (Kyushu University)
Ishii, Akira (Hiroshima University)
Iwao, Kosuke (The University of Tokyo)
Jinzenji, Masao (Hokkaido University)
Kajiura, Hiroshige (RIMS, Kyoto University)
Kajiwara, Takeshi (Yokohama National University)
Kanno, Hiroaki (Nagoya University)
Kato, Tsuyoshi (Kyoto University)
Kawaguchi, Shu (Osaka University)
Kawai, Toshiya (RIMS, Kyoto University)
Kim, Bumsing (Korea Institute for Advanced Study)
Kim, Hoil (Kyungpook Nat'l University)
Kimura, Yoshiyuki (Kyoto University)
Kobayashi Masanori (Tokyo Metoropolitan University)
Kondo, Satoshi (RIMS, Kyoto University)
Konishi, Yukiko (The University of Tokyo)
Kozako, Noriyasu (International Christian University)
Leung, Naichung Conan (Chinese University of Hong Kong)
Li, Jun (Stanford University)
Li, Weiping (Hong Kong University of Science and Technology)

Manabe, Masahide (Nagoya University)
Marelli, Giovanni (Korea Institute for Advanced Study)
Masuda, Tetsu (Kobe University)
Mckay, John (Concordia University)
Minabe, Satoshi (Hokkaido University)
Miyajima, Kimio (Kagoshima University)
Mochizuki, Takuro (Kyoto University)
Mori, Shigefumi (RIMS, Kyoto University)
Moriwaki, Atsushi (Kyoto University)
Moriya, Syunji (Kyoto University)
Moskovich, Daniel David (RIMS, Kyoto University)
Mukai, Shigeru (RIMS, Kyoto University)
Mulase, Motohico (University of California, Davis)
Nagao, Kentaro (Kyoto University)
Nakajima, Haruhisa (Josai University)
Nakayashiki, Atsushi (Kyushu University)
Naruse, Hiroshi (Okayama University)
Nasu, Hirokazu (RIMS, Kyoto University)
Nekrasov, Nikita (Institut des Hautes Études Scientifiques (IHÉS))
Nishinou, Takeo (Tohoku University)
Nohara, Yuichi (Tohoku University)
Ochiai, Hiroyuki (Nagoya University)
Ohkawa, Ryo (Tokyo Institute of Technology)
Ohta, Hiroshi (Nagoya University)
Okai, Takayuki (Osaka Sangyo University)
Ono, Kaoru (Hokkaido University)
Pandharipande, Rahul (Princeton University)
Pantev, Tony (University of Pennsylvania)
Sabbah, Claude (École Polytechnique)
Saito, Kyoji (RIMS, Kyoto University)
Saito, Masa-Hiko (Kobe University)
Satake, Ikuo (Osaka University)
Sato, Eiichi (Kyushu University)
Shimizu, Yuji (International Christian University)
Stienstra, Jan (Utrecht University)
Suzuki, Norio (Kitami Institute of Technology)
Suzuki, Sakie (RIMS, Kyoto University)
Tachiki, Hisato (Nagoya University)
Takahashi, Atsushi (Osaka University)
Takai, Hiroshi (Tokyo Metropolitan University)
Takasaki, Kanehisa (Kyoto University)
Tamagawa, Akio (RIMS, Kyoto University)

Toda, Yukinobu (The University of Tokyo)
Tokunaga, Hiroo (Tokyo Metropolitan University)
Tseng, Hsian-Hua (University of Wisconsin-Madison)
Tsuchiya, Akihiro (IPMU, The University of Tokyo)
Tsushima, Ryuji (Meiji University)
Ushio, Nozomi (Kobe University)
van Koert, Otto Ferdinand Bernard (Hokkaido University)
Vidunas, Raimundas (Kobe University)
Yamakawa, Daisuke (Kyoto University)
Yamasaki, Masahito (The University of Tokyo)
Yamasaki, Takao (Tohoku University)
Yanagida, Shintaro (Kobe University)
Yoshikawa, Kenichi (The University of Tokyo)
Yoshinaga, Masahiko (Kobe University)
Yoshioka, Kota (Kobe University)
Yotsutani, Naoto (Nagoya University)

CONTENTS

Advanced Studies in Pure Mathematics 59, 2010
New Developments in Algebraic Geometry,
Integrable Systems and Mirror Symmetry (Kyoto, 2008)
pp. 1–30

On SYZ mirror transformations

Kwokwai Chan and Naichung Conan Leung

Abstract.

In this expository paper, we discuss how Fourier–Mukai-type transformations, which we call SYZ mirror transformations, can be applied to provide a geometric understanding of the mirror symmetry phenomena for semi-flat Calabi–Yau manifolds and toric Fano manifolds. We also speculate the possible applications of these transformations to other more general settings.

§1. Introduction

In 1996, Strominger, Yau and Zaslow suggested, in their groundbreaking work [40], a geometric approach to the mirror symmetry for Calabi–Yau manifolds. Roughly speaking, the *Strominger–Yau–Zaslow (SYZ) Conjecture* asserts that any Calabi–Yau manifold X should admit a fibration by special Lagrangian tori and the mirror of X, which is another Calabi–Yau manifold Y, can be obtained by T-duality, i.e. dualizing the special Lagrangian torus fibration of X. Moreover, the symplectic geometry (A-model) of X should be interchanged with the complex geometry (B-model) of Y, and vice versa, through fiberwise Fourier–Mukai-type transformations, suitably modified by quantum corrections. These transformations are called *SYZ mirror transformations* and they will be the theme in this article.

Much work has been done on the SYZ Conjecture. Following the work of Hitchin [24], Leung–Yau–Zaslow [32] and Leung [31] explained successfully and neatly the mirror symmetry for semi-flat Calabi–Yau manifolds by using *semi-flat SYZ mirror transformations*. These are

Received July 16, 2008.
Revised May 5, 2009.
2000 *Mathematics Subject Classification.* Primary 14J32; Secondary 14N35, 14J45.
Key words and phrases. Mirror symmetry, SYZ conjecture, SYZ transformation, toric Fano manifold, Landau–Ginzburg model, quantum cohomology, Jacobian ring, Lagrangian submanifold, holomorphic vector bundle.

honest fiberwise real Fourier–Mukai transformations. The advantage in this case is the absence of quantum corrections by holomorphic curves and discs. This is due to the fact that the special Lagrangian torus fibrations on semi-flat Calabi–Yau manifolds do not admit singularities, and, accordingly, the bases are smooth affine manifolds.

To deal with general compact Calabi–Yau manifolds, however, one cannot avoid singularities in Lagrangian torus fibrations, and hence singularities in the base affine manifolds. Consequently, quantum corrections will come into play. This necessitates the study of moduli spaces of special Lagrangian submanifolds and affine manifolds with singularities, which makes the subject much more sophisticated and difficult. Nevertheless, the recent progress made by Gross and Siebert [21], after earlier works of Fukaya [13] and Kontsevich–Soibelman [30], was doubtlessly a significant step towards establishing the SYZ Conjecture for general compact Calabi–Yau manifolds.[1]

On the other hand, mirror symmetry phenomena have also been observed for Fano manifolds (and other classes of manifolds or orbifolds as well). The mirror of a Fano manifold $\bar{X}$ is predicted by Physicists to be given by a *Landau–Ginzburg model*, which is a pair (Y, W), consisting of a non-compact Kähler manifold Y and a holomorphic function $W : Y \to \mathbb{C}$ called the *superpotential*. A very important class of examples is provided by toric Fano manifolds. In this case, the mirror manifold Y is biholomorphic to (a bounded domain of) $(\mathbb{C}^*)^n$ and the superpotential W is a Laurent polynomial which can be written down explicitly. Ample evidences have been found in this toric Fano case; in particular, Cho and Oh [9] proved that the superpotential can be computed in terms of the counting of Maslov index two holomorphic discs in $\bar{X}$ with boundary in Lagrangian torus fibers. In [4], Auroux applied the SYZ philosophy to the study of the mirror symmetry for a compact Kähler manifold equipped with an anticanonical divisor. This is a generalization of the mirror symmetry for Fano manifolds, and, again, the mirror is given by a Landau–Ginzburg model. Auroux also made an attempt to compute the superpotential in terms of the counting of holomorphic discs, and analyzed the resulting wall-crossing phenomena. In [7], we studied the mirror symmetry for toric Fano manifolds, again through the SYZ approach, and we constructed and applied *SYZ mirror transformations for toric Fano manifolds* to explain various geometric results implied by mirror symmetry.

---

[1] We should mention that the Gross–Siebert program is expected to work for non-Calabi–Yau manifolds (e.g. Fano manifolds) as well.

A brief explanation of the results in [7] is now in order; for more details, see Section 3. Let $\bar{X}$ be a toric Fano manifold, i.e. a smooth projective toric variety such that the anticanonical line bundle $K_{\bar{X}}$ is ample. Let $\omega_{\bar{X}}$ be a toric Kähler structure on $\bar{X}$. The moment map $\mu_{\bar{X}} : \bar{X} \to \bar{P}$ of the Hamiltonian T^n-action on $(\bar{X}, \omega_{\bar{X}})$ is a natural Lagrangian torus fibration. Here $\bar{P} \subset \mathbb{R}^n$ is a polytope defining $(\bar{X}, \omega_{\bar{X}})$. The restriction of the moment map to the open dense T^n-orbit $X \cong (\mathbb{C}^*)^n \subset \bar{X}$ is a Lagrangian torus bundle $\mu_X = \mu_{\bar{X}}|_X : X \to P$, where P denotes the interior of the polytope $\bar{P}$. Our first result in [7] showed that the mirror manifold Y is nothing but the *SYZ mirror manifold* of X, i.e. the total space of the torus bundle dual to $\mu_X : X \to P$ (see Proposition 3.1).[2] Furthermore, the semi-flat SYZ transformation $\mathcal{F}^{\mathrm{sf}}$ takes the exponential of ($\sqrt{-1}$ times) the symplectic structure $\omega_X = \omega_{\bar{X}}|_X$ on X to the holomorphic volume form Ω_Y on Y.[3] Note that Ω_Y determines a complex structure on Y by declaring that a 1-form α is a $(1,0)$-form if and only if $\alpha \lrcorner \Omega_Y = 0$. This part of the mirror symmetry does not involve quantum corrections.

To get the superpotential W, however, we need to take into account the quantum corrections due to the anticanonical toric divisor $D_\infty = \bar{X} \setminus X$, which we have ignored above. Before doing that, we first take a digression to a well-known construction. For a simply connected symplectic manifold (M, ω), let $\mathfrak{L}M$ be the free loop space, i.e. the space of smooth maps $\gamma : S^1 \to M$. The symplectic structure on M induces a symplectic structure on $\mathfrak{L}M$ which will also be denoted by ω. The action functional defined by

$$H(\gamma) := \frac{1}{2\pi} \int_{D_\gamma} \omega,$$

where D_γ is a disk contracting γ, becomes a well-defined function on the universal covering $\widetilde{\mathfrak{L}M}$ of the free loop space $\mathfrak{L}M$. The group of deck transformations is $H_2(M, \mathbb{Z})$. It is not hard to see that H is the moment map for the built-in S^1-action on $\widetilde{\mathfrak{L}M}$, and the gradient flow lines of H are (pseudo-)holomorphic cylinders if we fix a compatible (almost) complex structure on M. Tentatively, the quantum cohomology (or Floer cohomology) is the S^1-equivariant Morse–Witten cohomology of the moment map H on $\widetilde{\mathfrak{L}M}$. However, the fact that $\widetilde{\mathfrak{L}M}$ is infinite dimensional poses severe difficulties in implementing this idea.

---

[2]More precisely, the SYZ mirror manifold is a bounded domain in the mirror manifold Y predicted by Physicists.

[3]Throughout this paper, we assume that the B-field is zero.

One of our discoveries in [7] was that a finite dimensional subspace of $\mathcal{L}M$ is enough to capture the quantum corrections and recover the quantum cohomology, in the case when $M = \bar{X}$ is a toric Fano manifold. Consider the subspace LX of $\mathcal{L}\bar{X}$ consisting of those loops which are *geodesic* in the Lagrangian torus fibers (with respect to the flat metrics) of the moment map $\mu_{\bar{X}} : \bar{X} \to \bar{P}$. We consider the function Ψ on LX defined by $\Psi(\gamma) = \exp(-H(\gamma))$ if γ bounds a Maslov index two holomorphic disc and $\Psi(\gamma) = 0$ otherwise. The function $\Psi : LX \to \mathbb{C}$, as an object in the A-model of $\bar{X}$, turns out to be the mirror of the superpotential W. In [7], we constructed the SYZ mirror transformation $\mathcal{F}$ for the toric Fano manifold $\bar{X}$, and showed that the SYZ mirror transformation of Ψ is precisely the B-model superpotential W. Moreover, by incorporating the symplectic structure ω_X and the holomorphic volume form Ω_Y, we proved that

$$\begin{aligned} \mathcal{F}(e^{\sqrt{-1}\omega_X+\Psi}) &= e^W\Omega_Y, \\ \mathcal{F}^{-1}(e^W\Omega_Y) &= e^{\sqrt{-1}\omega_X+\Psi}, \end{aligned}$$

where $\mathcal{F}^{-1}$ is the inverse SYZ mirror transformation (see Theorem 3.1). Hence, the *corrected* symplectic structure on X and the complex structure on (Y, W) are interchanged by the SYZ mirror transformation. On the other hand, we identified the small quantum cohomology ring $QH^*(\bar{X})$ of $\bar{X}$ with an algebra of functions on LX, and realized the quantum product as a *convolution product* (see Proposition 3.2). Then, we showed that the SYZ mirror transformation $\mathcal{F}$ exhibits a natural isomorphism between $QH^*(\bar{X})$ and the Jacobian ring $Jac(W)$ of the superpotential W, which *takes the quantum product (now as a convolution product) to the ordinary product of Laurent polynomials, just as what classical Fourier series do* (see Theorem 3.2). We conclude that the mirror symmetry for toric Fano manifolds is nothing but a Fourier transformation!

The main goal of this article is to popularize the use of SYZ mirror transformations in exploring mirror symmetry phenomena. In Section 2, we review the use of semi-flat SYZ mirror transformations in the study of the mirror symmetry for semi-flat Calabi–Yau manifolds, where quantum corrections are absent. This is the toy case which lays the basis for subsequent development in the investigation of the SYZ Conjecture. Section 3 discusses the mirror symmetry for toric Fano manifolds, where quantum corrections arise due to the anticanonical toric divisor. Following [7], we demonstrate how to construct and apply SYZ mirror transformations in this case. The final section contains a brief discussion of possible generalizations.

Acknowledgments. The authors are grateful to the organizers of the conference "New developments in Algebraic Geometry, Integrable Systems and Mirror Symmetry" held in Kyoto University in January 2008 for giving them an opportunity to participate in such a stimulating and fruitful event. Thanks are also due to Hiroshi Iritani and Cheol-Hyun Cho for many useful discussions. Finally, we thank the referee for several helpful comments. K.-W. C. was partially supported by Harvard University and the Croucher Foundation Fellowship. N.-C. L. was partially supported by RGC grants from the Hong Kong Government.

§2. SYZ mirror transformations without corrections

In this section, we review the construction of SYZ mirror transformations for semi-flat Calabi–Yau manifolds and see how they were applied in the study of semi-flat mirror symmetry.

2.1. Semi-flat SYZ mirror transformations

Denote by $N \cong \mathbb{Z}^n$ a rank-n lattice and $M = \mathrm{Hom}(N, \mathbb{Z})$ the dual lattice. Let $D \subset M_{\mathbb{R}} = M \otimes_{\mathbb{Z}} \mathbb{R}$ be a convex domain.[4] Then the tangent bundle $TD = D \times \sqrt{-1}M_{\mathbb{R}}$ is naturally a complex manifold with complex coordinates $x_j + \sqrt{-1}y_j$, $j = 1, \ldots, n$, where $x_1, \ldots, x_n \in \mathbb{R}$ and $y_1, \ldots, y_n \in \mathbb{R}$ are respectively the base coordinates on D and fiber coordinates on $M_{\mathbb{R}}$. We have the standard holomorphic volume form $\Omega_{TD} = d(x_1 + \sqrt{-1}y_1) \wedge \ldots \wedge d(x_n + \sqrt{-1}y_n)$ on TD. By taking fiberwise quotient by the lattice $M \subset M_{\mathbb{R}}$, we can compactify the fiber directions to give the complex manifold

$$Y = TD/M = D \times \sqrt{-1}T_M,$$

where T_M denotes the torus $M_{\mathbb{R}}/M$. The complex coordinates on Y are naturally given by $z_j = \exp(x_j + \sqrt{-1}y_j)$, $j = 1, \ldots, n$, where $y_1, \ldots, y_n \in \mathbb{R}/2\pi\mathbb{Z}$ are now coordinates on T_M. Note that Y is biholomorphic to an open part of $(\mathbb{C}^*)^n = TM_{\mathbb{R}}/M$. The projection to D is a torus bundle, which we denote by $\nu_Y : Y \to D$. The holomorphic n-form Ω_{TD} descends to give the holomorphic volume form

$$\Omega_Y = \frac{dz_1}{z_1} \wedge \ldots \wedge \frac{dz_n}{z_n}$$

[4]More generally, instead of a convex domain, one may consider a smooth affine manifold.

on Y. As mentioned in the introduction, Ω_Y in turn determines the complex structure on Y: a 1-form α is of $(1,0)$-type if and only if $\alpha \lrcorner \Omega_Y = 0$. Further, if ϕ is an elliptic solution of the *real Monge–Ampère equation*

$$\det\left(\frac{\partial^2\phi}{\partial x_j \partial x_k}\right) = const,$$

then the Kähler form

$$\omega_Y := \sqrt{-1}\partial\bar{\partial}\phi = \sum_{j,k} \phi_{jk} dx_j \wedge dy_k,$$

with ϕ_{jk} denoting $\frac{\partial^2\phi}{\partial x_j \partial x_k}$, gives a Calabi–Yau metric on Y, and

$$\nu_Y : Y \to D$$

becomes a special Lagrangian torus bundle (*SYZ fibration*). In summary, we have the following structures on the complex n-dimensional semi-flat Calabi–Yau manifold Y:

Riemannian metric	$g_Y = \sum_{j,k} \phi_{jk}(dx_j \otimes dx_k + dy_j \otimes dy_k)$
Holomorphic volume form	$\Omega_Y = \bigwedge_{j=1}^n (dx_j + \sqrt{-1} dy_j)$
Symplectic form	$\omega_Y = \sum_{j,k} \phi_{jk} dx_j \wedge dy_k$
SYZ fibration	$\nu_Y : Y \to D$

As suggested in the monumental work of Strominger–Yau–Zaslow [40], the mirror of Y, which is another Calabi–Yau manifold we denote by X, should be given by the moduli space of pairs (L, ∇), where L is a special Lagrangian torus fiber in Y, and ∇ is a flat $U(1)$-connection on the trivial complex line bundle $L \times \mathbb{C} \to L$. This is nothing but the total space of the torus fibration $\mu_X : X = D \times \sqrt{-1}T_N \to D$, where $T_N = N_\mathbb{R}/N = (T_M)^\vee$ and $N_\mathbb{R} = N \otimes_\mathbb{Z} \mathbb{R}$, which is dual to $\nu_Y : Y \to D$. This is called *T-duality* in physics. Furthermore, X can naturally be viewed as the fiberwise quotient of the cotangent bundle $T^*D = D \times \sqrt{-1}N_\mathbb{R}$ by the lattice $N \subset N_\mathbb{R}$. In particular, the standard symplectic form $\omega_{T^*D} = \sum_{j=1}^n dx_j \wedge du_j$ descends to give a symplectic form

$$\omega_X = \sum_{j=1}^n dx_j \wedge du_j$$

on $X = T^*D/N$, where $u_1, \ldots, u_n \in \mathbb{R}/2\pi\mathbb{Z}$ are coordinates on T_N. Through the metric

$$g_X = \sum_{j,k} (\phi_{jk} dx_j \otimes dx_k + \phi^{jk} du_j \otimes du_k),$$

where (ϕ^{jk}) is the inverse matrix of (ϕ_{jk}), we obtain a complex structure on X with complex coordinates given by $d\log(w_j) = \sum_{k=1}^n \phi_{jk}dx_k + \sqrt{-1}du_j$. There is a corresponding holomorphic volume form which can be written as

$$\Omega_X = \frac{dw_1}{w_1} \wedge \ldots \wedge \frac{dw_n}{w_n} = \bigwedge_{j=1}^{n} (\sum_{k=1}^{n} \phi_{jk}dx_k + \sqrt{-1}du_j).$$

The projection map

$$\mu_X : X \to D$$

now naturally becomes a special Lagrangian torus fibration. In summary, we have the following structures on X:

Riemannian metric	$g_X = \sum_{j,k}(\phi_{jk}dx_j \otimes dx_k + \phi^{jk}du_j \otimes du_k)$
Holomorphic volume form	$\Omega_X = \bigwedge_{j=1}^n(\sum_{k=1}^n \phi_{jk}dx_k + \sqrt{-1}du_j)$
Symplectic form	$\omega_X = \sum_{j=1}^n dx_j \wedge du_j$
SYZ fibration	$\mu_X : X \to D$

We remark that both Y and X admit natural Hamiltonian T^n-actions, but while $\mu : X \to D$ is a moment map for the T_N-action on X, $\nu : Y \to D$ is not a moment map for the T_M-action on Y. In fact, a moment map $\mu_Y : Y \to N_{\mathbb{R}}$ for the T_M-action on Y is given by

$$\mu_Y = L_\phi \circ \nu_Y,$$

where $L_\phi : D \to N_{\mathbb{R}}$ is the *Legendre transform* of ϕ defined by

$$L_\phi(x_1, \ldots, x_n) = d\phi_x = \Big(\frac{\partial\phi}{\partial x_1}, \ldots, \frac{\partial\phi}{\partial x_n}\Big).$$

Since ϕ is convex, the image $D^* = L_\phi(D)$ is an open convex subset of $(M_{\mathbb{R}})^* = N_{\mathbb{R}}$. (For this and other properties of the Legendre transform, see the book of Guillemin [22], Appendix 1.) In the action coordinates $x^1, \ldots, x^n$ of D^*, which are given by $\frac{\partial x^j}{\partial x_k} = \phi_{jk}$, the various structures on Y can be rewritten as:

Riemannian metric	$g_Y = \sum_{j,k}(\phi^{jk}dx^j \otimes dx^k + \phi_{jk}dy_j \otimes dy_k)$
Holomorphic volume form	$\Omega_Y = \bigwedge_{j=1}^n(\sum_{k=1}^n \phi^{jk}dx^k + \sqrt{-1}dy_j)$
Symplectic form	$\omega_Y = \sum_{j=1}^n dx^j \wedge dy_j$
SYZ fibration	$\mu_Y : Y \to D^*$

We call X the *SYZ mirror manifold* of Y (and vice versa) since the symplectic (resp. complex) geometry of X and the complex (resp. symplectic) geometry of Y are interchanged under the *semi-flat SYZ mirror transformation*, which is described as follows.

First recall that the dual torus $T_M = (T_N)^\vee$ can be interpreted as the moduli space of flat $U(1)$-connections on the trivial complex line bundle over T_N. More precisely, given $y = (y_1, \ldots, y_n) \in M_\mathbb{R} \cong \mathbb{R}^n$, we have a flat $U(1)$-connection

$$\nabla_y = d + \frac{\sqrt{-1}}{2} \sum_{j=1}^n y_j du_j$$

on $T_N \times \mathbb{C} \to \mathbb{C}$. The *holonomy* of ∇_y is given by the map

$$\mathrm{hol}_{\nabla_y} : N \to U(1), \ v \mapsto e^{-\sqrt{-1}\langle y,v \rangle}.$$

Hence, ∇_y is gauge equivalent to the trivial connection if and only if $y \in M \cong (2\pi\mathbb{Z})^n$. Moreover this construction gives all flat $U(1)$-connections on the trivial complex line bundle over T_N up to unitary gauge transformations. The universal $U(1)$-bundle, i.e. the Poincaré line bundle $\mathcal{P}$, is given by the trivial complex line bundle $(T_N \times T_M) \times \mathbb{C} \to T_N \times T_M$ equipped with the connection $d + \frac{\sqrt{-1}}{2} \sum_{j=1}^n (y_j du_j - u_j dy_j)$. The curvature of this connection is the two form

$$F = \sqrt{-1} \sum_{j=1}^n dy_j \wedge du_j.$$

Now consider the relative version of this picture. Let $X \times_D Y = D \times \sqrt{-1}(T_N \times T_M)$ be the fiber product of the dual torus bundles $\mu : X \to D$ and $\nu : Y \to D$. By abuse of notations, we still use $\mathcal{P}$ and $F = \sqrt{-1} \sum_{j=1}^n dy_j \wedge du_j \in \Omega^2(X \times_D Y)$ to denote the fiberwise universal line bundle and curvature two form respectively.

Definition 2.1. *The semi-flat SYZ mirror transformation*

$$\mathcal{F}^{sf} : \Omega^*(X) \to \Omega^*(Y)$$

is defined by

$$\begin{aligned} \mathcal{F}^{sf}(\alpha) &= \frac{1}{(2\pi\sqrt{-1})^n} \pi_{Y,*}(\pi_X^*(\alpha) \wedge e^{\sqrt{-1}F}) \\ &= \frac{1}{(2\pi\sqrt{-1})^n} \int_{T_N} \pi_X^*(\alpha) \wedge e^{\sqrt{-1}F}, \end{aligned}$$

where $\pi_X : X \times_D Y \to X$ and $\pi_Y : X \times_D Y \to Y$ are the two projections.

What is crucial is that this Fourier–Mukai-type transformation transforms the symplectic structure on X to the complex structure on Y in the sense of the following two propositions. These already appeared in [7], Proposition 3.2. We include their proofs, which are somewhat interesting, here for completeness.

Proposition 2.1.

$$\mathcal{F}^{sf}(e^{\sqrt{-1}\omega_X}) = \Omega_Y.$$

Proof.

$$\begin{aligned}
\mathcal{F}^{\mathrm{sf}}(e^{\sqrt{-1}\omega_X}) &= \frac{1}{(2\pi\sqrt{-1})^n}\int_{T_N}\pi_X^*(e^{\sqrt{-1}\omega_X})\wedge e^{\sqrt{-1}F} \\
&= \frac{1}{(2\pi\sqrt{-1})^n}\int_{T_N} e^{\sqrt{-1}\sum_{j=1}^n (dx_j+\sqrt{-1}dy_j)\wedge du_j} \\
&= \frac{1}{(2\pi\sqrt{-1})^n}\int_{T_N}\bigwedge_{j=1}^n\left(1+\sqrt{-1}(dx_j+\sqrt{-1}dy_j)\wedge du_j\right) \\
&= \frac{1}{(2\pi)^n}\int_{T_N}\left(\bigwedge_{j=1}^n (dx_j+\sqrt{-1}dy_j)\right)\wedge du_1\wedge\ldots\wedge du_n \\
&= \Omega_Y,
\end{aligned}$$

where we have $\int_{T_N} du_1\wedge\ldots\wedge du_n = (2\pi)^n$ in the final step. Q.E.D.

As a mirror transformation, $\mathcal{F}^{\mathrm{sf}}$ should have the *inversion property.* This is the following proposition.

Proposition 2.2. *If we define the inverse transform* $(\mathcal{F}^{sf})^{-1}: \Omega^*(Y)\to\Omega^*(X)$ *by*

$$\begin{aligned}
(\mathcal{F}^{sf})^{-1}(\alpha) &= \frac{1}{(2\pi\sqrt{-1})^n}\pi_{X,*}(\pi_Y^*(\alpha)\wedge e^{-\sqrt{-1}F}) \\
&= \frac{1}{(2\pi\sqrt{-1})^n}\int_{T_M}\pi_Y^*(\alpha)\wedge e^{-\sqrt{-1}F},
\end{aligned}$$

then we have

$$(\mathcal{F}^{sf})^{-1}(\Omega_Y) = e^{\sqrt{-1}\omega_X}.$$

Proof.

$$
\begin{aligned}
(\mathcal{F}^{\mathrm{sf}})^{-1}(\Omega_Y) &= \frac{1}{(2\pi\sqrt{-1})^n}\int_{T_M}\pi_Y^*(\Omega_Y)\wedge e^{-\sqrt{-1}F}\\
&= \frac{1}{(2\pi\sqrt{-1})^n}\int_{T_M}\left(\bigwedge_{j=1}^n(dx_j+\sqrt{-1}dy_j)\right)\wedge e^{\sum_{j=1}^n dy_j\wedge du_j}\\
&= \frac{1}{(2\pi\sqrt{-1})^n}\int_{T_M}\bigwedge_{j=1}^n\left((dx_j+\sqrt{-1}dy_j)\wedge e^{dy_j\wedge du_j}\right)\\
&= \frac{1}{(2\pi\sqrt{-1})^n}\int_{T_M}\bigwedge_{j=1}^n\left(dx_j+\sqrt{-1}dy_j+dx_j\wedge dy_j\wedge du_j\right)\\
&= \frac{1}{(2\pi)^n}\int_{T_M}\bigwedge_{j=1}^n\left((1+\sqrt{-1}dx_j\wedge du_j)\wedge dy_j\right)\\
&= \frac{1}{(2\pi)^n}\int_{T_M}\bigwedge_{j=1}^n\left(e^{\sqrt{-1}dx_j\wedge du_j}\wedge dy_j\right)\\
&= \frac{1}{(2\pi)^n}\int_{T_M}e^{\sqrt{-1}\sum_{j=1}^n dx_j\wedge du_j}\wedge dy_1\wedge\ldots\wedge dy_n\\
&= e^{\sqrt{-1}\omega_X}.
\end{aligned}
$$

Q.E.D.

By exactly the same arguments, one can also show that

$$\mathcal{F}^{\mathrm{sf}}(\Omega_X)=e^{\sqrt{-1}\omega_Y},\ (\mathcal{F}^{\mathrm{sf}})^{-1}(e^{\sqrt{-1}\omega_Y})=\Omega_X.$$

If we take into account the *B-fields*, then the semi-flat SYZ transformation will give an identification between the moduli space of complexified Kähler structures on X with the moduli space of complex structures on Y, and vice versa. For this and transformations of other geometric structures, we refer the reader to Leung [31].

2.2. Transformations of branes

Lying at the heart of the SYZ Conjecture is the basic but important observation that a point $z=\exp(x+\sqrt{-1}y)\in Y$ defines a flat $U(1)$-connection ∇_y on the trivial complex line bundle over the special Lagrangian torus fiber $L_x=\mu_X^{-1}(x)$. Now, the point $z\in Y$ together with its structure sheaf $\mathcal{O}_z$ can be considered as a *B-brane* on Y; while the pair $(L_x,\mathbb{L}_y)$, where $\mathbb{L}_y$ denotes the flat $U(1)$-bundle $(L_x\times\mathbb{C},\nabla_y)$, gives an *A-brane* on X. This implements the simplest case of correspondence between branes on mirror manifolds via SYZ transformations:

$$(L_x, \mathbb{L}_y) \longleftrightarrow (z, \mathcal{O}_z).$$

The space of infinitestimal deformations of the A-brane $(L_x, \mathbb{L}_y)$, which is given by $H^1(L_x, \mathbb{R}) \times H^1(L_x, \sqrt{-1}\mathbb{R}) = H^1(L_x, \mathbb{C})$, is canonically identified with the tangent space $T_z Y$, the space of infinitestimal deformations of the sheaf $\mathcal{O}_z$.

On the other hand, consider a section $L = \{(x, u(x)) \in X : x \in D\}$ of $\mu_X : X \to D$. The submanifold L is Lagrangian if and only if (locally) there exists a function f such that $u_j = \frac{\partial f}{\partial x_j}$. By the above observation (now used in the opposite way), a point $(x, u(x)) \in L$ determines a flat $U(1)$-connection $\nabla_{u(x)}$ on the trivial complex line bundle over the fiber $(L_x)^\vee = \nu_Y^{-1}(x)$. The family of points $\{(x, u(x)) : x \in D\}$ thus patch together to give the $U(1)$-connection

$$\nabla_L = d_Y - \frac{\sqrt{-1}}{2} \sum_{j=1}^{n} u_j(x) dy_j$$

on a certain complex line bundle over Y; its curvature two form is given by

$$F_L = d_Y \Big(- \frac{\sqrt{-1}}{2} \sum_{j=1}^{n} u_j(x) dy_j \Big) = -\frac{\sqrt{-1}}{2} \sum_{j,k} \frac{\partial u_j}{\partial x_k} dx_k \wedge dy_j,$$

and, in particular,

$$F_L^{2,0} = \frac{1}{8} \sum_{j<k} \Big(\frac{\partial u_j}{\partial x_k} - \frac{\partial u_k}{\partial x_j} \Big) \frac{dz_j}{z_j} \wedge \frac{dz_k}{z_k}.$$

We conclude that ∇_L is integrable, i.e. $F_L^{2,0} = 0$, if and only if L is Lagrangian. More generally, we can equip L with a flat $U(1)$-bundle $\mathbb{L} = (L \times \mathbb{C}, d_L + \alpha)$, where $\alpha \in \Omega^1(L, \mathbb{R})$ is a closed (and hence exact) one-form. The A-brane $(L, \mathbb{L})$ is then transformed to the $U(1)$-connection

$$\nabla_{L,\mathbb{L}} = \nabla_L + \alpha,$$

which again is integrable if and only if L is Lagrangian. Furthermore, one can prove that $\nabla_{L,\mathbb{L}}$ satisfies the *deformed Hermitian–Yang–Mills equations* if and only if L is special Lagrangian (see Leung–Yau–Zaslow [32] and Leung [31] for the detailed proofs). $\nabla_{L,\mathbb{L}}$ is a connection on the holomorphic line bundle over Y given by the semi-flat SYZ transformation of $\mathbb{L}$:

$$\mathcal{L}_{L,\mathbb{L}} = \pi_{Y,*}(\pi_X^*(\iota_*\mathbb{L}) \otimes \mathcal{P}),$$

where $\iota : L \hookrightarrow X$ is the inclusion map. In conclusion, the A-brane $(L, \mathbb{L})$ is transformed to the B-brane $(Y, \mathcal{L}_{L,\mathbb{L}})$ through semi-flat SYZ transformations:

$$(L, \mathbb{L}) \longleftrightarrow (Y, \mathcal{L}_{L,\mathbb{L}}).$$

§3. SYZ mirror transformations with corrections

In the previous section, we see that T-duality and SYZ mirror transformations can be applied successfully to give a geometric understanding of the mirror symmetry for semi-flat Calabi–Yau manifolds. However, no quantum corrections were involved in this case due to the absence of holomorphic curves and discs. The existence of quantum corrections is also closely related to the singularities of the Lagrangian torus fibrations, which again are not present in the semi-flat case. In this section, following [7], we are going to discuss how SYZ mirror transformations can be applied to a case where quantum corrections *do* exist, namely, the mirror symmetry for toric Fano manifolds.

3.1. Mirror symmetry for toric Fano manifolds

We begin with a more detailed description of the mirror picture for toric Fano manifolds [17], [29], [27]. Let $\bar{P} \subset M_\mathbb{R}$ be a smooth reflexive polytope given by the inequalities

$$\langle x, v_i \rangle \geq \lambda_i, \quad i = 1, \ldots, d,$$

where $v_1, \ldots, v_d \in N$ are primitive vectors and $\langle \cdot, \cdot \rangle : M_\mathbb{R} \times N_\mathbb{R} \to \mathbb{R}$ is the dual pairing. This determines a toric Fano manifold $\bar{X}$, together with a Kähler structure $\omega_{\bar{X}}$. Unlike the case of Calabi–Yau manifolds, the mirror of $\bar{X}$ is not another compact Kähler manifold, but a Landau–Ginzburg model: a pair (Y, W) consisting of a noncompact Kähler manifold Y, which (as a complex manifold) is biholomorphic to (a bounded domain of) $(\mathbb{C}^*)^n$, and the Laurent polynomial

$$W = e^{\lambda_1} z^{v_1} + \ldots + e^{\lambda_d} z^{v_d} : Y \to \mathbb{C},$$

which is called the superpotential. Here z^{v_i} denotes the monomial $z_1^{v_{i1}} \ldots z_n^{v_{in}}$ in the coordinates $z_1, \ldots, z_n$ of Y. For example, if $P = \{(x_1, x_2, x_3) \in \mathbb{R}^3 : x_1 \geq 0, x_2 \geq 0, x_1 + x_2 \leq t\}$, then $\bar{X} = \mathbb{C}P^2$ and the mirror Landau–Ginzburg model is given by the Laurent polynomial $W(z_1, z_2) = z_1 + z_2 + \frac{e^{-t}}{z_1 z_2}$ on $Y = (\mathbb{C}^*)^2$.

Among the many mirror symmetry predictions are the following conjectures:

Conjecture 3.1.

1. *The small quantum cohomology ring $QH^*(\bar{X})$ of $\bar{X}$ is isomorphic to the Jacobian ring $Jac(W)$ of W, where*

$$Jac(W) = \mathbb{C}[z_1^{\pm 1}, \ldots, z_n^{\pm 1}]/\langle \partial_1 W, \ldots, \partial_n W \rangle,$$

and ∂_j denotes $z_j \frac{\partial}{\partial z_j}$.

2. *(Homological mirror symmetry, see* [29], [39], [37]*) There are equivalences of triangulated categories*

$$\begin{aligned} D^b Coh(\bar{X}) &\cong D^\pi Fuk(Y, W) \\ D^\pi Fuk(\bar{X}) &\cong D_{Sing}(Y, W) \end{aligned}$$

where $D^\pi Fuk(Y, W)$ is (a suitably defined version of) the derived Fukaya category of the Landau–Ginzurg model (Y, W) and $D_{Sing}(Y, W)$ is the category of singularities of (Y, W).

Substantial evidences [19], [25], [39], [41], [5], [6], [1], [2], [9], [8] have been found for these conjectures, while evidence in the Calabi–Yau and other non-toric cases is much rarer. This is partly due to the fact that geometric structures on toric varieties are highly computable and explicit, making them an exceptionally fertile testing ground for techniques and conjectures.

One of these explicit structures: the Lagrangian torus fibration on $\bar{X}$ given by the moment map $\mu_{\bar{X}} : \bar{X} \to \bar{P}$ of the Hamiltonian T_N-action on $(\bar{X}, \omega_{\bar{X}})$, is particularly important in the SYZ approach and in the constructions of SYZ mirror transformations. Let

$$\mu_X : X \to P$$

be the restriction of the moment map to the open dense T_N-orbit $X = \bar{X} \setminus D_\infty$, where $D_\infty = \bigcup_{i=1}^d D_i$ is the anticanonical toric divisor, and P is the interior of $\bar{P}$. In the symplectic (or action-angle) coordinates,

$$X = T^*P/N = P \times \sqrt{-1}T_N$$

and the restriction of $\omega_{\bar{X}}$ to X is nothing but the standard symplectic structure

$$\omega_X = \sum_{j=1}^n dx_j \wedge du_j,$$

where $x_1, \ldots, x_n \in \mathbb{R}$ and $u_1, \ldots, u_n \in \mathbb{R}/2\pi\mathbb{Z}$ are respectively the base coordinates on P and fiber coordinates on T_N (see Abreu [3]). Now we are in exactly the same situation as in the previous section and it is tempting to assert that the mirror manifold Y predicted by Physicists is given by the SYZ mirror manifold of X, which is $TP/M = P \times \sqrt{-1}T_M$. This is indeed nearly the case.

Proposition 3.1 (Proposition 3.1 in [7]). *The mirror manifold $Y = (\mathbb{C}^*)^n$ predicted by Physicists contains the SYZ mirror manifold $TP/M = P \times \sqrt{-1}T_M$ of $X = \bar{X} \setminus D_\infty$ as a bounded domain*

$$\{(z_1, \ldots, z_n) \in Y : |e^{\lambda_i} z^{v_i}| < 1 \text{ for } i = 1, \ldots, d\}.$$

Equivalently, the SYZ mirror manifold is given by the preimage of $P \subset M_\mathbb{R} = \mathbb{R}^n$ under the Log map

$$Log : (\mathbb{C}^*)^n \to \mathbb{R}^n,\ (z_1, \ldots, z_n) \mapsto (\log|z_1|, \ldots, \log|z_n|).$$

The same result also appeared in Auroux's paper [4] (Proposition 4.2). Also included in his paper was a discussion of the issue that the SYZ mirror manifold (a bounded domain in $(\mathbb{C}^*)^n$) is "smaller" than Hori–Vafa's mirror manifold (the whole $(\mathbb{C}^*)^n$). There is evidence (say, in Abouzaid's works [1], [2]) showing that one should work with the SYZ mirror manifold, instead of the whole $(\mathbb{C}^*)^n$, in studying mirror symmetry. In any case, we will use and work with the SYZ mirror manifold, i.e. the bounded domain in $(\mathbb{C}^*)^n$, and denote it by Y henceforth.

In terms of the coordinates $z_1 = \exp(-x_1 - \sqrt{-1}y_1), \ldots, z_n = \exp(-x_n - \sqrt{-1}y_n) \in \mathbb{C}^*$ of $Y \subset (\mathbb{C}^*)^n$, the holomorphic volume form is given by the standard one on $(\mathbb{C}^*)^n$:

$$\Omega_Y = \frac{dz_1}{z_1} \wedge \ldots \wedge \frac{dz_n}{z_n}$$

and the torus fibration $\nu_Y : Y \to P$ is the restriction of the Log map. We remark that metrically we are *not* considering $X = \bar{X} \setminus D_\infty$ as a Calabi–Yau manifold; instead of the semi-flat Calabi–Yau metric, we use the T_N-invariant Kähler metric on $\bar{X}$ (and the corresponding dual metric on Y). These are defined (cf. Guillemin [22] and Abreu [3]) using the strictly convex function $\phi_P : P \to \mathbb{R}$ given by

$$\phi_P(x) = \frac{1}{2}\sum_{i=1}^{d} l_i(x) \log l_i(x),$$

where $l_i(x) = \langle x, v_i \rangle - \lambda_i$ for $i = 1, \ldots, d$, instead of a solution of the real Monge–Ampère equation. For example, this gives the standard Fubini–Study metric on $\bar{X} = \mathbb{C}P^n$. Using these metrics and the corresponding holomorphic volume forms, X and Y are almost Calabi–Yau manifolds and the torus fibers of μ_X and ν_Y are special Lagrangian submanifolds (also see Section 2 in Auroux [4]).

3.2. SYZ transformations for toric Fano manifolds

By applying the semi-flat SYZ mirror transformation or T-duality, we can obtain the mirror manifold Y. But where comes the superpotential $W : Y \to \mathbb{C}$? Recall that, in applying T-duality, we have completely ignored the compactification of X, which is given by adding the anti-canonical toric divisor $D_\infty = \bigcup_{i=1}^d D_i$. As suggested in the foundational work of Fukaya–Oh–Ohta–Ono [14], this has tremendous effect on the Floer theory of the Lagrangian torus fibers of $\mu_X : X \to P$, and this is indeed where quantum corrections by holomorphic discs come into play.

As have been discussed in the introduction, motivated by the idea of using Morse theory on the free loop space $\mathfrak{L}\bar{X}$ to construct the quantum cohomology $QH^*(\bar{X})$, we introduce the subspace $LX \subset \mathfrak{L}\bar{X}$ consisting of those loops which are geodesic in the Lagrangian torus fibers of the moment map $\mu_X : X \to P$, i.e.

$$LX = \{\gamma \in \mathfrak{L}\bar{X} : \gamma \text{ is a geodesic in } L_x = \mu_X^{-1}(x) \text{ for some } x \in P\}.$$

Concretely, we have

$$LX = X \times N = P \times \sqrt{-1}T_N \times N,$$

and we consider it as a (trivial) $\mathbb{Z}^n$-cover of X, $\pi : LX \to X$. Notice that, for each Lagrangian torus fiber L_x, $x \in P$, we have a canonical identification $\pi_1(L_x) \cong N$.

We are going to define a function Ψ on LX in terms of the counting of holomorphic discs in $\bar{X}$ of minimal Maslov index. This will recapture the information of the compactification of X by D_∞, which we have ignored previously, and Ψ serves as the object in the A-model of $\bar{X}$ mirror to the superpotential W. To do this, let's first recall the fundamental results of Cho–Oh [9] on the classification of holomorphic discs in $\bar{X}$ with boundary in Lagrangian torus fibers of $\mu_X : X \to P$.

Let $L_x = \mu_X^{-1}(x)$ be the Lagrangian torus fiber in X over a point $x \in P$. Then the relative homotopy group $\pi_2(\bar{X}, L_x)$ is generated by the Maslox index two classes $\beta_1, \ldots, \beta_d$, which are represented by holomorphic discs in $(\bar{X}, L_x)$. Note that we have, $\partial\beta_i = v_i$, for $i = 1, \ldots, d$, where $\partial : \pi_2(\bar{X}, L_x) \to \pi_1(L_x) \cong N$ is the natural boundary map. In [9], Cho and Oh proved that, for $i = 1, \ldots, d$ and for each point $p \in L_x$, there is a unique (up to automorphism of the domain) Maslov index two J-holomorphic disc $\varphi_i : (D^2, \partial D^2) \to (\bar{X}, L_x)$ in the class β_i which passes through p and intersects the toric divisor D_i at an interior point.[5] Here J is the complex structure on $\bar{X}$ determined by the fan Σ dual to $\bar{P}$.

Definition 3.1. *For $i = 1, \ldots, d$, define $\Psi_i : LX \to \mathbb{R}$ by*

$$\Psi_i(p, v) = \begin{cases} n_i(p)\exp(-\frac{1}{2\pi}\int_{\beta_i}\omega_{\bar{X}}) & \text{if } v = v_i \\ 0 & \text{if } v \neq v_i, \end{cases}$$

for $(p, v) \in LX = X \times N$, where $n_i(p)$ is the algebraic number of Maslov index two J-holomorphic discs in $(\bar{X}, L_{\mu_X(p)})$ in the class β_i which pass through p. Then set

$$\Psi = \Psi_1 + \ldots + \Psi_d : LX \to \mathbb{R}.$$

By their definitions, the T_N-invariant functions $\Psi_1, \ldots, \Psi_d$ carry enumerative meaning, although by Cho and Oh's result, we always have $n_i(p) = 1$, for all i and any p. One may think of the T_N-invariant function Ψ as recording which cycle $v \in N = \pi_1(L_x)$ collapses to a point as one goes towards the anticanonical toric divisor D_∞, or equivalently, which geodesic loop $\gamma \in LX$ bounds a holomorphic disc of Maslov index two.

Remark 3.1. Before showing how to transform Ψ to get the superpotential W, we remark that the T_N-invariant function $\Phi : LX \to \mathbb{R}$ introduced in [7], Definition 2.1, is nothing but the "exponential" of Ψ, i.e.

$$\Phi = \text{Exp}\,\Psi,$$

where Exp Ψ is defined as $\sum_{k=0}^{\infty}\frac{1}{k!}\underbrace{\Psi\star\ldots\star\Psi}_{k \text{ times}}$ in which $\star$ denotes the convolution product of a certain class of functions on LX with respect to the lattice N. Now each point $q = (q_1, \ldots, q_l)$ $(l = d - n)$ in the Kähler cone $\mathcal{K}(\bar{X}) \subset H^2(\bar{X}, \mathbb{R})$ determines a symplectic structure $\omega_{\bar{X}}$ on $\bar{X}$ and we can choose the polytope $\bar{P} = \{x \in M_{\mathbb{R}} : \langle x, v_i\rangle \geq \lambda_i, \quad i = 1, \ldots, d\}$ such that $v_1 = e_1, \ldots, v_n = e_n$ is the standard basis of $N = \mathbb{Z}^n$, $\lambda_1 = \ldots = \lambda_n = 0$ and $\lambda_{n+a} = \log q_a$ for $a = 1, \ldots, l$. We thus get

[5] Another way to state this result is the following. Let $\mathcal{M}_1(\beta_i)$ be the moduli space of J-holomorphic discs $\varphi : (D^2, \partial D^2) \to (\bar{X}, L_x)$ in the class β_i with 1 boundary marked point. Let $ev : \mathcal{M}_1(\beta_i) \to L_x$ be the evaluation map at the boundary marked point. Then the result of Cho and Oh says that $ev_*[\mathcal{M}_1(\beta_i)] = [L_x]$ as n-cycles in L_x. See also Sections 3.1 and 4 in Auroux [4].

two families of functions $\{\Psi_q\}_{q\in\mathcal{K}}$ and $\{\Phi_q\}_{q\in\mathcal{K}}$. By the symplectic area formula of Cho–Oh ([9], Theorem 8.1), we have

$$\int_{D^2}\varphi_i^*\omega_{\bar{X}} = \int_{\beta_i}\omega_{\bar{X}} = 2\pi(\langle x, v_i\rangle - \lambda_i),$$

for $i = 1, \ldots, d$. Hence, for any $(p, v) \in LX$,

$$\Psi_i(p, v) = \begin{cases} e^{-\langle x, v_i\rangle} & \text{if } v = v_i \\ 0 & \text{if } v \neq v_i, \end{cases}$$

for $i = 1, \ldots, n$, and

$$\Psi_{n+a}(p, v) = \begin{cases} q_a e^{-\langle x, v_{n+a}\rangle} & \text{if } v = v_{n+1} \\ 0 & \text{if } v \neq v_{n+a}, \end{cases}$$

for $a = 1, \ldots, l$, where $x = \mu_X(p)$. It follows that

$$q_a\frac{\partial\Phi_q}{\partial q_a} = \Phi_q \star \Psi_{n+a}$$

for $a = 1, \ldots, l$, which is the first part of Proposition 1.1 in [7].

On the other hand, the functions $\Psi_1, \ldots, \Psi_d$ are intimately related to the small quantum cohomology $QH^*(\bar{X})$ of $\bar{X}$, as was shown in the following

Proposition 3.2 (Second part of Proposition 1.1 in [7]). *Assume that $\bar{X}$ is a product of projective spaces. Then we have a natural isomorphism of $\mathbb{C}$-algebras*

$$QH^*(\bar{X}) \cong \mathbb{C}[\Psi_1^{\pm 1}, \ldots, \Psi_n^{\pm 1}]/\mathfrak{L}$$

where $\mathbb{C}[\Psi_1^{\pm 1}, \ldots, \Psi_n^{\pm 1}]$ is the polynomial algebra generated by $\Psi_1^{\pm 1}, \ldots,$ $\Psi_n^{\pm 1}$ with respect to the convolution product $\star$, and $\mathfrak{L}$ is the ideal generated by linear relations: $\sum_{i=1}^d a_i\Psi_i \sim \sum_{i=1}^d b_i\Psi_i$ if and only if the corresponding divisors $\sum_{i=1}^d a_iD_i$ and $\sum_{i=1}^d b_iD_i$ are linearly equivalent.

Remark 3.2. By employing Givental's mirror theorem [19], one can in fact show that the proposition holds for *all* toric Fano manifolds. See Remark 2.3 in [7] for details.

We need the assumption that $\bar{X}$ is a product of projective spaces as we are intended for a geometric understanding of the isomorphism in Proposition 3.2 by using *tropical geometry.* This is briefly described as follows (see Subsection 2.2 in [7] for details). One first defines a tropical version $QH^*_{trop}(\bar{X})$ of the small quantum cohomology ring of $\bar{X}$. Since $\bar{X}$

is a product of projective spaces, we have a one-to-one correspondences between the J-holomorphic curves in $\bar{X}$ which have contribution to the quantum product in $QH^*(\bar{X})$ and those tropical curves in $N_{\mathbb{R}}$ which have contribution to the tropical quantum product in $QH^*_{trop}(\bar{X})$, by the *correspondence theorem* of Mikhalkin [33] and Nishinou–Siebert [36]. From this follows the canonical isomorphism

$$QH^*(\bar{X}) \cong QH^*_{trop}(\bar{X}).$$

Then comes a simple but important observation: *Each tropical curve which has contribution to the tropical quantum product in $QH^*_{trop}(\bar{X})$ is obtained by gluing tropical discs in $N_{\mathbb{R}}$.*[6] On the other hand, these tropical discs are exactly corresponding to the families of Maslov index two J-holomorphic discs in $\bar{X}$ with boundary in Lagrangian torus fibers, which were used to define the functions $\Psi_1, \ldots, \Psi_d$. Hence, we naturally have another canonical isomorphism

$$QH^*_{trop}(\bar{X}) \cong \mathbb{C}[\Psi_1^{\pm 1}, \ldots, \Psi_n^{\pm 1}]/\mathfrak{L}.$$

For example, let us take a look at the case of $\bar{X} = \mathbb{C}P^2$. See Fig. 1 below.

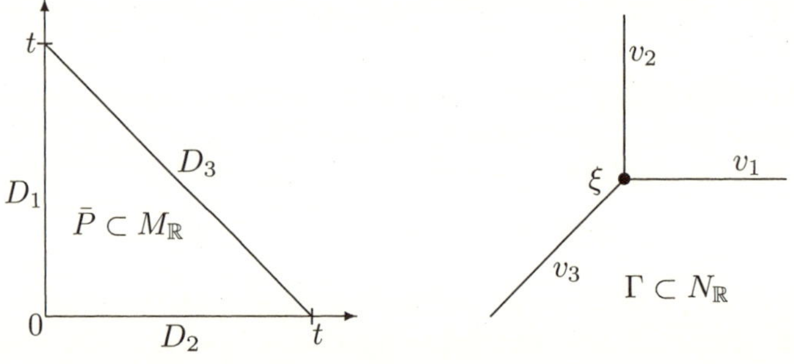

Fig. 1

Denote by $\{e_1, e_2\}$ the standard basis of $N = \mathbb{Z}^2$. We have $v_1 = (1,0), v_2 = (0,1), v_3 = (-1,-1)$, and the polytope $\bar{P} \subset M_{\mathbb{R}} \cong \mathbb{R}^2$ is defined by the inequalities

$$x_1 \geq 0,\ x_2 \geq 0,\ x_1 + x_2 \leq t,$$

[6]This idea was recently generalized by Gross [20] to understand tropically the big quantum cohomology and mirror symmetry of $\mathbb{C}P^2$.

where $t > 0$. There are three toric divisors D_1, D_2, D_3 corresponding to three functions $\Psi_1, \Psi_2, \Psi_3 \in C^\infty(LX)$ defined by

$$\begin{aligned}
\Psi_1(p, v) &= \begin{cases} e^{-x_1} & \text{if } v = (1, 0) \\ 0 & \text{otherwise,} \end{cases} \\
\Psi_2(p, v) &= \begin{cases} e^{-x_2} & \text{if } v = (0, 1) \\ 0 & \text{otherwise,} \end{cases} \\
\Psi_3(p, v) &= \begin{cases} e^{-(t-x_1-x_2)} & \text{if } v = (-1, -1) \\ 0 & \text{otherwise,} \end{cases}
\end{aligned}$$

for $(p, v) \in LX$ and $(x_1, x_2) = \mu_X(p) \in P$, respectively. The small quantum cohomology ring is given by

$$\begin{aligned}
QH^*(\mathbb{C}P^2) &= \mathbb{C}[D_1, D_2, D_3]/\langle D_1 - D_3, D_2 - D_3, D_1 * D_2 * D_3 - q\rangle \\
&= \mathbb{C}[H]/\langle H^3 - q\rangle,
\end{aligned}$$

where we have, by abuse of notations, also use $D_i \in H^2(\mathbb{C}P^2, \mathbb{C})$ to denote the cohomology class Poincaré dual to D_i, $H \in H^2(\mathbb{C}P^2, \mathbb{C})$ is the hyperplane class and $q = e^{-t}$. Fix any point $p \in \mathbb{C}P^2 \setminus D_\infty$, then the quantum corrections, which appear in the relation

$$D_1 * D_2 * D_3 = H^3 = q,$$

is due to the unique holomorphic curve $\varphi : (\mathbb{P}^1; x_1, x_2, x_3, x_4) \to \mathbb{C}P^2$ of degree 1 (i.e. a line) with 4 marked points such that $\varphi(x_4) = p$ and $\varphi(x_i) \in D_i$, for $i = 1, 2, 3$. Let $x = \mu_X(p) \in P$ and $L_x = \mu_X^{-1}(x)$ be the Lagrangian torus fiber containing p. Using tropical geometry, one

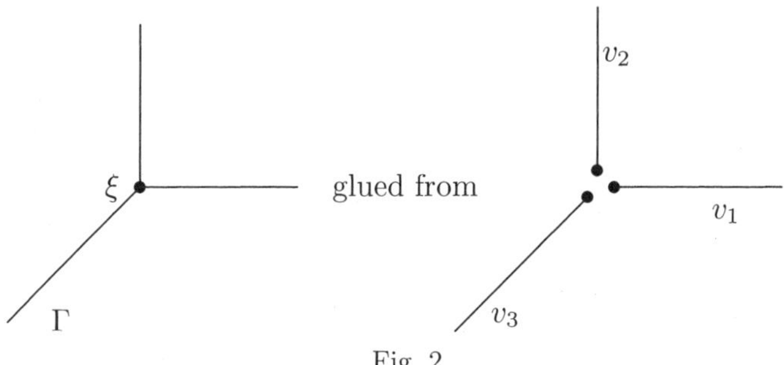

Fig. 2

sees that there is a tropical curve Γ in $N_\mathbb{R}$ with three unbounded edges in the directions v_1, v_2, v_3 and the vertex mapped to $\xi = \mathrm{Log}(p) \in N_\mathbb{R}$, which is corresponding to this holomorphic curve (see Fig. 1 above). Here, we identify X with $(\mathbb{C}^*)^2$, and $\mathrm{Log} : X = (\mathbb{C}^*)^2 \to N_\mathbb{R} = \mathbb{R}^2$ is the Log map we defined in Proposition 3.1. It is obvious that Γ can be obtained by gluing the three half lines emanating from the point $\xi \in N_{\mathbb{R}}$ in the directions v_1, v_2, v_3. See Fig. 2. These half lines are the tropical discs which are corresponding to the three families of Maslov index two holomorphic discs $\varphi_1, \varphi_2, \varphi_3$ respectively. We see that the above quantum relation corresponds exactly to the equation

$$\Psi_1 \star \Psi_2 \star \Psi_3 = q$$

in $\mathbb{C}[\Psi_1^{\pm 1}, \Psi_2^{\pm 1}]$.

Without the assumption that $\bar{X}$ is a product of projective spaces, the tropical interpretation will break down. This is because for general toric Fano manifolds, the holomorphic curves which contribute to the small quantum product may have components mapped into the anticanonical toric divisor D_∞. An example is provided by the exceptional curve in the blowup of $\mathbb{C}P^2$ at one T_N-invariant point (see Example 3 in Section 4 in [7]). Now the problem is that tropical geometry cannot be used to count these holomorphic curves. In other words, there are no tropical curves corresponding to such holomorphic curves (cf. Rau [38]).

Now it's time to return to the main theme of this section, namely, we can construct and apply SYZ mirror transformations to the study of mirror symmetry for toric Fano manifolds. First we equip $LX = X \times N$ with the symplectic structure $\pi^*(\omega_X)$, which we denote again by ω_X. Also let $\mu_{LX} : LX \to P$ be the composition map $\mu_X \circ \pi$. Analog to the semi-flat case, we consider the fiber product $LX \times_P Y = P \times N \times \sqrt{-1}(T_N \times T_M)$ of the fibrations $\mu_{LX} : LX \to P$ and $\nu_Y : Y \to P$. Note that we have a covering map $LX \times_P Y \to X \times_P Y$. Pulling back the universal curvature two-form $F = \sqrt{-1}\sum_{j=1}^n dy_j \wedge du_j \in \Omega^2(X \times_P Y)$, we get a two-form on $LX \times_P Y$, which we again denote by F. We further define the *holonomy function* $\mathrm{hol} : LX \times_P Y \to \mathbb{C}$ by

$$\mathrm{hol}(p, v, z) = \mathrm{hol}_{\nabla_y}(v) = e^{-\sqrt{-1}\langle y, v\rangle}$$

for $(p, v) \in LX$, $z = \exp(-x - \sqrt{-1}y) \in Y$ such that $\mu_X(p) = \nu_Y(z) = x$. The SYZ mirror transformation for toric Fano manifolds is constructed as *a combination of the semi-flat SYZ transformation $\mathcal{F}^{sf}$ and fiberwise Fourier series.*

Definition 3.2. *The SYZ mirror transformation $\mathcal{F} : \Omega^*(LX) \to \Omega^*(Y)$ for $\bar{X}$ is defined by*

$$\begin{aligned} \mathcal{F}(\alpha) &= (-2\pi\sqrt{-1})^{-n}\pi_{Y,*}(\pi_{LX}^*(\alpha) \wedge e^{\sqrt{-1}F} hol) \\ &= (-2\pi\sqrt{-1})^{-n} \int_{N \times T_N} \pi_{LX}^*(\alpha) \wedge e^{\sqrt{-1}F} hol, \end{aligned}$$

where $\pi_{LX} : LX \times_P Y \to LX$ and $\pi_Y : LX \times_P Y \to Y$ are the two natural projections.

The basic properties of $\mathcal{F}$ are similar to those of other Fourier-type transformations, and in particular, it satisfies the inversion property with the *inverse SYZ mirror transformation* $\mathcal{F}^{-1} : \Omega^*(Y) \to \Omega^*(LX)$ defined by

$$\begin{aligned}\mathcal{F}^{-1}(\alpha) &= (-2\pi\sqrt{-1})^{-n}\pi_{LX,*}(\pi_Y^*(\alpha) \wedge e^{-\sqrt{-1}F}\mathrm{hol}^{-1}) \\ &= (-2\pi\sqrt{-1})^{-n}\int_{T_M} \pi_Y^*(\alpha) \wedge e^{-\sqrt{-1}F}\mathrm{hol}^{-1}.\end{aligned}$$

In [7], the SYZ mirror transformation was, for the first time, used to study the appearance of the superpotential W as quantum corrections. More precisely, we showed that

Theorem 3.1 (First part of Theorem 1.1 in [7]). *The SYZ mirror transformation (or fiberwise Fourier series) of the function Ψ, defined in terms of the counting of Maslov index two J-holomorphic discs in the toric Fano manifold $\bar{X}$ with boundary in Lagrangian torus fibers, gives the superpotential $W : Y \to \mathbb{C}$ on the mirror manifold:*

$$\mathcal{F}(\Psi) = W.$$

Furthermore, we can incorporate the symplectic structure ω_X to give the holomorphic volume form of the Landau–Ginzburg model (Y, W) in the sense that

$$\mathcal{F}(e^{\sqrt{-1}\omega_X + \Psi}) = e^W \Omega_Y.$$

Conversely, we have

$$\mathcal{F}^{-1}(W) = \Psi,\ \mathcal{F}^{-1}(e^W \Omega_Y) = e^{\sqrt{-1}\omega_X + \Psi}.$$

Remark 3.3.

1. We shall mention that the fact that the superpotential W can be computed in terms of the counting of Maslov index two holomorphic discs in $\bar{X}$ with boundary in Lagrangian torus fibers was originally due to Cho and Oh [9]. The key point of our result is that there is an explicit Fourier–Mukai-type transformation, namely, the SYZ mirror transformation $\mathcal{F}$, that gives the superpotential W by transforming an object (the function Ψ) in the A-model of $\bar{X}$.
2. Apparently, the statements written here are slightly different from those in Theorem 1.1 in [7], but realizing that $\Phi = \mathrm{Exp}\ \Psi$, it is easy to see that they are in fact the same statements.

3. The complex oscillatory integrals

$$\int_\Gamma e^W \Omega_Y$$

of the n-form $e^W \Omega_Y$ over Lefschetz thimbles $\Gamma \subset Y$ (defined by the singularities of $W : Y \to \mathbb{C}$), which satisfy certain Picard–Fuchs equations, play the role of periods for Calabi–Yau manifolds. This is why we call $e^W \Omega_Y$ the holomorphic volume form of the Landau–Ginzburg model (Y, W).

On the other hand, we also showed that the SYZ mirror transformation (which, in this case, is fiberwise Fourier series) $\mathcal{F}(\Psi_i)$ of the function Ψ_i is nothing but the monomial $e^{\lambda_i} z^{v_i}$ on Y, for $i = 1, \ldots, d$. Since the Jacobian ring $Jac(W)$ of the superpotential W is generated by the monomials $e^{\lambda_1} z^{v_1}, \ldots, e^{\lambda_d} z^{v_d}$, by Proposition 3.2, the SYZ mirror transformation realizes a natural isomorphism between the small quantum cohomology $QH^*(\bar{X})$ and the Jacobian ring $Jac(W)$.

Theorem 3.2 (Second part of Theorem 1.1 in [7]). *The SYZ mirror transformation $\mathcal{F}$ induces a natural isomorphism of $\mathbb{C}$-algebras*

$$\mathcal{F} : QH^*(\bar{X}) \xrightarrow{\cong} Jac(W),$$

which takes the quantum product, now realized as a convolution product, to the ordinary product of Laurent polynomials, provided that $\bar{X}$ is a product of projective spaces.

In the example of $\bar{X} = \mathbb{C}P^2$, the superpotential is the Laurent polynomial $W(z_1, z_2) = z_1 + z_2 + \frac{q}{z_1 z_2}$ on $Y = (\mathbb{C}^*)^2$, where $q = e^{-t}$. Its logarithmic partial derivatives are given by

$$\partial_1 W = z_1 - \frac{q}{z_1 z_2}, \; \partial_2 W = z_2 - \frac{q}{z_1 z_2},$$

so that the Jacobian ring is given by

$$\begin{aligned} Jac(W) &= \mathbb{C}[z_1^{\pm 1}, z_2^{\pm 1}] / \langle z_1 - \frac{q}{z_1 z_2}, z_2 - \frac{q}{z_1 z_2} \rangle \\ &= \mathbb{C}[Z_1, Z_2, Z_3] / \langle Z_1 - Z_3, Z_2 - Z_3, Z_1 Z_2 Z_3 - q \rangle, \end{aligned}$$

where the monomials $Z_1 = z_1$, $Z_2 = z_2$ and $Z_3 = \frac{q}{z_1 z_2}$ are the SYZ mirror transformations (i.e. fiberwise Fourier series) of the functions Ψ_1, Ψ_2 and Ψ_3 respectively.

Remark 3.4.

1. In [10], Coates, Corti, Iritani and Tseng formulated the mirror symmetry conjecture for toric manifolds (and orbifolds) as an isomorphism of graded $\frac{\infty}{2}$VHS between the A-model $\frac{\infty}{2}$VHS associated to a toric manifold and the B-model $\frac{\infty}{2}$VHS associated to the mirror Landau–Ginzburg model (see also Iritani [28]). It is desirable to have this isomorphism, which contains more information than the isomorphism in the above theorem, realized by SYZ mirror transformations.
2. In [15] (and also [16]), Fukaya–Oh–Ohta–Ono applied the machinery developed in [14] to the case of toric manifolds. They considered Floer cohomology with coefficients in the Novikov ring, instead of $\mathbb{C}$ used here and in Auroux's paper [4]. They have results on the superpotential even in the non-Fano toric case. The isomorphism $QH^*(\bar{X}) \cong Jac(W)$ (over the Novikov ring) was also discussed and proved in their work (Theorem 1.9 in [15]). Their proof is combinatorial, using Batyrev's presentation of the small quantum cohomology ring for toric Fano manifolds, the validity of which in turn relies on Givental's mirror theorem. They claimed that a more conceptual and geometric proof for toric, not necessarily Fano, manifolds will appear in a sequel to their paper.

3.3. Transformation of branes

This subsection is an attempt to understand the correspondence between A-branes of the toric Fano manifold $\bar{X}$ and B-branes of the mirror Landau–Ginzburg model (Y, W) via SYZ mirror transformations.

We will deal with the simplest case of the correspondence. So let $L_x = \mu_{\bar{X}}^{-1}(x)$ be the Lagrangian torus fiber of $\bar{X}$ over a point $x \in P$. We equip L_x with a flat $U(1)$-bundle $\mathbb{L}_y = (L_x \times \mathbb{C}, \nabla_y)$, where ∇_y is the flat $U(1)$-connection corresponding to $y \in (L_x)^\vee$. The mirror of the A-brane $(L_x, \mathbb{L}_y)$ is given, according to SYZ, by the B-brane $(z = \exp(-x - \sqrt{-1}y) \in Y, \mathcal{O}_z)$. In other words, the correspondence on the level of objects is the same as in the semi-flat Calabi–Yau case. Quantum corrections will emerge and make a difference when we consider their endomorphisms.

According to Hori (see [26], Chapter 39), the endomorphism algebra $\mathrm{End}(z, \mathcal{O}_z)$ of the B-brane $(z, \mathcal{O}_z)$, as a $\mathbb{C}$-vector space, is given by the cohomology of the complex

$$(\bigwedge^* T_z Y, \delta = \iota_{\partial W(z)}),$$

where $\iota_{\partial W(z)}$ is contraction with the vector $\partial W(z) = \sum_{j=1}^{n} \partial_j W(z)(\partial_j)_z$ and here again ∂_j denotes $z_j \frac{\partial}{\partial z_j}$. The following elementary proposition shows that the introduction of the superpotential W "localizes" the category B-branes to the critical points of W.

Proposition 3.3. *The endomorphism End$(z, \mathcal{O}_z)$ is nontrivial if and only if $z \in Y$ is a critical point of the superpotential $W : Y \to \mathbb{C}$, and in which case, End$(z, \mathcal{O}_z)$ is isomorphic to $\bigwedge^* T_z Y$ as $\mathbb{C}$-vector spaces.*

On the other hand, the endomorphism algebra of the A-brane $(L_x, \mathbb{L}_y)$ in the (derived) Fukaya category is given by the *Floer cohomology ring* $HF(L_x, \mathbb{L}_y)$,[7] which in turn, as a $\mathbb{C}$-vector space, is given by the cohomology of the Floer complex

$$(C^*(L_x, \mathbb{C}), \delta = m_1)$$

where $m_1 = m_1(L_x, \mathbb{L}_y)$ denotes the Floer differential. In [9], [8], Cho and Oh explicitly computed the Floer differential m_1. Recall that $H^1(L_x, \mathbb{C})$, viewed as the space of infinitestimal deformations of the pair $(L_x, \mathbb{L}_y)$, is canonically isomorphic to $T_z Y$. Let $C_1, \ldots, C_n$ be the basis of $H^1(L_x, \mathbb{C})$ corresponding to $(\partial_1)_z, \ldots, (\partial_n)_z$. Then the results of Cho and Oh stated that $m_{1,\beta_i}(C_j) = C_j \cdot \partial\beta_i = v_i^j$ and

$$\begin{aligned} m_1(C_j) &= \sum_{i=1}^{d} m_{1,\beta_i}(C_j) \exp\left(-\frac{1}{2\pi}\int_{\beta_i} \omega_X\right) \mathrm{hol}_{\nabla_y}(\partial\beta_i) \\ &= \sum_{i=1}^{d} v_i^j z^{v_i} = \partial_j W(z). \end{aligned}$$

This shows that $m_1 = \iota_{\partial W(z)}$ on $H^1(L_x, \mathbb{C}) = T_z Y$, and $m_1 = 0$ on $H^1(L_x, \mathbb{C})$ if and only if z is a critical point of W. The following result proved by Cho–Oh in [9] is parallel to the above proposition.

Theorem 3.3 (Cho–Oh [9]). *The Floer cohomology $HF(L_x, \mathbb{L}_y)$ is nontrivial and isomorphic to $H^*(L_x, \mathbb{C})$ if and only if $m_1 = 0$ on $H^1(L_x, \mathbb{C})$.*

We conclude that

Theorem 3.4. *The Floer cohomology $HF(L_x, \mathbb{L}_y)$ of the A-brane $(L_x, \mathbb{L}_y)$ is isomorphic to the endomorphism algebra End$(z, \mathcal{O}_z)$ of the mirror B-brane $(z, \mathcal{O}_z)$ as $\mathbb{C}$-vector spaces.*

---

[7]We use $\mathbb{C}$, instead of the Novikov ring, as the coefficient ring.

It is intriguing to see whether this isomorphism can be realized by explicit SYZ mirror transformations.

Remark 3.5. In [8], Cho proved that the Floer cohomology ring $HF(L_x, \mathbb{L}_y)$, equipped with the product structure given by $m_2 = m_2(L_x, \mathbb{L}_y)$, is a Clifford algebra generated by $H^1(L_x, \mathbb{C})$ with the bilinear form given by the Hessian of W: $Q(C_j, C_k) = \partial_j \partial_k W(z)$. This implies that the isomorphism in Theorem 3.4 is in fact an isomorphism of $\mathbb{C}$-algebras. This confirms a prediction by Physicists. See the paper of Cho [8] for details.

§4. Further questions

The results described in this article represent the first step in our program which is aimed at exploring mirror symmetry via SYZ mirror transformations. In particular, they showed that these transformations can be applied successfully to explain the mirror symmetry for toric Fano manifolds, a case where quantum corrections do exist. However, we shall emphasize that the quantum corrections in the toric Fano case, which are due to the anticanonical toric divisor, are much simpler than those in the general case (Gross–Siebert [21], Auroux [4]), where quantum corrections may arise due to contributions from the proper singular Lagrangian fibers of the Lagrangian torus fibrations and complicated wall-crossing phenomena start to interfere. In terms of affine geometry, this means that the bases of the Lagrangian torus fibrations in the toric case are affine manifolds with boundary but without singularities, while in the general case, the bases are affine manifolds with *both* boundary and singularities (and in the semi-flat case, the bases are affine manifolds without boundary and singularities). Certainly much more work remains to be done in the future. In this final section, we will comment on several possible future research directions. The discussion is going to be rather speculative.

4.1. Toric Fano manifolds

We have seen that the simplest correspondence between A-branes on a toric Fano manifold $\bar{X}$ and B-branes on the mirror Landau–Ginzburg model (Y, W), namely

$$(L_x, \mathbb{L}_y) \longleftrightarrow (z, \mathcal{O}_z),$$

is compatible with the SYZ philosophy. It is desirable to see how other A-branes on X are transformed to the corresponding mirror B-branes on (Y, W). An interesting and important example would be the Lagrangian submanifold $\mathbb{R}P^n \subset \mathbb{C}P^n$ for odd n, which can be viewed as a multi-section of the moment map of $\mathbb{C}P^n$. Employing the SYZ approach, the mirror B-brane is expected to be a trivial rank-2^n holomorphic vector bundle over Y, equipped with some additional information related to W. A possible choice of this additional information would be a *matrix factorization* of W; currently, it is widely believed that the category of B-branes on (Y, W) is given by the category of matrix factorizations of W. This was first proposed by Kontsevich, see Orlov [37] for details. The relation between these matrix factorizations and the computation of Floer cohomology will be the key to a complete understanding of the correspondences of branes.

On the other hand, we have not even touched the correspondence between B-branes on $\bar{X}$ and A-branes on (Y, W). As we mentioned in the introduction, the results of Seidel [39], Ueda [41], Auroux–Katzarkov–Orlov [5], [6] and Abouzaid [1], [2] have provided substantial evidences for this half of the Homological Mirror Symmetry Conjecture. In particular, Abouzaid [2] made use of an idea originated from the SYZ conjecture, namely, the mirror of a Lagrangian section should be a holomorphic line bundle. His results also showed that the correspondence is in line with the SYZ picture. Recently, Fang [11] and Fang–Liu–Treumann–Zaslow [12] proved a version of Homological Mirror Symmetry for toric manifolds by explicitly using T-duality. It is an interesting question whether one can construct an explicit SYZ mirror transformation to realize the correspondence between B-branes on $\bar{X}$ and A-branes on (Y, W).

4.2. Toric non-Fano or non-toric Fano manifolds

As in the case of toric Fano manifolds, non-toric Fano manifolds such as Grassmannians and flag manifolds admit natural Lagrangian torus fibrations, provided by Gelfand–Cetlin integrable systems (see, for example, Guillemin–Sternberg [23]), which are convenient for applying SYZ mirror transformations. While mirror symmetry for these manifolds has been studied for some time by Givental [18] and others, new tools and new ideas are needed if we want to apply SYZ mirror transformations to these examples. The recent works of Nishinou–Nohara–Ueda [34], [35] have shed some light on this case. In particular, they obtained a classification of holomorphic discs in flag manifolds with boundary in Lagrangian torus fibers, which should be very useful in the constructions of SYZ mirror transformations.

On the other hand, the mirror symmetry for toric non-Fano manifolds is also not well understood too. As can be seen from the works of

Givental [19], the mirror map between the complexified Kähler and complex moduli spaces in this case is a nontrivial *coordinate change*, instead of an identity map as in the toric Fano case. In Auroux [4], nontrivial coordinate changes and wall-crossing phenomena were also observed in constructing the superpotentials for the mirrors of non-toric examples. Hence, the definitions and constructions of SYZ mirror transformations must have to be adjusted in order to incorporate the nontrivial mirror map and also wall-crossing phenomena. For this, we will have to make the construction of SYZ mirror transformations local. A very preliminary attempt to this is made in Section 5 in [7].

4.3. Calabi–Yau manifolds

The ultimate goal of our program is no doubt applying SYZ mirror transformations to get a better understanding of the mirror symmetry for Calabi–Yau manifolds and the SYZ Conjecture. Works of Fukaya [13], Kontsevich–Soibelman [30] and Gross–Siebert [21] have laid an important foundation for understanding the SYZ framework for both Calabi–Yau and non-Calabi–Yau manifolds. In view of the fact that toric varieties have played an important role in the constructions of Gross and Siebert, it would be nice if we can incorporate our methods with their new techniques to study SYZ mirror transformations for Calabi–Yau manifolds; and hopefully, this would let us reveal geometrically the secret of mirror symmetry.

References

[1] M. Abouzaid, Homogeneous coordinate rings and mirror symmetry for toric varieties, Geom. Topol., **10** (2006), 1097–1157; arXiv:math/0511644[math.SG].

[2] ———, Morse homology, tropical geometry, and homological mirror symmetry for toric varieties, Selecta Math. (N.S.), **15** (2009), 189–270; arXiv:math/0610004.

[3] M. Abreu, Kähler geometry of toric manifolds in symplectic coordinates, In: Symplectic and Contact Topology: Interactions and Perspectives, Toronto, ON/Montreal, QC, 2001, Fields Inst. Commun., **35**, Amer. Math. Soc., Providence, RI, 2003, pp. 1–24; arXiv:math/0004122[math.DG].

[4] D. Auroux, Mirror symmetry and T-duality in the complement of an anticanonical divisor, J. Gökova Geom. Topol. GGT, **1** (2007), 51–91; arXiv:0706.3207.

[5] D. Auroux, L. Katzarkov and D. Orlov, Mirror symmetry for weighted projective planes and their noncommutative deformations, Ann. of Math. (2), **167** (2008), 867–943; arXiv:math/0404281[math.AG].

[6] ———, Mirror symmetry for del Pezzo surfaces: vanishing cycles and coherent sheaves, Invent. Math., **166** (2006), 537–582, arXiv:math/0506166[math.AG].

[7] K.-W. Chan and N.-C. Leung, Mirror symmetry for toric Fano manifolds via SYZ transformations, Adv. Math., **223** (2010), 797–839; arXiv:0801.2830.

[8] C.-H. Cho, Products of Floer cohomology of torus fibers in toric Fano manifolds, Comm. Math. Phys., **260** (2005), 613–640; arXiv:math/0412414[math.SG].

[9] C.-H. Cho and Y.-G. Oh, Floer cohomology and disc instantons of Lagrangian torus fibers in Fano toric manifolds, Asian J. Math., **10** (2006), 773–814; arXiv:math/0308225[math.SG].

[10] T. Coates, A. Corti, H. Iritani and H.-H. Tseng, Wall-crossings in toric Gromov–Witten theory. I. Crepant examples, Geom. Topol., **13** (2009), 2675–2744; arXiv:math/0611550[math.AG].

[11] B. Fang, Homological mirror symmetry is T-duality for $\mathbb{P}^n$, Commun. Number Theory Phys., **2** (2008), 719–742; arXiv:0804.0646.

[12] B. Fang, C.-C. M. Liu, D. Treumann and E. Zaslow, A categorification of Morelli's theorem and homological mirror symmetry for toric varieties, preprint, 2008, arXiv:0811.1228.

[13] K. Fukaya, Multivalued Morse theory, asymptotic analysis and mirror symmetry, In: Graphs and Patterns in Mathematics and Theoretical Physics, Proc. Sympos. Pure Math., **73**, Amer. Math. Soc., Providence, RI, 2005, pp. 205–278.

[14] K. Fukaya, Y.-G. Oh, H. Ohta and K. Ono, Lagrangian intersection Floer theory: Anomaly and obstruction, preprint, 2000.

[15] ———, Lagrangian Floer theory on compact toric manifolds I, Duke Math. J., **151** (2009), 23–174; arXiv:0802.1703.

[16] ———, Lagrangian Floer theory on compact toric manifolds II: Bulk deformations, preprint, 2008, arXiv:0810.5654.

[17] A. Givental, Homological geometry and mirror symmetry, In: Proceedings of the International Congress of Mathematicians, Vol. 1, 2, Zürich, 1994, Birkhäuser, Basel, 1995, pp. 472–480.

[18] ———, Stationary phase integrals, quantum Toda lattices, flag manifolds and the mirror conjecture, In: Topics in Singularity Theory, Amer. Math. Soc. Transl. Ser. 2, **180**, Amer. Math. Soc., Providence, RI, 1997, pp. 103–115; arXiv:alg-geom/9612001.

[19] ———, A mirror theorem for toric complete intersections, In: Topological Field Theory, Primitive Forms and Related Topics, Kyoto, 1996, Progr. Math., **160**, Birkhäuser Boston, Boston, MA, 1998, pp. 141–175; arXiv:alg-geom/9701016.

[20] M. Gross, Mirror symmetry for $\mathbb{P}^2$ and tropical geometry, to appear in Adv. Math., arXiv:0903.1378.

[21] M. Gross and B. Siebert, From real affine geometry to complex geometry, preprint, 2007, arXiv:math/0703822[math.AG].

[22] V. Guillemin, Moment Maps and Combinatorial Invariants of Hamiltonian T^n-Spaces, Progr. Math., **122**, Birkhäuser Boston, Boston, MA, 1994.

[23] V. Guillemin and S. Sternberg, The Gelfand–Cetlin system and quantization of the complex flag manifolds, J. Funct. Anal., **52** (1983), 106–128.

[24] N. Hitchin, The moduli space of special Lagrangian submanifolds, Ann. Scuola Norm. Sup. Pisa Cl. Sci. (4), **25** (1997), 503–515; arXiv:dg-ga/9711002.

[25] K. Hori, A. Iqbal and C. Vafa, D-branes and mirror symmetry, preprint, 2000, arXiv:hep-th/0005247.

[26] K. Hori, S. Katz, A. Klemm, R. Pandharipande, R. Thomas, C. Vafa, R. Vakil and E. Zaslow, Mirror Symmetry, Clay Math. Monogr., **1**, Amer. Math. Soc., Providence, RI; Clay Math. Instit., Cambridge, MA, 2003.

[27] K. Hori and C. Vafa, Mirror symmetry, preprint, 2000, arXiv:hep-th/0002222.

[28] H. Iritani, Real and integral structures in quantum cohomology I: toric orbifolds, preprint, 2007, arXiv:0712.2204.

[29] M. Kontsevich, Lectures at ÉNS, Paris, Spring 1998, notes taken by J. Bellaiche, J.-F. Dat, I. Marin, G. Racinet and H. Randriambololona.

[30] M. Kontsevich and Y. Soibelman, Affine structures and non-Archimedean analytic spaces, In: The Unity of Mathematics, Progr. Math., **244**, Birkhäuser Boston, Boston, MA, 2006, pp. 321–385; arXiv:math/0406564[math.AG].

[31] N.-C. Leung, Mirror symmetry without corrections, Comm. Anal. Geom., **13** (2005), 287–331; arXiv:math/0009235[math.DG].

[32] N.-C. Leung, S.-T. Yau and E. Zaslow, From special Lagrangian to Hermitian–Yang–Mills via Fourier–Mukai transform. Adv. Theor. Math. Phys., **4** (2000), 1319–1341; arXiv:math/0005118[math.DG].

[33] G. Mikhalkin, Enumerative tropical algebraic geometry in $\mathbb{R}^2$, J. Amer. Math. Soc., **18** (2005), 313–377; arXiv:math/0312530[math.AG].

[34] T. Nishinou, Y. Nohara and K. Ueda, Toric degenerations of Gelfand–Cetlin systems and potential functions, to appear in Adv. Math., arXiv:0810.3470.

[35] ———, Potential functions via toric degenerations, preprint, 2008, arXiv:0812.0066.

[36] T. Nishinou and B. Siebert, Toric degenerations of toric varieties and tropical curves, Duke Math. J., **135** (2006), 1–51; arXiv:math/0409060[math.AG].

[37] D. Orlov, Triangulated categories of singularities and D-branes in Landau–Ginzburg models, Proc. Steklov Inst. Math., **246** (2004), 227–248; arXiv:math/0302304[math.AG].

[38] J. Rau, Intersections on tropical moduli spaces, preprint, 2008, arXiv:0812.3678.

[39] P. Seidel, More about vanishing cycles and mutation, In: Symplectic Geometry and Mirror Symmetry, Seoul, 2000, World Sci. Publ., River Edge, NJ, 2001, pp. 429–465; arXiv:math/0010032[math.SG].

[40] A. Strominger, S.-T. Yau and E. Zaslow, Mirror symmetry is T-duality, Nuclear Phys. B, **479** (1996), 243–259; arXiv:hep-th/9606040.

[41] K. Ueda, Homological mirror symmetry for toric del Pezzo surfaces, Comm. Math. Phys., **264** (2006), 71–85; arXiv:math/0411654[math.AG].

Kwokwai Chan
Department of Mathematics
Harvard University
1 Oxford Street, Cambridge, MA 02138
U.S.A.
E-mail address: kwchan@math.harvard.edu

Naichung Conan Leung
The Institute of Mathematical Sciences and Department of Mathematics
The Chinese University of Hong Kong
Shatin
Hong Kong
E-mail address: leung@ims.cuhk.edu.hk

Advanced Studies in Pure Mathematics 59, 2010
New Developments in Algebraic Geometry,
Integrable Systems and Mirror Symmetry (Kyoto, 2008)
pp. 31–77

Hitchin integrable systems, deformations of spectral curves, and KP-type equations

Andrew R. Hodge[1] **and Motohico Mulase**[2]

Abstract.

An effective family of spectral curves appearing in Hitchin fibrations is determined. Using this family the moduli spaces of stable Higgs bundles on an algebraic curve are embedded into the Sato Grassmannian. We show that the Hitchin integrable system, the natural algebraically completely integrable Hamiltonian system defined on the Higgs moduli space, coincides with the KP equations. It is shown that the Serre duality on these moduli spaces corresponds to the formal adjoint of pseudo-differential operators acting on the Grassmannian. From this fact we then identify the Hitchin integrable system on the moduli space of Sp_{2m}-Higgs bundles in terms of a reduction of the KP equations. We also show that the dual Abelian fibration (the SYZ mirror dual) to the Sp_{2m}-Higgs moduli space is constructed by taking the symplectic quotient of a Lie *algebra* action on the moduli space of GL-Higgs bundles.

Contents

Received January 24, 2008.

2000 *Mathematics Subject Classification.* Primary 14H60; Secondary 14H70, 35Q53, 37J35.

Key words and phrases. Hitchin integrable system, spectral curve, Higgs bundle, Sato Grassmannian, symplectic KP equation.

[1]Research supported by NSF grants DMS-0135345 (VIGRE) and DMS-0406077.

[2]Research supported by NSF grant DMS-0406077, JSPS grant (S-19104002) and UC Davis.

§1. Introduction

The purpose of this paper is to determine the relation between the KP-type equations defined on the Sato Grassmannians and the Hitchin integrable systems defined on the moduli spaces of stable Higgs bundles. The results established are the following:

(1) We determine the *effective* family of spectral curves appearing in the Hitchin fibration of the moduli spaces of stable Higgs bundles.
(2) We embed the effective family of Jacobian varieties of the spectral curves into the Sato Grassmannian and show that the KP flows are tangent to each fiber of the Hitchin fibration.
(3) The moduli space of Higgs bundles of rank n and degree $n(g-1)$ on an algebraic curve of genus $g \geq 2$ is embedded into the *relative* Grassmannian of [2, 4, 23]. Using this embedding we show that the Hitchin integrable system is exactly the restriction of the KP equations on the Grassmannian to the image of this embedding.
(4) It is shown that the Krichever construction transforms the Serre duality of the geometric data consisting of algebraic curves and vector bundles on them to the formal adjoint of pseudo-differential operators acting on the Grassmannian. By identifying the fixed-point-set of the Serre duality and the formal adjoint operation we determine the KP-type equations that are equivalent to the Hitchin integrable system defined on the moduli space of Sp_{2m}-Higgs bundles.
(5) There are two ways to *reduce* an algebraically completely integrable Hamiltonian system: one by restriction and the other by taking a quotient of a Lie algebra action that is similar to the symplectic quotient. When applied to the moduli spaces of Higgs bundles, these constructions yield SYZ-mirror pairs. We interpret the SL-PGL and Sp_{2m}-SO_{2m+1} dualities in this way.

Let $\mathcal{G}_{\mathbb{C}}$ be the category of complex Lie groups, and $\mathcal{CY}$ the category of Calabi–Yau spaces. For a compact oriented surface Σ of genus $g \geq 2$, the functor

$$\mathrm{Hom}(\hat{\pi}_1(\Sigma),\ \cdot\ )/\!\!/\ \cdot\ : \mathcal{G}_{\mathbb{C}} \longrightarrow \mathcal{CY}$$

assigns to each complex Lie group G its character variety

$$\mathrm{Hom}\big(\hat{\pi}_1(\Sigma), G\big)/\!\!/G,$$

where $\hat{\pi}_1(\Sigma)$ is the central extension of the fundamental group of Σ. The quotient by conjugation is the geometric invariant theory quotient of Mumford [21]. An amazing discovery of Hausel and Thaddeus [8], and its generalizations by [5, 14] and others, is that the character variety functor transforms the Langlands duality in $\mathcal{G}_{\mathbb{C}}$ to the mirror symmetry of Calabi–Yau spaces in the sense of Strominger–Yau–Zaslow [28]:

$$\begin{array}{ccc}
\mathcal{G}_{\mathbb{C}} & \xrightarrow{\mathrm{Hom}(\hat{\pi}_1(\Sigma),\ \cdot\ )/\!\!/\ \cdot} & \mathcal{CY} \\
{\scriptstyle \text{Langlands Dual}}\big\downarrow & & \big\downarrow{\scriptstyle \text{Mirror Dual}} \\
\mathcal{G}_{\mathbb{C}} & \xrightarrow[\mathrm{Hom}(\hat{\pi}_1(\Sigma),\ \cdot\ )/\!\!/\ \cdot]{} & \mathcal{CY}.
\end{array}$$

The character variety $\mathrm{Hom}\big(\hat{\pi}_1(\Sigma), G\big)/\!\!/G$ has many distinct complex structures [8, 14]. To understand the SYZ mirror symmetry among the character varieties, it is most convenient to realize them as *Hitchin integrable systems.* In his seminal papers [10, 11], Hitchin identifies the character variety with the moduli space of stable *G-Higgs bundles*, which has the structure of an *algebraically completely integrable Hamiltonian system.*

An algebraically completely integrable Hamiltonian system [4, 31] is a holomorphic symplectic manifold (X, ω) of dimension $2N$ together with a holomorphic map $H : X \rightarrow \mathfrak{g}^*$ such that

(1) a general fiber $H^{-1}(s)$, $s \in \mathfrak{g}^*$, is an Abelian variety of dimension N,
(2) $\mathfrak{g}^*$ is the dual Lie algebra of a general fiber $H^{-1}(s)$ considered as a Lie group, and
(3) the coordinate components of the map H are Poisson commutative with respect to the symplectic structure ω.

The notion corresponding to an algebraically completely integrable Hamiltonian system in *real* symplectic geometry is the cotangent bundle of a torus. The procedure of symplectic quotient is to remove the effect of this cotangent bundle from a given symplectic manifold. In the holomorphic context, it is often useful to take the quotient by a *family* of groups that have the same Lie algebra. Suppose we have a Lie algebra direct sum decomposition $\mathfrak{g} = \mathfrak{g}_1 \oplus \mathfrak{g}_2$. If $\mathfrak{g}_1$-action on X is integrable to a group $G_{1,s}$-action in each fiber $H^{-1}(s)$ for $s \in \mathfrak{g}_2^*$, then we can define a *quotient* $X /\!/ \mathfrak{g}_1$ as the family of quotients $H^{-1}(s)/G_{1,s}$ over $\mathfrak{g}_2^*$. We can also construct a *reduction* of (X, ω, H) by restricting the fibration to $\mathfrak{g}_2^*$ and considering the family of $\mathfrak{g}_2$-orbits in $H^{-1}(s)$, if the $\mathfrak{g}_2$-action is integrated to a group action over $\mathfrak{g}_2^*$. When applied to the moduli space of Higgs bundles, these two constructions yield Abelian fibrations that are dual to one another, producing an SYZ mirror pair. We examine these constructions for the SL-PGL and Sp_{2m}-SO_{2m+1} dualities.

From the results established in [4, 16, 17, 19], we know that linear integrable evolution equations on the Jacobians or Prym varieties are realized as the restriction of KP-type equations defined on the Sato Grassmannians through a generalization of Krichever construction. Since the Hitchin integrable systems are defined on a *family* of Jacobian varieties or Prym varieties, we need to embed the whole family into the Sato Grassmannian to compare the Hitchin systems and the KP equations. To deal with families, we use two different approaches in this article. One approach is to utilize the theory of Sato Grassmannians defined over an arbitrary scheme developed in [2, 4, 22, 23]. In this way we can directly compare the integrable Hamiltonian systems on the Higgs moduli spaces and the KP equations. The other approach is to examine the deformations of spectral curves that appear in the Hitchin Hamiltonian systems. Once we identify the effective family of spectral curves, we can embed the whole family into a single Sato Grassmannian over $\mathbb{C}$, using the method developed in [16].

The second approach has an unexpected application: we can identify the effect of Serre duality operation on the algebro-geometric data in terms of the language of Grassmannians. Note that Sato Grassmannians are constructed from pseudo-differential operators [24, 25]. We will show, using Abel's theorem, that the Serre duality is simply the formal adjoint operation on the pseudo-differential operators. Since the Sp-Hitchin system is the fixed-point-set of the Serre duality on the GL-Hitchin system, we can determine the integrable equations corresponding to the Sp case as a reduction of the KP equations on the fixed-point-set of the formal adjoint action on pseudo-differential operators. Since our formal adjoint is slightly different from what has been studied in the literature [13, 27, 29], the equations coming up for the symplectic groups are *not* BKP or CKP equations. Let

$$P = \sum a_i(x) \left(\frac{\partial}{\partial x}\right)^i$$

be a formal pseudo-differential operator, where $a_i(x)$ is a matrix valued function. We define the *formal adjoint* by

$$P^* = \sum \left(\frac{\partial}{\partial x}\right)^i \cdot a_i(-x)^t.$$

The reduction of the KP equations that corresponds to the Sp-Hitchin system is the $2m$-component KP equations that preserve the algebraic condition

$$\mathbf{L}^* = \begin{bmatrix} & I_m \\ I_m & \end{bmatrix} \cdot \mathbf{L} \cdot \begin{bmatrix} & I_m \\ I_m & \end{bmatrix} \tag{1.1}$$

for a $2m \times 2m$ matrix Lax operator $\mathbf{L}$ with the leading term $I_{2m} \cdot \partial/\partial x$. Several authors have proposed integrable systems with Sp_{2m}-symmetry (cf. [30]). It would be interesting to study our reduction (1.1) from the point of view of integrable systems and to investigate the relation with the other Sp integrable systems.

The fundamental literature of the algebro-geometric study of the Hitchin integrable systems and related topics is the book [4] by Donagi and Markman. Our present paper employs a slightly different perspective, that leads to the discovery of the KP-type equations corresponding to the Sp Hitchin system.

The relation between the Hitchin integrable systems and the KP equations was also studied in [15]. The treatment there was limited to the study of the Hitchin system on a single fiber. The present article extends the result therein.

We also note that many topics of this paper have been studied by the Salamanca school of algebraic geometers from yet another point of view [2, 7, 9, 22].

The paper is organized as follows. The first section is devoted to reviewing the Hitchin integrable systems of [3, 11]. We then determine an effective family of spectral curves in Section 3. Section 4 is devoted to giving two constructions of reduced integrable systems from a Hitchin system: one is a straightforward specialization, and the other is a kind of symplectic reduction by a Lie subalgebra. These two constructions give rise to an Abelian fibration and its dual Abelian fibration. We show that the GL-Hitchin integrable system is equivalent to the KP equations in Section 5. The identification of the Serre duality in terms of Grassmannians and pseudo-differential operators is carried out in Section 6. Finally we determine the KP-type equations for the Sp Hitchin system.

§2. Hitchin integrable systems

In this section we review the algebraically completely integrable Hamiltonian systems defined on the moduli spaces of Higgs bundles, following [3, 4, 10, 11].

Throughout the paper we denote by C a non-singular algebraic curve of genus $g \geq 2$. The moduli space of semistable algebraic vector bundles on C of rank n and degree d is denoted by $\mathcal{U}_C(n,d)$. When n and d are relatively prime, a semistable bundle is automatically stable, and the moduli space is a non-singular projective algebraic variety of dimension $n^2(g-1)+1$. We denote by

$$\mathcal{U}_C(n) = \coprod_{d \in \mathbb{Z}} \mathcal{U}_C(n,d) \tag{2.1}$$

the space of all stable vector bundles. A *Higgs bundle* is a pair (E, ϕ) consisting of an algebraic vector bundle E on C and a global section

$$\phi \in H^0(C, \mathrm{End}(E) \otimes K_C) \tag{2.2}$$

of the endomorphism sheaf of E twisted by the canonical sheaf K_C of C. A Higgs bundle is *stable* if $\frac{\deg F}{\operatorname{rank} F} < \frac{\deg E}{\operatorname{rank} E}$ for every ϕ-invariant proper holomorphic vector subbundle F. An endomorphism of a Higgs bundle (E, ϕ) is an endomorphism ψ of E that commutes with ϕ:

$$\begin{array}{ccc} E & \xrightarrow{\psi} & E \\ \phi\downarrow & & \downarrow\phi \\ E \otimes K_C & \xrightarrow[\psi \otimes 1]{} & E \otimes K_C. \end{array}$$

It is known that $H^0(C, \mathrm{End}(E, \phi)) = \mathbb{C}$ for a stable Higgs bundle, and one can define the moduli space of stable objects. We denote by $\mathcal{H}_C(n,d)$ the moduli space of stable Higgs bundles of rank n and degree d on C, and

$$\mathcal{H}_C(n) = \coprod_{d \in \mathbb{Z}} \mathcal{H}_C(n,d). \tag{2.3}$$

Note that the *Serre duality* induces an involution on $\mathcal{H}_C(n)$ defined by

$$\mathcal{H}_C(n,d) \ni (E, \phi) \longmapsto (E^* \otimes K_C, -\phi^*) \in \mathcal{H}_C(n, -d + 2n(g-1)). \tag{2.4}$$

The dual of the Higgs field $\phi : E \to E \otimes K_C$ is a homomorphism $\phi^* : E^* \otimes K_C^{-1} \to E^*$. We use the same notation for the homomorphism $E^* \otimes K_C \to E^* \otimes K_C^{\otimes 2}$ induced by ϕ^*.

If E is stable, then (E, ϕ) is stable for any ϕ of (2.2). And if $\phi = 0$, then the stability of (E, ϕ) simply means E is stable. Therefore, the Higgs moduli space contains the total space of the holomorphic cotangent bundle

$$T^*\mathcal{U}_C(n,d) \subset \mathcal{H}_C(n,d), \tag{2.5}$$

since

$$H^0(C, \mathrm{End}(E) \otimes K_C) \cong H^1(C, \mathrm{End}(E))^* \cong T_E^*\mathcal{U}_C(n,d).$$

Note that $p^*\Lambda^1(\mathcal{U}_C(n,d)) \subset \Lambda^1(T^*\mathcal{U}_C(n,d))$ has a tautological section

$$\eta \in H^0(T^*\mathcal{U}_C(n,d), p^*\Lambda^1(\mathcal{U}_C(n,d))), \tag{2.6}$$

where $p : T^*\mathcal{U}_C(n,d) \to \mathcal{U}_C(n,d)$ is the projection, and Λ^r denotes the sheaf of holomorphic r-forms. The differential $\omega = -d\eta$ of the tautological section defines the canonical holomorphic symplectic form on $T^*\mathcal{U}_C(n,d)$. The restriction of ω on $\mathcal{U}_C(n,d)$, which is the 0-section of the cotangent bundle, is identically 0. Therefore the 0-section is a Lagrangian submanifold of this holomorphic symplectic space.

Let us denote by

$$V_{GL}^* = V_{GL_n(\mathbb{C})}^* = \bigoplus_{i=1}^{n} H^0(C, K_C^{\otimes i}). \tag{2.7}$$

As a vector space V_{GL}^* has the same dimension of $H^0(C, \mathrm{End}(E) \otimes K_C)$. The Higgs field $\phi : E \to E \otimes K_C$ induces a homomorphism of the i-th anti-symmetric tensor product spaces

$$\wedge^i(\phi) : \wedge^i(E) \longrightarrow \wedge^i(E \otimes K_C) = \wedge^i(E) \otimes K_C^{\otimes i},$$

or equivalently $\wedge^i(\phi) \in H^0(C, \mathrm{End}(\wedge^i(E)) \otimes K_C^{\otimes i})$. Taking its natural trace, we obtain

$$\mathrm{tr} \wedge^i (\phi) \in H^0(C, K_C^{\otimes i}).$$

This is exactly the i-th characteristic coefficient of the twisted endomorphism ϕ:

$$\det(x - \phi) = x^n + \sum_{i=1}^{n} (-1)^i \mathrm{tr} \wedge^i (\phi) \cdot x^{n-i}. \tag{2.8}$$

By assigning the coefficients of (2.8), Hitchin [11] defines a holomorphic map, now known as the *Hitchin fibration* or *Hitchin map*,

$$H : \mathcal{H}_C(n,d) \ni (E, \phi) \longmapsto \det(x - \phi) \in \bigoplus_{i=1}^{n} H^0(C, K_C^{\otimes i}) = V_{GL}^*. \tag{2.9}$$

The map H to the vector space V_{GL}^* is a collection of $N = n^2(g-1)+1$ globally defined holomorphic functions on $\mathcal{H}_C(n,d)$. The 0-fiber of the Hitchin fibration is the moduli space $\mathcal{U}_C(n,d)$.

To determine the generic fiber of H, the notion of *spectral curves* is introduced in [11]. The total space of the canonical sheaf $K_C = \Lambda^1(C)$ on C is the cotangent bundle T^*C. Let

$$\pi : T^*C \longrightarrow C \tag{2.10}$$

be the projection, and

$$\tau \in H^0(T^*C, \pi^* K_C) \subset H^0(T^*C, \Lambda^1(T^*C))$$

be the tautological section of π^*K_C on T^*C. Here again $\omega_{T^*C} = -d\tau$ is the holomorphic symplectic form on T^*C. The tautological section τ satisfies that $\sigma^*\tau = \sigma$ for every section $\sigma \in H^0(C, K_C)$ viewed as a holomorphic map $\sigma : C \to T^*C$. The characteristic coefficients (2.8) of ϕ give a section

(2.11)

$$s = \det(\tau - \phi) = \tau^{\otimes n} + \sum_{i=1}^{n} (-1)^i \mathrm{tr} \wedge^i (\phi) \cdot \tau^{\otimes n-1} \in H^0(T^*C, \pi^* K_C^{\otimes n}).$$

We define the spectral curve C_s associated with a Higgs pair (E, ϕ) as the divisor of 0-points of the section $s = \det(\tau - \phi)$ of the line bundle $\pi^* K_C^{\otimes n}$:

$$C_s = (s)_0 \subset T^*C. \tag{2.12}$$

The projection π of (2.10) defines a ramified covering map $\pi : C_s \to C$ of degree n.

There is yet another construction of the spectral curve C_s. Since the section $s = \det(\tau - \phi)$ is determined by the characteristic coefficients of ϕ, by abuse of notation we identify s as an element of V_{GL}^*:

$$s = (s_1, s_2, \dots, s_n) = (-\mathrm{tr}\,\phi, \mathrm{tr} \wedge^2 (\phi), \dots, (-1)^n \mathrm{tr} \wedge^n (\phi)) \in \bigoplus_{i=1}^{n} H^0(C, K_C^{\otimes i}).$$

It defines an $\mathcal{O}_C$-module $(s_1 + s_2 + \cdots + s_n) \cdot K_C^{\otimes -n}$. Let $\mathcal{I}_s$ denote the ideal generated by this module in the symmetric tensor algebra $\mathrm{Sym}(K_C^{-1})$. Since K_C^{-1} is the sheaf of linear functions on T^*C, the scheme associated to this tensor algebra is $\operatorname{Spec}(\operatorname{Sym}(K_C^{-1})) = T^*C$. The spectral curve as the divisor of 0-points of s is then defined by

$$C_s = \operatorname{Spec}\left(\frac{\operatorname{Sym}(K_C^{-1})}{\mathcal{I}_s}\right) \subset \operatorname{Spec}(\operatorname{Sym}(K_C^{-1})) = T^*C. \tag{2.13}$$

We denote by U_{reg} the set consisting of points s for which C_s is non-singular. It is an open dense subset of V_{GL}^* [3]. We note that since $\deg(K_C) = 2g-2 > 0$, every divisor of T^*C intersects with the 0-section C. Therefore, if C_s is non-singular, then it has only one component, and is therefore irreducible. The genus of C_s is $g(C_s) = n^2(g-1)+1$, which follows from an isomorphism

$$\pi_*\mathcal{O}_{C_s} = \operatorname{Sym}(K_C^{-1})/\mathcal{I}_s \cong \bigoplus_{i=0}^{n-1} K_C^{\otimes -i}$$

as an $\mathcal{O}_C$-module.

The Higgs field $\phi \in H^0(C, \operatorname{End}(E) \otimes K_C)$ gives a homomorphism

$$\varphi : K_C^{-1} \longrightarrow \operatorname{End}(E),$$

which induces an algebra homomorphism, still denoted by the same letter,

$$\varphi : \operatorname{Sym}(K_C^{-1}) \longrightarrow \operatorname{End}(E).$$

Since $s \in V_{GL}^*$ is the characteristic coefficients of φ, by the Cayley–Hamilton theorem, the homomorphism φ factors through

$$\operatorname{Sym}(K_C^{-1}) \longrightarrow \operatorname{Sym}(K_C^{-1})/\mathcal{I}_s \longrightarrow \operatorname{End}(E).$$

Hence E is a module over $\operatorname{Sym}(K_C^{-1})/\mathcal{I}_s$ of rank 1. The rank is 1 because the ranks of E and $\operatorname{Sym}(K_C^{-1})/\mathcal{I}_s$ are the same as $\mathcal{O}_C$-modules. In this way a Higgs pair (E, ϕ) gives rise to a line bundle $\mathcal{L}_E$ on the spectral curve C_s, if it is nonsingular. Since $\mathcal{L}_E$ being an $\mathcal{O}_{C_s}$-module is equivalent to E being a $\operatorname{Sym}(K_C^{-1})/\mathcal{I}_s$-module, we recover E from $\mathcal{L}_E$ simply by $E = \pi_*\mathcal{L}_E$, which has rank n because π is a covering of degree n. From the equality $\chi(C, E) = \chi(C_s, \mathcal{L}_E)$ and Riemann–Roch, we find that $\deg \mathcal{L}_E = d + n(n-1)(g-1)$. To summarize, the above construction defines an inclusion map

$$H^{-1}(s) \subset \operatorname{Pic}^{d+n(n-1)(g-1)}(C_s) \cong \operatorname{Jac}(C_s),$$

if C_s is non-singular.

Conversely, suppose we have a line bundle $\mathcal{L}$ of degree $d + n(n-1)(g-1)$ on a non-singular spectral curve C_s. Then $\pi_*\mathcal{L}$ is a module over $\pi_* \mathcal{O}_{C_s} = \mathrm{Sym}(K_C^{-1})/\mathcal{I}_s$, which defines a homomorphism $K_C^{-1} \to \mathrm{End}(\pi_* \mathcal{L})$, or equivalently, $\psi : \pi_* \mathcal{L} \to \pi_* \mathcal{L} \otimes K_C$. It is easy to see that the Higgs pair $(\pi_* \mathcal{L}, \psi)$ is stable. Suppose we had a ψ-invariant subbundle $F \subset \pi_* \mathcal{L}$ of rank $k < n$. Since $(F, \psi|_F)$ is a Higgs pair, it gives rise to a spectral curve $C_{s'}$ that is a degree k covering of C. Note that the characteristic polynomial $s' = \det(x - \psi)$ is a factor of the full characteristic polynomial $s = \det(x - \phi)$. Therefore, $C_{s'}$ is a component of C_s, which contradicts to our assumption that C_s is irreducible. Therefore, $\pi_* \mathcal{L}$ has no ψ-invariant proper subbundle. Thus we have established that

(2.14)
$$H^{-1}(s) = \mathrm{Pic}^{d+n(n-1)(g-1)}(C_s) \cong \mathrm{Jac}(C_s), \qquad s \in U_{\mathrm{reg}} \subset V_{GL}^*.$$

The vector bundle $\pi_* \mathcal{L}$ itself is not necessarily stable. It is proved in [3] that the locus of $\mathcal{L}$ in $\mathrm{Pic}^{d+n(n-1)(g-1)}(C_s)$ that gives unstable $\pi_* \mathcal{L}$ has codimension two or more.

Recall that the tautological section

$$\eta \in H^0(T^* \mathcal{U}_C(n,d), p^* \Lambda^1(\mathcal{U}_C(n,d)))$$

is a holomorphic 1-form on $T^* \mathcal{U}_C(n,d) \subset \mathcal{H}_C(n,d)$. Its restriction to the fiber $H^{-1}(s)$ of $s \in U_{\mathrm{reg}}$ for which C_s is non-singular extends to a holomorphic 1-form on the whole fiber $H^{-1}(s) \cong \mathrm{Jac}(C_s)$ since η is undefined only on a codimension 2 subset. Hence η extends as a holomorphic 1-form on $H^{-1}(U_{\mathrm{reg}})$. Thus η is well defined on $T^* \mathcal{U}_C(n,d) \cup H^{-1}(U_{\mathrm{reg}})$. The complement of this open subset in $\mathcal{H}_C(n,d)$ consists of such Higgs pairs (E, ϕ) that E is unstable *and* C_s is singular. Since the stability of E and the non-singular condition for C_s are both open conditions, this complement has codimension at least two. Consequently, both the tautological section η and the holomorphic symplectic form $\omega = -d\eta$ extend holomorphically to the whole Higgs moduli space $\mathcal{H}_C(n,d)$.

We note that there are no holomorphic 1-forms other than constants on a Jacobian variety since its cotangent bundle is trivial. It implies that

$$\omega|_{H^{-1}(s)} = -d(\eta|_{H^{-1}(s)}) \equiv 0$$

for $s \in U_{\mathrm{reg}}$. Therefore, a generic fiber of the Hitchin fibration is a Lagrangian subvariety of the holomorphic symplectic variety $(\mathcal{H}_C(n,d), \omega)$. The *Poisson structure* on $H^0(\mathcal{H}_C(n,d), \mathcal{O}_{\mathcal{H}_C(n,d)})$ is defined by

$$\{f, g\} = \omega(X_f, X_g), \qquad f, g \in H^0(\mathcal{H}_C(n,d), \mathcal{O}_{\mathcal{H}_C(n,d)}),$$

where X_f denotes the Hamiltonian vector field defined by the relation $df = \omega(X_f, \cdot)$. Since ω vanishes on a generic fiber of H, the holomorphic functions on $\mathcal{H}_C(n,d)$ coming from the coordinates of the Hitchin fibration are *Poisson commutative* with respect to the holomorphic symplectic structure ω. Therefore, $(\mathcal{H}_C(n,d), \omega, H)$ is an algebraically completely integrable Hamiltonian system [4, 31], called the *Hitchin integrable system.*

Theorem 2.1 (Hitchin [11]). *The Hitchin fibration*

$$H : \mathcal{H}_C(n,d) \longrightarrow V^*_{GL}$$

is a generically Lagrangian Jacobian fibration that defines an algebraically completely integrable Hamiltonian system $(\mathcal{H}_C(n,d), \omega, H)$. A generic fiber $H^{-1}(s)$ is a Lagrangian with respect to the holomorphic symplectic structure ω and is isomorphic to the Jacobian variety of a spectral curve C_s.

The Hitchin map H restricted to $T^*_E\mathcal{U}_C(n,d)$ for a stable E is a polynomial map

$$T^*_E\mathcal{U}_C(n,d) = H^0(C, \mathrm{End}(E) \otimes K_C) \ni \phi \longmapsto \det(x - \phi) \in \bigoplus_{i=1}^{n} H^0(C, K_C^{\otimes i})$$

consisting of g linear components, $3g-3$ quadratic components, $5g-5$ cubic components, etc., and $(2n-1)(g-1)$ components of degree n. Thus the inverse image

$$H^{-1}(s) \cap T^*_E\mathcal{U}_C(n,d)$$

for a generic (E,ϕ) consists of

$$\delta = \prod_{i=1}^{n} i^{(2i-1)(g-1)} \tag{2.15}$$

points [3]. Consequently, the map

$$p : \mathrm{Jac}(C_s) \cong H^{-1}(s) \longrightarrow H^{-1}(0) = \mathcal{U}_C(n,d) \tag{2.16}$$

is a covering morphism of degree δ.

Let us now determine the Hamiltonian vector field associated to each coordinate function of H. Take a point $(E,\phi) \in T^*\mathcal{U}_C(n,d) \cap H^{-1}(U_{\mathrm{reg}})$. The tangent space of the Higgs moduli space at this point is given by

$$T_{(E,\phi)}\mathcal{H}_C(n,d) = H^1(C, \mathrm{End}(E)) \oplus H^0(C, \mathrm{End}(E) \otimes K_C). \tag{2.17}$$

Since the symplectic form $\omega = -d\eta$ is the exterior derivative of the tautological 1-form η of (2.6) on $T^*\mathcal{U}_C(n,d)$, the evaluation of ω at (E,ϕ) is given by

$$\omega\big((a_1,b_1),(a_2,b_2)\big) = \langle a_1, b_2\rangle - \langle a_2, b_1\rangle, \tag{2.18}$$

where $(a_1, b_1), (a_2, b_2) \in H^1(C, \mathrm{End}(E)) \oplus H^0(C, \mathrm{End}(E) \otimes K_C)$, and

$$\langle\cdot,\cdot\rangle : H^1(C, \mathrm{End}(E)) \oplus H^0(C, \mathrm{End}(E) \otimes K_C) \longrightarrow \mathbb{C}$$

is the Serre duality pairing. This expression is the standard Darboux form of the symplectic form ω. Choose an open neighborhood Y_1 of E in $\mathcal{U}_C(n,d)$ on which the cotangent bundle $T^*\mathcal{U}_C(n,d)$ is trivial. Then H is a polynomial map

$$H : Y_1 \times H^0(C, \mathrm{End}(E) \otimes K_C) \ni (E', \phi) \longmapsto s = \det(x - \phi) \in \bigoplus_{i=1}^{n} H^0(C, K_C^{\otimes i})$$

that depends only on the second factor. The differential $dH_{(E,\phi)}$ at the point (E, ϕ) gives a linear isomorphism

$$dH_{(E,\phi)} : H^0(C, \mathrm{End}(E) \otimes K_C) \xrightarrow{\sim} \bigoplus_{i=1}^{n} H^0(C, K_C^{\otimes i}). \tag{2.19}$$

As to the first factor of the tangent space of (2.17), we use the differential of the covering map p of (2.16):

$$dp_{(E,\phi)} : H^1(C_s, \mathcal{O}_{C_s}) \xrightarrow{\sim} H^1(C, \mathrm{End}(E)), \tag{2.20}$$

where $s = H(E, \phi)$. The dual of (2.19), together with (2.20), gives an isomorphism

$$V_{GL} \stackrel{\mathrm{def}}{=} \bigoplus_{i=0}^{n-1} H^1(C, K_C^{-\otimes i}) = H^1(C, \pi_* \mathcal{O}_{C_s}) \xrightarrow{\sim} H^1(C, \mathrm{End}(E)) \cong H^1(C_s, \mathcal{O}_{C_s}). \tag{2.21}$$

From (2.19) and (2.21), we see that the tangent space to the Higgs moduli is given by

$$\begin{aligned} T_{(E,\phi)}\mathcal{H}_C(n,d) &\cong \bigoplus_{i=1}^{n} H^1(C, K_C^{-\otimes(i-1)}) \oplus \bigoplus_{i=1}^{n} H^0(C, K_C^{\otimes i}) \\ &= V_{GL} \oplus V_{GL}^*. \end{aligned} \tag{2.22}$$

Therefore, the symplectic form has a decomposition into n pieces $\omega = \omega_1 + \omega_2 + \cdots + \omega_n$, and in each factor $H^1(C, K_C^{-\otimes(i-1)}) \oplus H^0(C, K_C^{\otimes i})$, ω_i takes the standard Darboux form

$$\omega_i\big((a_1^i, b_1^i), (a_2^i, b_2^i)\big) = \langle a_1^i, b_2^i\rangle_i - \langle a_2^i, b_1^i\rangle_i, \tag{2.23}$$

where $(a_1^i, b_1^i), (a_2^i, b_2^i) \in H^1(C, K_C^{-\otimes(i-1)}) \oplus H^0(C, K_C^{\otimes i})$, and $\langle \cdot, \cdot \rangle_i$ is the duality pairing of $H^1(C, K_C^{-\otimes(i-1)})$ and $H^0(C, K_C^{\otimes i})$.

Since $s = H(E, \phi) \in U_{\text{reg}}$ is a regular value of H, there is an open subset $Y_2 \subset U_{\text{reg}} \subset V_{GL}^*$ around s such that $H^{-1}(Y_2)$ is locally the product of the fiber $H^{-1}(s)$ and Y_2. By taking Y_1 smaller if necessary, we thus obtain a local product neighborhood

$$p^{-1}(Y_1) \cap H^{-1}(Y_2) \cong Y_1 \times Y_2 = Y$$

of (E, ϕ) in $\mathcal{H}_C(n, d)$. By construction, the projections to the first and the second factors coincide with the projection $p : T^*\mathcal{U}_C(n, d) \to \mathcal{U}_C(n, d)$ and the Hitchin map H:

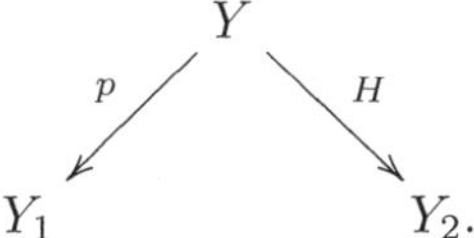

The neighborhood Y and these projections provide the Darboux coordinate system for ω, and its expression (2.18, 2.23) globally holds on Y. In particular, the Hamiltonian vector fields corresponding to the components of H on Y are constant vector fields that are determined by elements of $H^1(C, K_C^{-\otimes(i-1)})$ for each $i = 1, \ldots, n$. Let $(h_1, \ldots, h_N)$ be a linear coordinate system of V_{GL}^*, where $N = n^2(g-1)+1$. Since $V_{GL}^* = \bigoplus_{i=1}^n H^0(C, K_C^{\otimes i})$, each coordinate function is an element of V_{GL}:

$$h_1, \ldots, h_N \in V_{GL} = \bigoplus_{i=0}^{n-1} H^1(C, K_C^{-\otimes i}). \tag{2.24}$$

Since $H^0(C_s, \mathcal{O}_{C_s}) = \mathbb{C}$, the Hitchin map $H : \mathcal{H}_C(n, d) \to V_{GL}^*$ is actually the pull back of the coordinate functions $H^{-1}(h_1), \ldots, H^{-1}(h_N)$ on $H^{-1}(U_{\text{reg}}) \subset \mathcal{H}_C(n, d)$. The identification of V_{GL} as the first factor of the tangent space in (2.22) gives the Hamiltonian vector fields corresponding to the coordinate components of the Hitchin map. We have therefore established that the Hamiltonian vector fields are linear flows with respect to the linear coordinate of the Jacobian $\text{Jac}(C_s)$ determined by (2.21). Of course each of these linear flows extends globally on $\text{Jac}(C_s)$ since the tangent bundle of a Jacobian is trivial and a local linear function extends globally as a linear function.

Theorem 2.2 (Hitchin [11]). *The Hamiltonian vector fields corresponding to the Hitchin fibration are linear Jacobian flows on a generic fiber* $H^{-1}(s) \cong \text{Jac}(C_s)$.

How does the construction of the *spectral data* $(\mathcal{L}_E, C_s)$ from a Higgs pair (E, ϕ) change when we consider the Serre dual $(E^* \otimes K_C, -\phi^*)$? To answer this question, first we note

$$\begin{aligned} \operatorname{tr} \wedge^i (\phi) = \operatorname{tr} \wedge^i (\phi^*) &\in H^0(C, \operatorname{End}(\wedge^i E) \otimes K_C^i) \\ &= H^0(C, \operatorname{End}(\wedge^i (E^* \otimes K_C)) \otimes K_C^i). \end{aligned} \tag{2.25}$$

For $s = (s_1, s_2, \ldots, s_n) \in V_{GL}^*$, we write $s^* = (-s_1, s_2, \ldots, (-1)^n s_n)$. By definition, the spectral curves C_s and C_{s^*} are isomorphic. As divisors of T^*C, their isomorphism $C_s \cong \epsilon(C_{s^*})$ is induced by the involution of T^*C

$$\epsilon : T^*C \ni (p, x) \longmapsto (p, -x) \in T^*C, \tag{2.26}$$

where $p \in C$ and $x \in T_p^*C$.

Proposition 2.3 (Hitchin [12]). *The spectral data* $(\mathcal{L}_{E^* \otimes K_C}, C_{s^*})$ *corresponding to the Serre dual* $(E^* \otimes K_C, -\phi^*)$ *of the Higgs pair* (E, ϕ) *is given by*

$$\begin{cases} \mathcal{L}_{E^* \otimes K_C} = \epsilon^*(\mathcal{L}_E^* \otimes K_{C_s}) \\ C_{s^*} = \epsilon(C_s). \end{cases}$$

The degree of these isomorphic line bundles is $-\deg(E) + (n^2+n)(g-1)$.

Proof. We use the exact sequence of [3, 12] that characterizes $\mathcal{L}_E$:

$$0 \longrightarrow \mathcal{L}_E \otimes K_{C_s}^{-1} \otimes \pi^* K_C \longrightarrow \pi^* E \longrightarrow \pi^*(E \otimes K_C) \longrightarrow \mathcal{L}_E \otimes \pi^* K_C \longrightarrow 0.$$

The dual of this sequence is then

$$\begin{aligned} 0 \longrightarrow \mathcal{L}_E^* \otimes \pi^* K_C \longrightarrow \pi^*(E^* \otimes K_C) &\longrightarrow \pi^*(E^* \otimes K_C^{\otimes 2}) \\ &\longrightarrow \mathcal{L}_E^* \otimes K_{C_s} \otimes \pi^* K_C \longrightarrow 0. \end{aligned}$$

Thus the spectral data of the Higgs pair $(E^* \otimes K_C, \phi^*)$ is $(\mathcal{L}_E^* \otimes K_{C_s}, C_s)$. The involution ϵ comes in here when we consider the Higgs pair $(E^* \otimes K_C, -\phi^*)$. Q.E.D.

Proposition 2.4. *Hitchin's integrable system for the group* $Sp_{2m}(\mathbb{C})$ *is realized as the fixed-point-set of the Serre duality involution* $(E, \phi) \mapsto (E^* \otimes K_C, -\phi^*)$ *defined on the Higgs moduli space* $\mathcal{H}_C(2m, 2m(g-1))$.

Proof. The fixed-point-set consists of Higgs pairs (E, ϕ) such that $E \cong E^* \otimes K_C$ and $\phi = -\phi^*$. Choose a square root of K_C and define $F = E \otimes K_C^{-1/2}$. The Higgs field $\phi = -\phi^*$ naturally acts on this bundle, and the pair (F, ϕ) forms the moduli space of Sp_{2m}-Higgs bundles [12]. The characteristic coefficients satisfy the relation $s = s^*$, hence

$$s \in V_{Sp}^* \stackrel{\text{def}}{=} \bigoplus_{i=1}^{m} H^0(C, K_C^{\otimes 2i}). \tag{2.27}$$

The spectral curve C_s has a non-trivial involution $\epsilon : C_s \to C_s$. From the exact sequence

$$0 \longrightarrow \mathcal{L}_E \otimes K_{C_s}^{-1} \otimes \pi^* K_C^{1/2} \longrightarrow \pi^*(E \otimes K_C^{-1/2}) \longrightarrow \pi^*(E \otimes K_C^{1/2}) \longrightarrow \mathcal{L}_E \otimes \pi^* K_C^{1/2} \longrightarrow 0,$$

we see that $\mathcal{L}_F \cong \mathcal{L}_E \otimes \pi^* K_C^{-1/2}$ and $\mathcal{L}_{F^*} \cong \epsilon^* \mathcal{L}_F^* \otimes K_{C_s} \otimes \pi^* K_C^{-1/2}$. If we define $\mathcal{L}_0 = \mathcal{L}_F \otimes \pi^* K_C^{-m+1/2}$ following [12], then $\mathcal{L}_0^* \cong \epsilon^* \mathcal{L}_0$. Q.E.D.

§3. Deformations of spectral curves

The Hitchin fibration (2.9) is a family of deformations of Jacobians, but it is not *effective*. In this section we determine the natural effective family associated with the Hitchin fibration.

An obvious action on $\mathcal{H}_C(n, d)$ that preserves the spectral curves is the scalar multiplication of Higgs fields $\phi \mapsto \lambda \cdot \phi$ by $\lambda \in \mathbb{C}^*$. Although this $\mathbb{C}^*$-action is *not* symplectomorphic because it changes the symplectic form $\omega \mapsto \lambda \cdot \omega$, from the point of view of constructing an effective family of Jacobians, we need to quotient it out. We note that the $\mathbb{C}^*$-action on V_{GL}^* defined by

$$V_{GL}^* = \bigoplus_{i=1}^{n} H^0(C, K_C^{\otimes i}) \ni s = (s_1, s_2, \ldots, s_n) \longmapsto \lambda \cdot s = (\lambda s_1, \lambda^2 s_2, \ldots, \lambda^n s_n) \in V_{GL}^* \tag{3.1}$$

makes the Hitchin fibration $\mathbb{C}^*$-equivariant:

$$\begin{array}{ccc} \mathcal{H}_C(n,d) & \xrightarrow{\quad\lambda\quad} & \mathcal{H}_C(n,d) \\ {\scriptstyle H}\big\downarrow & & \big\downarrow{\scriptstyle H} \\ \bigoplus_{i=1}^n H^0(C, K_C^{\otimes i}) & \xrightarrow[\operatorname{diag}(\lambda,\ldots,\lambda^n)]{} & \bigoplus_{i=1}^n H^0(C, K_C^{\otimes i}), \end{array} \qquad \lambda \in \mathbb{C}^*. \tag{3.2}$$

Since C_s is defined by the equation (2.11), the natural $\mathbb{C}^*$-action on the cotangent bundle T^*C gives the isomorphism $\lambda : C_s \to C_{\lambda \cdot s}$ that commutes with the projection $\pi : T^*C \to C$. The line bundles $\mathcal{L}_{(E,\phi)}$ on C_s

and $\mathcal{L}_{(E,\lambda\phi)}$ on $C_{\lambda\cdot s}$ corresponding to E are related by this isomorphism by

$$\lambda^*\mathcal{L}_{(E,\lambda\phi)} \xrightarrow{\sim} \mathcal{L}_{(E,\phi)}.$$

Another group action on the Higgs moduli space that leads to trivial deformations of the spectral curves is the Jacobian action. The $\mathrm{Jac}(C)$-action on $\mathcal{H}_C(n,d)$ is defined by $(E,\phi) \longmapsto (E \otimes L, \phi)$, where $L \in \mathrm{Jac}(C) = \mathrm{Pic}^0(C)$ is a line bundle on C of degree 0. The Higgs field is preserved in this action because $(E \otimes L)^* \otimes (E \otimes L) = E^* \otimes E$ is unchanged, hence

$$\phi \in H^0(C, \mathrm{End}(E) \otimes K_C) = H^0(C, \mathrm{End}(E \otimes L) \otimes K_C).$$

Thus the cotangent bundle $T^*\mathcal{U}_C(n,d)$ is trivial along every orbit of the $\mathrm{Jac}(C)$-action on $\mathcal{U}_C(n,d)$, and each $\mathrm{Jac}(C)$-orbit in $\mathcal{H}_C(n,d)$ lies in the same fiber of the Hitchin fibration. Let us identify the $\mathrm{Jac}(C)$-action on a generic fiber $H^{-1}(s) \cong \mathrm{Jac}(C_s)$. The covering map $\pi : C_s \to C$ induces an injective homomorphism $\pi^* : \mathrm{Jac}(C) \ni L \longmapsto \pi^*L \in \mathrm{Jac}(C_s)$. This is injective because if $\pi^*L \cong \mathcal{O}_{C_s}$, then by the projection formula we have

$$\pi_*(\pi^*L) \cong \pi_*\mathcal{O}_{C_s} \otimes L \cong \bigoplus_{i=0}^{n-1} L \otimes K_C^{\otimes -i},$$

which has a nowhere vanishing section. Hence $L \cong \mathcal{O}_C$. Take a point $(E,\phi) \in H^{-1}(s)$ and let $\mathcal{L}_E$ be the corresponding line bundle on C_s. Since $\pi_*(\mathcal{L}_E \otimes \pi^*L) \cong E \otimes L$, the action of $\mathrm{Jac}(C)$ on $H^{-1}(s) \cong \mathrm{Jac}(C_s)$ is through the canonical action of the subgroup

$$\mathrm{Jac}(C) \cong \pi^*\mathrm{Jac}(C) \subset \mathrm{Jac}(C_s)$$

on $\mathrm{Jac}(C_s)$.

On the open subset $T^*\mathcal{U}_C(n,d)$ of $\mathcal{H}_C(n,d)$, the $\mathrm{Jac}(C)$-action is symplectomorphic because it is induced by the action on the base space $\mathcal{U}_C(n,d)$. On the other open subset $H^{-1}(U_{\mathrm{reg}})$ the action is also symplectomorphic because it preserves each fiber which is a Lagrangian. Since the symplectic form ω is defined by extending the canonical form $\omega = -d\eta$ to $\mathcal{H}_C(n,d)$, the $\mathrm{Jac}(C)$-action is globally holomorphic symplectomorphic on $\mathcal{H}_C(n,d)$. This action is actually a Hamiltonian action and the first component of the Hitchin map

$$H_1 : \mathcal{H}_C(n,d) \ni (E,\phi) \longmapsto \mathrm{tr}(\phi) \in H^0(C, K_C) \tag{3.3}$$

is the moment map. Note that $H^1(C, \mathcal{O}_C)$ is the Lie algebra of the Abelian group $\mathrm{Jac}(C)$, and $H^0(C, K_C)$ is its dual Lie algebra. Since the infinitesimal action of $H^1(C, \mathcal{O}_C)$ on the Higgs moduli space defines a vector field which is obtained by identifying $H^1(C, \mathcal{O}_C)$ with the first component of (2.21), the symplectic dual to this vector field is the map to the first component of second factor in (2.22), i.e, H_1.

We can therefore construct the symplectic quotient $\mathcal{H}_C(n,d)/\!/\mathrm{Jac}(C)$, which we will do in Section 4. Here our interest is to determine an effective family of spectral curves. The moment map H_1 of (3.3) being the trace of ϕ, it is natural to define

$$V^*_{SL} = V^*_{SL_n(\mathbb{C})} = \bigoplus_{i=2}^{n} H^0(C, K_C^{\otimes i}) \subset V^*_{GL}. \tag{3.4}$$

This is a vector space of dimension $(n^2-1)(g-1)$. Now consider a partial projective space

$$\mathbb{P}(\mathcal{H}_C^{SL}(n,d)) = \left(H^{-1}(V^*_{SL}) \setminus H^{-1}(0)\right)/\mathbb{C}^*. \tag{3.5}$$

This is no longer a holomorphic symplectic manifold, yet the Hitchin fibration naturally descends to a generically Jacobian fibration

$$PH : \mathbb{P}(\mathcal{H}_C^{SL}(n,d)) \longrightarrow \mathbb{P}_w(V^*_{SL}) \tag{3.6}$$

over the weighted projective space of V^*_{SL} defined by the restriction of (3.1) on V^*_{SL}. We now claim

Theorem 3.1 (Effective Jacobian fibration)**.** *The Jacobian fibration*

$$PH : \mathbb{P}(\mathcal{H}_C^{SL}(n,d)) \longrightarrow \mathbb{P}_w(V^*_{SL})$$

is generically effective.

The rest of this section is devoted to the proof of this theorem. The statement is equivalent to the claim that the family of deformations of spectral curves parametrized by $\mathbb{P}_w(V^*_{SL})$ is effective. Let $s \in V^*_{SL}$ be a point for which the spectral curve C_s is non-singular, and $\bar{s}$ the corresponding point of $\mathbb{P}_w(V^*_{SL})$. We wish to show that the Kodaira-Spencer map

$$T_{\bar{s}}\mathbb{P}_w(V^*_{SL}) \longrightarrow H^1(C_s, \mathcal{T}C_s)$$

is injective, where $\mathcal{T}X$ denotes the tangent sheaf of X.

The spectral curve C_s of (2.11) is the divisor of T^*C defined by the section

$$\det(\tau - \phi) \in H^0(T^*C, \pi^* K_C^{\otimes n}),$$

where τ is the tautological section of $\pi^* K_C$ on T^*C. Let $\mathcal{N}_s$ denote the normal sheaf of C_s in T^*C. Since $K_{T^*C} = \Lambda^2(T^*C) \cong \mathcal{O}_{T^*C}$ by the holomorphic symplectic form $\omega_C = -d\tau$, we have

$$\mathcal{N}_s \cong K_{C_s} \cong \pi^* K_C^{\otimes n}.$$

From

$$0 \longrightarrow \mathcal{T}C_s \longrightarrow \mathcal{T}T^*C|_{C_s} \longrightarrow \mathcal{N}_s \longrightarrow 0,$$

we obtain
(3.7)

$$\begin{aligned} 0 \longrightarrow H^0(C_s, \mathcal{T}T^*C) &\xrightarrow{\ \iota\ } H^0(C_s, \mathcal{N}_s) \xrightarrow{\ \kappa\ } H^1(C_s, \mathcal{T}C_s) \\ \longrightarrow H^1(C_s, \mathcal{T}T^*C) &\longrightarrow H^1(C_s, \mathcal{N}_s). \end{aligned}$$

Since

$$H^0(C_s, \mathcal{N}_s) \cong H^0(C, \pi_* K_{C_s}) \cong \bigoplus_{i=1}^{n} H^0(C, K_C^{\otimes i}) = V_{GL}^*,$$

the homomorphism κ is the Kodaira–Spencer map for the deformations of spectral curves on V_{GL}^*. We thus need to identify the image of ι. We note that the tangent sheaf and the cotangent sheaf are isomorphic on the total space of the cotangent bundle T^*C, i.e., $\mathcal{T}T^*C \cong \Lambda^1(T^*C)$.

Proposition 3.2. *We have the following isomorphisms:*

$$\begin{aligned} &H^0(T^*C, \Lambda^0(T^*C)) \cong \mathbb{C} \\ &H^0(T^*C, \Lambda^1(T^*C)) \cong H^0(C, K_C) \oplus \mathbb{C} \cdot \tau \qquad (3.8) \\ &H^0(T^*C, \Lambda^2(T^*C)) \cong \mathbb{C}. \end{aligned}$$

Proof. The first isomorphism of (3.8) asserts that every globally defined holomorphic function f on T^*C is a constant. A section $\sigma \in H^0(C, K_C)$ is a map $\sigma : C \to T^*C$. Take an arbitrary pair of points (x, y) in T^*C. If they are not on the same fiber, then there is a section σ such that both x and y are on the image $\sigma(C)$. Since f is constant on $\sigma(C)$, $f(x) = f(y)$. If they are on the same fiber, then choose a point $z \in T^*C$ not on this fiber and use the same argument.

Since $\Lambda^2(T^*C) \cong \mathcal{O}_{T^*C}$, the third isomorphism follows from the first one.

The second isomorphism of (3.8) asserts that every holomorphic 1-form α on T^*C is either the pull-back of a holomorphic 1-form on C via $\pi : T^*C \to C$, the tautological 1-form τ, or a linear combination of them. Let $\sigma : C \to T^*C$ be an arbitrary section. Since the normal sheaf of $\sigma(C)$ in T^*C is $K_{\sigma(C)}$, we have on $\sigma(C)$

$$0 \longrightarrow K_{\sigma(C)}^{-1} \longrightarrow \Lambda^1(T^*C) \otimes \mathcal{O}_{\sigma(C)} \longrightarrow K_{\sigma(C)} \longrightarrow 0.$$

Therefore, identifying $C \cong \sigma(C)$, we obtain

$$\begin{aligned} 0 \longrightarrow H^0(C, \Lambda^1(T^*C) \otimes \mathcal{O}_C) \longrightarrow H^0(C, K_C) \xrightarrow{\kappa_0} H^1(C, K_C^{-1}) \\ \longrightarrow H^1(C, \Lambda^1(T^*C) \otimes \mathcal{O}_C) \longrightarrow H^1(C, K_C) \longrightarrow 0. \end{aligned}$$

Here κ_0 is the Kodaira–Spencer map assigning a deformation of C to a *displacement* of C in T^*C through a section of K_C. But since $\sigma(C)$ is always isomorphic to C, we do not obtain any deformation of C in this way. Hence κ_0 is the 0-map. Therefore,

$$H^0(\sigma(C), \Lambda^1(T^*C) \otimes \mathcal{O}_{\sigma(C)}) \cong H^0(\sigma(C), K_{\sigma(C)}) \cong H^0(C, K_C). \tag{3.9}$$

Now consider an exact sequence on T^*C

$$\begin{aligned} 0 \longrightarrow \Lambda^1(T^*C) \otimes \mathcal{O}_{T^*C}(-\sigma(C)) \longrightarrow \Lambda^1(T^*C) \\ \longrightarrow \Lambda^1(T^*C) \otimes \mathcal{O}_{\sigma(C)} \longrightarrow 0, \end{aligned}$$

which produces

$$\begin{aligned} 0 \longrightarrow H^0(T^*C, \Lambda^1(T^*C) \otimes \mathcal{O}_{T^*C}(-\sigma(C))) \longrightarrow H^0(T^*C, \Lambda^1(T^*C)) \\ \xrightarrow{r} H^0(\sigma(C), \Lambda^1(T^*C) \otimes \mathcal{O}_{\sigma(C)}). \end{aligned}$$

Because of (3.9), the homomorphism r is equal to the pull-back

$$r = \sigma^* : H^0(T^*C, \Lambda^1(T^*C)) \longrightarrow H^0(C, K_C)$$

by the section $\sigma : C \to T^*C$. It is surjective because $\sigma^* \circ \pi^* = id_{H^0(C,K_C)}$. Therefore, we have a splitting exact sequence

$$\begin{aligned} 0 \longrightarrow H^0(T^*C, \Lambda^1(T^*C) \otimes \mathcal{O}(-\sigma(C))) \longrightarrow H^0(T^*C, \Lambda^1(T^*C)) \\ \xrightarrow{\sigma^*} H^0(C, K_C) \longrightarrow 0. \end{aligned} \tag{3.10}$$

The tautological 1-form $\tau \in H^0(T^*C, \pi^*K_C)$ defines

$$0 \longrightarrow \pi^*K_C \longrightarrow \Lambda^1(T^*C) \xrightarrow{\wedge\tau} \Lambda^2(T^*C) \otimes \mathcal{O}_{T^*C}(-C) \longrightarrow 0,$$

noting that τ vanishes along the divisor $C \subset T^*C$. Since

$$H^0(T^*C, \Lambda^2(T^*C) \otimes \mathcal{O}_{T^*C}(-C)) \cong H^0(T^*C, \mathcal{O}_{T^*C}(-C)) = 0,$$

we obtain

$$H^0(T^*C, \pi^* K_C) \cong H^0(T^*C, \Lambda^1(T^*C)).$$

Take $\alpha \in H^0(T^*C, \Lambda^1(T^*C) \otimes \mathcal{O}_{T^*C}(-C)) \cong H^0(T^*C, \pi^* K_C(-C))$. Then

$$\alpha/\tau \in H^0(T^*C, \pi^* K_C(-C) \otimes \pi^* K_C^{-1}(C)) \cong H^0(T^*C, \mathcal{O}_{T^*C}) \cong \mathbb{C}.$$

Therefore, α is a constant multiple of τ, and we have obtained

$$H^0(T^*C, \Lambda^1(T^*C) \otimes \mathcal{O}_{T^*C}(-C)) \cong \mathbb{C}. \tag{3.11}$$

From (3.10) and (3.11), we conclude that

$$H^0(T^*C, \Lambda^1(T^*C)) \cong H^0(C, K_C) \oplus \mathbb{C} \cdot \tau.$$

Q.E.D.

Lemma 3.3. *Let $s \in V_{GL}^*$ be a point such that the spectral curve C_s is nonsingular. Then we have an isomorphism*

$$H^0(C_s, TT^*C) \cong H^0(C_s, \Lambda^1(T^*C)) \cong H^0(T^*C, \Lambda^1(T^*C)). \tag{3.12}$$

Proof. The line bundle on T^*C that corresponds to the divisor $C_s \subset T^*C$ is $\pi^* K_C^{\otimes n}$. Thus

$$\mathcal{O}_{T^*C}(-C_s) \cong \pi^* K_C^{\otimes(-n)} \cong \mathcal{O}_{T^*C}(-nC).$$

As above, let us consider an exact sequence

$$0 \longrightarrow \Lambda^1(T^*C) \otimes \mathcal{O}(-nC) \longrightarrow \Lambda^1(T^*C) \longrightarrow \Lambda^1(T^*C) \otimes \mathcal{O}_{C_s} \longrightarrow 0$$

and its cohomology sequence

$$0 \longrightarrow H^0(T^*C, \Lambda^1(T^*C) \otimes \mathcal{O}(-nC)) \longrightarrow H^0(T^*C, \Lambda^1(T^*C)) \xrightarrow{q} H^0(C_s, \Lambda^1(T^*C) \otimes \mathcal{O}_{C_s}).$$

From (3.11), we have

$$H^0(T^*C, \Lambda^1(T^*C) \otimes \mathcal{O}_{T^*C}(-nC)) = 0$$

for $n \geq 2$, hence q is injective.

Take $\beta \in H^0(C_s, \Lambda^1(T^*C) \otimes \mathcal{O}_{C_s})$, and extend it as a meromorphic 1-form on T^*C. Since $\deg K_C = 2g - 2 > 0$, every divisor of T^*C intersects with the 0-section C, and since $C_s \sim nC$ as a divisor, it also intersects with C_s. If $D \subset T^*C$ is the pole divisor of β, then it cannot intersect with C_s, hence $D = \emptyset$. Therefore, q is surjective. Q.E.D.

Let us go back to the Kodaira–Spencer map (3.7). We now know from (3.8) and (3.12) that

$$(3.13)\quad 0 \longrightarrow H^0(C, K_C) \oplus \mathbb{C}\cdot\tau \stackrel{\iota}{\longrightarrow} \bigoplus_{i=1}^{n} H^0(C, K_C^{\otimes i}) \stackrel{\kappa}{\longrightarrow} H^1(C_s, K_{C_s}^{-1}).$$

The $H^0(C, K_C)$-factor of the second term of (3.13) maps to the first component of the third term via the injective homomorphism ι. The tautological 1-form τ on T^*C that appears as the second factor of the second term is mapped to a diagonal ray in the third term.

To see this fact, we recall that the tangent and cotangent sheaves are isomorphic on T^*C through the symplectic form $\omega_C = -d\tau$. Let v be the vector field on T^*C corresponding to τ through this isomorphism, i.e., $\tau = \omega_C(\cdot, v)$. This vector field v represents the $\mathbb{C}^*$-action on T^*C along fiber. In terms of a local coordinate system, these correspondences are clearly described. Choose a local coordinate z on the algebraic curve C around a point $p \in C$, and denote by x the linear coordinate on T_p^*C with respect to the basis dz_p. Then at the point $(p, x dz_p) \in T^*C$ we have the following expressions:

$$\begin{cases} \tau = x\, dz \\ \omega_C = dz \wedge dx \\ v = x\, \frac{\partial}{\partial x}\, . \end{cases}$$

The $\lambda \in \mathbb{C}^*$ action on T^*C generated by the vector fields v produces a displacement of $C_s \subset T^*C$ to $C_{\lambda\cdot s}$, which corresponds to the equivariant action of λ on $\mathcal{H}_C(n, d)$ as described in (3.2). In terms of the holomorphic 1-form τ, its restriction on C_s gives an element

$$\begin{aligned}\tau|_{C_s} \in H^0(C_s, K_{C_s}) \cong H^0(C, \pi_* \mathcal{O}_{C_s} \otimes K_C^{\otimes n}) \cong H^0(C, \oplus_{i=1}^n K_C^{\otimes i}) \\ = \bigoplus_{i=1}^{n} H^0(C, K_C^{\otimes i}).\end{aligned}$$

This is the image $\iota(\tau)$ of (3.13).

Summing up, we have constructed an injective homomorphism

$$(3.14)\quad V_{SL}^*/\mathbb{C} \cong \Big(\bigoplus_{i=1}^{n} H^0(C, K_C^{\otimes i})\Big)/\iota\big(H^0(C, K_C) \oplus \mathbb{C}\cdot\tau\big) \stackrel{\bar\kappa}{\longrightarrow} H^1(C_s, K_{C_s}^{-1}).$$

The image of $\bar\kappa$ represents the generic Kodaira–Spencer class of the Hitchin fibration PH of (3.6). This completes the proof of Theorem 3.1.

§4. Symplectic quotient of the Higgs moduli space and Prym fibrations

The Hamiltonian vector fields corresponding to the coordinate components of the Hitchin map $\mathcal{H}_C(n,d) \to V_{GL}^*$ are constant Jacobian flows along each fiber of the map. Suppose we have a direct sum decomposition of V_{GL} into two Lie subalgebras

$$V_{GL} \stackrel{\text{def}}{=} \bigoplus_{i=0}^{n-1} H^1(C, K_C^{\otimes -i}) = \mathfrak{g}_1 \oplus \mathfrak{g}_2.$$

Then there are two possible ways to construct new algebraically completely integrable Hamiltonian systems. If the $\mathfrak{g}_1$-action on $\mathcal{H}_C(n,d)$ is integrated to a group action, then it is Hamiltonian by definition and we can construct the symplectic quotient. Or if the $\mathfrak{g}_2$-action is integrated to a group action instead, then we may find a family of $\mathfrak{g}_2$-orbits in $\mathcal{H}_C(n,d)$ fibered over the dual Lie algebra $\mathfrak{g}_2^*$. An important difference between real symplectic geometry and holomorphic symplectic geometry is that in the latter case integrations of the same Lie algebra may generate different (non-isomorphic) Lie groups. Consequently, the idea of symplectic quotient has to be generalized so that we can allow a *family* of groups acting on a symplectic manifold. The discovery of Hausel and Thaddeus in [8] is that the above two constructions lead to *mirror symmetric* pairs of Calabi–Yau spaces in the sense of Strominger–Yau–Zaslow [28]. In this section we consider two cases, the SL-PGL duality and the Sp_{2m}-SO_{2m+1} duality.

The SL-PGL duality comes from the decomposition

$$V_{GL} = H^1(C, \mathcal{O}_C) \oplus \left(\bigoplus_{i=i}^{n-1} H^1(C, K_C^{\otimes -i}) \right).$$

Obviously, the vector fields generated by the $H^1(C, \mathcal{O}_C)$-action are integrable to the $\mathrm{Jac}(C)$-action everywhere on $\mathcal{H}_C(n,d)$. Therefore, we do have the usual symplectic quotient mod $\mathrm{Jac}(C)$. On the other hand, the integration of the other Lie algebra

$$V_{SL} \stackrel{\text{def}}{=} \bigoplus_{i=i}^{n-1} H^1(C, K_C^{\otimes -i})$$

produces different Lie groups, called *Prym varieties*, along each fiber of the Hitchin fibration.

The spectral covering $\pi : C_s \to C$ induces two group homomorphisms dual to one another, the pull-back π^* and the *norm map* Nm_π

defined by

$$\tag{4.1} \begin{aligned} \pi^* &: \mathrm{Jac}(C) \ni L \longmapsto \pi^* L \in \mathrm{Jac}(C_s) \\ \mathrm{Nm}_\pi &: \mathrm{Jac}(C_s) \ni \mathcal{L} \longmapsto \det(\pi_* \mathcal{L}) \otimes \det(\pi_* \mathcal{O}_{C_s})^{-1} \in \mathrm{Jac}(C). \end{aligned}$$

In terms of divisors the norm map can be defined alternatively by

$$\mathrm{Pic}(C_s) \ni \sum_{p \in C_s} m(p) \cdot p \longmapsto \sum_{p \in C_s} m(p) \cdot \pi(p) \in \mathrm{Jac}(C).$$

The *Prym variety* and the *dual Prym variety*

$$\tag{4.2} \begin{aligned} \mathrm{Prym}(C_s/C) &\overset{\mathrm{def}}{=} \mathrm{Ker}(\mathrm{Nm}_\pi) \\ \mathrm{Prym}^*(C_s/C) &\overset{\mathrm{def}}{=} \mathrm{Jac}(C_s)/\pi^* \mathrm{Jac}(C) \end{aligned}$$

constructed by using these homomorphisms are Abelian varieties of dimension $g(C_s) - g(C)$ and are *dual* to one another. The algebraically completely integrable Hamiltonian systems with these Abelian fibrations, that are naturally constructed from $\mathcal{H}_C(n,d)$, then become SYZ-mirror symmetric.

We have shown in Section 3 that the $\mathrm{Jac}(C)$-action on $\mathcal{H}_C(n,d)$ is Hamiltonian. So we can define the symplectic quotient

$$\tag{4.3} \mathcal{PH}_C(n,d) \overset{\mathrm{def}}{=} \mathcal{H}_C(n,d) /\!\!/ \mathrm{Jac}(C) = H_1^{-1}(0)/\mathrm{Jac}(C).$$

This is a symplectic space of dimension $2(n^2-1)(g-1)$ modeled by the moduli space of stable principal $PGL_n(\mathbb{C})$-bundles on C [10, 11]. Since the $\mathrm{Jac}(C)$-action on $\mathcal{H}_C(n,d)$ preserves the Hitchin fibration, we have an induced Lagrangian fibration

$$\tag{4.4} H_{PGL} : \mathcal{PH}_C(n,d) \longrightarrow V_{SL}^* = \bigoplus_{i=2}^{n} H^0(C, K_C^{\otimes i}).$$

It's 0-fiber is $H_{PGL}^{-1}(0) = \mathcal{U}_C(n,d)/\mathrm{Jac}(C)$. Following [21] we denote by $\mathcal{SU}_C(n,d)$ the moduli space of stable vector bundles with a fixed determinant line bundle. This is a fiber of the determinant map

$$\tag{4.5} \mathcal{U}_C(n,d) \ni E \longmapsto \det E \in \mathrm{Pic}^d(C),$$

and is independent of the choice of the value of the determinant. Note that (4.5) is a non-trivial fiber bundle. The equivariant $\mathrm{Jac}(C)$-action on (4.5) is given by

$$\begin{array}{ccc} \mathcal{U}_C(n,d) & \xrightarrow{\otimes L} & \mathcal{U}_C(n,d) \\ {\scriptstyle \det}\downarrow & & \downarrow{\scriptstyle \det} \\ \mathrm{Pic}^d(C) & \xrightarrow[\otimes L^{\otimes n}]{} & \mathrm{Pic}^d(C) \end{array} \qquad L \in \mathrm{Jac}(C). \tag{4.6}$$

The isotropy subgroup of the $\mathrm{Jac}(C)$-action on $\mathcal{U}_C(n,d)$ is the group of n-torsion points

$$J_n(C) \overset{\text{def}}{=} \{L \in \mathrm{Jac}(C) \mid L^{\otimes n} = \mathcal{O}_C\} \cong H^1(C, \mathbb{Z}/n\mathbb{Z}), \tag{4.7}$$

since $E \otimes L \cong E$ implies $\det(E) \otimes L^{\otimes n} \cong \det(E)$. Choose a reference line bundle $L_0 \in \mathrm{Pic}^d(C)$ and consider a degree n covering

$$\nu : \mathrm{Pic}^d(C) \ni L \otimes L_0 \longmapsto L^{\otimes n} \otimes L_0 \in \mathrm{Pic}^d(C), \qquad L \in \mathrm{Jac}(C).$$

Then the pull-back bundle $\nu^* \mathcal{U}_C(n,d)$ on $\mathrm{Pic}^d(C)$ becomes trivial:

$$\nu^* \mathcal{U}_C(n,d) = \mathrm{Pic}^d(C) \times \mathcal{SU}_C(n,d).$$

The quotient of this product by the diagonal action of $J_n(C)$ is the original moduli space:

$$\big(\mathrm{Pic}^d(C) \times \mathcal{SU}_C(n,d)\big)/J_n(C) \cong \mathcal{U}_C(n,d). \tag{4.8}$$

It is now clear that

$$\mathcal{U}_C(n,d)/\mathrm{Jac}(C) \cong \mathcal{SU}_C(n,d)/J_n(C).$$

What are the other fibers of (4.4)? Let $s \in V_{SL}^* \cap U_{\mathrm{reg}}$ be a point such that C_s is non-singular. We have already noted that the covering map $\pi : C_s \to C$ induces an injective homomorphism $\pi^* : \mathrm{Jac}(C) \ni L \longmapsto \pi^* L \in \mathrm{Jac}(C_s)$. Therefore, the fiber $H_{PGL}^{-1}(s)$ is isomorphic to the dual Prym variety $\mathrm{Prym}^*(C_s/C)$. Similarly to the equivariant action (4.6), we have

$$\begin{array}{ccc} \mathrm{Jac}(C_s) & \xrightarrow{\otimes L} & \mathrm{Jac}(C_s) \\ {\scriptstyle \mathrm{Nm}}\downarrow & & \downarrow{\scriptstyle \mathrm{Nm}} \\ \mathrm{Jac}(C) & \xrightarrow[\otimes L^{\otimes n}]{} & \mathrm{Jac}(C) \end{array} \qquad L \in \mathrm{Jac}(C). \tag{4.9}$$

By the same argument of (4.8), we obtain

$$\big(\mathrm{Prym}(C_s/C) \times \mathrm{Jac}(C)\big)/J_n(C) \cong \mathrm{Jac}(C_s). \tag{4.10}$$

From (4.2) and (4.10), it follows that

$$\mathrm{Prym}^*(C_s/C) = \mathrm{Prym}(C_s/C)/J_n(C).$$

We have thus established

Theorem 4.1 ([11, 12]). *The natural fibration*

$$H_{PGL} : \mathcal{PH}_C(n,d) \longrightarrow V_{SL}^* = \bigoplus_{i=2}^{n} H^0(C, K_C^{\otimes i})$$

of (4.4) *is a Lagrangian dual Prym fibration with respect to the canonical holomorphic symplectic form* $\bar{\omega}$ *on* $\mathcal{PH}_C(n,d)$.

We recall that the Higgs moduli space $\mathcal{H}_C(n,d)$ contains the cotangent bundle $T^*\mathcal{U}_C(n,d)$ as an open dense subspace, and that the holomorphic symplectic form ω is the canonical symplectic form on this cotangent bundle. Similarly, we can show the following

Proposition 4.2. *The symplectic form* $\bar{\omega}$ *on* $\mathcal{PH}_C(n,d)$ *given by the symplectic quotient is the canonical cotangent symplectic form on the cotangent bundle*

$$T^*\big(\mathcal{SU}_C(n,d)/J_n(C)\big) \subset \mathcal{PH}_C(n,d).$$

Proof. Let E be a stable vector bundle on C. The exact sequence

$$0 \longrightarrow \mathcal{O}_C \longrightarrow \mathrm{End}(E) \longrightarrow \mathcal{Q} = \mathrm{End}(E)/\mathcal{O}_C \longrightarrow 0$$

induces a cohomology sequence

$$0 \longrightarrow H^1(C, \mathcal{O}_C) \longrightarrow H^1(C, \mathrm{End}(E)).$$

Therefore,

$$T_E\big(\mathcal{SU}_C(n,d)/J_n(C)\big) \cong H^1(C, \mathrm{End}(E))/H^1(C, \mathcal{O}_C).$$

Dualizing the situation, we have

$$0 \longrightarrow \mathrm{End}_0(E) \otimes K_C \longrightarrow \mathrm{End}(E) \otimes K_C \xrightarrow{\mathrm{tr}} K_C \longrightarrow 0,$$

where $\mathrm{End}_0(E)$ is the sheaf of traceless endomorphisms of E. We then have

$$0 \longrightarrow H^0(C, \mathrm{End}_0(E) \otimes K_C) \longrightarrow H^0(C, \mathrm{End}(E) \otimes K_C) \xrightarrow{\mathrm{tr}} H^0(C, K_C) \longrightarrow 0.$$

The trace homomorphism is globally surjective because $\mathcal{O}_C$ is contained in End(E). Hence

$$T_E^*(\mathcal{SU}_C(n,d)/J_n(C)) \cong H^0(C, \mathrm{End}_0(E) \otimes K_C).$$

Since the moment map H_1 of (3.3) is the trace map, we conclude that the symplectic form $\bar{\omega}$ on the symplectic quotient $\mathcal{PH}_C(n,d)$ is the canonical cotangent symplectic form on the cotangent bundle

$$T^*(\mathcal{SU}_C(n,d)/J_n(C)).$$

Q.E.D.

The other reduction of $\mathcal{H}_C(n,d)$ consisting of the V_{SL}-orbits is the moduli space $\mathcal{SH}_C(n,d)$ of stable Higgs bundles (E,ϕ) with a fixed determinant $\det(E) \cong L$ and traceless Higgs fields

$$\phi \in H^0(C, \mathrm{End}_0(E) \otimes K_C).$$

This moduli space is modeled by the moduli space of stable principal $SL_n(\mathbb{C})$-bundles on C and has dimension $2(n^2-1)(g-1)$. The cotangent bundle $T^*\mathcal{SU}_C(n,d)$ is an open dense subspace of $\mathcal{SH}_C(n,d)$. Since $\mathrm{tr}(\phi) = 0$, the Hitchin fibration H of (2.9) naturally restricts to

$$H_{SL} : \mathcal{SH}_C(n,d) \longrightarrow V_{SL}^* = \bigoplus_{i=2}^{n} H^0(C, K_C^{\otimes i}). \tag{4.11}$$

The 0-fiber is $H_{SL}^{-1}(0) = \mathcal{SH}_C(n,d)$. For a generic $s \in V_{SL}^*$ such that C_s is non-singular, the fiber $H_{SL}^{-1}(s)$ is the subset of $H^{-1}(s) \cong \mathrm{Jac}(C_s)$ consisting of Higgs bundles (E,ϕ) such that $\det(E) = L$ is fixed and $\det(x-\phi) = s$. Therefore, $H_{SL}^{-1}(s) \cong \mathrm{Prym}(C_s/C)$.

By comparing $\mathcal{PH}_C(n,d)$ and $\mathcal{SH}_C(n,d)$, we find that

$$\mathcal{SH}_C(n,d)/J_n(C) \cong \mathcal{PH}_C(n,d), \tag{4.12}$$

where the $J_n(C)$-action on $\mathcal{SH}_C(n,d)$ is defined by $E \mapsto E \otimes L$ for $L \in J_n(C)$. Since $J_n(C)$ is a finite group and $\mathcal{PH}_C(n,d)$ is a holomorphic symplectic variety, we can define a holomorphic symplectic form $\hat{\omega}$ on $\mathcal{SH}_C(n,d)$ via the pull-back of the projection

$$\mathcal{SH}_C(n,d) \longrightarrow \mathcal{SH}_C(n,d)/J_n(C).$$

Obviously $\hat{\omega}$ agrees with the canonical cotangent symplectic form on $T^*\mathcal{SU}_C(n,d)$. Therefore,

Theorem 4.3 ([11, 12]). *The fibration*

$$H_{SL} : \mathcal{SH}_C(n,d) \longrightarrow V_{SL}^* = \bigoplus_{i=2}^{n} H^0(C, K_C^{\otimes i})$$

is a Lagrangian Prym fibration with respect to the canonical holomorphic symplectic form $\hat{\omega}$ *on* $\mathcal{SH}_C(n,d)$.

Hausel and Thaddeus [8] shows that

Theorem 4.4 (Theorem (3.7) in [8]). *The two fibrations*

$$\begin{array}{ccc} \mathcal{SH}_C(n,d) & & \mathcal{PH}_C(n,d) \\ & {}_{H_{SL}}\searrow \quad \swarrow_{H_{PGL}} & \\ & V_{SL}^* & \end{array}$$

are mirror symmetric in the sense of Strominger–Yau–Zaslow [28].

The effectiveness of the family of Prym and dual Prym varieties can be established by the same method of Section 3. Let us define partial projective moduli spaces

$$\mathbb{P}(\mathcal{PH}_C(n,d)) \stackrel{\text{def}}{=} \left(\mathcal{PH}_C(n,d) \setminus H_{PGL}^{-1}(0)\right)/\mathbb{C}^*$$

and

$$\mathbb{P}(\mathcal{SH}_C(n,d)) \stackrel{\text{def}}{=} \left(\mathcal{SH}_C(n,d) \setminus H_{SL}^{-1}(0)\right)/\mathbb{C}^*.$$

The induced Hitchin maps are denoted by

$$\begin{aligned} PH_{PGL} &: \mathbb{P}(\mathcal{PH}_C(n,d)) \longrightarrow \mathbb{P}_w(V_{SL}^*) \\ PH_{SL} &: \mathbb{P}(\mathcal{SH}_C(n,d)) \longrightarrow \mathbb{P}_w(V_{SL}^*). \end{aligned} \tag{4.13}$$

We have the following

Theorem 4.5. *The Prym and dual Prym fibrations*

$$\begin{aligned} PH_{SL} &: \mathbb{P}(\mathcal{SH}_C(n,d)) \longrightarrow \mathbb{P}_w(V_{SL}^*) \\ PH_{PGL} &: \mathbb{P}(\mathcal{PH}_C(n,d)) \longrightarrow \mathbb{P}_w(V_{SL}^*) \end{aligned}$$

are generically effective.

For the case of $Sp_{2m}(\mathbb{C})$-Hitchin systems, we consider Lie subalgebras

$$\begin{aligned} \mathfrak{g} &= \bigoplus_{i=0}^{m-1} H^1(C, K_C^{\otimes -2i}) \\ V_{Sp} &= \bigoplus_{i=0}^{m-1} H^1(C, K_C^{\otimes -2i-1}), \end{aligned} \tag{4.14}$$

and a direct sum decomposition

$$V_{GL_{2m}} = \mathfrak{g} \oplus V_{Sp}. \tag{4.15}$$

This time the Hamiltonian flows on $\mathcal{H}_C(n,d)$ generated by elements of $\mathfrak{g}$ do not form a group action. However, if we restrict our attention to points of $U_{\text{reg}} \cap V_{Sp}^*$ as in (2.27), then the integral of the Lie algebra action becomes a group action. Recall that the spectral curve C_s has an involution induced by ϵ of (2.26). We denote by

$$r : C_s \rightarrow C'_s = C_s/\langle\epsilon\rangle \tag{4.16}$$

the natural projection. It is ramified at the intersection of C_s with the 0-section of T^*C, which is the divisor of π^*K_C on C_s of degree $4m(g-1)$. Therefore, we find the genus of C'_s by the Riemann–Hurwitz formula:

$$g(C'_s) = m(2m-1)(g-1) + 1 = \dim \mathfrak{g}. \tag{4.17}$$

Since $H^0(C'_s, r_*\mathcal{O}_{C_s})$ has a nowhere vanishing section, we have

$$0 \longrightarrow \mathcal{O}_{C'_s} \longrightarrow r_*\mathcal{O}_{C_s} \longrightarrow N^{-1} \longrightarrow 0 \tag{4.18}$$

with a line bundle N on C'_s of degree $2m(g-1)$. Note that

$$N^{\otimes 2} = \det(r_*\mathcal{O}_{C_s})^{\otimes -2} \cong \operatorname{Nm}_r(\pi^*K_C),$$

hence N is a square root of the *branch divisor* $\operatorname{Nm}_r(\pi^*K_C)$ of the covering r. The exact sequence (4.18) gives

$$H^1(C_s, \mathcal{O}_{C_s}) \cong H^1(C'_s, \mathcal{O}_{C'_s}) \oplus H^1(C'_s, N^{-1}), \tag{4.19}$$

and the projection to the first factor is the differential of the norm map

$$\operatorname{Nm}_r : \operatorname{Jac}(C_s) \ni \mathcal{L} \longmapsto \det(r_*\mathcal{L}) \otimes \det(r_*\mathcal{O}_{C_s})^{-1} \in \operatorname{Jac}(C'_s).$$

The construction of the Sp-Hitchin system of Proposition 2.4 is to reduce the GL-Hitchin system $\mathcal{H}_C(2m, 2m(g-1))$ by finding the right fibration of groups. Along the fixed-point-set of the Serre duality on this Higgs moduli space, the action of V_{Sp} generates the Prym fibration $\{\operatorname{Prym}(C_s/C'_s)\}_{s\in V_{Sp}^*}$ since the condition $\mathcal{L}_0^* \cong \epsilon^*\mathcal{L}_0$ on C_s is the same as $\mathcal{L}_0 \in \operatorname{Prym}(C_s/C'_s)$. Comparing (4.15) and (4.19), we have

$$H^1(C'_s, \mathcal{O}_{C'_s}) \cong \mathfrak{g} \qquad \text{and} \qquad H^1(C'_s, N^{-1}) \cong V_{Sp}.$$

The *dual* fibration is the result of a kind of symplectic quotient of $\mathcal{H}_C(2m, 2m(g-1))$ by $\mathfrak{g}$. In this quotient we restrict the Hitchin fibration to the 0-fiber of

$$\mathfrak{g}^* = \bigoplus_{i=1}^{m} H^0(C, K_C^{\otimes 2i-1}),$$

and then take the quotient of each fiber by the Lie group $\mathrm{Jac}(C'_s)$ of $\mathfrak{g}$. The result is the dual Prym fibration with a fiber

$$\mathrm{Prym}^*(Cs/C'_s) = \mathrm{Jac}(C_s)/\mathrm{Jac}(C'_s) = \mathrm{Prym}(Cs/C'_s)/J_2(C'_s)$$

over each $s \in V^*_{Sp}$, where $J_2(C'_s)$ denotes the group of 2-torsion points of $\mathrm{Jac}(C'_s)$.

§5. Hitchin's integrable systems and the KP equations

The partial projective Hitchin fibration

$$PH : \mathbb{P}(\mathcal{H}_C^{SL}(n,d)) \to \mathbb{P}_w(V^*_{SL})$$

is a generically effective Jacobian fibration. In this section we embed $\mathbb{P}(\mathcal{H}_C^{SL}(n,d))$ into a quotient of the Sato Grassmannian. There is also a natural embedding of $\mathcal{H}_C(n, n(g-1))$ into a *relative* Grassmannian of [2, 4, 22, 23]. We show that the linear Jacobian flows on the Hitchin integrable system $(\mathcal{H}_C(n, n(g-1)), \omega, H)$ are exactly the KP equations on the Grassmannian via this second embedding.

Following Quandt [23], we define

Definition 5.1 (Sato Grassmannian). *Let n be a positive integer. The Sato Grassmannian Gr_n is a functor from the category of schemes to the small category of sets. It assigns to every scheme S a set $Gr_n(S)$ consisting of quasi-coherent $\mathcal{O}_S$-submodules W of $\mathcal{O}_S((z))^{\oplus n}$ such that both the kernel and the cokernel of the natural homomorphism*

(5.1)

$$0 \longrightarrow \mathrm{Ker}(\gamma_W) \longrightarrow W \xrightarrow{\gamma_W} \mathcal{O}_S((z))^{\oplus n}/\mathcal{O}_S[[z]]^{\oplus n} \longrightarrow \mathrm{Coker}(\gamma_W) \longrightarrow 0$$

are coherent $\mathcal{O}_S$ modules. We refer to this condition simply the Fredholm condition. *Here we denote by $\mathcal{O}_S[[z]]$ the ring of formal power series in z with coefficients in $\mathcal{O}_S$, and $\mathcal{O}_S((z)) = \mathcal{O}_S[[z]] + \mathcal{O}_S[z^{-1}]$.*

Remark 5.1. A more general and powerful theory of relative Sato Grassmannians has been recently established by Plaza Martín [22]. It would be an interesting project to study possible relations between [22] and the geometric Langlands correspondence.

Remark 5.2. The Grassmannian $Gr_n(\mathbb{C})$ defined over the point scheme Spec($\mathbb{C}$) has the structure of an infinite-dimensional pro-ind scheme over $\mathbb{C}$. If S is irreducible, then $Gr_n(S)$ is a disjoint union of an infinite number of components *indexed* by the Grothendieck group $K(S)$:
(5.2)

$$\text{index} : Gr_n(S) \ni W \longmapsto \text{index}(\gamma_W) \overset{\text{def}}{=} \text{Ker}(\gamma_W) - \text{Coker}(\gamma_W) \in K(S).$$

Remark 5.3. The *big cell* is the open subscheme of the index 0 piece of the Grassmannian $Gr_n^0(S)$ consisting of W's such that γ_W is an isomorphism. This is the stage where the Lax and Zakharov–Shabat formalisms of integrable nonlinear partial differential equations, such as KP, KdV, and many other equations, are interpreted as infinite-dimensional dynamical systems [24, 25]. The fundamental result is the identification (6.17) of the big cell with the group of 0-th order monic pseudodifferential operators [16, 23, 24, 25].

Remark 5.4. The above W is a closed subset of $\mathcal{O}_S((z))^{\oplus n}$ with respect to the topology defined by the filtration

$$\cdots \supset z^{\nu-1} \cdot \mathcal{O}_S[[z]]^{\oplus n} \supset z^{\nu} \cdot \mathcal{O}_S[[z]]^{\oplus n} \supset z^{\nu+1} \cdot \mathcal{O}_S[[z]]^{\oplus n} \supset \cdots .$$

One of the consequences of the Fredholm condition is that

$$W \cap z^{\nu} \cdot \mathcal{O}_S[[z]]^{\oplus n} = 0 \qquad \text{for } \nu >> 0. \tag{5.3}$$

Consider the group
(5.4)

$$\Gamma_n(S) = \mathcal{O}_S^{\times} \cdot \begin{bmatrix} 1 & & & \\ & 1 & & \\ & & \ddots & \\ & & & 1 \end{bmatrix} + z \cdot \begin{bmatrix} \mathcal{O}_S[[z]] & & & \\ & \mathcal{O}_S[[z]] & & \\ & & \ddots & \\ & & & \mathcal{O}_S[[z]] \end{bmatrix}$$

consisting of $n \times n$ invertible diagonal matrices with entries in the ring of formal power series whose constant term is scalar diagonal. It acts on $Gr_n(S)$ by left-multiplication without fixed points. Indeed, let $W \in Gr_n(S)$ and $c+\gamma \in \mathcal{O}_S^{\times} \cdot I_n + z \cdot \mathcal{O}_S[[z]]^{\oplus n}$ satisfy that $W = (c+\gamma) \cdot W$. This means that $\gamma \cdot w \in W$ for every $w \in W$ since W is an $\mathcal{O}_S$-module. Since $\gamma \in z \cdot \mathcal{O}_S[[z]]^{\oplus n}$, it then contradicts to (5.3) unless $\gamma = 0$. The *quotient Grassmannian*

$$Z_n \overset{\text{def}}{=} Gr_n/\Gamma_n \tag{5.5}$$

is a smooth infinite-dimensional scheme. We denote by $\overline{W}$ the point of Z_n corresponding to $W \in Gr_n$.

The tangent space $T_W Gr_n$ of the Sato Grassmannian at W is given by the space of continuous $\mathcal{O}_S$-homomorphisms

$$T_W Gr_n = \mathrm{Hom}_{\mathcal{O}_S}\big(W, \mathcal{O}_S((z))^{\oplus n}/W\big).$$

The tangent space to Z_n is then given by

$$T_{\overline{W}} Z_n \cong \mathrm{Hom}_{\mathcal{O}_S}\big(W, \mathcal{O}_S((z))^{\oplus n}/(\Gamma_n \cdot W)\big). \tag{5.6}$$

This expression does not depend on the choice of the lift $W \in Gr_n$ of $\overline{W} \in Z_n$.

Every element $a \in \mathcal{O}_S((z))^{\oplus n}$ defines a homomorphism

$$KP(a)_W : W \ni w \longmapsto \overline{a \cdot w} \in \mathcal{O}_S((z))^{\oplus n}/(\Gamma_n \cdot W)$$

through the left multiplication as a diagonal matrix, which in turn determines a global vector field

$$\mathcal{O}_S((z))^{\oplus n} \ni a \longmapsto KP(a) \in H^0(Z_n, TZ_n).$$

We call $KP(a)$ the n-component *KP flow* associated with a. As explained in [16, 17, 19, 23], the quotient of the index zero Grassmannian Z_n^0 is naturally identified with the set of *Lax operators* (6.18), and the action of $\mathcal{O}_S((z))^{\oplus n}$ on a Lax operator is written as an infinite system of nonlinear partial differential equations called *Lax equations.* This system for the case of $S = \mathrm{Spec}(\mathbb{C})$ is the n-component *Kadomtsev–Petviashvili hierarchy* [16].

Associated to the Hitchin fibration $H : \mathcal{H}_C(n,d) \to V_{GL}^*$ we have a family of spectral curves:

$$\begin{array}{ccc} \mathfrak{C}_{V_{GL}^*}(n,d) & \xrightarrow{\text{inclusion}} & T^*C \times V_{GL}^* \\ & {\scriptstyle F}\searrow \quad \swarrow{\scriptstyle p_2} & \\ & V_{GL}^*. & \end{array} \tag{5.7}$$

Here the fiber $F^{-1}(s)$ of $s \in V_{GL}^*$ is the spectral curve $C_s \subset T^*C \times \{s\}$, and p_2 is the projection to the second factor. We note that the ramification points of the covering $\pi : C_s \to C$ are determined by the resultant of the defining equation

$$x^n + s_1 x^{n-1} + \cdots + s_n = 0 \tag{5.8}$$

of C_s and its derivative

$$nx^{n-1} + (n-1)s_1 x^{n-1} + \cdots + s_{n-1} = 0. \tag{5.9}$$

For every $s \in V_{GL}^*$, we denote by $\mathrm{Res}(s)$ this resultant. The Sylvester matrix of these polynomials (5.8) and (5.9) show that

$$\mathrm{Res}(s) \in H^0(C, K_C^{\otimes n(n-1)}).$$

Since the linear system $|K_C|$ is base-point-free, for every choice of $p \in C$, the subset

(5.10)

$$U_p = \{s \in V_{GL}^* \mid C_s \text{ is non-singular and } \pi : C_s \to C \text{ is unramified at } p\}$$

is Zariski open in V_{GL}^*.

Theorem 5.5. *There is a rational map*

$$\mu : \mathbb{P}(\mathcal{H}_C^{SL}(n,d)) \longrightarrow Z_n^{d-n(g-1)}(\mathbb{C})$$

of the partial projective moduli space of Higgs bundles into the quotient Grassmannian of index $d - n(g-1)$. This map is generically injective. At a general point of the image of the embedding the n-component KP flows defined on $Z_n(\mathbb{C})$ are tangent to the Hitchin fibration $PH : \mathbb{P}(\mathcal{H}_C^{SL}(n,d)) \longrightarrow \mathbb{P}_w(V_{SL}^*)$.

Proof. A point of the partial projective Higgs moduli space represents an isomorphism class of spectral data $(\pi : C_s \to C, \mathcal{L})$, where $\mathcal{L}$ is a line bundle on C_s of degree $d + (n^2 - n)(g-1)$. Choose a point $p \in C$, a coordinate z of the formal completion $\hat{C}_p$ of C at p, and $s \in U_p \cap V_{SL}^*$ so that C_s is non-singular and π is unramified over p. The formal coordinate z defines an identification $\mathcal{O}_{\hat{C}_p} = \mathbb{C}[[z]]$. We also choose a local trivialization of $\mathcal{L}$ around $\pi^{-1}(p)$, i.e., an isomorphism

$$\beta : \mathcal{L}|_{\hat{C}_{s,\pi^{-1}(p)}} \xrightarrow{\sim} \mathcal{O}_{\hat{C}_{s,\pi^{-1}(p)}} = \mathbb{C}[[z]]^{\oplus n}. \tag{5.11}$$

Since the formal completion $\hat{C}_{s,\pi^{-1}(p)}$ is the disjoint union of n copies of $\hat{C}_p$, the formal coordinate z also defines an identification $\mathcal{O}_{\hat{C}_{s,\pi^{-1}(p)}} = \mathbb{C}[[z]]^{\oplus n}$.

Now define

$$W = \beta(H^0(C_s \setminus \pi^{-1}(p), \mathcal{L})) \subset \mathbb{C}((z))^{\oplus n},$$

which is the set of meromorphic sections of $\mathcal{L}$ that are holomorphic on $C_s \setminus \pi^{-1}(p)$ and have finite poles at $\pi^{-1}(p)$. Since we have

$$\begin{cases} \mathrm{Ker}(\gamma_W) \cong H^0(C_s, \mathcal{L}) \\ \mathrm{Coker}(\gamma_W) \cong H^1(C_s, \mathcal{L}), \end{cases}$$

W is a point of the Grassmannian $Gr_n^{d-n(g-1)}$ of index $d-n(g-1)$. The different choice of the local trivialization β' in (5.11) leads to an element $\beta' \circ \beta^{-1} \in \Gamma_n$. Therefore, the point $\overline{W} \in Z_n^{d-n(g-1)}(\mathbb{C})$ is uniquely determined by $(\pi : C_s \to C, p, z, \mathcal{L})$. Conversely this set of geometric data is uniquely determined by $\overline{W}$ (see Section 5 of [16]). Thus the rational map μ is generically one-to-one. Notice that the Γ_n action on the Grassmannian is inessential from the geometric point of view because it simply changes the local trivialization of (5.11).

The tangent space at a general point of $\mathbb{P}(\mathcal{H}_C^{SL}(n,d))$ is

$$H^1(C_s, \mathcal{O}_{C_s}) \oplus V_{SL}^*/\mathbb{C}.$$

Since the Kodaira–Spencer map (3.14) is injective, it suffices to show injectivity of the natural map

$$\begin{aligned} d\mu : H^1(C_s, \mathcal{O}_{C_s}) \oplus H^1(C_s, K_{C_s}^{-1}) &\longrightarrow \mathrm{Hom}(W, \mathbb{C}((z))^{\oplus n}/W) \\ &= T_W Gr_n(\mathbb{C}), \end{aligned}$$

which is induced by the local trivialization of $\mathcal{O}_{C_s}$ and K_C coming from the choice of the local coordinate z. Using z, we define

$$A_W = H^0(C_s \setminus \pi^{-1}(p), \mathcal{O}_{C_s}) \subset \mathbb{C}((z))^{\oplus n},$$

which is a point of the Grassmannian satisfying

$$\begin{cases} \mathrm{Ker}(\gamma_{A_W}) \cong H^0(C_s, \mathcal{O}_{C_s}) \\ \mathrm{Coker}(\gamma_{A_W}) \cong H^1(C_s, \mathcal{O}_{C_s}). \end{cases}$$

We note that A_W is a ring and W is an A_W-module. Since

$$\mathrm{Coker}(\gamma_{A_W}) = \frac{\mathbb{C}((z))^{\oplus n}}{A_W + \mathbb{C}[[z]]^{\oplus n}},$$

$H^1(C_s, \mathcal{O}_{C_s})$ is injectively mapped to $\mathrm{Hom}(W, \mathbb{C}((z))^{\oplus n}/W)$.

The local coordinate z determines a local trivialization of K_C

$$K_C|_{\hat{C}_p} \xrightarrow{\sim} \mathcal{O}_{\hat{C}_p} \cdot dz \cong \mathcal{O}_{\hat{C}_p},$$

and hence that of $K_{C_s}^{-1}$

$$K_{C_s}^{-1}|_{\hat{C}_{s,\pi^{-1}(p)}} \xrightarrow{\sim} \left(\mathcal{O}_{\hat{C}_p}\right)^{\oplus n} \cdot \frac{\partial}{\partial z} \cong \mathbb{C}[[z]]^{\oplus n} \cdot \frac{\partial}{\partial z}.$$

This trivialization gives

$$H^1(C_s, K_{C_s}^{-1}) \cong \frac{\mathbb{C}((z))^{\oplus n} \cdot \partial/\partial z}{D_W + \mathbb{C}[[z]]^{\oplus n} \cdot \partial/\partial z},$$

where

$$D_W = H^0(C_s \setminus \pi^{-1}(p), K_{C_s}^{-1}) \subset \mathbb{C}((z))^{\oplus n} \cdot \partial/\partial z.$$

Note that $s \in V_{GL}^*$ determines an element of $H^0(C_s, K_{C_s})$ because

$$s \in V_{GL}^* = \bigoplus_{i=1}^{n} H^0(C, K_C^{\otimes i}) \cong H^0(C, \pi_* \pi^* K_C^{\otimes n}) \cong H^0(C_s, K_{C_s}).$$

This holomorphic 1-form on C_s induces a homomorphism

$$\mathcal{L} \to \mathcal{L} \otimes K_{C_s},$$

or equivalently,

$$K_{C_s}^{-1} \otimes \mathcal{L} \longrightarrow \mathcal{L}.$$

In terms of the local coordinate z, this homomorphism gives an action of D_W on W. Since $D_W \cdot W \subset W$, we conclude that $H^1(C_s, K_{C_s}^{-1})$ is injectively mapped to $T_W Gr_n(\mathbb{C})$. In this construction $H^1(C_s, \mathcal{O}_{C_s})$ and $H^1(C_s, K_{C_s}^{-1})$ have no common element in $T_W Gr_n(\mathbb{C})$ except for 0. This establishes the injectivity of $d\mu$.

In [16], it is proved that the orbit of the n-component KP flows starting from $\overline{W}$ is isomorphic to $\mathrm{Pic}^{d+(n^2-n)(g-1)}(C_s)$ (Theorem 5.8, [16]), which is the fiber of the Hitchin map PH. This completes the proof of the theorem. Q.E.D.

The above theorem does not say anything about the Hitchin integrable system, because the partial projective moduli space is not a symplectic manifold and we do not have any integrable systems on it. To directly compare the Jacobian flows of the Hitchin systems and the KP flows, we use the *relative* Grassmannian of [23] defined on the scheme U_p of (5.10). So let $\pi^{-1}(p) = \{p_1, \ldots, p_n\}$, and denote by $z_i = \pi^*(z)$ the formal coordinate of $\hat{C}_{s,p_i}$. Choose a linear coordinate system $(h_1, \ldots, h_N)$ for V_{GL}^* as in (2.24). Recall that $\pi_* \mathcal{O}_{C_s} \cong \oplus_{i=0}^{n-1} K_C^{\otimes -i}$, and from (2.21) we have

$$H^1(C_s, \mathcal{O}_{C_s}) \cong \bigoplus_{i=0}^{n-1} H^1(C, K_C^{\otimes -i}) = V_{GL}.$$

Using the Čech cohomology computation based on the covering $C = (C \setminus \{p\}) \cup \hat{C}_p$, we expand each $h_k \in V_{GL}$ as

$$h_k = \sum_{i=1}^{n} \sum_{j \geq 1} t_{ijk} z_i^{-j}, \qquad t_{ijk} \in \mathbb{C}. \tag{5.12}$$

The n-component KP flows on the Grassmannian $Gr_n(U_p)$ is defined by a formal action of

$$\exp\left(\sum_{i=1}^{n}\sum_{j\geq 1}\sum_{k=1}^{N} t_{ijk} z_i^{-j}\right). \tag{5.13}$$

For degree $d = n(g-1)$, we have the following:

Theorem 5.6. *Let us choose a point $p \in C$, a formal coordinate z of $\hat{C}_p$, and a square root of the canonical sheaf $K_C^{1/2}$. Then the fibration*

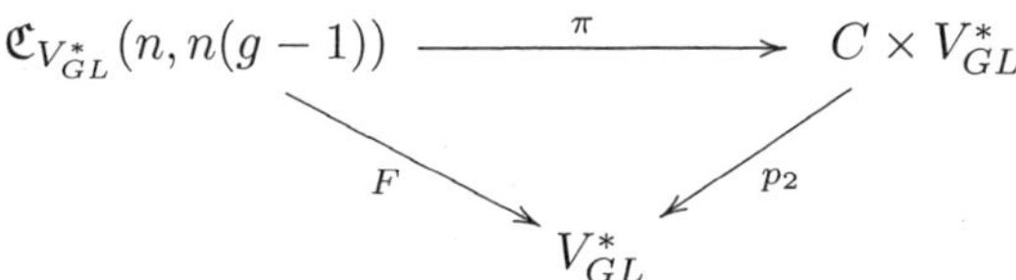

determines a point $W \in Gr_n^0(U_p)$. There is a birational map from $\mathcal{H}_C(n, n(g-1))$ to the orbit of the n-component KP flows on the quotient Grassmannian $Z_n^0(U_p)$ starting from $\overline{W}$. The Hitchin integrable system on the Higgs moduli space $\mathcal{H}_C(n, n(g-1))$ is the pull-back of the n-component KP equations via this birational map.

Proof. We regard C, p, z, and $K_C^{1/2}$ being defined over the trivial family $C \times V_{GL}^*$. There is a natural choice of a local trivialization of $K_C^{1/2}$ on $\hat{C}_p$ determined by the coordinate z. Note that

$$\mathcal{K}^{1/2} \stackrel{\text{def}}{=} \pi^*(K_C^{n/2}) \tag{5.14}$$

is a square root of the relative dualizing sheaf $\mathcal{K} = \pi^*(K_C^n)$ on

$$\mathfrak{C}_{V_{GL}^*}(n, n(g-1)).$$

Take an arbitrary point $s \in U_p$. Since $p \in C$ is a point of C at which π is unramified, the formal coordinate z determines a local trivialization of $\mathcal{K}^{1/2}$ along $\pi^{-1}(p) \subset \mathfrak{C}_{V_{GL}^*}(n, n(g-1))$. Define

$$W = H^0(\mathfrak{C}_{V_{GL}^*}(n, n(g-1)) \setminus \pi^{-1}(p), \mathcal{K}^{1/2}) \subset \mathcal{O}_{U_p}((z))^{\oplus n}, \tag{5.15}$$

using this local trivialization. Since

$$\begin{cases} \mathrm{Ker}(\gamma_{W_s}) \cong H^0(C_s, K_{C_s}^{1/2}) \\ \mathrm{Coker}(\gamma_{W_s}) \cong H^1(C_s, K_{C_s}^{1/2}) \end{cases}$$

for each $s \in U_p$, we have $W \in Gr_n^0(U_p)$. Let us analyze the action of (5.13) at W. On each C_s, $h_k = \sum t_{ijk} z_i^{-j}$ defines an element of $H^1(C_s, \mathcal{O}_{C_s})$. The exponential of (5.13) is the Lie integration map

$$(5.16) \quad V_{GL} \cong H^1(C_s, \mathcal{O}_{C_s}) \xrightarrow{\exp} H^1(C_s, \mathcal{O}_{C_s})/H^1(C_s, \mathbb{Z}) = \mathrm{Jac}(C_s).$$

Therefore, the action of (5.13) at the point W produces simultaneous deformations of the line bundle

$$(5.17) \qquad \mathcal{K}^{1/2} \longmapsto \mathcal{K}^{1/2} \otimes \exp\left(\sum_{i=1}^{n} \sum_{j\geq 1} \sum_{k=1}^{N} t_{ijk} z_i^{-j}\right)$$

on $\mathfrak{C}_{V_{GL}^*}(n, n(g-1))$. Consequently, the orbit of the n-component KP flows on $Z_n^0(U_p)$ starting at $\overline{W}$ is naturally identified with the family of $\mathrm{Pic}^{n^2(g-1)}(C_s)$ on U_p. Since the Hitchin fibration $H : \mathcal{H}_C(n, n(g-1)) \to V_{GL}^*$ is also a family of $\mathrm{Pic}^{n^2(g-1)}(C_s)$ on V_{GL}^*, we have a rational map from $H : \mathcal{H}_C(n, n(g-1))$ to the n-component KP orbit of $\overline{W}$ in $Z_n^0(U_p)$.

Recall that the Hitchin integrable system on $H : \mathcal{H}_C(n, n(g-1))$ is the linear Jacobian flows defined by the coordinate functions $(h_1, \ldots, h_N)$. We note that this is exactly what (5.17) gives. Q.E.D.

Remark 5.7. So far we have assumed that C_s is non-singular. Although it will not be an Abelian variety, we can still define the Jacobian $\mathrm{Jac}(C_s)$ using (5.16) when the spectral curve C_s is singular. As long as we choose $p \in C$ so that $\pi^{-1}(p)$ avoids the singular locus of C_s, the KP flows are well defined. Moreover, the theory of *Heisenberg KP flows* introduced in [1, 16] allows us to consider the covering $\pi : C_s \to C$ right at a ramification point. It is more desirable to define the Grassmannian over the whole V_{GL}^* and to deal with the entire family $F : \mathfrak{C}_{V_{GL}^*} \to V_{GL}^*$ together with the moduli stack of Higgs bundles instead of the stable moduli we have considered here, since the framework of the Sato Grassmannian allows us to consider all vector bundles on C and degenerated spectral curves. We refer to [4] for a study in this direction.

§6. Serre duality and formal adjoint of pseudo-differential operators

In this section we analyze the Serre duality of Higgs bundles in terms of the language of the Sato Grassmannian, and identify the involution on the corresponding pseudo-differential operators.

The Krichever construction ([17, 19, 25]) assigns a point W of the Sato Grassmannian,

$$Gr_n(\mathbb{C}) \ni W = \beta(H^0(C \setminus \{p\}, F)) \subset \mathbb{C}((z))^{\oplus n}, \tag{6.1}$$

to a set of geometric data (C, p, z, F, β), where C is an irreducible algebraic curve, $p \in C$ is a non-singular point, z is a formal parameter of the completion $\hat{C}_p$, F is a torsion-free sheaf of rank n on C, and

$$\beta : F|_{\hat{C}_p} \xrightarrow{\sim} \mathcal{O}_{\hat{C}_p}^{\oplus n} = \mathbb{C}[[z]]^{\oplus n}$$

is a local trivialization of F around p. We continue to assume that C is non-singular, hence F is locally-free on C. As noted in Section 5, the choice of a local coordinate z automatically determines a local trivialization of K_C:

$$\alpha : K_C|_{\hat{C}_p} \xrightarrow{\sim} \mathcal{O}_{\hat{C}_p} \cdot dz \cong \mathcal{O}_{\hat{C}_p}. \tag{6.2}$$

For every homomorphism $\xi : F \to K_C$, we have an element

$$\beta^*(\xi) \overset{\text{def}}{=} \alpha \circ \xi \circ \beta^{-1} \in \left(\mathcal{O}_{\hat{C}_p}^{\oplus n}\right)^* = \mathcal{O}_{\hat{C}_p}^{\oplus n}$$

that is determined by the commutative diagram

$$\begin{array}{ccc} F|_{\hat{C}_p} & \xrightarrow{\xi} & K_C|_{\hat{C}_p} \\ \beta \downarrow \wr & & \wr \downarrow \alpha \\ \mathcal{O}_{\hat{C}_p}^{\oplus n} & \xrightarrow{\beta^*(\xi)} & \mathcal{O}_{\hat{C}_p}. \end{array} \tag{6.3}$$

Thus we have the canonical choice of the local trivialization

$$\beta^* : F^* \otimes K_C|_{\hat{C}_p} \xrightarrow{\sim} \mathcal{O}_{\hat{C}_p}^{\oplus n}.$$

Definition 6.1 (Serre Dual). *The* Serre dual *of the geometric date* (C, p, z, F, β) *is the set of data* $(C, p, z, F^* \otimes K_C, \beta^*)$.

To identify the counterpart of the Serre duality on the Sato Grassmannian, we introduce a non-degenerate symmetric paring in $\mathbb{C}((z))$ by (6.4)

$$\langle a(z), b(z) \rangle = \frac{1}{2\pi i} \oint a(z)b(z)dz = \text{ the coefficient of } z^{-1} \text{ in } a(z)b(z),$$

and define

$$\langle f(z), g(z) \rangle_n = \sum_{i=1}^{n} \langle f_i(z), g_i(z) \rangle, \qquad f(z), g(z) \in \mathbb{C}((z))^{\oplus n}. \tag{6.5}$$

Lemma 6.1. *Let $W \in Gr_n(\mathbb{C})$ be a point of the Sato Grassmannian of index d. Then*

$$W^{\perp} = \{f \in \mathbb{C}((z))^{\oplus n} \mid \langle f, w \rangle_n = 0 \text{ for all } w \in W\}$$

is a point of $Gr_n(\mathbb{C})$ of index $-d$. Moreover, we have

$$\begin{cases} \mathrm{Ker}(\gamma_W)^* \cong \mathrm{Coker}(\gamma_{W^{\perp}}) \\ \mathrm{Coker}(\gamma_W)^* \cong \mathrm{Ker}(\gamma_{W^{\perp}}), \end{cases} \tag{6.6}$$

where γ_W is defined by the exact sequence

$$0 \longrightarrow \mathrm{Ker}(\gamma_W) \longrightarrow W \xrightarrow{\gamma_W} \mathbb{C}((z))^{\oplus n}/\mathbb{C}[[z]]^{\oplus n} \longrightarrow \mathrm{Coker}(\gamma_W) \longrightarrow 0.$$

Proof. Take an element $f \in W^{\perp} \cap \mathbb{C}[[z]]^{\oplus n}$. Then

$$\langle f, W + \mathbb{C}[[z]]^{\oplus n} \rangle_n = 0.$$

Thus it induces a linear map

$$f : \mathbb{C}((z))/(W + \mathbb{C}[[z]]^{\oplus n}) \ni g \longmapsto \langle f, g \rangle_n \in \mathbb{C},$$

defining a natural inclusion

$$\mathrm{Ker}(\gamma_{W^{\perp}}) \subset \mathrm{Coker}(\gamma_W)^*. \tag{6.7}$$

Now let $h^* \in \mathrm{Coker}(\gamma_W)^*$ be an arbitrary element. Choose a sequence $\{g_1^i, g_2^i, g_3^i \dots\}_{i=1,\dots,n}$ of elements of $\mathbb{C}((z))^{\oplus n}$ such that

(1) $g_j^i = \mathbf{e}^i z^{-j} + \sum_{\ell=1}^n \sum_{k \geq 1} g_{\ell jk} \mathbf{e}^{\ell} z^{-j+k}$;
(2) if W has an element whose leading term is $\mathbf{e}^i z^{-j}$ for $j > 0$, then $g_j^i \in W$.

Here $\mathbf{e}^i = (0, \dots, 0, 1, 0, \dots, 0)^T$ is the i-th standard basis vector for $\mathbb{C}^n$. We denote by $\bar{g}_j^i$ the projection image of g_j^i in $\mathbb{C}((z))^{\oplus n}/(W + \mathbb{C}[[z]]^{\oplus n})$. Consider the set of equations

$$\langle f, g_j^i \rangle_n = h^*(\bar{g}_j^i), \qquad i = 1, \dots, n; \quad j = 1, 2, 3, \dots \tag{6.8}$$

for $f \in \mathbb{C}[[z]]^{\oplus n}$. If we write $f(z) = \sum_{i=1}^n \sum_{j\geq 0} a_{ij}\mathbf{e}^i z^j$, then (6.8) is equivalent to

$$
\begin{aligned}
a_{i0} &= h^*(g_1^i)\\
a_{i1} &= h^*(g_2^i) - \sum_{\ell=1}^n a_{\ell 0} g_{\ell 21}\\
a_{i2} &= h^*(g_3^i) - \left(\sum_{\ell=1}^n a_{\ell 1} g_{\ell 31} + a_{\ell 0} g_{\ell 32}\right)\\
a_{i3} &= h^*(g_4^i) - \left(\sum_{\ell=1}^n a_{\ell 2} g_{\ell 41} + a_{\ell 1} g_{\ell 42} + a_{\ell 0} g_{\ell 43}\right)\\
&\vdots
\end{aligned}
$$

It is obvious that (6.8) has a unique solution. By construction we have $\langle f, W\rangle_n = 0$, hence $f \in \mathrm{Ker}(\gamma_{W^\perp})$. Therefore, the inclusion (6.7) is indeed a surjective map.

The application of the same argument to $W \cap \mathbb{C}[[z]]^{\oplus n}$ establishes that

$$\mathrm{Ker}(\gamma_W) \cong \mathrm{Coker}(\gamma_{W^\perp})^*.$$

Q.E.D.

Remark 6.2. We note that $W^{\perp\perp} = W$. It is obvious that $W \subset W^{\perp\perp}$. The relation (6.6) then makes the inclusion relation actually the equality.

Theorem 6.3. *Let $W \in Gr_n(\mathbb{C})$ be the point of the Grassmannian corresponding to a set of geometric data (C, p, z, F, β). Then the point corresponding to its Serre dual $(C, p, z, F^* \otimes K_C, \beta^*)$ is $W^\perp$.*

Proof. Let W' be the point of the Grassmannian corresponding to $(C, p, z, F^* \otimes K_C, \beta^*)$. Note that

$$\mathrm{index}(\gamma_W) = \dim H^0(C, F) - \dim H^1(C, F) = -\mathrm{index}(\gamma_{W'}).$$

Take an arbitrary $f \in H^0(C \setminus \{p\}, F)$ and $g \in H^0(C \setminus \{p\}, F^* \otimes K_C)$. Then

$$g \cdot f \in H^0(C \setminus \{p\}, K_C).$$

From (6.3) we see that $\beta^*(g) \cdot \beta(f) = \alpha(g \cdot f)$. Abel's theorem tells us that the sum of the residues of the meromorphic 1-form $g \cdot f$ on C is 0. Since $g \cdot f$ is holomorphic everywhere on C except for p, we have

$$0 = \mathrm{res}_p(g \cdot f) = \langle \beta^*(g), \beta(f)\rangle_n.$$

Therefore, $W' \subset W^{\perp}$. Since $\gamma_{W'}$ and $\gamma_{W^{\perp}}$ have the isomorphic kernels and cokernels, we conclude that $W' = W^{\perp}$. Q.E.D.

The ring of *ordinary differential operators* is defined as

$$\mathcal{D} = \mathbb{C}[[x]]\left[\frac{d}{dx}\right]. \tag{6.9}$$

Extending the powers of the differentiation to negative integers, we define the ring of *pseudo-differential operators* by

$$\mathcal{E} = \mathcal{D} + \mathbb{C}[[x]]\left[\left[\left(\frac{d}{dx}\right)^{-1}\right]\right]. \tag{6.10}$$

Let $\mathcal{E}x$ be the left maximal ideal of $\mathcal{E}$ generated by x. Then $\mathcal{E}/\mathcal{E}x$ is a left $\mathcal{E}$-module. Let us denote $\partial = d/dx$. As a $\mathbb{C}$-vector space we identify

$$\mathcal{E}/\mathcal{E}x = \mathbb{C}((\partial^{-1})). \tag{6.11}$$

Two useful formulas for calculating pseudo-differential operators are

$$\partial^n \cdot a(x) = \sum_{i\geq 0}\binom{n}{i} a^{(i)}(x)\partial^{n-i}, \tag{6.12}$$

where $a^{(i)}(x)$ is the i-th derivative of $a(x)$, and

$$a(x)\cdot\partial^n = \sum_{i\geq 0}(-1)^i\binom{n}{i}\partial^{n-i}\cdot a^{(i)}(x). \tag{6.13}$$

A pseudo-differential operator has thus two expressions

$$\begin{aligned} P = \sum_{n\in\mathbb{Z}} a_n(x)\partial^n &= \sum_{n\in\mathbb{Z}}\sum_{i\geq 0}(-1)^i\binom{n}{i}\partial^{n-i}\cdot a_n^{(i)}(x) \\ &= \sum_{m\in\mathbb{Z}}\sum_{i\geq 0}(-1)^i\binom{m+i}{i}\partial^m\cdot a_{m+i}^{(i)}(x). \end{aligned}$$

The natural projection $\mathcal{E} \to \mathcal{E}/\mathcal{E}x$ is given by

$$\begin{aligned} \rho : \mathcal{E} \ni P = \sum_n a_n(x)\partial^n & \\ \longmapsto \sum_m\sum_{i\geq 0}(-1)^i\binom{m+i}{i} a_{m+i}^{(i)}(0)\partial^m &\in \mathbb{C}((\partial^{-1})). \end{aligned} \tag{6.14}$$

The relation to the Grassmannian comes from the identification

$$\mathbb{C}((z)) \ni f(z) \longmapsto f(\partial^{-1}) \cdot \partial^{-1} \in \mathbb{C}((\partial^{-1})). \tag{6.15}$$

Then $\mathbb{C}((z))$ becomes an $\mathcal{E}$-module. The action of $P \in \mathcal{E}$ on $f(z) \in \mathbb{C}((z))$ is given by the formula

$$P \cdot f(z) = \rho(P \cdot f(\partial^{-1})\partial^{-1}) \in \mathbb{C}((z)).$$

Definition 6.2 (Adjoint). *The* adjoint *of* $P = \sum_\ell a_\ell(x)\partial^\ell$ *is defined by*

$$P^* = \sum_\ell \partial^\ell \cdot a_\ell(-x).$$

Remark 6.4. Our adjoint is slightly different from the *formal adjoint* of [13, 29], which is defined to be $\sum_\ell (-\partial)^\ell \cdot a_\ell(x)$. This is due to the fact that we are considering the action of pseudo-differential operators on the function space $\mathbb{C}((z))$ through *Fourier transform.* Thus ∂ acts as the multiplication of z^{-1}, and x acts as the differentiation.

Let us compute the (m,n)-matrix entry of P. We find

$$\begin{aligned}
&\langle z^m, P \cdot z^n \rangle \\
&= \left\langle z^m, \rho\left(\sum_\ell a_\ell(x)\partial^{\ell-n-1}\right)\right\rangle \\
&= \left\langle z^m, \rho\left(\sum_\ell \sum_i (-1)^i \binom{\ell-n-1}{i} \partial^{\ell-n-1-i} \cdot a_\ell^{(i)}(x)\right)\right\rangle \\
&= \left\langle z^m, \sum_\ell \sum_i (-1)^i \binom{\ell-n-1}{i} a_\ell^{(i)}(0) \cdot z^{-\ell+n+i}\right\rangle \\
&= \sum_i (-1)^i \binom{m+i}{i} a^{(i)}_{m+n+i+1}(0).
\end{aligned}$$

Similarly, we find

$$\begin{aligned}
&\langle z^n, P^* \cdot z^m \rangle \\
&= \left\langle z^n, \rho\left(\sum_\ell \partial^\ell a_\ell(-x)\cdot\partial^{-m-1}\right)\right\rangle \\
&= \left\langle z^n, \rho\left(\sum_\ell\sum_i (-1)^i(-1)^i\binom{-m-1}{i}\partial^{\ell-m-1-i}\cdot a_\ell^{(i)}(-x)\right)\right\rangle \\
&= \left\langle z^n, \sum_\ell\sum_i \binom{-m-1}{i} a_\ell^{(i)}(0)\cdot z^{-\ell+m+i}\right\rangle \\
&= \sum_i \binom{-m-1}{i} a_{m+n+i+1}^{(i)}(0).
\end{aligned}$$

Following [16], we introduce the ring of differential and pseudo-differential operators with matrix coefficients and denote them by $gl_n(\mathcal{D})$ and $gl_n(\mathcal{E})$, respectively. The adjoint of a matrix pseudo-differential operator is defined in an obvious way:

$$P = [P_{ij}]_{i,j=1,\ldots,n} \Longrightarrow P^* \overset{\text{def}}{=} [P_{ji}^*]_{i,j=1,\ldots,n}. \tag{6.16}$$

Proposition 6.5. *For arbitrary $f(z), g(z) \in \mathbb{C}((z))^{\oplus n}$ and $P \in gl_n(\mathcal{E})$ we have the adjoint formula*

$$\langle f(z), P\cdot g(z)\rangle_n = \langle P^*\cdot f(z), g(z)\rangle_n.$$

Proof. The statement follows from the above computations, noting the identity

$$(-1)^i\binom{m+i}{i} = \binom{-m-1}{i},$$

and the usual matrix transposition. Q.E.D.

A more direct relation between pseudo-differential operators and the Sato Grassmannian is that the identification of the *big cell* of $Gr_n(\mathbb{C})$ with the group

$$G = I_n + gl_n(\mathbb{C}[[x]])[[\partial^{-1}]]\cdot\partial^{-1} \tag{6.17}$$

of monic pseudo-differential operators of order 0 ([16], Theorem 6.5).

Theorem 6.6. *Let $W \in Gr_n(\mathbb{C})$ be in the big cell, and correspond to a pseudo-differential operator $S \in G$. Then $W^\perp$ corresponds to $(S^*)^{-1} \in G$.*

Proof. The correspondence of $S \in G$ and W in the big cell is the equation

$$W = S^{-1} \cdot \mathbb{C}[z^{-1}]^{\oplus n} \cdot z^{-1}.$$

For $f(z), g(z) \in \mathbb{C}[z^{-1}]^{\oplus n} \cdot z^{-1}$, we have

$$\langle S^* \cdot f(z), S^{-1} \cdot g(z) \rangle_n = \langle f(z), SS^{-1} \cdot g(z) \rangle_n = 0.$$

Q.E.D.

The *Lax operator* is defined by

$$\mathbf{L} = S \cdot I_n \cdot \partial \cdot S^{-1}. \tag{6.18}$$

Lax operators bijectively correspond to points of the big cell of the quotient Grassmannian $Z_n(\mathbb{C})$. If $\overline{W}$ corresponds to $\mathbf{L}$, then $\overline{W}^{\perp}$ corresponds to $\mathbf{L}^*$. Therefore, the Serre duality indeed corresponds to the adjoint action of the pseudo-differential operators.

§7. The Hitchin integrable system for Sp_{2m} and corresponding KP-type equations

In this section we identify the KP-type equations that are equivalent to the Hitchin integrable system for Sp_{2m}. Since the moduli space of the Sp_{2m}-Higgs bundles is identified as the fixed-point-set of the Serre duality involution on $\mathcal{H}_C(2m, 2m(g-1))$ (Proposition 2.4), we will see that the evolution equations are the KP equations restricted to a certain type of self-adjoint Lax operators.

Let $(E, \phi) = (E^* \otimes K_C, -\phi^*)$ be a fixed point of the Serre duality operation on the Higgs moduli space $\mathcal{H}_C(2m, 2m(g-1))$, and $(\mathcal{L}_E, C_s)$ the corresponding spectral data. We choose a point $p \in C$ and a local coordinate z as in Section 5. We assume that the spectral cover $\pi : C_s \to C$ is unramified at p. This time we choose a local trivialization

$$\beta : \mathcal{L}_E|_{\hat{C}_{s,\pi^{-1}(p)}} \xrightarrow{\sim} \mathcal{O}_{\hat{C}_{s,\pi^{-1}(p)}} \cong \mathbb{C}[[z]]^{\oplus 2m}.$$

Then the Serre dual of the data $(\pi_s : C_s \to C, p, z, \mathcal{L}_E, \phi, \beta)$ is

$$(\pi_{s^*} : C_{s^*} \to C, p, z, \epsilon^*(\mathcal{L}_E^* \otimes K_{C_s}), -\phi^*, \epsilon^*\beta^*).$$

In general the dual trivialization is defined by

$$\begin{array}{ccc}
\mathcal{L}_E|_{\hat{C}_{s,\pi_s^{-1}(p)}} & \xrightarrow{\ \xi\ } & K_{C_s}|_{\hat{C}_{s,\pi_s^{-1}(p)}} \\
\beta \big\downarrow \wr & & \wr \big\downarrow \alpha \\
\mathcal{O}_{\hat{C}_{s,\pi_s^{-1}(p)}} & \xrightarrow{\ \beta^*(\xi)\ } & \mathcal{O}_{\hat{C}_{s,\pi_s^{-1}(p)}} \\
\epsilon \big\downarrow \wr & & \wr \big\downarrow \epsilon \\
\mathcal{O}_{\hat{C}_{s^*,\pi_{s^*}^{-1}(p)}} & \xrightarrow{\ \epsilon^*\beta^*(\xi)\ } & \mathcal{O}_{\hat{C}_{s^*,\pi_{s^*}^{-1}(p)}}
\end{array}.$$

Note that in the current case we have $s = s^*$ and $\epsilon : C_s \to C_s$ is a non-trivial involution. Since ϵ commutes with π, it induces an involution of the set $\pi^{-1}(p) = \{p_1, \ldots, p_{2m}\}$. Let us number the $2m$ distinct points so that

$$\epsilon(p_1, \ldots, p_{2m}) = (p_1, \ldots, p_{2m}) \cdot A,$$

where

$$A = \begin{bmatrix} 0 & I_m \\ I_m & 0 \end{bmatrix}. \tag{7.1}$$

By the same argument we used in Theorem 6.3, if we define

$$W = \beta(H^0(C_s \setminus \pi^{-1}(p), \mathcal{L}_E)) \subset \mathbb{C}((z))^{\oplus 2m},$$

then we have

$$AW^\perp = \epsilon^*\beta^*(H^0(C_{s^*} \setminus \pi^{-1}(p), \epsilon^*(\mathcal{L}_E^* \otimes K_{C_s}))) \subset \mathbb{C}((z))^{\oplus 2m}. \tag{7.2}$$

Let us assume that E is not on the theta divisor of [3] so that $H^0(C, E) = H^1(C, E) = 0$. Then $W \in Gr_{2m}(\mathbb{C})$ is on the big cell, and hence corresponds to a monic 0-th order pseudo-differential operator $S \in G$. Since $W = AW^\perp$, we have

$$S = A \cdot (S^*)^{-1} \cdot A. \tag{7.3}$$

The Lax operator

$$\mathbf{L} = S \cdot I_{2m} \cdot \partial \cdot S^{-1} = \begin{bmatrix} \mathbf{L}_1 & \mathbf{L}_2 \\ \mathbf{L}_3 & \mathbf{L}_4 \end{bmatrix}$$

then satisfies that

$$\mathbf{L}^* = A \cdot \mathbf{L} \cdot A, \tag{7.4}$$

or equivalently

$$\mathbf{L}_1^* = \mathbf{L}_4, \qquad \mathbf{L}_2^* = \mathbf{L}_2, \qquad \mathbf{L}_3^* = \mathbf{L}_3.$$

The time evolution of S and the Lax operator $\mathbf{L}$ is given by the formula established in Section 6 of [16]. Let $S(t)$ be the solution of the n-component KP equations with the initial data $S(0) = S$. Then $S(t)$ is given by the generalized Birkhoff decomposition of [16, 18]

$$\exp\left(\begin{bmatrix} D_1 & \\ & D_2 \end{bmatrix}\right) \cdot S(0)^{-1} = S(t)^{-1} \cdot Y(t), \tag{7.5}$$

where D_1 and D_2 are diagonal matrices of the shape

$$D_i = \begin{bmatrix} \sum_{j\geq 1} t_{ij1}\partial^j & & \\ & \ddots & \\ & & \sum_{j\geq 1} t_{ijm}\partial^j \end{bmatrix}$$

corresponding to (5.13), and $Y(t)$ is an invertible infinite order differential operator introduced in [18]. We impose that the time evolution $S(t)$ satisfies the same twisted self-adjoint condition (7.3). Applying the adjoint-inverse operation and conjugation by A of (7.1) to (7.5), we obtain

$$\exp\left(A\begin{bmatrix} -D_1 & \\ & -D_2 \end{bmatrix}A\right) \cdot S(0)^{-1} = S(t)^{-1} \cdot (A(Y(t)^*)^{-1}A).$$

Therefore, the time evolution (7.5) with the condition $D_2 = -D_1$ preserves the twisted self-adjointness (7.3). We have thus established

Theorem 7.1. *The KP-type equations that generate the Hitchin integrable systems on the moduli spaces of Sp_{2m}-Higgs bundles are the reduction of the $2m$-component KP equations that preserve the twisted self-adjointness* (7.3) *for the operator S, or* (7.4) *for the Lax operator $\mathbf{L}$. The time evolution $S \longmapsto S(t)$ preserving the condition is given by the generalized Birkhoff decomposition*

$$\exp\left(\begin{bmatrix} D & \\ & -D \end{bmatrix}\right) \cdot S(0)^{-1} = S(t)^{-1} \cdot Y(t), \tag{7.6}$$

where

$$D = \begin{bmatrix} \sum_{j\geq 1} t_{j1}\partial^j & & \\ & \ddots & \\ & & \sum_{j\geq 1} t_{jm}\partial^j \end{bmatrix}.$$

Remark 7.2. Since the time evolution of (7.6) is given by a traceless matrix $\mathrm{diag}(D, -D)$, every finite-dimensional orbit of this system is a Prym variety by the general theory of [16], which is expected from the Sp Hitchin fibration.

References

[1] M. R. Adams and M. J. Bergvelt, The Krichever map, vector bundles over algebraic curves, and Heisenberg algebras, Commun. Math. Phys., **154** (1993), 265–305.

[2] A. Álvarez Vázquez, J. M. Muñoz Porras and F. J. Plaza Martín, The algebraic formalism of soliton equations over arbitrary base fields, In: Workshop on Abelian Varieties and Theta Functions, Aportaciones Mat. Investig., **13**, Soc. Mat. Mexicana, 1998, pp. 3–40.

[3] A. Beauville, M. S. Narasimhan and S. Ramanan, Spectral curves and the generalised theta divisor, J. Reine Angew. Math., **398** (1989), 169–179.

[4] R. Donagi and E. Markman, Spectral covers, algebraically completely integrable Hamiltonian systems, and moduli of bundles, Lecture Notes in Math., **1620**, Springer-Verlag, 1996.

[5] R. Donagi and T. Pantev, Langlands duality for Hitchin systems, arXiv:math/0604617v2.

[6] E. Frenkel, Lectures on the Langlands program and conformal field theory, arXiv:hep-th/0512172.

[7] E. Gómez González, J. M. Muñoz Porras and F. J. Plaza Martín, Prym varieties, curves with automorphisms and the Sato Grassmannian, arXiv:math.AG/0207202[math.AG].

[8] T. Hausel and M. Thaddeus, Mirror symmetry, Langlands duality, and the Hitchin system, Invent. Math., **153** (2003), 197–229.

[9] D. Hernandez-Serrano, J. M. Muñoz Porras and F. J. Plaza Martín, Equations of the moduli of Higgs pairs and infinite Grassmannian, Internat. J. Math., to appear.

[10] N. J. Hitchin, The self-duality equations on a Riemann surface, Proc. London Math. Soc. (3), **55** (1987), 59–126.

[11] N. J. Hitchin, Stable bundles and integrable systems, Duke Math. J., **54** (1987), 91–114.

[12] N. J. Hitchin, Langlands duality and G_2 spectral curves, arXiv:math.AG/0611524.

[13] M. Jimbo and T. Miwa, Solitons and infinite-dimensional Lie algebras, Publ. Res. Inst. Math. Sci., **19** (1983), 943–1001.

[14] A. Kapustin and E. Witten, Electric-magnetic duality and the geometric Langlands program, arXiv:hep-th/0604151v3.

[15] Y. Li and M. Mulase, Hitchin systems and KP equations, Internat. J. Math., **7** (1996), 277–244.

[16] Y. Li and M. Mulase, Prym varieties and integrable systems, Comm. Anal. Geom., **5** (1997), 279–332.

[17] M. Mulase, Cohomological structure in soliton equations and Jacobian varieties, J. Differential Geom., **19** (1984), 403–430.

[18] M. Mulase, Solvability of the super KP equation and a generalization of the Birkhoff decomposition, Invent. Math., **92** (1988), 1–46.

[19] M. Mulase, Category of vector bundles on algebraic curves and infinite dimensional Grassmannians, Internat. J. Math., **1** (1990), 293–342.

[20] M. Mulase, Geometry of character varieties of surface groups, RIMS Kôkyûroku, **1605** (2008), 1–21.

[21] D. Mumford, J. Fogarty and F. Kirwan, Geometric Invariant Theory, Third ed., Springer-Verlag, 1994.

[22] F. Plaza Martín, Families of infinite Grassmannians and the induced central extensions, in preparation.

[23] I. Quandt, On a relative version of the Krichever correspondence, Bayreuth. Math. Schr., **52** (1997), 1–74.

[24] M. Sato and Y. Sato, Soliton equations as dynamical systems on infinite-dimensional Grassmann manifold, In: Nonlinear Partial Differential Equations in Applied Science, North-Holland Math. Stud., **81**, 1983, pp. 259–271.

[25] G. Segal and G. Wilson, Loop groups and equations of KdV type, Inst. Hautes Études Sci. Publ. Math., **61** (1985), 5–65.

[26] T. Shiota, Characterization of Jacobian varieties in terms of soliton equations, Invent. Math., **83** (1986), 333–382.

[27] T. Shiota, Prym varieties and soliton equations, In: Infinite-Dimensional Lie Algebras and Groups, Adv. Ser. Math. Phys., **7**, 1989, pp. 407–448.

[28] A. Strominger, S.-T. Yau and E. Zaslow, Mirror symmetry is T-duality, Nuclear Phys. B, **479** (1996), 243–259.

[29] K. Takasaki, The World of Integrable Systems—The Fellowship of the Toda Lattice (in Japanese), Kyoritsu Shuppan Publ. Co., 2001.

[30] T. Tsuchida, New reductions of integrable matrix PDEs—$Sp(2m)$-invariant systems, arXiv:0712.4373[nlin.SI].

[31] P. Vanhaecke, Integrable systems in the realm of algebraic geometry, Lecture Notes in Math., **1638**, Springer-Verlag, 1996.

Andrew R. Hodge
Department of Defense
Fort George G. Meade, MD 20755-6000
U.S.A.

Motohico Mulase
Department of Mathematics
University of California
Davis, CA 95616–8633
U.S.A.
E-mail address: mulase@math.ucdavis.edu

Advanced Studies in Pure Mathematics 59, 2010
New Developments in Algebraic Geometry,
Integrable Systems and Mirror Symmetry (Kyoto, 2008)
pp. 79–110

BCOV ring and holomorphic anomaly equation

Shinobu Hosono

Abstract.

We study certain differential rings over the moduli space of Calabi–Yau manifolds. In the case of an elliptic curve, we observe a close relation to the differential ring of quasi-modular forms due to Kaneko–Zagier[23].

§1. Introduction

Since the pioneering work by Candelas, de la Ossa, Green and Parkes [8] in 1991, the theory of variation of Hodge structures has been one of the indispensable tools in the study of mirror symmetry of Calabi–Yau manifolds and its application to Gromov–Witten theory or enumerative geometry on Calabi–Yau manifolds. In particular, in the generalization due to Bershadsky, Cecotti, Ooguri and Vafa (BCOV) to higher genus Gromov–Witten potentials, the theory of variation of Hodge structures was combined with a framework which is called t-t^* geometry [9]. t-t^* geometry is a deformation theory of $N = 2$ supersymmetric quantum field theory in two dimensions, and there is a natural hermitian (real) structure in the space of observables. BCOV identifies this hermitian structure with the hermitian structure over the moduli space of Calabi–Yau manifolds given by the Weil–Petersson metric, and have proposed a profound recursive relation, called *holomorphic anomaly equation*, for higher genus Gromov–Witten potentials.

In case of dimension one, i.e. for elliptic curves, the counting problem and higher genus Gromov–Witten potentials have been determined by Dijkgraaf [11] in 1995, where it was remarked that BCOV theory

---

Received October 27, 2008.
2000 *Mathematics Subject Classification.* Primary 53Z50; Secondary 53B50, 33C90.
Partly supported by JSPS Grant-in-Aid for Scientific Research (C-18540014, S-19104002).

is closely related to the theory of elliptic quasi-modular forms. In the same proceedings volume as [11], Kaneko and Zagier have presented a general theory of quasi-modular forms introducing the (differential) ring of almost holomorphic modular forms. It is also found in [21] that, for an rational elliptic surface, the higher genus Gromov–Witten potentials are expressed by quasi-modular forms, extending the genus zero result by [25], [26].

For Calabi–Yau threefolds, it has been expected that the BCOV holomorphic anomaly equation is defined over a certain differential ring which generalizes the almost holomorphic elliptic modular forms due to Kaneko and Zagier. Recently, in physics literatures, Yamaguchi and Yau [32] and later Alim and Länge [2] have made important developments toward the structure of the expected differential ring of Calabi–Yau threefolds (see also [1], [15]). In this paper, to make a parallel argument to the theory due to Kaneko and Zagier, we introduce three different differential rings $\mathcal{R}^0_{BCOV}$, $\mathcal{R}^\Gamma_{BCOV}$ and $\mathcal{R}^{hol}_{BCOV}$ over the moduli space of Calabi–Yau threefolds. We call these differential rings simply as *BCOV rings.* Our BCOV rings $\mathcal{R}^\Gamma_{BCOV}$ and $\mathcal{R}^{hol}_{BCOV}$, should be regarded as a natural generalization of the ring of almost holomorphic modular forms and quasi-modular forms, respectively, and may be recognized in the original work [5] and more explicitly in recent physics literatures [32], [2], [1], [15]. We will introduce another form of the BCOV ring $\mathcal{R}^0_{BCOV}$ and observe that, with this ring, our parallelism to Kaneko–Zagier theory becomes complete.

Here we summarize briefly the theory of quasi-modular forms due to Kaneko and Zagier. Kaneko–Zagier [23] starts from a ring

$$\mathbf{C}[[\tau]][\frac{1}{\tau-\bar{\tau}}] \ , \tag{1.1}$$

with τ in the upper-half plane, and the standard modular group action, $\tau \mapsto \frac{a\tau+b}{c\tau+d}$. Inside this large ring, one first considers the almost modular forms $\widehat{M}(\Gamma)_k$ of weight k as almost holomorphic functions $F(\tau,\bar{\tau})$ on the upper-half plane which transforms like a modular form of weight k;

$$F(\frac{a\tau+b}{c\tau+d},\overline{\frac{a\tau+b}{c\tau+d}}) = (c\tau+d)^k F(\tau,\bar{\tau}) \ .$$

Then the ring of the almost holomorphic modular forms $\widehat{M}(\Gamma) = \oplus_{k\geq 0}\widehat{M}(\Gamma)_k$ becomes a differential ring under $D_\tau : \widehat{M}(\Gamma)_k \to \widehat{M}(\Gamma)_{k+2}$, with $D_\tau := \frac{1}{2\pi i}\big(\frac{\partial}{\partial\tau} + \frac{k}{\tau-\bar{\tau}}\big)$. Elements of $\widehat{M}(\Gamma)$ have an expansion $F = \sum_{m\geq 0} c_m(\tau)\big(\frac{1}{\tau-\bar{\tau}}\big)^m$, and by taking the first coefficient $c_0(\tau)$ we obtain holomorphic objects. Kaneko–Zagier shows that this map defines a (differential) ring isomorphism $\varphi : \widehat{M}(\Gamma) \to \widetilde{M}(\Gamma)$, where $\widetilde{M}(\Gamma) = \mathbf{C}[E_2(\tau), E_4(\tau), E_6(\tau)]$ is the ring of the quasi-modular forms with the differential $\partial_\tau := \frac{1}{2\pi i}\frac{\partial}{\partial \tau}$. Our observation here is that the ring (1.1) is defined by the Kähler geometry on the upper-half plane, and has a natural generalization to the Weil–Petersson geometry on the moduli space of Calabi–Yau manifolds. Based on this, we will introduce our BCOV ring $\mathcal{R}^0_{BCOV}$ in terms of purely geometric data, and subsequently introduce other forms of the ring, $\mathcal{R}^\Gamma_{BCOV}$, $\mathcal{R}^{hol}_{BCOV}$. We may schematically write our parallelism of the BCOV rings to the relevant rings in Kaneko–Zagier theory:

$$\begin{array}{ccccc} \mathbf{C}[[\tau]][\frac{1}{\tau-\bar\tau}] & \supset & \widehat{M}(\Gamma) & \xrightarrow{\varphi} & \widetilde{M}(\Gamma) \\ \mathcal{R}^0_{BCOV} & \to & \mathcal{R}^\Gamma_{BCOV} & \xrightarrow{\bar t \to \infty} & \mathcal{R}^{hol}_{BCOV} \end{array} .$$

As one see in the arrow $\mathcal{R}^0_{BCOV} \to \mathcal{R}^\Gamma_{BCOV}$, instead of $\supset$, the relations shown in this diagram are not exact correspondences but should be understood simply as parallelism. In fact, in our BCOV ring, following [5], we work with meromorphic sections of certain bundles instead of (almost) holomorphic forms in Kaneko–Zagier theory. Details will be described in the text, however it should be helpful to have this schematic diagram in mind.

The main result of this paper is the introduction of the BCOV ring $\mathcal{R}^0_{BCOV}$ (Definition 3.3, Theorem 3.5) and making the parallelism to Kaneko–Zagier theory of quasi-modular forms complete.

Construction of this paper is as follows. To make the paper self-contained, in Section 2, we review the geometry of the moduli space of Calabi–Yau manifold, which is called special Kähler geometry, and set up our notations. In Section 3, we define our BCOV rings. In subsection (3-1), we introduce the first form of our BCOV (differential) ring $\mathcal{R}^0_{BCOV}$ based on the special Kähler geometry (Definition 3.1, Theorem 3.5). We remark that the BCOV ring $\mathcal{R}^0_{BCOV}$ is infinitely generated, however there is a natural reduction $\mathcal{R}^{0,red}_{BCOV}$ to a finitely generated ring. In (3-2), we consider modular (monodromy) property of the ring and we will define a differential ring $\mathcal{R}^\Gamma_{BCOV}$ as the ring of monodromy invariants. We, then, understand in our framework the process of 'fixing a holomorphic (meromorphic) ambiguities in the propagator functions' given in the Section 6.3 of [5]. In (3-3), BCOV rings $\mathcal{R}^0_{BCOV}$ and $\mathcal{R}^\Gamma_{BCOV}$ will be determined explicitly for an elliptic curve. There we provide a precise relation to the theory of quasi-modular forms (Propositions 3.11, 3.12). In (3-4), the final form $\mathcal{R}^{hol}_{BCOV}$ is defined under a choice of a symplectic basis of the middle dimensional homology group. In Section 4, toward an application to the holomorphic anomaly equation, we introduce the holomorphic anomaly equation in the form appeared in [32], [2]. Considering holomorphic anomaly equation in the reduced ring $\mathcal{R}^{0,red}_{BCOV}$ we note that the equation simplifies to a system of linear differential equation (Proposition 4.3) which is easy to handle. Conclusions and discussion are given in Section 5. There, the equivalence of the modular anomaly equation in [21] to BCOV holomorphic anomaly equation is also announced.

Acknowledgments. The main result of this paper was announced in the workshop "Number Theory and Physics at the Crossroads", Sep. 21–26, 2008 at Banff International Research Station. The author would like to thank the organizers for providing a wonderful research environment there. He also would like to thank M.-H. Saito for valuable discussions. This work is supported in part by Grant-in Aid Scientific Research (C 18540014).

§2. Special Kähler geometry on deformation spaces

(2-1) Period integrals. Let us consider a family of Calabi–Yau 3-folds $\mathcal{Y} = \{Y_x\}$ over a small complex domain B with fibers Y_x $(x \in B)$. As such a family, we will consider hypersurfaces or complete intersections in projective toric varieties, and assume that the local family eventually extends to a family over a toric variety $\mathcal{M}$ $(= \mathbf{P}_{Sec(\Sigma)}$ with the secondary fan, see eg. [12]). We also assume that $\dim\mathcal{M} = \dim H^{2,1}(Y_{x_0})$ for a smooth Y_{x_0}.

We fix a smooth $Y_{x_0} (x_0 \in B)$ as above. We denote the cohomology $H^3(Y_{x_0}, \mathbf{Z})$ by H_{x_0} and write the symplectic form there by

$$\langle u, v\rangle := \sqrt{-1}\int_{Y_{x_0}} u \cup v \ . \tag{2.1}$$

We choose a symplectic basis $\{\alpha_I, \beta^J\}_{0\le I,J\le r}$ satisfying $\langle \alpha_I, \beta^J\rangle = \delta_I^{\ J}$, $\langle \alpha_I, \alpha_J\rangle = \langle \beta^I, \beta^J\rangle = 0$, and we denote its dual homology basis by $\{A^I, B_J\}_{0\le I,J\le r}$ $(r := \dim H^{2,1}(Y_{x_0}))$. With respect to this basis, we write the symplectic form

$$Q = \begin{pmatrix} 0 & E \\ -E & 0 \end{pmatrix},$$

where $E = E_{r+1}$ represents the unit matrix of size $(r+1)$. We define the period domain

$$\mathcal{D} = \{[\omega] \in \mathbf{P}(H_{x_0} \otimes \mathbf{C}) \,|\, \langle \omega, \omega\rangle = 0, \langle \omega, \overline{\omega}\rangle > 0\} \ .$$

We choose a holomorphic three form of the fiber Y_x ($x \in B$) and denote it by $\Omega_x := \Omega(Y_x)$. Then the period map $\mathcal{P}_0 : B \to \mathcal{D}$ is defined by

$$\mathcal{P}_0(x) = [\sum_I \left(\int_{A^I} \Omega_x \right) \alpha_I + \sum_J \left(\int_{B_J} \Omega_x \right) \beta^J] \ .$$

Using path-dependent identification $H_3(Y_x, \mathbf{Z}) \cong H_3(Y_{x_0}, \mathbf{Z})$, we may globalize this period map on B to $\mathcal{M}$ by introducing a covering space $\tilde{\mathcal{M}}$. We denote the resulting period map $\mathcal{P} : \tilde{\mathcal{M}} \to \mathcal{D}$ and assume $\Gamma \subset Sp(2r+2, \mathbf{Z})$ as the covering group. In this paper, we will write the period map $\mathcal{P}(x) = [\vec{\omega}(x)]$ (or $\mathcal{P}_0(x) = [\vec{\omega}(x)]$ ($x \in B$)) with the notations for the period integrals,

$$\vec{\omega}(x) = \sum_I X^I(x) \alpha_I + \sum_J P_J(x) \beta^J \ .$$

We write by $\mathcal{U}$ the restriction to $\mathcal{D}$ of the tautological line bundle $\mathcal{O}(-1)$ over $\mathbf{P}(H_{x_0} \otimes \mathbf{C})$. We then set $\mathcal{L} = \mathcal{P}^* \mathcal{U}$, i.e., the pullback to $\tilde{\mathcal{M}}$. Complex conjugate of $\mathcal{L}$ will be denoted by $\overline{\mathcal{L}}$. The sections of $\mathcal{L}^{\otimes n} \otimes \overline{\mathcal{L}}^{\otimes m}$ will be often referred to as 'sections of weight (n,m)'. The period integral $\vec{\omega}(x)$ may be considered as a section of $\mathcal{L}$, and thus has weight $(1,0)$.

(2-2) Prepotential. The symplectic form (2.1) naturally induces one form θ on $\mathcal{D}$ by $\theta := \langle d\omega, \omega \rangle$. With this one form, $(\mathcal{D}, \theta)$ becomes a holomorphic contact manifold of dimension $2r+1$. Since locally, the period map $\mathcal{P}_0 : B \to \mathcal{D}$ is an embedding [6], [30], [31] and also $\theta|_{\mathcal{P}_0(B)} = 0$ due to Griffiths transversality, we know that the image of the period map $\mathcal{P}_0$ is a Legendre submanifold. Combining this with Gauss correspondence in projective geometry, it is found in general [7] that the image of the period map $\mathcal{P}_0$ can be recovered by the half of the period integrals $X^I(x) = \int_{A^I} \Omega_x$. More concretely, it is known that:

1) The map $x \mapsto [X^0(x), \cdots, X^r(x)] \in \mathbf{P}^r$ is an local isomorphism $B \to \mathbf{P}^r$,
2) Integrating $\theta|_{\mathcal{P}_0(B)} = 0$ on B, we can write the other half of the period integrals,

$$P_J(x) = \frac{\partial \mathcal{F}(X^I)}{\partial X^J} \quad (J = 0, 1, \cdots, r),$$

in terms of a holomorphic function $\mathcal{F}(X)$ called *prepotential.*

The function $\mathcal{F}(X)$ is an holomorphic function of $X^0(x), \cdots, X^r(x)$ and has the following homogeneous property

$$\sum_{I=0}^{r} X^I(x) \frac{\partial \mathcal{F}}{\partial X^I} = 2\mathcal{F}(X) \ .$$

This potential function exists locally for the small domain B. When globalizing the above local arguments to $\tilde{\mathcal{M}}$, we naturally see that the monodromy group Γ plays a role for the definition of $\mathcal{F}(X)$ (see below). The group action of Γ is referred to as *duality transformation* in physics (see e.g. [10] and references therein). Here we remark that the holomorphic prepotential has a simple relation to the so-called Griffiths–Yukawa coupling of the family $\{Y_x\}_{x \in B}$;

$$\text{(2.2)} \quad -\int_{Y_x} \Omega_x \cup \frac{\partial}{\partial x^i} \frac{\partial}{\partial x^j} \frac{\partial}{\partial x^k} \Omega_x = \sum_{I,J,K=0}^{r} \frac{\partial X^I}{\partial x^i} \frac{\partial X^J}{\partial x^j} \frac{\partial X^K}{\partial x^k} \frac{\partial^3 \mathcal{F}(X)}{\partial X^I \partial X^J \partial X^K} \ ,$$

where the l.h.s. is will be written by $C_{ijk}(x)$ hereafter. We denote $\overline{C}_{\bar{i}\bar{j}\bar{k}}$ the complex conjugate of $C_{ijk}(x)$.

(2-3) Period matrix (1). The most important object to introduce the BCOV anomaly equation is the classical period matrix. For simplicity, let us assume that our family $\{Y_x\}$ is given by a family of hypersurfaces in a (smooth) toric variety $\mathbf{P}^4_\Sigma$, with the parameter $x = (x^1, x^2, \cdots, x^r)$ compactified to a toric variety $\mathcal{M}$. We then consider $\mathcal{A}^q_k$, the set of rational q forms on $\mathbf{P}^4_\Sigma$ with a pole order less than k along Y_x, and set the cohomology group

$$\mathcal{H}_k = \mathcal{A}^4_k / d\mathcal{A}^3_{k-1} \ .$$

Obviously we have $\mathcal{H}_1 \subset \mathcal{H}_2 \subset \cdots$, and, in fact, this stabilizes at $\mathcal{H}_4$ ($= \mathcal{H}_5 = \cdots$) to the rational 4 forms $\mathcal{H}$ of poles along Y_x. We note that, for 3-folds, the 'tubular' map $\tau : H_3(Y_x, \mathbf{Z}) \tilde{\to} H_4(\mathbf{P}_\Sigma \setminus Y_x, \mathbf{Z})$ is an isomorphism ([Gr, Proposition 3.5]), and is related to the Poincaré residue map $R : \mathcal{H} \to H^3(Y_x, \mathbf{C})$ along Y_x by

$$\int_{\tau(\gamma)} \omega = \int_\gamma R(\omega) \ \ (\omega \in \mathcal{H}_k).$$

This residue map is an isomorphism, and more precisely, maps the filtration $\mathcal{H}_1 \subset \mathcal{H}_2 \subset \mathcal{H}_3 \subset \mathcal{H}_4$ to the Hodge filtration

$$F^{3,0} \subset F^{3,1} \subset F^{3,2} \subset F^{3,3} \ .$$

The holomorphic three form $\Omega_x = R(\omega_0)$ may then be given by a basis ω_0 of $\mathcal{H}_1 \cong F^{3,0}$. We take a basis $\omega_0, \omega_1, \cdots, \omega_r$ of $\mathcal{H}_2$, and define the period matrix

$$(2.3)\quad \mathbf{\Omega} = \begin{pmatrix} \int_{\tau(A_0)} \omega_0 & \cdots & \int_{\tau(A_r)} \omega_0 & \int_{\tau(B_0)} \omega_0 & \cdots & \int_{\tau(B_r)} \omega_0 \\ & \vdots & & & \vdots & \\ \int_{\tau(A_0)} \omega_r & \cdots & \int_{\tau(A_r)} \omega_r & \int_{\tau(B_0)} \omega_r & \cdots & \int_{\tau(B_r)} \omega_r \end{pmatrix}$$

as $(r+1, 2r+2)$ matrix. The first row of this matrix coincides with the period integral $\vec{\omega}(x) = \sum_I X^I(x)\alpha_I + \sum_J P_J(x)\beta^J$, and the monodromy group Γ acts on this period matrix from the right.

The following properties of $\mathbf{\Omega}$ are consequences of the filtration (2) and the Hodge–Riemann bilinear relations;

$$(2.4)\quad \begin{array}{ll} 1) & \mathbf{\Omega}\, Q\, {}^t\mathbf{\Omega} = 0 \\ 2) & \sqrt{-1}\mathbf{\Omega}\, Q\, {}^t\overline{\mathbf{\Omega}} > 0\ . \end{array}$$

Here 2) means that when we decompose the $(r+1) \times (r+1)$ hermitian matrix $\sqrt{-1}\mathbf{\Omega} Q {}^t\overline{\mathbf{\Omega}}$ into the block form compatible with the filtration $\mathcal{H}_1 \subset \mathcal{H}_2$, then the first diagonal block (1×1 matrix) is positive definite and the whole $(r+1) \times (r+1)$ hermitian matrix has 1 positive eigenvalue and r negative eigenvalues.

In our case of hypersurfaces (or complete intersections) in toric varieties $\mathbf{P}^4_\Sigma$, the basis ω_0 for the rational differential $\mathcal{H}_1$ can be given explicitly by the defining equation of Y_x with the deformations $x^1, \cdots, x^r$. Then our assumption hereafter for the family $\{Y_x\}_{x \in B}$ is that the derivatives

$$(2.5)\quad \partial_1\omega_0, \cdots, \partial_r\omega_0 \quad (\partial_i\omega_0 := \frac{\partial}{\partial x^i}\omega_0)$$

span $\mathcal{H}_2$ together with the basis ω_0 of $\mathcal{H}_1$ for each small complex domain $B \subset \mathcal{M}$. It will be useful to define a notation $\partial_0\omega_0 \equiv \omega_0$. Now using $\int_{\tau(\gamma)} \partial_i\omega_0 = \int_\gamma R(\partial_i\omega_0) = \partial_i \int_\gamma R(\omega_0)$, we can write the period matrix simply by

$$(2.6)\quad \mathbf{\Omega} = \begin{pmatrix} \partial_i X^I & \partial_i P_J \end{pmatrix} = \begin{pmatrix} \partial_i X^I & \partial_i X^J \tau_{IJ} \end{pmatrix}_{0 \le i, I \le r}$$

where we define $\tau_{IJ} = \frac{\partial^2}{\partial X^I \partial X^J}\mathcal{F}(X)$, and set $\partial_0 X^I \equiv X^I, \partial_0 P_J \equiv P_J$. In the above formula, and also hereafter, the repeated indices are assumed to be summed over (Einstein's convention) unless otherwise mentioned. Using the property 1) in the previous paragraph (2-2), we may assume

$$\det(\partial_i X^I)_{0\le i,I\le r} = (X^0)^{r+1} \det\left(\partial_i\left(\frac{X^I}{X^0}\right)\right)_{1\le i,I\le r} \neq 0$$

for x such that $X^0(x) \neq 0$. Hence, at least locally, we can normalize the period matrix in the form

$$(\partial_i X^I)^{-1}\mathbf{\Omega} = (E\ \tau) \ .$$

In this normalized form, the bilinear relation 1) in (2.4) is trivial since $\tau_{IJ} = \tau_{JI}$, while 2) entails

$$\sqrt{-1}(\bar{\tau} - \tau) > 0 \ ,$$

which means the matrix $\mathrm{Im}\,\tau$ has one positive and r negative eigenvalues. Here we note a similarity to the period matrix of genus g curves, however the mixed property of the eigenvalues is a new feature in higher dimensions.

Finally, we note that the monodromy group Γ acts on the normalized period matrix from the right, and for $\left(\begin{smallmatrix} D & B \\ C & A \end{smallmatrix}\right) \in \Gamma$ we have

$$\tau \mapsto (C\tau + D)^{-1}(A\tau + B) \ . \tag{2.7}$$

Since $\frac{\partial}{\partial X^I}\frac{\partial}{\partial X^J}\mathcal{F} = \tau_{IJ}$, this describes the transformation property of the prepotential which is defined locally for the family over B.

(2-4) Period matrix (2). Period matrix (2.6) has been defined entirely in the holomorphic category, since it is based on the Hodge filtration. One may modify the Hodge filtration to Hodge decomposition if we incorporate a hermitian structure coming from the Kähler geometry on $\mathcal{M}$. Let us first note that over the small domain B, and hence on $\mathcal{M}$ except the degeneration loci, there exists a Kähler metric $g_{i\bar{j}} = \partial_i \partial_{\bar{j}} K(x,\bar{x})$ $(1 \le i, \bar{j} \le r)$ with the Kähler potential

$$K(x,\bar{x}) = -\log\{\langle \Omega_x, \overline{\Omega}_x \rangle\} = -\log\{\langle R(\omega_0), \overline{R(\omega_0)} \rangle\} \ .$$

This Kähler metric is called Weil–Petersson metric on $\mathcal{M}$. The bases $\partial_0\omega_0 \equiv \omega_0; \partial_1\omega_0, \cdots, \partial_r\omega_0$ which are compatible with the filtration $\mathcal{H}_1 \subset \mathcal{H}_2$, or the Hodge filtration $F^{3,0} \subset F^{3,1}$, may now be modified to

$$D_0\omega_0 \equiv \omega_0; D_1\omega_0, \cdots, D_r\omega_0 \ \ (D_i = \partial_i + K_i, i = 1, \cdots, r),$$

where $K_i = \partial_i K(x,\bar{x})$. One should observe that, since $\langle R(\omega_0), R(D_i\omega_0) \rangle = 0$ holds for $i = 1, \cdots, r$, these bases are compatible to the Hodge decomposition $H^{3,0}(Y_x)\oplus H^{2,1}(Y_x)$. Correspondingly, the period matrix (2.6) may be modified to

$$\boldsymbol{\Omega} = \begin{pmatrix} D_iX^I & D_iP_J \end{pmatrix} = \begin{pmatrix} D_iX^I & D_iX^J\tau_{IJ} \end{pmatrix}_{0\le i,I\le r} \ , \tag{2.8}$$

where we set $D_0X^I \equiv X^I, D_0P_J \equiv P_J$. Since we have $\det(D_iX^I)_{0\le i,I\le r} = \det(\partial_iX^I)_{0\le i,I\le r} \ne 0$, the normalized period integral has a similar form as before;

$$(D_iX^I)^{-1}\boldsymbol{\Omega} = \begin{pmatrix} E & \tau \end{pmatrix} \ .$$

The same monodromy group Γ acts from the right on the period matrix. In contrast to this, the left action shows a nice connection to the Kähler geometry on $\mathcal{M}$.

(2-5) Special Kähler geometry on $\mathcal{M}$. After some algebra, we have for the Kähler metric

$$g_{i\bar{j}} = \partial_i\partial_{\bar{j}}K(x,\bar{x}) = -e^{K(x,\bar{x})}\langle R(D_i\omega_0), \overline{R(D_j\omega_0)}\rangle.$$

With respect to this, we introduce the metric connections

$$\Gamma^k_{ij} = g^{k\bar{k}}\partial_i g_{j\bar{k}} \ , \ \Gamma^{\bar{k}}_{\bar{i}\bar{j}} = g^{\bar{k}k}\partial_{\bar{i}} g_{\bar{j}k} \ ,$$

for the holomorphic tangent bundle $T\mathcal{M}$ and the anti-holomorphic tangent bundle $T'\mathcal{M}$, respectively. The curvature tensor for these connections are given by

$$R_{i\bar{j}\ l}^{\ \ k} = -\partial_{\bar{j}}\Gamma^k_{il} \ , \ R_{i\bar{j}\ \bar{l}}^{\ \ \bar{k}} = \partial_i\Gamma^{\bar{k}}_{\bar{j}\bar{l}} \ .$$

In addition to these, we have the Griffiths–Yukawa couplings $C_{ijk}(x)$ and $\overline{C}_{\bar{i}\bar{j}\bar{k}}(\bar{x})$ on the moduli space $\mathcal{M}$. These tensors define the so-called *special Kähler geometry* on $\mathcal{M}$, which can be summarized into a property of the period matrix (2.8).

Let us introduce the following $(2r+2)\times(2r+2)$ matrix

$$\begin{pmatrix} \boldsymbol{\Omega} \\ \overline{\boldsymbol{\Omega}} \end{pmatrix} = \begin{pmatrix} D_iX^I & D_iP_J \\ \overline{D_iX^I} & \overline{D_iP_J} \end{pmatrix}_{0\le i,I,J\le r} \ . \tag{2.9}$$

By definition of the period matrix, the row vectors represent the Hodge decomposition $H^{3,0}(Y_x)\oplus H^{2,1}(Y_x)\oplus H^{1,2}(Y_x)\oplus H^{0,3}(Y_x)$ in terms of the bases

$$R(\omega_0), \ \ R(D_i\omega_0) \ , \ \ \overline{R(D_i\omega_0)} \ , \ \ \overline{R(\omega_0)} \ . \tag{2.10}$$

By simply writing (2.9), we implicitly understand the row vectors are ordered according to the Hodge decomposition above. With these explicit bases in mind, we introduce the covariant derivatives acting on the column vectors $\mathbf{\Omega}$ and $\overline{\mathbf{\Omega}}$;

$$D_i = \begin{cases} \partial_i + K_i - \Gamma^*_{i*} & \text{on } \mathbf{\Omega} \\ \partial_i & \text{on } \overline{\mathbf{\Omega}} \end{cases}, \quad \overline{D_{\bar{i}}} = \begin{cases} \partial_{\bar{i}} & \text{on } \mathbf{\Omega} \\ \partial_{\bar{i}} + K_{\bar{i}} - \Gamma^*_{\bar{i}*} & \text{on } \overline{\mathbf{\Omega}} \end{cases},$$

where Γ^*_{i*} and $\Gamma^*_{\bar{i}*}$ represents the conventional form of the contraction via the metric connection.

Theorem 2.1. *The period matrix satisfies*

$$(2.11) \qquad (D_i + \mathcal{A}_i)\begin{pmatrix} \mathbf{\Omega} \\ \overline{\mathbf{\Omega}} \end{pmatrix} = \begin{pmatrix} \mathbf{0} \\ \mathbf{0} \end{pmatrix}, \quad (\overline{D_{\bar{i}}} + \bar{\mathcal{A}}_{\bar{i}})\begin{pmatrix} \mathbf{\Omega} \\ \overline{\mathbf{\Omega}} \end{pmatrix} = \begin{pmatrix} \mathbf{0} \\ \mathbf{0} \end{pmatrix},$$

with

$$\mathcal{A}_i = \begin{array}{c} \\ 0 \\ m \\ \bar{m} \\ \bar{0} \end{array} \begin{array}{c} \begin{array}{cccc} 0 & n & \bar{n} & \bar{0} \end{array} \\ \begin{pmatrix} 0 & -\delta_i^n & 0 & 0 \\ 0 & 0 & C_{im}^{\bar{n}} & 0 \\ 0 & 0 & 0 & -g_{i\bar{m}} \\ 0 & 0 & 0 & 0 \end{pmatrix} \end{array}, \quad \bar{\mathcal{A}}_{\bar{i}} = \begin{array}{c} \\ 0 \\ m \\ \bar{m} \\ \bar{0} \end{array} \begin{array}{c} \begin{array}{cccc} 0 & n & \bar{n} & \bar{0} \end{array} \\ \begin{pmatrix} 0 & 0 & 0 & 0 \\ -g_{\bar{i}m} & 0 & 0 & 0 \\ 0 & \overline{C}_{\bar{i}\bar{m}}^{n} & 0 & 0 \\ 0 & 0 & -\delta_{\bar{i}}^{\bar{n}} & 0 \end{pmatrix} \end{array},$$

where we set $C_{im}^{\bar{n}} = \sqrt{-1}e^K C_{imn} g^{n\bar{n}}$ *and* $\overline{C}_{\bar{i}\bar{m}}^{n} = -\sqrt{-1}e^K \overline{C}_{\bar{i}\bar{m}\bar{n}} g^{n\bar{n}}$.

Proof. Let us consider, the three from $R(D_j\omega_0)$ in the Hodge decomposition (2.10). Then $D_iR(D_j\omega_0)$ is a three form and may be expressed in terms of the basis (2.10) as

$$D_iR(D_j\omega_0) = c_0R(\omega_0) + c_mR(D_m\omega_0) + d_m\overline{R(D_m\omega_0)} + d_0\overline{R(\omega_0)}.$$

Now, using $\langle R(D_n\omega_0), \overline{R(D_m\omega_0)}\rangle = -e^{-K}g_{n\bar{m}}$ and other orthogonal relations, it is easy to see $c_0 = c_m = d_0 = 0$ and

$$\begin{aligned} -d_me^{-K}g_{k\bar{m}} &= \langle R(D_k\omega_0), D_iR(D_j\omega_0)\rangle \\ &= -\langle R(\omega_0), D_kD_iD_jR(\omega_0)\rangle = \sqrt{-1}C_{ijk}, \end{aligned}$$

where C_{ijk} is the Griffiths–Yukawa coupling (2.2) ($R(\omega_0) = \Omega_x$). One can continue similar arguments for other bases of the Hodge decomposition (2.10). Integrating three forms over the cycles $\{A^I, B_J\}$, we obtain the claimed linear relations for the row vectors of the period matrix. Q.E.D.

The connection matrix $\mathcal{A}_i, \mathcal{A}_{\bar{i}}$ was first determined by Strominger in [28]. The existence of the first order differential operator is due to the fact that the Hodge decomposition is 'flat' over $\mathcal{M}$, and we should have the compatibility relations for the first order system. Indeed one can see that

$$[D_i + \mathcal{A}_i, D_j + \mathcal{A}_j] = 0 = [\overline{D}_{\bar{i}} + \overline{\mathcal{A}}_{\bar{i}}, \overline{D}_{\bar{j}} + \overline{\mathcal{A}}_{\bar{j}}],$$

are ensured by the existence of the prepotential (2.2), and another mixed-type compatibility condition imposes a rather strong constraint on the Kähler geometry [28], which is called special Kähler geometry (see, e.g., [10] and references therein).

Theorem 2.2. *The compatibility condition*

$$[D_i + \mathcal{A}_i, \overline{D}_{\bar{j}} + \overline{\mathcal{A}}_{\bar{j}}] = 0$$

is equivalent to

$$-R_{i\bar{j}\ l}^{\ \ k} = g_{i\bar{j}}\delta_l^{\ k} + g_{\bar{j}l}\delta_i^{\ k} - e^{2K} C_{ilm} \overline{C}_{\bar{j}\bar{k}\bar{m}} g^{m\bar{m}} g^{k\bar{k}} \ . \tag{2.12}$$

In Section (3-1), we will derive the relation (2.12) directly evaluating the metric connection. The both equations (2.11) and (2.12) are often referred to as special Kähler geometry relations, and will play central roles in solving BCOV anomaly equation.

(2-6) Notations. As we have summarized above, the special Kähler geometry relations, in the holomorphic local coordinate $x^i (i = 1, \cdots, r)$, on $\mathcal{M}$ arises from the flat property of the period matrix $\boldsymbol{\Omega}$ and its complex conjugate. Since the row vectors of $\boldsymbol{\Omega}$ correspond to the decomposition $H^{3,0}(Y_x) \oplus H^{2,1}(Y_x)$, it is convenient to introduce the (Greek letter) notation $\alpha = (0, i)$ and $\bar{\alpha} = (\bar{0}, \bar{i})$ for the indices of the row vectors and their complex conjugates, respectively. Then the period matrix may be written simply by

$$\boldsymbol{\Omega} = (D_\alpha X^I \, D_\alpha P_J) = (D_\alpha X^I) \, (\, E \quad \tau_{IJ} \,) \ .$$

As remarked in subsection (2-4), $(r+1) \times (r+1)$ matrix $(D_\alpha X^I)$ is invertible. We set $\xi_\alpha^{\ I} := D_\alpha X^I$, $\bar{\xi}_{\bar{\alpha}}^{\ I} := \overline{D_\alpha X^I}$ and define $\xi_I^{\ \alpha}$ and $\bar{\xi}_I^{\ \bar{\alpha}}$, respectively, by

$$(\xi_I^{\ \alpha}) = (D_\alpha X^I)^{-1} \ , \ \ (\bar{\xi}_I^{\ \bar{\alpha}}) = (\overline{D_\alpha X^I})^{-1} \ .$$

Note that the i-th row of $\boldsymbol{\Omega}$ represents period integrals of the three form $R(D_i \omega_0) \in H^{2,1}(Y_x)$ for a symplectic basis $\{A^I, B_J\}$. Then we can see the Weil–Petersson metric in the following $(r+1)\times(r+1)$ hermitian matrix,

$$\begin{pmatrix} g_{0\bar{0}} & 0 \\ 0 & g_{i\bar{j}} \end{pmatrix} = -e^{K(x,\bar{x})}\sqrt{-1}\boldsymbol{\Omega} Q {}^t\overline{\boldsymbol{\Omega}} \quad (g_{0\bar{0}} \equiv -1).$$

Equivalently one can write this matrix relation by

$$g_{\alpha\bar{\beta}} = \sqrt{-1}e^{K(x,\bar{x})}\,\xi_\alpha^{\ I}\,(\tau-\bar{\tau})_{IJ}\,\bar{\xi}_{\bar{\beta}}^{\ J} \ .$$

For the tensor analysis in later sections, we introduce a "metric" by

$$(\mathcal{G}_{IJ}) = \sqrt{-1}e^{K(x,\bar{x})}\big(\tau-\bar{\tau}\big) \ , \ (\mathcal{G}^{IJ}) = -\sqrt{-1}e^{-K(x,\bar{x})}\big(\tau-\bar{\tau}\big)^{-1} .$$

Then we have $g_{\alpha\bar{\beta}} = \xi_\alpha^{\ I}\mathcal{G}_{IJ}\bar{\xi}_{\bar{\beta}}^{\ J}, g^{\alpha\bar{\beta}} = \xi_I^{\ \alpha}\mathcal{G}^{IJ}\bar{\xi}_J^{\ \bar{\beta}}$. With these metrics we will raise and lower the indices $I, J, \cdots$ as well as the Greek indices.

Now the special Kähler geometry relations in Theorem2.1 may be expressed by

$$(2.13) \quad \begin{cases} D_i\xi_0^{\ I} = \xi_i^{\ I} \\ D_i\xi_j^{\ I} = -\sqrt{-1}e^K C_{ijk}g^{k\bar{k}}\bar{\xi}_{\bar{k}}^{\ I} \\ D_i\bar{\xi}_{\bar{j}}^{\ I} = g_{i\bar{j}}\bar{\xi}_{\bar{0}}^{\ I} \\ D_i\bar{\xi}_{\bar{0}}^{\ I} = 0 \end{cases} , \begin{cases} \overline{D}_{\bar{i}}\xi_0^{\ I} = 0 \\ \overline{D}_{\bar{i}}\xi_j^{\ I} = g_{\bar{i}j}\xi_0^{\ I} \\ \overline{D}_{\bar{i}}\bar{\xi}_{\bar{j}}^{\ I} = \sqrt{-1}e^K\overline{C}_{\bar{i}\bar{j}\bar{k}}g^{k\bar{k}}\xi_k^{\ I} \\ \overline{D}_{\bar{i}}\bar{\xi}_{\bar{0}}^{\ I} = \bar{\xi}_{\bar{i}}^{\ I} \end{cases} .$$

It will be useful to note that the following relation holds by definition;

$$(2.14) \qquad D_i\xi_j^{\ J} = C_{ijm}S^{mn}\xi_n^{\ I} \ ,$$

where S^{mn} is the *propagator* that will be introduced in the next section.

§3. BCOV rings

(3-1) BCOV ring $\mathcal{R}^0_{BCOV}$. Based on the special Kähler geometry relations summarized in the previous section, we introduce a differential ring over the meromorphic sections of a certain vector bundle over $\tilde{\mathcal{M}}$. Let us first introduce the so-called propagators;

Definition 3.1.

$$S_{\bar{\alpha}\bar{\beta}} = e^{2K}(\tau-\bar{\tau})_{IJ}\bar{\xi}_{\bar{\alpha}}^{\ I}\bar{\xi}_{\bar{\beta}}^{\ J} \ , \ S^{\alpha\beta} = g^{\alpha\bar{\alpha}}g^{\beta\bar{\beta}}S_{\bar{\alpha}\bar{\beta}} \ .$$

Obviously $S_{\bar{\alpha}\bar{\beta}}$ and $S^{\alpha\beta}$ are symmetric with respect to the indices. Note that, since the factor e^{2K} has weight $(-2,-2)$, both $S^{\alpha\beta}$ and $S_{\bar{\alpha}\bar{\beta}}$ have weight $(-2,0)$ with respect to the line bundle $\mathcal{L}$ (see Section (2-1)).

Following [5], we will often use the notation S, S^i, S^{ij}, which are related to $S^{\alpha\beta}$ by

$$\begin{pmatrix} S^{00} & S^{0i} \\ S^{0i} & S^{ij} \end{pmatrix} = \begin{pmatrix} 2S & -S^i \\ -S^i & S^{ij} \end{pmatrix} .$$

If we use the property of τ and the prepotential introduced in (2-2), we have

$$S = \frac{1}{2} e^{2K} (\tau - \bar{\tau})_{IJ} \bar{X}^I \bar{X}^J = e^{2K} \left(\frac{1}{2} \tau_{IJ} \bar{X}^I \bar{X}^J - \overline{\mathcal{F}}(\bar{X}) \right) .$$

Proposition 3.2. *The following relations hold,*

$$1) \quad \overline{D}_{\bar{i}} S = S_{\bar{0}\bar{i}} \qquad 2) \quad \overline{D}_{\bar{i}} \overline{D}_{\bar{j}} S = S_{\bar{i}\bar{j}} \qquad 3) \quad \overline{D}_{\bar{i}} \overline{D}_{\bar{j}} \overline{D}_{\bar{k}} S = e^{2K} \overline{C}_{\bar{i}\bar{j}\bar{k}} .$$

These are equivalent to 1) $\partial_{\bar{i}} S = g_{\bar{i}j} S^j$, 2) $\partial_{\bar{i}} S^j = g_{\bar{i}i} S^{ij}$ *and* 3) $\partial_{\bar{i}} S^{jk} = e^{2K} \overline{C}_{\bar{i}\bar{j}\bar{k}} g^{j\bar{j}} g^{k\bar{k}}$.

Proof. Acting $D_{\bar{i}} (= \partial_{\bar{i}})$ on $S = \frac{1}{2} e^{2K} (\tau - \bar{\tau})_{IJ} \bar{X}^I \bar{X}^J$, we obtain

$$\partial_{\bar{i}} S = e^{2K} (\tau - \bar{\tau})_{IJ} D_{\bar{i}} \bar{X}^I \bar{X}^J = S_{\bar{0}\bar{i}} ,$$

where we use $(\frac{\partial}{\partial \bar{X}^K} \bar{\tau}_{IJ}) \bar{X}^J = \bar{X}^J (\frac{\partial}{\partial \bar{X}^J} \bar{\tau}_{IK}) = 0$. By similar calculations with the special Kähler geometry relation (2.13), the properties 2) and 3) above follow. Q.E.D.

S^{kl}, S^k, S are called propagators in physics literatures, and contain the anti-holomorphic prepotential $\bar{\mathcal{F}}(\bar{X})$ in their definitions. On the other hand, the holomorphic prepotential $\mathcal{F}(X)$ defines the so-called n-point functions on a Riemann sphere by

$$C_{i_1 i_2 \cdots i_n} = D_{i_1} D_{i_2} \cdots D_{i_n} \mathcal{F}(X) \quad (n \geq 3),$$

where the covariant derivative $D_i = \partial_i + 2\partial_i K(x, \bar{x}) - \Gamma^*_{i*}$ acts on $\mathcal{F}(X)$ by the standard contractions of the holomorphic indices (see (3.4) in general). By Kähler geometry, the holomorphic covariant derivatives commute with each other, so the n-point functions are symmetric tensor, and in particular C_{ijk} coincides with the Griffith–Yukawa coupling (2.2).

Definition 3.3. As an symmetric algebra, we define

$$\mathcal{R}^0_{BCOV} = \mathbf{Q}[S, S^i, S^{ij}, K_i, C_{i_1 i_2 i_3}, \cdots, C_{i_1 i_2 i_3 \cdots i_n}, \cdots] ,$$

where $K_i = \frac{\partial}{\partial x^i} K(x, \bar{x})$, We call this symmetric algebra *BCOV ring*.

In general this BCOV ring is infinitely generated. However we will see in the next section that if we consider the corresponding ring for an elliptic curve, the ring is finitely generated. In case of Calabi–Yau threefolds, as it turns out later that this problem of finiteness is related to the explicit form of the holomorphic function $h_{ijkl}(x)$ derived in (3.7) below. For convenience, we will often abbreviate the infinite series of the generators $C_{jkl}, D_{i_1}C_{jkl}, \cdots$ by $\{C_{ijk}\}$. With this convention, the BCOV ring may be written simply by

$$\mathcal{R}^0_{BCOV} = \mathbf{Q}[S, S^i, S^{ij}, K_i, \{C_{ijk}\}] \ . \tag{3.1}$$

The propagator $S^{\alpha\beta}$ and $C_{i_1 i_2 \cdots i_n}$ have respective weights $(-2,0)$ and $(2,0)$, and also K_i has weight $(0,0)$. Therefore the symmetric algebra is graded and defined in the set of global sections of

$$\bigoplus_{m,n\geq 0} \bigoplus_{k=-\infty}^{\infty} \left(\pi^*(T^*\mathcal{M})\right)^{\otimes m} \otimes \left(\pi^*(T\mathcal{M})\right)^{\otimes n} \otimes \mathcal{L}^k \ , \tag{3.2}$$

where $\pi : \tilde{\mathcal{M}} \to \mathcal{M}$ is the covering map and $\mathcal{L} \to \tilde{\mathcal{M}}$ is the line bundle introduced in (2-1). Precisely, K_i is a connection of the holomorphic line bundle $\mathcal{L}$ and therefore this is not a one form. However we regard K_i as a one form taking a global trivialization of $\mathcal{L}$ over $\tilde{\mathcal{M}}$. By the following lemma, we see how the BCOV ring $\mathcal{R}^0_{BCOV}$ depends on the sheets of the covering.

Lemma 3.4. *When we change the symplectic basis by* $\mathbf{\Omega} \to \mathbf{\Omega}\left(\begin{smallmatrix} D & B \\ C & A \end{smallmatrix}\right)$, *we have the corresponding change of the generators,*

$$S^{\alpha\beta} \to S^{\alpha\beta} + \xi_I^{\ \alpha}\left[C(D+\tau C)\right]^{IJ} \xi_J^{\ \beta} \ . \tag{3.3}$$

Proof. We keep our convention of the contraction by writing the indices of the symplectic matrix by $(X^I P_J) \to (X^I P_J)\left(\begin{smallmatrix} D_I^{\ J} & B_{IJ} \\ C^{IJ} & A^I_{\ J} \end{smallmatrix}\right)$. Then we have

$$\xi_\alpha^{\ I} \to \xi_\alpha^{\ J}(D+\tau C)_J^{\ I} \ , \quad \tau_{IJ} \to \left[(D+\tau C)^{-1}(B+\tau A)\right]_{IJ} \ ,$$

and also, using $\left(\begin{smallmatrix} D & B \\ C & A \end{smallmatrix}\right) Q \, {}^t\!\left(\begin{smallmatrix} D & B \\ C & A \end{smallmatrix}\right) = Q$, we have

$$(\tau - \bar\tau) \to (D+\tau C)^{-1} (\tau - \bar\tau) \, \overline{{}^t(D+\tau C)}^{-1} \ .$$

After some algebra, the claimed transformation property follows directly from the definitions. Q.E.D.

We note that the metric connection Γ^k_{ij} and K_i define the covariant derivative D_i on the sections (3.2). Thus, for example, for the section $V_i^{\ jl}$ of weight $(k,0)$ we have

$$(3.4)\quad D_n V_i^{\ jl} = \frac{\partial}{\partial x^n} V_i^{\ jl} - \Gamma^m_{ni} V_m^{\ jl} + \Gamma^j_{nm} V_i^{\ ml} + \Gamma^l_{nm} V_i^{\ jm} + k K_n V_i^{\ jl} \ .$$

Theorem 3.5. *The BCOV ring $\mathcal{R}^0_{\rm BCOV}$ is a graded, differential, symmetric algebra with the (commuting) differentials D_i $(i = 1, \cdots, r)$.*

This theorem is a direct consequence of the following proposition.

Proposition 3.6. *The covariant derivative acts on the generators of $\mathcal{R}^0_{\rm BCOV}$ by*

$$(3.5)\quad \begin{aligned} D_i S^{kl} &= \delta^k_i S^l + \delta^l_i S^k - C_{imn} S^{mk} S^{nl} \ , \\ D_i S^k &= -C_{imn} S^m S^{nk} + 2\delta^k_i S \ , \\ D_i S &= -\frac{1}{2} C_{imn} S^m S^n \ , \\ D_i K_j &= -K_i K_j + C_{ijm} S^{mn} K_n - C_{ijm} S^m \ . \end{aligned}$$

Proof. The first three equations follow from the definitions and the special Kähler geometry relations (2.12), (2.13). There, it is useful to write $S^{\alpha\beta} = \frac{1}{\sqrt{-1}} e^K \mathcal{G}^{IJ} \xi_I^{\ \alpha} \xi_J^{\ \beta}$ and use the following relations,

$$D_i \mathcal{G}_{LM} = \sqrt{-1} e^K C_{ikl} \xi_L^{\ k} \xi_M^{\ l} \ , \quad D_i \xi_I^{\ \alpha} = -C_{imn} S^{n\alpha} \xi_I^{\ m} - \xi_I^{\ 0} \delta_i^{\ \alpha} \ .$$

For the fourth relation, we formulate the following two lemmas and use the relation (3.6) below. Q.E.D.

Lemma 3.7.

$$\partial_{\bar{k}} \xi_I^{\ i} = 0 \, , \quad \partial_{\bar{k}} \xi_I^{\ 0} = -\partial_{\bar{k}} (K_m \xi_I^{\ m}) \ .$$

Proof. The matrix $(\xi_I^{\ \alpha})$ is the inverse of $(\xi_\alpha^{\ I})$ by definition. From this, we have $\partial_{\bar{k}} \xi_I^{\ \alpha} = -\xi_I^{\ \beta} \partial_{\bar{k}} \xi_\beta^{\ J} \xi_J^\alpha$. Now using $\partial_{\bar{k}} \xi_\beta^{\ J} = \partial_{\bar{k}} D_\beta X^J = g_{\bar{k}\beta} \xi_0^{\ J}$, we obtain

$$\partial_{\bar{k}} \xi_I^{\ \alpha} = -\xi_I^{\ \beta} g_{\bar{k}\beta} \xi_0^{\ J} \xi_J^\alpha = -g_{\bar{k}\beta} \xi_I^{\ \beta} \delta_0^{\ \alpha} .$$

The first claimed relation is the case when $\alpha = i$. The second relation follows from the case $\alpha = 0$ together with the first relation. Q.E.D.

Lemma 3.8. *Define* $f_{ij}^k = (\partial_i\partial_j X^I)\xi_I{}^k$, *then* f_{ij}^k *is holomorphic and we have*

$$\partial_i K_j - K_i K_j = -C_{ijk}S^k + f_{ij}^m K_m + h_{ij} \ ,$$

where $h_{ij} = -(\partial_i\partial_j X^I)h_I$ *with a holomorphic function* $h_I = h_I(x)$.

Proof. When differentiating twice the defining relation of the Kähler potential $e^{-K} = \langle \Omega_x, \bar{\Omega}_x\rangle = \sqrt{-1}X^I(\bar{\tau}-\tau)_{IJ}\bar{X}^J$, we have

$$e^{-K}(-\partial_i K_j + K_i K_j) = \sqrt{-1}\partial_i\partial_j X^I(\bar{\tau}-\tau)_{IJ}\bar{X}^J - \sqrt{-1}\partial_i X^I\partial_j X^L\tau_{ILJ}\bar{X}^J,$$

where we set $\tau_{IJK} = \frac{\partial}{\partial X^K}\tau_{IJ}$. On the other hand, by definition of S^m, we have

$$\begin{aligned} C_{ijm}S^m &= -e^{2K}C_{ijm}(\tau-\bar{\tau})_{IJ}\bar{\xi}_{\bar{0}}^{\,I}\bar{\xi}_{\bar{m}}^{\,J}g^{m\bar{m}}g^{0\bar{0}} \\ &= -\sqrt{-1}e^K C_{ijm}\mathcal{G}_{IJ}\bar{X}^I\bar{\xi}_{\bar{m}}^{\,J}g^{m\bar{m}} \\ &= -\sqrt{-1}e^K C_{ijm}\bar{X}^L\xi_L{}^m = -\sqrt{-1}e^K\tau_{IJL}\partial_i X^I\partial_j X^J\bar{X}^L \ , \end{aligned}$$

where we use $X^M\tau_{IJM} = 0$ which follows from the homogeneity property of $\mathcal{F}(X)$. Using $\mathcal{G}_{IJ} = \xi_I{}^\alpha g_{\alpha\bar{\beta}}\bar{\xi}_J^{\bar{\beta}}$ and $\bar{\xi}_{\bar{0}}^J = \bar{X}^J$, we also have

$$\sqrt{-1}e^K\partial_i\partial_j X^I(\bar{\tau}-\tau)_{IJ}\bar{X}^J = -\partial_i\partial_j X^I\mathcal{G}_{IJ}\bar{X}^J = \partial_i\partial_j X^I\xi_I{}^0 \ ,$$

where, due to Lemma 3.7, we may use $\xi_I{}^0 = -\xi_I{}^m K_m + h_I$ with some holomorphic function h_I. Substituting all these relations into the first equation, we obtain the claimed formula. The holomorphicity of $f_{ij}^k = (\partial_i\partial_j X^I)\xi_I{}^k$ follows from the same Lemma 3.7. Q.E.D.

From the above lemma, and $\partial_{\bar{m}}S^k = g_{m\bar{m}}S^{mk}$, we obtain

$$\Gamma_{ij}^k = g_i^k K_j + g_j^k K_i - C_{ijm}S^{mk} + f_{ij}^k \ , \tag{3.6}$$

which derives the special Kähler relation (2.12) directly from the definitions.

The connection (3.6) contains non-geometric object $f_{ij}^k(x)$, however this does not appear in the formulas (3.5) since the first three equations follows directly from the special Kähler relations as we have already seen. For the the fourth equation, one observes that $f_{ij}^k(x)$ cancels in the evaluation of $D_i K_j$.

Remark 3.9. The BCOV ring $\mathcal{R}^0_{BCOV}$ is not finitely generated in general, however it is very 'close' to this property. This can be observed in the following formula

$$\begin{aligned} D_l C_{ijk} &= \sum_{\{a,b\}\cup\{c,d\}=\mathcal{I}} C_{abm} S^{mn} C_{ncd} + \tau_{IJKL} \xi_i{}^I \xi_j{}^J \xi_k{}^K \xi_l{}^L \\ &= \sum_{\{a,b\}\cup\{c,d\}=\mathcal{I}} C_{abm} S^{mn} C_{ncd} - \sum_{a\in\mathcal{I}} K_a C_{\mathcal{I}\setminus\{a\}} + h_{ijkl} \end{aligned} \tag{3.7}$$

where we set $\mathcal{I} = \{i,j,k,l\}$ and $h_{ijkl} := \tau_{IJKL} \partial_i X^I \partial_j X^J \partial_k X^K \partial_l X^L$ with $\tau_{IJKL} = \frac{\partial}{\partial X^I} \frac{\partial}{\partial X^J} \frac{\partial}{\partial X^K} \frac{\partial}{\partial X^L} \mathcal{F}(X)$. Eq.(3.7) follows from $C_{ijk} = \tau_{IJK} \xi_i{}^I \xi_j{}^J \xi_k{}^K$ with the relations (2.14) and $D_k \tau_{IJK} = (\partial_k - K_k)\tau_{IJK} = \tau_{IJKL} \xi_k{}^L$. If the holomorphic term h_{ijkl} were zero, then the BCOV ring $\mathcal{R}^0_{BCOV}$ reduces to a finitely generated ring. This finite generating property can be realized in general by considering the following quotient: First, let us note that under the relation (3.7), the BCOV ring may be written as

$$\mathcal{R}^0_{BCOV} = \mathbf{Q}[S, S^i, S^{ij}, K_i, C_{ijk}, \{h_{ijkl}\}] \ ,$$

where $\{h_{ijkl}\}$ means the infinite sequence of the covariant derivations of h_{ijkl}. Considering a differential ideal $\mathbf{Q}[\{h_{ijkl}\}]$, we may reduce the ring $\mathcal{R}^0_{BCOV}$ to the quotient

$$\mathcal{R}^{0,red}_{BCOV} = \mathcal{R}^0_{BCOV} / \mathbf{Q}[\{h_{ijkl}\}] \ . \tag{3.8}$$

We call this quotient ring *reduced BCOV ring.*

(3-2) BCOV ring $\mathcal{R}^\Gamma_{BCOV}$. As we have remarked in the previous section, the BCOV ring is defined in the algebra of global (meromorphic) sections of the bundle (3.2) over the covering space $\tilde{\mathcal{M}}$. Since there is a natural action of the covering group $\Gamma \subset Sp(2r+2, \mathbf{Z})$ on the sections, one may consider the invariants under this group action. Elements in $\mathcal{R}^0_{BCOV}$, in general, are not invariant under this group action, however they can be 'lifted' to define Γ-invariants by specifying a 'lift' for each propagator (see below). We then define $\mathcal{R}^\Gamma_{BCOV}$ the minimal differential ring of Γ-invariants which contains those Γ-invariants from $\mathcal{R}^0_{BCOV}$. We may call $\mathcal{R}^\Gamma_{BCOV}$ as a Γ-completion of the BCOV ring $\mathcal{R}^0_{BCOV}$.

Let us first note that the generators $K_i(x,\bar{x})$ and $C_{ijk}(x)$ are invariant under the group Γ by their definitions. Since the generators $S^{\alpha\beta}$ are transformed according to (3.3), we modify them to Γ-invariants $\tilde{S}^{\alpha\beta}$.

Let us assume $\tilde{S}^{kl} = S^{kl} + \Delta S^{kl}$ for a Γ-invariant lift. Then it may be determined simply by writing the equation (3.6) as

$$\Gamma^k_{ij} = \delta^k_i K_j + \delta^k_i K_j - C_{ijm} \tilde{S}^{mk} + \tilde{f}^k_{ij} \ , \quad (\tilde{f}^k_{ij} = f^k_{ij} + C_{ijm} \Delta S^{mk}) \ , \tag{3.9}$$

and requiring Γ-invariance of $\tilde{f}^k_{ij}$. We will show, in the example of an elliptic curve, the simplest way to impose the invariance is to require $\tilde{f}^k_{ij}$ to be a rational function (section) on $\mathcal{M}$. In the original paper by BCOV, this process is referred to as 'fixing holomorphic (meromorphic) ambiguity'(Section 6.3 of [5]).

Once $\tilde{S}^{kl}$ is determined in this way, the form of other Γ-invariant propagators $\tilde{S}^k$, $\tilde{S}$ may be restricted, by requiring the relations $\partial_{\bar{k}}\tilde{S}^l = g_{k\bar{k}}\tilde{S}^{kl}$ and $g_{k\bar{k}}\tilde{S}^k = \partial_{\bar{k}}\tilde{S}$ given in Proposition 3.2, to

$$(3.10)\quad \tilde{S}^k = S^k + \Delta S^{kl}K_l + \Delta S^k \ , \quad \tilde{S} = S + \frac{1}{2}\Delta S^{kl}K_kK_l + \Delta S^k K_k + \Delta S \ ,$$

where ΔS^k and ΔS are suitable meromorphic sections. The form of ΔS^k can also be determined, in a similar way to (3.9), from

$$(3.11)\quad \partial_i K_j - K_i K_j = -C_{ijk}\tilde{S}^k + \tilde{f}^m_{ij}K_m + \tilde{h}_{ij} \ , \quad (\tilde{h}_{ij} = h_{ij} + C_{ijk}\Delta S^k),$$

by requiring that $\tilde{h}_{ij}$ is a rational function (section) on $\mathcal{M}$.

Proposition 3.10. *For the Γ-invariant propagators, we have*

$$(3.12)\quad \begin{aligned} D_i\tilde{S}^{kl} &= \delta^k_i\tilde{S}^l + \delta^l_i\tilde{S}^k - C_{imn}\tilde{S}^{mk}\tilde{S}^{nl} + \mathcal{E}^{kl}_i \ , \\ D_i\tilde{S}^k &= -C_{imn}\tilde{S}^m\tilde{S}^{nk} + 2\delta^k_i\tilde{S} + \mathcal{E}^{km}_i K_m + \mathcal{E}^k_i \ , \\ D_i\tilde{S} &= -\frac{1}{2}C_{imn}\tilde{S}^m\tilde{S}^n + \frac{1}{2}\mathcal{E}^{kl}_i K_kK_l + \mathcal{E}^m_i K_m + \mathcal{E}_i \ , \\ D_iK_j &= -K_iK_j + C_{ijm}\tilde{S}^{mn}K_n - C_{ijm}\tilde{S}^m + C_{ijm}\kappa^m \ , \end{aligned}$$

where we set $\kappa^m = \Delta S^m$ and

$$\mathcal{E}^{kl}_i = \mathcal{D}^f_i\Delta S^{kl} - \delta^k_i\Delta S^l - \delta^l_i\Delta S^k + C_{imn}\Delta S^{mk}\Delta S^{nl},$$

$$\mathcal{E}^k_i = \mathcal{D}^f_i\Delta S^k - 2\delta^k_i\Delta S + C_{imn}\Delta S^n\Delta S^{nl} \ , \quad \mathcal{E}_i = \mathcal{D}^f_i\Delta S + \frac{1}{2}C_{imn}\Delta S^m\Delta S^n \ .$$

We also define $\mathcal{D}^f_i := \partial_i + f^*_{i*}$ a (covariant) derivative with $f^k_{ij}(x)$ in (3.6) being treated as a connection.

Proof. Use the definitions $\tilde{S}^{kl}, \tilde{S}^k, \tilde{S}, \Gamma^k_{ij} = \delta^k_i K_j + \delta^k_j K_i - C_{ijm}S^{mk} + f^k_{ij}$ and Proposition 3.6 for the evaluations. After some algebra, the claimed formulas follow. Q.E.D.

If we define

$$(3.13)\quad \hat{\mathcal{E}}^{kl}_i = \mathcal{E}^{kl}_i \ , \quad \hat{\mathcal{E}}^k_i = \mathcal{E}^{km}_i K_m + \mathcal{E}^k_i \ , \quad \hat{\mathcal{E}}_i = \frac{1}{2}\mathcal{E}^{kl}_i K_kK_l + \mathcal{E}^m_i K_m + \mathcal{E}_i$$

then these are sections of weight $(-2,0)$ which are invariant under the action Γ. Considering all covariant derivatives of these tensors, and also κ^m, we have the minimal ring of Γ-invariants

$$\mathcal{R}^{\Gamma}_{BCOV} = \mathbf{Q}[\tilde{S}^{kl}, \tilde{S}^k, \tilde{S}, K_i, \{C_{ijk}\}, \{\hat{\mathcal{E}}^{kl}_i\}, \{\hat{\mathcal{E}}^k_i\}, \{\hat{\mathcal{E}}_i\}, \{\kappa^m\}] \ ,$$

where the bracket notation is used for the infinite set of the generators as before. We should note that the explicit forms of the new generators $\hat{\mathcal{E}}^{kl}_i, \hat{\mathcal{E}}^k_i, \hat{\mathcal{E}}_i$ and κ^m depend on the 'lifts' of the propagators $\tilde{S}^{ij}, \tilde{S}^k, \tilde{S}$. Hence the ring $\mathcal{R}^{\Gamma}_{BCOV}$ also depends on the lifts.

In case of elliptic curves, we will observe that the ring $\mathcal{R}^{\Gamma}_{BCOV}$ has a close similarity to the ring of almost holomorphic modular forms studied in Kaneko–Zagier [23].

(3-3) Example (elliptic curve). We have introduced the BCOV ring for Calabi–Yau threefolds, however if we replace the special Kähler geometry by the geometry of upper-half plane, it naturally reduces to the rather standard theory of (almost holomorphic) modular forms[23].

Let us consider a family of elliptic curves over $\mathcal{M}$ and its period integrals following (2-1). We consider a family of hypersurfaces $Y_{\vec{a}}$

$$\mathtt{W}(a) := a_0 + a_1 U + a_2 V + a_3 \frac{1}{U^3 V^2} = 0 \ \subset (\mathbf{C}^*)^2$$

in the torus $(\mathbf{C}^*)^2$. Compactifying $(\mathbf{C}^*)^2$ to a suitable toric variety $\mathbf{P}_\Sigma$, we obtain our family of elliptic curves. The moduli space $\mathcal{M}$ arises as the parameter space of the defining equation. Because of the natural torus actions on the parameters, it is easy to see that $\mathcal{M}$ is given by $\mathbf{P}^1$, and we have

$$\Omega_x = R\Big(\frac{a_0}{\mathtt{W}(a)} \frac{dU}{U} \frac{dV}{V}\Big) \quad \Big(x = \frac{a_1^3 a_2^2 a_3}{a_0^6} \in \mathbf{P}^1\Big).$$

Taking a symplectic basis A, B with $Q = \left(\begin{smallmatrix} 0 & 1 \\ -1 & 0 \end{smallmatrix}\right)$, we define the period integral $\vec{\omega} = (w_0(x), w_1(x))$. The period integrals satisfy the Picard–Fuchs differential equation of the form,

$$\{\theta_x^2 - 12x(6\theta_x + 1)(6\theta_x + 5)\}\omega_i(x) = 0 \ ,$$

where $\theta_x = x\frac{d}{dx}$. The 'Griffiths–Yukawa coupling' in this case is simply defined by

$$C_x := -\int_{Y_x} \Omega_x \cup \frac{d}{dx}\Omega_x = \frac{1}{(1-432x)x} \ .$$

Let us fix (uniquely) the A cycle by the condition that the corresponding period integral $w_0(x)$ is regular at $x = 0$ and normalized by $\omega_0(x) =$

$1+\cdots$. Then we take a dual cycle B to A. With this choice of the basis, the period matrix takes the form $\mathbf{\Omega}=(\omega_0(x)\,\omega_1(x))=\omega_0(x)(\,1\ t\,)$ with $t\sim\frac{1}{2\pi i}\log x+\cdots$ near $x=0$. We invert the relation $t=\frac{\omega_1(x)}{\omega_0(x)}$ as $x=x(t)$. Then it is standard to obtain the following identities (see eg. [24]);

$$\frac{1}{2\pi i}\frac{1}{\omega_0(x)^2}C_x\frac{dx}{dt}=1\ ,\ \ \omega_0(x(t))^4=E_4(t)\ ,\ \ C_x(x(t))=j(t)\ , \tag{3.14}$$

where $E_4(t)$ is the Eisenstein series, $j(t)$ is the normalized j-function with their Fourier expansion given by $q=e^{2\pi i t}$ (and $\frac{1}{2\pi i}\frac{d}{dt}=q\frac{d}{dq}$). We also have the following useful relation

$$\frac{\omega_0(x(t))^{12}}{C_x(x(t))}=\eta(t)^{24}\ , \tag{3.15}$$

in terms of the Dedekind η-function. Note that the first identity of (3.14) simply represents the fact that there is no quantum correction to the Griffiths–Yukawa coupling.

(3-3.a) The BCOV ring $\mathcal{R}^0_{BCOV}=\mathbf{Q}[S,K_x,C_x]$. For an elliptic curve, Definition 3.1 of $S=S^{00}$ should be read as

$$S=\frac{1}{2\pi i}g^{0\bar{0}}g^{0\bar{0}}e^{2K}(t-\bar{t})\bar{\xi}_{\bar{0}}\bar{\xi}_{\bar{0}}=\frac{1}{2\pi i}e^{2K}(t-\bar{t})\bar{\omega}_0\bar{\omega}_0\ .$$

with the period matrix $\mathbf{\Omega}=\omega_0(\,1\,t\,)$. Here, for elliptic curves, we introduce the factor $\frac{1}{2\pi i}$ in the definition of S. For the Kähler potential, we have $e^{-K}=i\,\mathbf{\Omega}Q\,{}^t\overline{\mathbf{\Omega}}$. Then it is straightforward to obtain

$$S=\frac{1}{2\pi i}\frac{1}{\omega_0^2}\frac{1}{\bar{t}-t}\ ,\quad K_x=\frac{dt}{dx}\frac{1}{\bar{t}-t}-\frac{d}{dx}\log(\omega_0(x))\quad .$$

Proposition 3.11. *The BCOV ring $\mathcal{R}^0_{BCOV}$ is finitely generated by S,K_x and C_x. The covariant differential D_x acts on the generators by*

$$D_xS=-C_xSS\,,\ \ D_xK_x=-K_xK_x-60\,C_x\,,\ \ D_xC_x=0\ . \tag{3.16}$$

Proof. It is sufficient to derive the differentials of generators (3.16). The metric connection Γ^x_{xx} may be determined from the relation;

$$\begin{aligned}&(-\partial_xK_x+K_xK_x)e^{-K}\\&=i\frac{d^2}{dx^2}\mathbf{\Omega}\,Q\,{}^t\overline{\mathbf{\Omega}}=\Big(-\frac{C_x'}{C_x}K_x+60\,C_x\Big)e^{-K}\ ,\end{aligned} \tag{3.17}$$

where we use the Picard–Fuchs equation

$$\frac{d^2}{dx^2}\vec{\omega} - \frac{C_x'}{C_x}\frac{d}{dx}\vec{\omega} - 60C_x\vec{\omega} = \vec{0} \ ,$$

to rewrite $i\frac{d^2}{dx^2}\mathbf{\Omega}\, Q\, {}^t\overline{\mathbf{\Omega}} = i\frac{d^2}{dx^2}\vec{\omega}Q^t\overline{\vec{\omega}}$ as above. After differentiating (3.17) by $\partial_{\bar{x}}$, we have $\Gamma^x_{xx} = 2K_x + \frac{C_x'}{C_x}$. Now using (3.17) again, we have

$$D_xK_x = (\partial_x + \Gamma^x_{xx})K_x = -K_xK_x - 60C_x \ .$$

Similarly, noting the generators S, C_x have their weights $(-2,0)$ and $(2,0)$, respectively, and using the relations (3.14), it is straightforward to obtain the claimed relations. Q.E.D.

It will be useful to have the following expression for the connection $\Gamma^x_{xx} = 2K_x + \frac{C_x'}{C_x}$;

$$\Gamma^x_{xx} = 2\frac{dt}{dx}\frac{1}{\bar{t}-t} + \frac{d}{dx}\log\left(\frac{C_x}{\omega_0(x)^2}\right) = 2\frac{dt}{dx}\frac{1}{\bar{t}-t} + \frac{dx}{dt}\frac{d}{dx}\frac{dt}{dx} \ , \tag{3.18}$$

where we use the identify $\frac{dt}{dx} = \frac{1}{2\pi i}\frac{C_x}{\omega_0^2}$ in (3.14). In particular, writing the first identity of (3.18) as

$$\Gamma^x_{xx} = 2C_xS + \frac{d}{dx}\log\left(\frac{C_x}{\omega_0(x)^2}\right) = 2C_xS + f^x_{xx} \ ,$$

one may regard this as the corresponding relation to (3.6).

(3-3.b) The BCOV ring $\mathcal{R}^\Gamma_{BCOV}$. In our example, the covering group Γ is given by the (genus one) modular subgroup $\langle \left(\begin{smallmatrix} 1 & 0 \\ -1 & 1 \end{smallmatrix}\right), \left(\begin{smallmatrix} 1 & 1 \\ 0 & 1 \end{smallmatrix}\right)\rangle \subset SL(2,\mathbf{Z})$. S is not invariant under the Γ action, however it is clear from the form $\frac{1}{\bar{t}-t}$ that S can be lifted to a Γ-invariant by

$$S \mapsto \tilde{S} = \frac{1}{\omega_0^2(x)}\left\{\frac{1}{2\pi i}\frac{1}{\bar{t}-t} - \frac{1}{12}E_2(t)\right\}$$

in terms of the Eisenstein series $E_2(t)$. $\tilde{S}$ is invariant since $E_2(t) - \frac{1}{2\pi i}\frac{12}{\bar{t}-t} = E_2^*(t)$ is the almost holomorphic (elliptic) modular form of weight 2 and $\omega_0(x)^2 = \sqrt{E_4(t)}$ for the denominator.

In our general formulation based on (3.9), the invariance arises in a rather weak form as follows: We first start with the 'shift';

$$\Gamma^x_{xx} = 2C_xS + f^x_{xx} = 2C_x\tilde{S} + \tilde{f}^x_{xx} \quad (\tilde{f}^x_{xx} := f^x_{xx} - 2C_x\Delta S) \ ,$$

where $\tilde{S} = S + \Delta S$. Accordingly, the formula $D_x S$ changes to

$$D_x \tilde{S} = -C_x \tilde{S}\tilde{S} + \mathcal{E}_x \quad \left(\mathcal{E}_x := \partial_x \Delta S + C_x \Delta S \Delta S + 2\frac{\omega_0'}{\omega_0}\Delta S \right) .$$

Proposition 3.12. *$\tilde{f}_{xx}^x$ is a rational function of x if and only if we set*

$$\Delta S = -\frac{1}{C_x}\frac{\omega_0'}{\omega_0} + r(x) \ ,$$

with some rational function $r(x)$. In terms of $r(x)$, $\mathcal{E}_x$ is given by $r'(x) + C_x r^2(x) - 60$. When $\mathcal{E}_x = \lambda C_x$ with some constant $\lambda \in \mathbf{Q}$, then the BCOV ring $\mathcal{R}_{BCOV}^{\Gamma}$ is finitely generated by $\tilde{S}, K_x, C_x$ with the following differentials,

$$D_x \tilde{S} = -C_x \tilde{S}\tilde{S} + \lambda C_x \, , \quad D_x K_x = -K_x K_x - 60\, C_x \, , \quad D_x C_x = 0 \ .$$

Proof. We evaluate $\tilde{f}_{xx}^x$ as

$$\tilde{f}_{xx}^x = \partial_x \log\left(\frac{C_x}{\omega_0^2}\right) - 2C_x \Delta S = -2\frac{\omega_0'}{\omega_0} + \frac{C_x'}{C_x} - 2C_x \Delta S \ ,$$

from which the first claim is clear. For the evaluation of $\mathcal{E}_x$, we use the Picard–Fuchs equation satisfied the period integral $\omega_0(x)$. The third claim is clear since the differentials closes among the generators. Q.E.D.

The differential equation $\mathcal{E}_x = \lambda C_x$ for $r(x)$ may be solved by hypergeometric series. From the solution, one may observe that there are infinitely many λ for which $r(x)$ becomes rational. The simplest result is given by

$$\lambda = \frac{1}{144} \, , \quad r(x) = \frac{1}{12}\frac{C_x'}{C_x^2} \, , \quad \tilde{S} = -\frac{1}{12}\frac{E_2^*}{\omega_0^2} \ ,$$

where we evaluate $S + \Delta S = S + \frac{1}{12}\frac{1}{C_x}\partial_x \log \frac{C_x}{\omega_0^{12}} = S - \frac{1}{12}\frac{1}{C_x}\partial_x \log \eta(t)^{12}$ for $\tilde{S}$. Similarly, for $\lambda = \frac{25}{144}, \frac{49}{144}, \frac{121}{144}$, for example, we obtain $r(x) = \frac{5}{12}\frac{C_x'}{C_x^2}, \frac{1}{12}\frac{C_x'}{C_x^2} - \frac{1}{2}\frac{1}{1-864x}, \frac{5}{12}\frac{C_x'}{C_x^2} - \frac{1}{2}\frac{1}{1-864x}$ and

$$\tilde{S} = \frac{-1}{12\omega_0^2}\left\{E_2^* + 4\frac{E_6}{E_4}\right\}, \ \frac{-1}{12\omega_0^2}\left\{E_2^* + 6\frac{E_4^2}{E_6}\right\}, \ \frac{-1}{12\omega_0^2}\left\{E_2^* + 4\frac{E_6}{E_4} + 6\frac{E_4^2}{E_6}\right\},$$

respectively.

We note that, when the ring is finitely generated, the BCOV ring is very close to the ring of almost holomorphic modular forms studied in Kaneko–Zagier [23]. For comparison, it might be useful to write our generators (for the case $\lambda = \frac{1}{144}$) in terms of the elliptic modular forms;

$$(3.19) \quad \tilde{S} = \frac{-1}{12} \frac{E_2^*(t)}{\omega_0(x)^2}, \; C_x = j(t), \; K_x = \frac{-1}{12} \frac{j(t)}{\omega_0(x)^2} \Big\{ E_2^*(t) - \frac{E_6(t)}{E_4(t)} \Big\}.$$

One should note, however, that the weight assignment in the BCOV ring is different from that of almost holomorphic modular forms. Also, in the BCOV ring, we have additional indices of the cotangents $\big(\pi^*(T^*\mathcal{M})\big)^{\otimes m}$.

(3-4) BCOV ring $\mathcal{R}^{hol}_{BCOV}$. For the applications to Gromov–Witten theory of Calabi–Yau manifolds, the most relevant form of the BCOV ring is the holomorphic limits of the invariants $\mathcal{R}^{\Gamma}_{BCOV}$, which is often referred to as "$\bar{t} \to \infty$" limit in physics literatures. For the above example of an elliptic curve, the meaning "$\bar{t} \to \infty$" should be clear as the 'limit' taking the constant term of $\sum_{n \ge 0} a_m(t) \big(\frac{1}{\bar{t}-t}\big)^n$. We need to formulate a precise meaning for Calabi–Yau threefolds. However the idea of the limit should be clear from the structure $\mathcal{R}^{\Gamma}_{BCOV}$ with the differentials (3.12). Namely, all the differentials are with respect to holomorphic coordinate, and therefore "throwing away" the anti-holomorphic dependence, at the cost of Γ-invariance, should be compatible with the differentiations.

To describe the holomorphic limit in more detail, let us introduce the so-called flat coordinate. We first fix a symplectic basis $\{A^I, B_J\}$ and denote the corresponding period integrals $(X^I(x), P_J(x))$. By the property 1) in Section (2-2), the (half) period maps $x \in B(\subset \mathcal{M}) \mapsto [X^I(x)] \in \mathbf{P}^r$ provides a local isomorphism. Due to this property we may introduce the so-called *flat* coordinate $(t^a)_{a=1,\cdots r}$ by the relation

$$(X^0(x), X^1(x), \cdots, X^r(x)) = X^0(x)(1, t^1, \cdots, t^r) \; ,$$

near $X^0(x) \neq 0$. In this flat coordinate we have for the Kähler potential $e^{-K(x,\bar{x})} = iX^0(x)\overline{X^0(x)}e^{-\mathcal{K}(t,\bar{t})}$ with

$$e^{-\mathcal{K}(t,\bar{t})} = 2\overline{F(t)} - 2F(t) + (t^a - \bar{t}^a)\big(\frac{\partial F}{\partial t^a} + \overline{\frac{\partial F}{\partial t^a}}\big),$$

and $F(t) = \frac{1}{(X^0)^2}\mathcal{F}(X) = \mathcal{F}(\frac{X^a}{X^0})$. Connections of the bundles in these two local coordinates (x^i) and (t^a) are related by

$$K_i = -\partial_i \log X^0(x) + \frac{\partial t^a}{\partial x^i}\mathcal{K}_{t^a} \; , \; \Gamma^k_{ij} = \frac{\partial x^k}{\partial t^c}\Gamma^{t^c}_{t^a t^b}\frac{\partial t^a}{\partial x^i}\frac{\partial t^b}{\partial x^j} + \frac{\partial x^k}{\partial t^a}\frac{\partial}{\partial x^i}\frac{\partial t^b}{\partial x^j} \; .$$

As we see in the formula $e^{-\mathcal{K}(t,\bar{t})}$, holomorphic and anti-holomorphic dependences are not separated by a factor like $f(t)g(\bar{t})$ (or $f(t) + g(\bar{t})$

in $\log \mathcal{K}$). We assume that the 'constant terms' against to the anti-holomorphic dependences are selected simply by setting to zero those expressions written by $\mathcal{K}_{t^a}(t,\bar{t})$ and $\Gamma^{t^c}_{t^a t^b}$ (and also their holomorphic derivatives).

Definition 3.13. Choose a symplectic basis $\mathcal{B} := \{A^I, B_J\}$. Then we define the holomorphic limit of the elements in $\mathcal{R}^{\Gamma}_{BCOV}$, with respect to $\mathcal{B}$, by the following replacements of the connections:

$$K_i \to \mathtt{K}_i := -\partial_i \log X^0(x) \ , \ \Gamma^k_{ij} \to \mathtt{\Gamma}^k_{ij} := \frac{\partial x^k}{\partial t^a}\frac{\partial}{\partial x^i}\frac{\partial t^a}{\partial x^j} \ .$$

As remarked above, the holomorphic limit commutes with the holomorphic differentials D_i, and hence we have the same differentials as (3.12). We denote the holomorphic limit of the generators $\tilde{S}^{ij}, \tilde{S}^k, \tilde{S}$, respectively by $\mathtt{S}^{ij}, \mathtt{S}^k, \mathtt{S}$. Also by $\mathtt{D}_i (= \partial_i \pm k\mathtt{K}_i \pm \mathtt{\Gamma}^*_{i*})$, we represent the holomorphic limit of the differential D_i. Accordingly, $\mathtt{K}_i$ should be assumed in the definitions (3.13) of $\hat{\mathcal{E}}^k_i, \hat{\mathcal{E}}_i$, although we use the same notation for these. Thus, taking the holomorphic limit of the Γ-invariant BCOV ring, $\mathcal{R}^{\Gamma}_{BCOV}$, we will have *holomorphic* BCOV ring,

$$\mathcal{R}^{hol}_{BCOV} = \mathbf{Q}[\mathtt{S}^{ij}, \mathtt{S}^k, \mathtt{S}, \mathtt{K}_i, \{C_{ijk}\}, \{\hat{\mathcal{E}}^{kl}_i\}, \{\hat{\mathcal{E}}^k_i\}, \{\hat{\mathcal{E}}_i\}, \{\kappa^m\}] \ .$$

The concrete form of the generators $\mathtt{S}^{ij}$ may be determined from the holomorphic limit of the relation (3.9);

$$\mathtt{\Gamma}^k_{ij} = \delta^k_i \mathtt{K}_j + \delta^k_j \mathtt{K}_i - C_{ijm}\mathtt{S}^{mk} + \tilde{f}^k_{ij} \ . \tag{3.20}$$

Similarly, for $\mathtt{S}^k$, we can use (3.11),

$$\partial_i \mathtt{K}_j - \mathtt{K}_i \mathtt{K}_j = -C_{ijk}\mathtt{S}^k + \tilde{f}^m_{ij}\mathtt{K}_m + \tilde{h}_{ij} \ . \tag{3.21}$$

These equations are used to determine the propagators in [5]. As noted there, when $r \geq 2$, the first relation (3.20) provides an overdetermined system for $\mathtt{S}^{ij}$, and the form of $\tilde{f}^k_{ij}$ should be restricted so that there exist solutions $\mathtt{S}^{ij}$. Similarly, $\tilde{h}_{ij}$ should be restricted so that the relation (3.21) has a solution $\mathtt{S}^k$. If we find a set of solutions $\mathtt{S}^{ij}, \mathtt{S}^k$, the first equation of (3.12) determines $\mathcal{E}^{kl}_i$, and the second relation of (3.12) determines $\mathtt{S}$ up to $\mathcal{E}^k_i$. As argued in [5], the possible forms of rational functions (sections) $\tilde{f}^k_{ij}, \tilde{h}_{ij}$ may be restricted, to some extent, by imposing regularity (or singularity) of $\mathtt{K}_i$ at certain degeneration loci of the family, see [5].

Example 1 (elliptic curve): When we take the symplectic basis $\mathcal{B} = \{A, B\}$ as in the previous section, the holomorphic limit of $\mathcal{R}^{\Gamma}_{BCOV}$ is exactly the map taking the constant term of $\sum c_m\left(\frac{1}{\bar{t}-t}\right)^m$. From the example in the previous section, it is immediate to obtain (for $\lambda = \frac{1}{144}$) that

$$\mathtt{S} = -\frac{1}{12}\frac{E_2(t)}{\omega_0(x)^2}\ ,\ \mathtt{K}_x = -\partial_x \log(\omega_0(x))\ ,\ \Gamma^x_{xx} = \frac{\partial x}{\partial t}\frac{\partial}{\partial x}\frac{\partial x}{\partial t}\ .$$

As for the differentials, we have the same form as those in $\mathcal{R}^{\Gamma}_{BCOV}$, i.e.,

$$\mathtt{D}_x\mathtt{S} = -C_x\mathtt{S}\mathtt{S} + \frac{C_x}{144}\ ,\ \mathtt{D}_x\mathtt{K}_x = -\mathtt{K}_x\mathtt{K}_x - 60\,C_x\ ,\ \mathtt{D}_x C_x = 0\ .$$

These relations define the BCOV ring $\mathcal{R}^{hol}_{BCOV} = \mathbf{Q}[\mathtt{S}, \mathtt{K}_x, C_x]$. The form of the generators are given simply by $E_2^*(t) \to E_2(t)$ in (3.19).

Example 2 (mirror quintic Calabi–Yau threefold): The construction of a symplectic basis $\mathcal{B}$ about the so-called large complex structure limit has been done in [8] (see also [17] and references therein for its combinatorial construction). We consider the holomorphic limit with respect to this basis. To fix the propagators $\mathtt{S}^{ij}, \mathtt{S}^k$, we have to solve the equations (3.20) and (3.21) finding suitable choices for the rational functions $\tilde{f}^x_{xx}, \tilde{h}_{xx}$. In [5], it has been found that these unknowns are uniquely fixed by requiring expected properties for the higher genus Gromov–Witten potential, $\mathcal{F}_g$, which comes from the anomaly equation. Here we simply translate their results into our conventions. First, the propagators $\mathtt{S}^{xx}, \mathtt{S}^x$ are determined by the choice $\tilde{f}^x_{xx} = -\frac{8}{5}\frac{1}{x}, \tilde{h}_{ij} = \frac{2}{25}\frac{1}{x^2}$ in

$$\Gamma^x_{xx} = 2\mathtt{K}_x - C_{xxx}\mathtt{S}^{xx} - \frac{8}{5}\frac{1}{x},\ \partial_x\mathtt{K}_x - \mathtt{K}_x\mathtt{K}_x = -C_{xxx}\mathtt{S}^x - \frac{8}{5}\frac{1}{x}\mathtt{K}_x + \frac{2}{25}\frac{1}{x^2},$$

where $C_{xxx} = \frac{5}{x^3(1-5^5x)}$ and x is related to ψ in [4], [5] by $x = \frac{1}{\psi^5}$. Then the differentials are evaluated to be

$$\begin{aligned}
\mathtt{D}_x\mathtt{S}^{xx} &= 2\,\mathtt{S}^x - C_{xxx}\mathtt{S}^{xx}\mathtt{S}^{xx} + \frac{x}{25}\ ,\\
\mathtt{D}_x\mathtt{S}^x &= 2\,\mathtt{S} - C_{xxx}\mathtt{S}^x\mathtt{S}^{xx} + \frac{x}{25}\mathtt{K}_x - \frac{1}{125}\ ,\\
\mathtt{D}_x\mathtt{S} &= -\frac{1}{2}C_{xxx}\mathtt{S}^x\mathtt{S}^x + \frac{x}{50}\mathtt{K}_x\mathtt{K}_x - \frac{1}{125}\mathtt{K}_x + \frac{2}{3125}\frac{1}{x}\ ,\\
\mathtt{D}_x\mathtt{K}_x &= -\mathtt{K}_x\mathtt{K}_x + C_{xxx}\mathtt{S}^{xx}\mathtt{K}_x - C_{xxx}\mathtt{S}^x + \frac{2}{25}\frac{1}{x^2}\ ,
\end{aligned}$$

from which we read $\mathcal{E}^{xx}_x = \frac{x}{25}, \mathcal{E}^x_x = -\frac{1}{125}, \mathcal{E}_x = \frac{2}{3125}\frac{1}{x}$ and $\kappa^x = \frac{2}{25}\frac{1}{x^2}\frac{1}{C_{xxx}}$. With these, the BCOV ring is determined by

$$\mathcal{R}^{hol}_{BCOV} = \mathbf{Q}[\mathtt{S}^{xx}, \mathtt{S}^x, \mathtt{S}, \mathtt{K}_x, \{C_{xxx}\}, \{\hat{\mathcal{E}}^{xx}_x\}, \{\hat{\mathcal{E}}^x_x\}, \{\hat{\mathcal{E}}_x\}, \{\kappa^x\}]\ .$$

Example 3: The Calabi–Yau manifolds whose higher genus Gromov–Witten invariants are studied in [19] are not complete intersections in toric varieties, but has an interesting property: there exist two different large complex structure limits (cusps) in the deformation space [27]. By mirror symmetry, this phenomenon is related to non-birational Calabi–Yau manifolds whose derived categories of coherent sheaves are equivalent [27], [3], [22]. The cusps of the example in [27], [19] are located at $x = 0$ and $z = \frac{1}{x} = 0$. The BCOV ring $\mathcal{R}^{hol}_{BCOV}$ with respect to a symplectic basis $\mathcal{B}_0$ at $x = 0$ has a similar form as in Example 2 with

$$\mathcal{E}^{xx}_x = \frac{-1}{14}\frac{xp(x)}{(x-3)^2}, \ \mathcal{E}^x_x = \frac{1}{14}\frac{p(x)}{(x-3)^2}, \ \mathcal{E}_x = \frac{-1}{28}\frac{p(x)(x+14)+q(x)}{(x-3)^3},$$

and $\kappa^x = \frac{2}{x^2}\frac{1}{C_{xxx}}$, where $p(x) = x^4 - 716x^3 + 422x^2 + 452x - 15$, $q(x) = 12374x^3 - 7166x^2 - 7630x + 246$. The BCOV ring $\mathcal{R}^{hol}_{BCOV}$ at the other cusp ($z = 0$) is defined with respect to a different symplectic basis $\mathcal{B}_\infty$. However we verify from the results in [19] that the ring is determined with $\mathcal{E}^{zz}_z = \mathcal{E}^{xx}_x\left(\frac{dz}{dx}\right)$, $\mathcal{E}^z_z = \mathcal{E}^x_x$, $\mathcal{E}_z = \mathcal{E}_x\left(\frac{dx}{dz}\right)$ and $\kappa^z = \kappa^x\left(\frac{dz}{dx}\right)$. From this, we observe that the BCOV ring $\mathcal{R}^{\Gamma}_{BCOV}$ is invariant under the symplectic transformation which connects $\mathcal{B}_0$ and $\mathcal{B}_\infty$.

§4. BCOV holomorphic anomaly equation in $\mathcal{R}^{0,red}_{BCOV}$

(4-1) BCOV holomorphic anomaly equation. The original form of the BCOV anomaly equation has been formulated based on the special Kähler geometry over the moduli space $\mathcal{M}$. Although mathematical ground of this anomaly equation has not yet been established, up to now, this equation provides the only way to a systematic calculation of higher genus Gromov–Witten potential $\mathtt{F}_g(t)$ for Calabi–Yau complete intersections and some cases beyond them. About this equation, recently, several important progress has been made in physics literatures [32], [1], [16], [19], [2]. In particular, the polynomial property found by Yamaguchi and Yau [32] and also in [2] is the one which we followed for our definition of the BCOV ring $\mathcal{R}^{\Gamma}_{BCOV}$.

To summarize the recursive procedure given in [5], let us write the anomaly equation in the following form, which appeared in [32] and [2],

$$\begin{aligned} \frac{\partial \mathcal{F}^{(g)}}{\partial \tilde{S}^{ij}} &= \frac{1}{2} D_i D_j \mathcal{F}^{(g-1)} + \frac{1}{2}\sum_{h=1}^{g-1} D_i \mathcal{F}^{(g-h)} D_j \mathcal{F}^{(h)} , \\ 0 &= \tilde{S}^{jk}\frac{\partial \mathcal{F}^{(g)}}{\partial \tilde{S}^k} + \tilde{S}^j \frac{\partial \mathcal{F}^{(g)}}{\partial \tilde{S}} + \frac{\partial \mathcal{F}^{(g)}}{\partial \tilde{K}_j} \qquad (g \geq 2). \end{aligned} \tag{4.1}$$

Then the recursion proceeds as follows, with $\tilde{f}_{ij}^k, \tilde{h}_{ij}, \mathcal{E}_i^k$ in (3.9), (3.11), (3.12), respectively, being unknown:

Step 1. We start with the fact that there exists a polynomial $f_0(x)$ and a choice $\tilde{f}_{ij}^k(x)$ such that

$$D_i \mathcal{F}^{(1)} = \frac{1}{2} C_{imn} \tilde{S}^{mn} - \left(\frac{\chi}{24} - 1\right) K_i + f_{1,i}(x) \ , \quad (f_{1,i} := \partial_i \log f_0)$$

gives the genus one Gromov–Witten potential $\mathtt{F}_1(t)$ of the mirror Calabi–Yau manifold X (with its Euler number χ) when we take the holomorphic limit. We refer [4], [5] for details of $\mathtt{F}_1(t)$. The polynomial $f_0(x)$ is essentially given by the discriminant of the family. We define, using the bracket notation,

$$\mathcal{R}_{BCOV}^{\Gamma,1} = \mathcal{R}_{BCOV}^{\Gamma}[\{f_{1,i}(x)\}] \ ,$$

and regard $D_i\mathcal{F}^{(1)}$ (and $f_{1,i}$) as an element of weight zero in $\mathcal{R}_{BCOV}^{\Gamma,1}$.

Step 2. Suppose we have $D_i\mathcal{F}^{(1)}$ and $\mathcal{R}_{BCOV}^{\Gamma,1}$ as above. Consider the anomaly equation (4.1) for $g = 2$ in the ring $\mathcal{R}_{BCOV}^{\Gamma,1}$ to find a (unique) solution $\mathcal{F}_0^{(2)}$ of weight $(-2,0)$ under the condition $\mathcal{F}_0^{(2)}|_{\tilde{S}^{ij}=\tilde{S}^k=\tilde{S}=0} = 0$. Then the observation made in [5] is that there exist a rational section $f_2(x)$ of $(\mathcal{L}^{-1})^{\otimes 2}$ and suitable choices $\tilde{h}_{ij}(x)$ and $\mathcal{E}_i^k(x)$ such that

$$\mathcal{F}^{(2)} = \mathcal{F}_0^{(2)} + f_2(x) \ ,$$

gives the Gromov–Witten potential $\mathtt{F}_2(t)$ under the holomorphic limit. ($\tilde{f}_{ij}^k(x)$ in step 1 and $\tilde{h}_{ij}(x), \mathcal{E}_i^k(x)$ in step 2 fix the lifting $S^{\alpha\beta}$ to $\tilde{S}^{\alpha\beta}$, and thus $\mathcal{R}_{BCOV}^{\Gamma}$.) We extend our BCOV ring to $\mathcal{R}_{BCOV}^{\Gamma,2} = \mathcal{R}_{BCOV}^{\Gamma,1}[\{f_2(x)\}]$.

For $g \geq 3$, this procedure continues genus by genus enlarging the BCOV ring by some rational section $f_g(x)$ of the line bundle $\mathcal{L}^{2-2g}$ to $\mathcal{R}_{BCOV}^{\Gamma,g+1} = \mathcal{R}_{BCOV}^{\Gamma,g}[\{f_g\}]$. When we define a notation

$$\mathcal{R}_{BCOV}^{\Gamma,\infty} = \varinjlim \mathcal{R}_{BCOV}^{\Gamma,g} \ ,$$

the solutions $\mathcal{F}_g$ are (scalar) elements in this ring of weight $(2-2g,0)$.

Remark 4.1. In general, the ring $\mathcal{R}_{BCOV}^{\Gamma,\infty}$ is a large ring. However, we note that, since $\mathcal{R}_{BCOV}^{\Gamma,\infty}$ consists of Γ-invariants, working in an affine coordinate $U \subset \mathcal{M}$ of the toric variety $\mathcal{M} = \mathbf{P}_{Sec(\Sigma)}$ makes sense. Let us consider an affine coordinate of U and consider the field of rational functions $\mathbf{Q}(x)$ on U. We assume that the unknowns $\mathcal{E}_i^{kl}, \mathcal{E}_i^k, \mathcal{E}_i, \kappa^m$ are rational functions on U (as seen in Example 2 and 3 of (3-4)). Then, due to the rationality of C_{ijk} and $f_g(x)$, we observe

$$\mathcal{R}^{\Gamma,\infty}_{BCOV}\big|_U \subset \mathbf{Q}(x)[\tilde{S}^{ij}, \tilde{S}^k, \tilde{S}, K_i] \ . \tag{4.2}$$

The ring in the r.h.s. of the inclusion is the local form which we see the BCOV ring $\mathcal{R}^{\Gamma,\infty}_{BCOV}$ in physics literatures, for example [32], [16], [19]. In reference [16], in particular, following the idea of [29], [13], an efficient way to impose certain boundary conditions to determine $f_g(x)$ has been found.

(4-2) BCOV anomaly equation in $\mathcal{R}^{0,red}_{BCOV}$. As briefly sketched above, solving BCOV anomaly equation (4.1) contains a process finding a suitable $f_g(x)$ at each genus. This is the main problem to determine the Gromov–Witten potential $\mathbf{F_g}$ from the anomaly equation. Apart from this important problem, we can extract some algebraic (combinatorial) structure of the equation by considering the same BCOV anomaly equation (4.1) in the reduced ring $\mathcal{R}^{0,red}_{BCOV}$ defined in (3.8).

Let us first note that $\mathcal{R}^{0,red}_{BCOV}$ is generated by S^{ij}, S^k, S, K_i and C_{ijk} over $\mathbf{Q}$ in the quotient ring, with the 'reduced' differential D_i (see Remark 3.9). Hereafter all manipulations should be understood in this quotient ring, although we abuse the same notations. The BCOV anomaly equation has the same form as (4.1) with obvious replacements of the generators, e.g. $\tilde{S}^{ij}$ by S^{ij}. Then the following property is due to [2]:

Proposition 4.2. *Define new generators by*

$$\hat{S}^{ij} = S^{ij} \ , \ \hat{S}^k = S^k - S^{km}K_m \ , \ \hat{S} = S - S^m K_m + \frac{1}{2} S^{mn} K_m K_n \ ,$$

then the second equation of (4.1) implies simply $\frac{\partial \mathcal{F}^{(g)}}{\partial K_m} = 0$, *namely*

$$\mathcal{F}^{(g)} \in \mathbf{Q}[\hat{S}^{ij}, \hat{S}^k, \hat{S}, C_{ijk}] \subset \mathbf{Q}[S^{ij}, S^k, S, K_m, C_{ijk}] \ (= \mathcal{R}^{0,red}_{BCOV}) \ .$$

Using the new generators above, the l.h.s of the first equation of (4.1) may be written as

$$\frac{\partial \mathcal{F}^{(g)}}{\partial \hat{S}^{ij}} - K_j \frac{\partial \mathcal{F}^{(g)}}{\partial \hat{S}^i} + \frac{1}{2} K_i K_j \frac{\partial \mathcal{F}^{(g)}}{\partial \hat{S}} \ ,$$

while the r.h.s. of that equation has the following expansion,

$$\begin{aligned}(4.3)\qquad &\frac{1}{2}D_iD_j\mathcal{F}^{(g-1)}+\frac{1}{2}\sum_{h=1}^{g-1}D_i\mathcal{F}^{(g-h)}D_j\mathcal{F}^h\\ &=Q_{ij}^{(g-1)}+Q_iK_j^{(g-1)}+Q_j^{(g-1)}K_i+\frac{1}{2}Q^{(g-1)}K_iK_j\ .\end{aligned}$$

Here one should note that by the above Proposition, the dependence on K_i comes only from the covariant derivatives. Now comparing each coefficient of $1, K_i,\ K_iK_j$, we have,

Proposition 4.3. *The BCOV anomaly equation in $\mathcal{R}^{0,red}_{BCOV}$ is equivalent to the following first order system of linear differential equations;*

$$(4.4)\quad \frac{\partial\mathcal{F}^{(g)}}{\partial\hat{S}^{ij}}=Q_{ij}^{(g-1)}\ ,\ \frac{\partial\mathcal{F}^{(g)}}{\partial\hat{S}^{i}}=-Q_{i}^{(g-1)}\ ,\ \frac{\partial\mathcal{F}^{(g)}}{\partial\hat{S}}=Q^{(g-1)}\ \ (g\geq 2).$$

With the initial data $Q_{ij}^{(1)},Q_i^{(1)},Q^{(1)}$ which follow from (4.3) with $D_i\mathcal{F}^{(1)} = \frac{1}{2}C_{ijk}\hat{S}^{jk}-(\frac{\chi}{24}-1)K_i$, this equation has a unique solution $\mathcal{F}^{(g)}\in \mathbf{Q}[\hat{S}^{ij},\hat{S}^k,\hat{S},C_{ijk}]$ of weight $(2-2g,0)$.

The uniqueness of the solution above follows from the weight consideration for the possible 'constants of integration' in the ring $\mathcal{R}^{0,red}_{BCOV}$.

For the application to Gromov–Witten potential $\mathtt{F}_g(t)$, the BCOV anomaly equation should be considered in the ring $\mathcal{R}^{\Gamma,\infty}_{BCOV}$, as we have summarized briefly in the previous section. However, the simple structure (4.4) extracted above in $\mathcal{R}^{0,red}_{BCOV}$ is still valid for the BCOV anomaly equation in the ring $\mathcal{R}^{\Gamma,\infty}_{BCOV}$. In fact, the form (4.4) of the BCOV equation has appeared first in [[19], Section (3-4)] to make the solutions $\mathcal{F}^{(g)}$ $(\mathtt{F}_g(t))$.

§5. Conclusions and discussions

After a self-contained introduction to the special Kähler geometry, we have introduced the differential ring $\mathcal{R}^0_{BCOV}$, which is geometric in nature. Combined with the modular property, we considered the Γ-invariant 'lifts' $\tilde{S}^{ij},\tilde{S}^k,\tilde{S}$ of the propagators. The 'lifting' process has been identified with that of fixing 'meromorphic ambiguities' in [5]. With a choice of Γ-invariant lifts of the propagators, we defined the ring $\mathcal{R}^{\Gamma}_{BCOV}$, which depends on the choice of the lifts. After taking a symplectic basis $\mathcal{B}$, we defined the holomorphic limit $\mathcal{R}^{hol}_{BCOV}$ following [5].

In case of an elliptic curve, we have shown a close relation of our BCOV rings to the theory of quasi-modular forms due to Kaneko–Zagier [23].

Considering a suitable quotient, we have reduced the ring $\mathcal{R}^0_{BCOV}$ to a finitely generated differential ring $\mathcal{R}^{0,red}_{BCOV}$. In this reduced ring, we have extracted a simple algebraic structure of the BCOV holomorphic anomaly equation which still exists before the reduction.

As briefly summarized in Section 4, our construction of the BCOV ring is an abstraction of the important progress made in [32] and [2] for the solutions of BCOV holomorphic anomaly equation. In 1999, in case of a rational elliptic surface $\frac{1}{2}K3$, M.-H. Saito, A. Takahashi and the present author [21] found a similar recursion formula for Gromov–Witten potentials,

$$\frac{\partial Z_{g;n}}{\partial E_2} = \frac{1}{24} \sum_{\substack{g'+g''=g \\ g',g''\geq 0}} \sum_{s=1}^{n-1} s(n-s) Z_{g';s} Z_{g'';n-s} + \frac{n(n+1)}{24} Z_{g-1;n} \ , \tag{5.1}$$

in terms of quasi-modular forms $Z_{g;n}{=}P_{g;n}\frac{q^{\frac{n}{2}}}{\eta(\tau)^{12n}}$ $(P_{g;n} \in \mathbf{Q}[E_2, E_4, E_6])$ with the initial data $Z_{0;1} = \frac{q^{\frac{1}{2}} E_4}{\eta(\tau)^{12}}$ (this generalized a previous result in [25], [26] for $g = 0$ case). Later it has been conjectured that the above recursion relation (5.1) is equivalent to the BCOV holomorphic anomaly equation evaluating $\mathcal{F}^{(g)}$ for $g \leq 3$ [18]. Due to recent progress made in [32] and [2], we have now the BCOV anomaly equation of the form,

$$\frac{\partial \mathcal{F}^{(g)}}{\partial S^{yy}} = \frac{1}{2}\Big(\sum_{h=1}^{g-1} D_y \mathcal{F}^{(g-h)} D_y \mathcal{F}^{(h)} + D_y D_y \mathcal{F}^{(g-1)} \Big) \ ,$$

which is defined over a suitable BCOV ring $\mathcal{R}^{\Gamma,\infty}_{BCOV}$. In this form, we can prove the equivalence of the 'modular anomaly equation' (5.1) to the BCOV holomorphic anomaly equation. It is almost clear that the close relationship between the ring of the quasi-modular forms and our BCOV ring presented in Section 3 plays a central role for the equivalence. The detailed results will be reported elsewhere [20].

References

[1] M. Aganagic, V. Bouchard and A. Klemm, Topological strings and (almost) modular forms, Comm. Math. Phys., **277** (2008), 771–819; arXiv:hep-th/0609123.

[2] M. Alim and J. D. Länge, Polynomial structure of the (open) topological string partition function, J. High Energy Phys., **2007** (2007), 045.

[3] L. Borisov and A. Caldararu The Pfaffian–Grassmannian derived equivalence, arXiv:math/0608404.

[4] M. Bershadsky, S. Cecotti, H. Ooguri and C. Vafa, Holomorphic anomalies in topological field theories, (with an appendix by S. Katz), Nuclear Phys. B, **405** (1993), 279–304.

[5] M. Bershadsky, S. Cecotti, H. Ooguri and C. Vafa, Kodaira–Spencer theory of gravity and exact results for quantum string amplitudes, Comm. Math. Phys., **165** (1994), 311–428.

[6] F. A. Bogomolov, Hamiltonian Kähler manifolds, Soviet Math. Dokl., **19** (1978), 1462–1465.

[7] R. L. Bryant and P. A. Griffiths, Some observations on the infinitesimal period relations for regular threefolds with trivial canonical bundle, In: Arithmetic and Geometry, Vol. II, Progr. Math., **36**, Birkhäuser Boston, Boston, MA, 1983, pp. 77–102.

[8] P. Candelas, X. C. de la Ossa, P. S. Green and L. Parkes, A pair of Calabi–Yau manifolds as an exactly soluble superconformal theory, Nuclear Phys. B, **356** (1991), 21–74.

[9] S. Cecotti and C. Vafa, On classification of $N = 2$ supersymmetric theories, Comm. Math. Phys., **158** (1993), 569–644.

[10] B. Craps, F. Roose, W. Troost and A. V. Proeyen, What is special Kähler geometry, Nuclear Phys. B, **503** (1997), 565–613.

[11] R. Dijkgraaf, Mirror symmetry and elliptic curves, In: The Moduli Space of Curves, Progr. Math., **129**, Birkhäuser Boston, Boston, MA, 1995, pp. 149–163.

[12] I. M. Gel'fand, M. M. Kapranov and A. V. Zelevinski, Discriminants, Resultants, and Multidimensional Determinants, Birkhäuser Boston, Inc., Boston, MA, 1994.

[13] D. Ghoshal and C. Vafa, c=1 String as the Topological Theory of the Conifold, Nuclear Phys. B, **453** (1995), 121–128.

[14] P. A. Griffiths, On the periods of certain rational integrals. I, II, Ann. of Math. (2), **90** (1969), 460–495; ibid., **90** (1969), 496–541.

[15] T. W. Grimm, A. Klemm, M. Marino and M. Weiss, Direct Integration of the Topological String, J. High Energy Phys., **2007** (2007), 058; arXiv:hep-th/0702187.

[16] M. X. Huang, A. Klemm and S. Quackenbush, Topological string theory on compact Calabi–Yau: Modularity and boundary conditions, arXiv:hep-th/0612125.

[17] S. Hosono, Central charges, symplectic forms, and hypergeometric series in ocal mirror symmetry, In: Mirror Symmetry. V, (eds. N. Yui, S.-T. Yau and J. Lewis), AMS/IP Stud. Adv. Math., **38**, Amer. Math. Soc., Providence, RI, 2006, pp. 405–439.

[18] S. Hosono, Counting BPS states via holomorphic anomaly equations, In: Calabi–Yau Varieties and Mirror Symmetry, (eds. N. Yui and J. Lewis),

Fields Inst. Commun., **38**, Amer. Math. Soc., Providence, RI, 2003, pp. 57–86.

[19] S. Hosono and Y. Konishi, Higher genus Gromow–Witten invariants of the Grassmannian, and the Pfaffian Calabi–Yau threefolds, to appear in Adv. Theor. Math. Phys., **13** (2009); arXiv:0704.2928[math.AG].

[20] S. Hosono and M.-H. Saito, Modular and holomorphic anomaly equations of a rational elliptic surface, to appear.

[21] S. Hosono, M.-H. Saito and A. Takahashi, Holomorphic anomaly equation and BPS state counting of rational elliptic surface, Adv. Theor. Math. Phys., **3** (1999), 177–208; arXiv:hep-th/9901151.

[22] A. Kuznetsov, Homological projective duality for Grassmannians of lines, arXiv:math/0610957.

[23] M. Kaneko and D. Zagier, A generalized Jacobi theta function and quasi-modular forms, In: The Moduli Space of Curves, Prog. Math., **129**, Birkhäuser Boston, MA, 1995, pp. 165–172.

[24] B. Lian and S.-T. Yau, Arithmetic properties of mirror map and quantum coupling, Comm. Math. Phys., **176** (1996), 163–192.

[25] J. A. Minahan, D. Nemeschansky and N. P. Warner, Partition functions for the BPS states of the E8 non-critical string, Adv. Theor. Math. Phys., **1** (1998), 167–183; arXiv:hep-th/9707149.

[26] J. A. Minahan, D. Nemeschansky, C. Vafa and N. P. Warner, E-strings and N=4 topological Yang–Mills theories, Nuclear Phys. B, **527** (1998), 581–623; arXiv:hep-th/9802168.

[27] E. A. Rødland, The Pfaffian Calabi–Yau, its Mirror and their link to the Grassmannian $G(2,7)$, Compositio Math., **122** (2000), 135–149; arXiv:math/9801092[math.AG].

[28] A. Strominger, Special geometry, Comm. Math. Phys., **133** (1990), 163–180.

[29] A. Strominger, Massless black holes and conifolds in string theory, Nuclear Phys. B, **451** (1995), 96–108.

[30] G. Tian, Smoothness of the universal deformation space of compact Calabi–Yau manifolds and its Peterson–Weil metric, In: Mathematical Aspects of String Theory, (ed. S.-T. Yau), World Sci. Publ., Singapore, 1987, pp. 629–646.

[31] A. N. Todorov, The Weil–Petersson geometry of the moduli space of SU($n \geq 3$) (Calabi–Yau) manifolds. I, Comm. Math. Phys., **126** (1989), 325–346.

[32] S. Yamaguchi and S.-T. Yau, Topological string partition functions as polynomials, J. High Energy Phys., **2004** (2004), 047.

Graduate School of Mathematical Sciences
University of Tokyo
Meguro-ku, Tokyo 153-8914
Japan
E-mail address: hosono@ms.u-tokyo.ac.jp

Advanced Studies in Pure Mathematics 59, 2010
New Developments in Algebraic Geometry,
Integrable Systems and Mirror Symmetry (Kyoto, 2008)
pp. 111–166

Ruan's conjecture and integral structures in quantum cohomology

Hiroshi Iritani

Abstract.

This is an expository article on the recent studies [27, 28, 45, 46, 23] of Ruan's crepant resolution/flop conjecture [62, 63] and its possible relations to the K-theory integral structure [45, 46, 49] in quantum cohomology.

§1. Introduction

The small quantum cohomology is a family $(H^*(X), \circ_\tau)$ of commutative ring structures on $H^*(X)$ parametrized by $\tau \in H^{1,1}(X)$. In the large radius limit, *i.e.* $-\Re\left(\int_C \tau\right) \to \infty$ for every effective curve $C \subset X$, the quantum product $\circ_\tau$ goes to the cup product. For a pair (X_1, X_2) of birational varieties in a "crepant" relationship (such as flops or crepant resolutions), Yongbin Ruan's conjecture states that the small quantum cohomologies $(H^*(X_1), \circ_{\tau_1})$ and $(H^*(X_2), \circ_{\tau_2})$ are isomorphic under analytic continuation of the parameter τ. The conjectural space where the quantum product $\circ_\tau$ is analytically continued is known as a Kähler moduli space $\mathcal{M}$ in physics. (See Figure 1.) In our situation, the space $\mathcal{M}$ has two limit points (which we call cusps) $\mathbf{0}_1$, $\mathbf{0}_2$ corresponding to the large radius limit points of X_1 and X_2 respectively. A neighborhood V_i of $\mathbf{0}_i$ is identified with an open subset of $H^{1,1}(X_i)$. A weak form of Ruan's conjecture asserts that *there exists a family $(F, \circ_\tau)$ of commutative rings over $\mathcal{M}$ such that its restriction to V_i is isomorphic to the small quantum cohomology of X_i.* In particular, the cohomology rings $H^*(X_1)$, $H^*(X_2)$ are connected via quantum deformations.

Received July 5, 2008.
Revised March 9, 2009.
2000 *Mathematics Subject Classification.* Primary 14N35; Secondary 53D45.
Partly supported by JSPS Grant-in-Aid for Scientific Research (S-19104002) and EPSRC (EP/E022162/1).

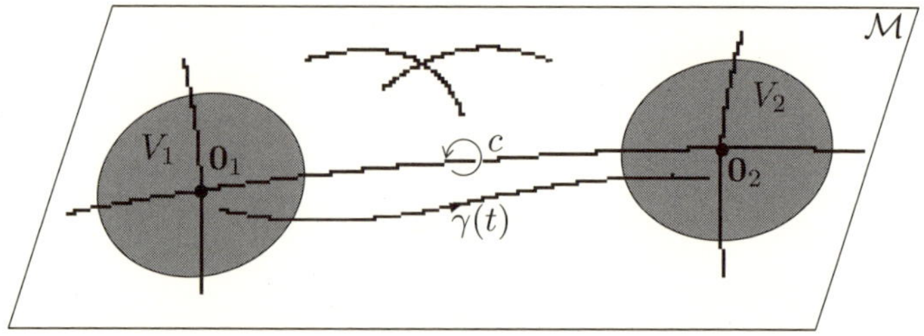

Fig. 1. Kähler moduli space $\mathcal{M}$ containing cusp neighborhoods $V_i \subset H^{1,1}(X_i, \mathbb{C})$, $i = 1, 2$. The *global quantum D-module* over $\mathcal{M}$ develops singularities along thick lines.

In a more precise picture, the family of rings should come from a meromorphic flat connection (F, ∇) over $\mathcal{M}$, which we call a *global quantum D-module.* The D-module (F, ∇) restricted to V_i is identified with the quantum D-module given by the Dubrovin connection (6):

$$\nabla_\alpha = \frac{\partial}{\partial t^\alpha} + \frac{1}{z}\phi_\alpha \circ_\tau, \quad \text{where } z \in \mathbb{C}^* \text{ is a parameter.}$$

It is clear that the operator $z\nabla_\alpha$ recovers the quantum product $\phi_\alpha \circ_\tau$ in the limit $z \to 0$, but the D-module contains much more information than a family of rings. In fact, the global quantum D-module (F, ∇) together with additional data—opposite subspace and dilaton shift—yields a Kyoji Saito's *flat structure* [66] on the extended Kähler moduli space[1]. Moreover, the local monodromy around each cusp determines a canonical choice of the opposite subspace and recovers the flat structure on V_i induced from the identification with the vector space $H^{1,1}(X_i)$. Here, as the example in [27] suggests, the flat structures from the different cusps $\mathbf{0}_1$ and $\mathbf{0}_2$ does not necessarily coincide.

In this article, we postulate furthermore that the global quantum D-module is underlain by an integral local system. We also conjecture that, over V_i, the integral local system in question comes from the K-theory of X_i. This has the following physical explanation. Quantum cohomology is part of the A-model topological string theory. A chiral field in the A-model (*i.e.* a section of the quantum D-module) should have a pairing

[1] A flat structure is also referred to as a Frobenius manifold structure due to Dubrovin [31]. In the later formulation, the parameter τ will be extended to the total cohomology group $H^*(X_i)$ and $\mathcal{M}$ is also extended accordingly.

with a B-type D-brane (*i.e.* an object of the derived category $D^b_{\mathrm{coh}}(X_i)$) (see *e.g.* [39]). This suggests that *a vector bundle on X_i should give a flat section of the quantum D-module.* Under mirror symmetry, this corresponds to the fact that a holomorphic n-form has a pairing with a (real) Lagrangian n-cycle by integration. Based on mirror symmetry for toric orbifolds, the author [45, 46] proposed a formula (13) which assigns a flat section of the quantum D-module to an element of the K-group. Katzarkov–Kontsevich–Pantev [49] also found a similar formula for a rational structure independently. The flat sections arising from the K-group define an integral local system over V_i. Via the analytic continuation of K-theory flat sections along a path $\gamma(t)$ connecting V_1 and V_2 (see Figure 1), our picture gives an isomorphism of K-groups:

$$\mathbb{U}_{K,\gamma} \colon K(X_1) \xrightarrow{\cong} K(X_2).$$

The isomorphism $\mathbb{U}_{K,\gamma}$ here contains complete information on the relationships between the genus zero Gromov–Witten theories (quantum cohomology) of X_1 and X_2. We hope that $\mathbb{U}_{K,\gamma}$ is given by a certain Fourier–Mukai transformation.

The paper is structured as follows. In Section 2, we review orbifold quantum cohomology/D-module and introduce the K-theory integral structure on it. In Section 3.1, we formulate a precise picture (Assumption 3.1) of the global quantum D-module sketched above. In Sections 3.2–3.7, we discuss what follows from the picture *without* using integral structures. The main observation here is the fact that *each cusp determines a possibly different flat structure on $\mathcal{M}$.* The Hard Lefschetz condition in Section 3.7 is a sufficient condition for the Frobenius structures from different cusps to match. These facts were found in [27], but the present article contains a complete proof of the characterization of Frobenius structures at cusps (Theorem 3.13, announced in [27]) and a *generalized* Hard Lefschetz condition (Theorem 3.22). In Sections 3.8, 3.9, we use integral structures to study the crepant resolution conjecture for Calabi–Yau orbifolds and give an explicit prediction (Conjecture 3.31) for the change of co-ordinates in local examples. Readers who want to know a role of integral structures in Ruan's conjecture can safely skip Sections 3.2–3.7 and go directly to Sections 3.8 or 3.9.

Finally, we remark that we restrict our attentions to the even parity part of the (orbifold) cohomology group throughout this paper.

Acknowledgments. Thanks are due to Masa-Hiko Saito for the invitation to the conference "New developments in Algebraic Geometry, Integrable Systems and Mirror symmetry" held at RIMS, January 2008. The author is grateful to Tom Coates, Alessio Corti, Hsian-Hua Tseng for helpful conversations during joint works [27, 24, 25] and in many other occasions. He is also grateful to Samuel Boissiere for explaining the McKay correspondence for a finite subgroup of $SO(3)$ and to Jim Bryan, Martin Guest and Yongbin Ruan for helpful comments on the present article. He also thanks Claus Hertling, Shinobu Hosono, Tony Pantev, Shunsuke Takagi for helpful discussions. This paper was written while the author was visiting Imperial College London. He would like to thank Imperial College for its hospitality.

§2. K-theory integral structure in quantum cohomology

In this section, we review the orbifold quantum cohomology for smooth Deligne–Mumford stacks and introduce the K-theory integral structure on it. Assuming the convergence of structure constants, quantum cohomology defines a flat connection, called the *Dubrovin connection*, on some cohomology bundle over a neighborhood of the "large radius limit point". This is called the *quantum D-module.* We will see that the K-group defines an integral lattice in the space of (multi-valued) flat sections of the quantum D-module. The key definition will be given in Definition 2.11. The true origin of this integral structure is yet to be known, but it has a number of good properties:

- This is invariant under every local monodromy around the large radius limit point (Proposition 2.12 (iii)).
- The pairing on quantum cohomology is translated into the Mukai pairing on the K-group (Proposition 2.12 (ii)).
- This gives a real structure which is *pure and polarized* in a neighborhood of the large radius limit point [45, Theorem 3.7]. In particular, we have *tt^*-geometry* [18, 37, 47] on quantum cohomology.
- This looks compatible with many computations done in the context of mirror symmetry [43, 7]. In particular, this matches with the integral structure on the Landau–Ginzburg mirror in the case of toric orbifolds [45, 46] (see Example 2.14 (iii)).
- Thus in toric case, this integral structure is compatible also with the *Stokes structure.*

In this article, we will not explain the last three items. See [45, 38, 49] for the properties "pure and polarized" or "compatibility with Stokes structure".

2.1. Orbifold quantum cohomology

We start with some notation concerning orbifolds. Let $\mathcal{X}$ be a smooth Deligne–Mumford stack with projective coarse moduli space X.

Let $I\mathcal{X}$ be the inertia stack of $\mathcal{X}$. A point on $I\mathcal{X}$ is given by a pair (x,g) of a point $x \in \mathcal{X}$ and an element g of the automorphism group (local group) $\mathrm{Aut}_{\mathcal{X}}(x)$ at x. The element $g \in \mathrm{Aut}_{\mathcal{X}}(x)$ is also called a stabilizer. Let

$$I\mathcal{X} = \bigsqcup_{v \in \mathsf{T}} \mathcal{X}_v = \mathcal{X}_0 \sqcup \bigsqcup_{v \in \mathsf{T}'} \mathcal{X}_v$$

be the decomposition of $I\mathcal{X}$ into connected components. Here T is a finite set parametrizing connected components of $I\mathcal{X}$. T contains a distinguished element $0 \in \mathsf{T}$ which corresponds to the trivial stabilizer $g = 1$ and we set $\mathsf{T} = \{0\} \cup \mathsf{T}'$. Then $\mathcal{X}_0$ is isomorphic to $\mathcal{X}$. At each point (x,g) in $I\mathcal{X}$, we can define a rational number $\iota_{(x,g)}$ called *age*. The element g of the automorphism group acts on the tangent space $T_x\mathcal{X}$ and decomposes it into eigenspaces:

$$T_x\mathcal{X} = \bigoplus_{0 \le f < 1} (T_x\mathcal{X})_f$$

where g acts on $(T_x\mathcal{X})_f$ by $\exp(2\pi \mathtt{i} f)$. The age $\iota_{(x,g)}$ is defined to be

$$\iota_{(x,g)} = \sum_{0 \le f < 1} f \dim_{\mathbb{C}} (T_x\mathcal{X})_f.$$

The age $\iota_{(x,g)}$ is constant along the connected component $\mathcal{X}_v$ of $I\mathcal{X}$, so we denote by ι_v the age $\iota_{(x,g)}$ at any point (x,g) in $\mathcal{X}_v$. The *orbifold or Chen–Ruan cohomology group* $H^*_{\mathrm{CR}}(\mathcal{X})$ is a $\mathbb{Q}$-graded vector space defined by

$$H^p_{\mathrm{CR}}(\mathcal{X}) = \bigoplus_{v \in \mathsf{T}} H^{p-2\iota_v}(\mathcal{X}_v, \mathbb{C}), \quad p \in \mathbb{Q}.$$

This is the same as $H^*(I\mathcal{X}, \mathbb{C})$ as a vector space, but the grading is shifted by the age. In this paper, we only consider the *even parity part* of $H^*_{\mathrm{CR}}(\mathcal{X})$, *i.e.* the summands satisfying $p - 2\iota_v \equiv 0 \mod 2$ in the above decomposition. Unless otherwise stated, we denote by $H^*_{\mathrm{CR}}(\mathcal{X})$ the even parity part. The inertia stack has an involution $\mathrm{inv}\colon I\mathcal{X} \to I\mathcal{X}$ which sends (x,g) to (x,g^{-1}). This induces an involution $\mathrm{inv}\colon \mathsf{T} \to \mathsf{T}$ on the index set T and $\mathrm{inv}^*\colon H^*_{\mathrm{CR}}(\mathcal{X}) \to H^*_{\mathrm{CR}}(\mathcal{X})$ on the cohomology. The *orbifold Poincaré pairing* on $H^*_{\mathrm{CR}}(\mathcal{X})$ is defined by

$$(\alpha, \beta)_{\mathrm{orb}} = \int_{I\mathcal{X}} \alpha \cup \mathrm{inv}^*(\beta) = \sum_{v \in \mathsf{T}} \int_{\mathcal{X}_v} \alpha_v \cup \beta_{\mathrm{inv}(v)},$$

where α_v, β_v are the v-components of α, β. This pairing is symmetric, non-degenerate and of degree $-2\dim_{\mathbb{C}} \mathcal{X}$.

Gromov–Witten theory for manifolds has been extended to the class of symplectic orbifolds or smooth Deligne–Mumford stacks. This was done by Chen–Ruan [21] in the symplectic category and by Abramovich–Graber–Vistoli [1] in the algebraic category. The formal properties of the genus zero Gromov–Witten theory hold in orbifold theory as well: the genus zero orbifold Gromov–Witten theory defines a cohomological field theory (see *e.g.* [57]) on the metric vector space $(H^*_{\mathrm{CR}}(\mathcal{X}), (\cdot,\cdot)_{\mathrm{orb}})$. In particular, we have the following correlation functions (Gromov–Witten invariants):

$$\langle \cdot, \dots, \cdot \rangle_{0,m,d} : (H^*_{\mathrm{CR}}(\mathcal{X}))^{\otimes m} \to \mathbb{C} \tag{1}$$

defined for $m \geq 0$ and $d \in H_2(X, \mathbb{Z})$. This is zero when d is not in the semigroup $\mathrm{Eff}_{\mathcal{X}} \subset H_2(X, \mathbb{Z})$ generated by classes of effective curves or $m \leq 2$ and $d = 0$. Also these correlation functions satisfy the so-called *WDVV equation* or the *splitting axiom* (see [1, Theorem 6.4.3]). The genus zero Gromov–Witten invariant is homogeneous with respect to the grading of $H^*_{\mathrm{CR}}(\mathcal{X})$. More precisely, $\langle \alpha_1, \dots, \alpha_m \rangle_{0,m,d} = 0$ unless $p_1 + \cdots + p_m = 2(\dim_{\mathbb{C}} \mathcal{X} + \langle c_1(\mathcal{X}), d \rangle + m - 3)$, where $\alpha_i \in H^{p_i}_{\mathrm{CR}}(\mathcal{X})$.

The genus zero Gromov–Witten invariants define a quantum product $\bullet_\tau$ on $H^*_{\mathrm{CR}}(\mathcal{X})$ parametrized by $\tau \in H^*_{\mathrm{CR}}(\mathcal{X})$:

$$(\alpha \bullet_\tau \beta, \gamma)_{\mathrm{orb}} = \sum_{d \in \mathrm{Eff}_{\mathcal{X}}, m \geq 0} \frac{1}{m!} \Big\langle \alpha, \beta, \gamma, \overbrace{\tau, \dots, \tau}^{m \text{ times}} \Big\rangle_{0,m+3,d} Q^d. \tag{2}$$

Here Q^d denotes the element of the group ring $\mathbb{C}[\mathrm{Eff}_{\mathcal{X}}]$ corresponding to $d \in \mathrm{Eff}_{\mathcal{X}} \subset H_2(X, \mathbb{Z})$. The right-hand side belongs to $\mathbb{C}[\![\tau]\!][\![\mathrm{Eff}_{\mathcal{X}}]\!]$ (a certain completion[2] of $\mathbb{C}[\![\tau]\!] \otimes \mathbb{C}[\mathrm{Eff}_{\mathcal{X}}]$) and defines the element $\alpha \bullet_\tau \beta$ in $H^*_{\mathrm{CR}}(\mathcal{X}) \otimes \mathbb{C}[\![\tau]\!][\![\mathrm{Eff}_{\mathcal{X}}]\!]$ because the orbifold Poincaré pairing is non-degenerate. By extending $\bullet_\tau$ as a $\mathbb{C}[\![\tau]\!][\![\mathrm{Eff}_{\mathcal{X}}]\!]$-bilinear map, we have an associative commutative ring $(H^*_{\mathrm{CR}}(\mathcal{X}) \otimes \mathbb{C}[\![\tau]\!][\![\mathrm{Eff}_{\mathcal{X}}]\!], \bullet_\tau)$. Here the associativity of the product $\bullet_\tau$ follows from the WDVV equation. This is the *orbifold quantum cohomology* of $\mathcal{X}$.

Using the *Divisor equation* (see [1, Theorem 8.3.1]), we can write

(3)
$$(\alpha \bullet_\tau \beta, \gamma)_{\mathrm{orb}} = \sum_{d \in \mathrm{Eff}_{\mathcal{X}}, m \geq 0} \frac{1}{m!} \Big\langle \alpha, \beta, \gamma, \overbrace{\tau', \dots, \tau'}^{m \text{ times}} \Big\rangle_{0,m+3,d} e^{\langle \tau_{0,2}, d \rangle} Q^d,$$

---

[2] For example, one completion is given by the additive valuation v on $\mathbb{C}[\mathrm{Eff}_{\mathcal{X}}]$ defined by $v(Q^d) = \int_d \omega$, where ω is a Kähler form on $\mathcal{X}$.

where we put

$$(4)\quad \tau = \tau_{0,2} + \tau', \quad \tau_{0,2} \in H^2(\mathcal{X}_0), \quad \tau' \in \bigoplus_{p\neq 2} H^p(\mathcal{X}_0) \oplus \bigoplus_{v \in \mathsf{T}'} H^*(\mathcal{X}_v).$$

This shows that the parameters τ and Q in the product $\bullet_\tau$ are redundant. In fact $\bullet_\tau$ depends only on τ' and $e^{\tau_{0,2}}Q$. We put

$$\circ_\tau := \bullet_\tau|_{Q=1}.$$

The product $\circ_\tau$ is a formal power series in τ' and a formal Fourier series in $\tau_{0,2}$. The original product $\bullet_\tau$ can be recovered from $\circ_\tau$ by substituting $e^{\langle \tau_{0,2}, d\rangle} Q^d$ for $e^{\langle \tau_{0,2}, d\rangle}$. In what follows, we will study $\circ_\tau$ instead of $\bullet_\tau$ and assume that *the product $\circ_\tau$ is convergent in some open set U of $H^*_{\mathrm{CR}}(\mathcal{X})$*.

Assumption 2.1. *The orbifold quantum product $\circ_\tau$ is convergent on a simply-connected open set U containing the following set*

$$\{\tau \in H^*_{\mathrm{CR}}(\mathcal{X}) \,;\, \Re(\langle d, \tau_{0,2}\rangle) < -M, \ \forall d \in \mathrm{Eff}_{\mathcal{X}} \setminus \{0\}, \ \|\tau'\| \le e^{-M}\},$$

where $\tau = \tau_{0,2} + \tau'$ is the decomposition in (4), $M > 0$ is a sufficiently big real number and $\|\cdot\|$ is a suitable norm on $H^*_{\mathrm{CR}}(\mathcal{X})$.

Remark 2.2. Working over a certain formal power series ring, we could discuss the K-theory integral structure without this assumption. However, when considering Ruan's conjecture later, we cannot avoid the convergence problem of quantum cohomology.

The open set U above is considered to be a neighborhood of the "*large radius limit point*" which is the limit point of the sequence

$$(5)\qquad \tau = \tau_{0,2} + \tau' \colon \quad \Re(\langle d, \tau_{0,2}\rangle) \to -\infty, \quad \tau' \to 0.$$

(This notion will be made more precise later.) In this limit, the orbifold quantum product $\circ_\tau$ goes to the *Chen–Ruan orbifold cup product* $\cup_{\mathrm{CR}}$. This product $\cup_{\mathrm{CR}}$ is the same as the cup product when $\mathcal{X}$ is a manifold, but in the orbifold case, this is different from the cup product on $I\mathcal{X}$.

2.2. Quantum D-modules with Galois actions

Let $\{\phi_i\}$ be a homogeneous $\mathbb{C}$-basis of $H^*_{\mathrm{CR}}(\mathcal{X})$ and $\{t^i\}$ be the linear co-ordinate system on $H^*_{\mathrm{CR}}(\mathcal{X})$ dual to the basis $\{\phi_i\}$. Denote by $\tau = \sum_{i=1}^N t^i \phi_i$ a general point on $H^*_{\mathrm{CR}}(\mathcal{X})$. The quantum D-module is a meromorphic flat connection on the trivial $H^*_{\mathrm{CR}}(\mathcal{X})$-bundle over $U \times \mathbb{C}$. Denote by (τ, z) a general point on the base space $U \times \mathbb{C}$. Let $(-)\colon U \times \mathbb{C} \to U \times \mathbb{C}$ be the map sending (τ, z) to $(\tau, -z)$.

Definition 2.3. The *quantum D-module* $QDM(\mathcal{X}) = (F, \nabla, (\cdot,\cdot)_F)$ is the trivial holomorphic vector bundle $F := H^*_{\mathrm{CR}}(\mathcal{X}) \times (U \times \mathbb{C}) \to (U \times \mathbb{C})$ endowed with the meromorphic flat connection ∇:

$$
\begin{aligned}
\nabla_i &= \nabla_{\frac{\partial}{\partial t^i}} = \frac{\partial}{\partial t^i} + \frac{1}{z}\phi_i \circ_\tau, \\
\nabla_{z\partial_z} &= z\frac{\partial}{\partial z} - \frac{1}{z} E \circ_\tau + \mu,
\end{aligned}
\tag{6}
$$

and the ∇-flat pairing

$$
(\cdot,\cdot)_F \colon (-)^*\mathcal{O}(F) \otimes \mathcal{O}(F) \to \mathcal{O}_{U\times\mathbb{C}}
$$

induced from the orbifold Poincaré pairing $F_{(\tau,-z)} \times F_{(\tau,z)} = H^*_{\mathrm{CR}}(\mathcal{X}) \times H^*_{\mathrm{CR}}(\mathcal{X}) \to \mathbb{C}$. Here E is the *Euler vector field* on U given by

$$
E := c_1(T\mathcal{X}) + \sum_{i=1}^{N} (1 - \frac{1}{2}\deg\phi_i) t^i \phi_i \tag{7}
$$

and $\mu \in \mathrm{End}(H^*_{\mathrm{CR}}(\mathcal{X}))$ is the *Hodge grading operator* defined by

$$
\mu(\phi_i) := (\frac{1}{2}\deg\phi_i - \frac{n}{2})\phi_i, \quad n = \dim_{\mathbb{C}} \mathcal{X}. \tag{8}
$$

The flat connection ∇ is called the *Dubrovin connection* or *the first structure connection.* Note that ∇_i has a pole of order 1 along $z = 0$ and ∇_{∂_z} has a pole of order 2 along $z = 0$. The flatness of ∇ follows from the WDVV equations and the homogeneity of Gromov–Witten invariants.

Remark 2.4. By *D-module* one means a module over the ring of differential operators. In our case, the ring $\mathcal{O}_{\mathcal{M}\times\mathbb{C}^*}\langle \partial_{t^i}, z\partial_z\rangle$ of differential operators on $\mathcal{M} \times \mathbb{C}^*$ acts on the space of sections of F via the flat connection: $\partial_{t^i} \mapsto \nabla_i$, $z\partial_z \mapsto \nabla_{z\partial_z}$. This explains the name "quantum D-module".

The quantum D-module admits certain discrete symmetries (Galois actions). Firstly, since $\circ_\tau$ depends only on $e^{\tau_{0,2}}$ and τ', it is clear that $\circ_\tau$ is invariant under the following translation:

$$
\tau_{0,2} \mapsto \tau_{0,2} - 2\pi\mathtt{i}\xi, \quad \xi \in H^2(X,\mathbb{Z}).
$$

This is a consequence of the Divisor equations and is familiar in ordinary Gromov–Witten theory. Interestingly, we have a finer symmetry for orbifold theory. Let $H^2(\mathcal{X},\mathbb{Z})$ be the sheaf cohomology of the constant sheaf $\mathbb{Z}$ on the stack $\mathcal{X}$ (not on the coarse moduli space X). This group is identified with the set of isomorphism classes of topological orbifold line bundles on $\mathcal{X}$. Then $H^2(X,\mathbb{Z})$ is identified with the subset of $H^2(\mathcal{X},\mathbb{Z})$ consisting of line bundles which are pulled back from the coarse moduli space X. For $\xi \in H^2(\mathcal{X},\mathbb{Z})$, let L_ξ be the corresponding topological orbifold line bundle on $\mathcal{X}$ and $\xi_0 := c_1(L_\xi) \in H^2(X,\mathbb{Q})$ be the first Chern class. For $v \in \mathsf{T}$, define $0 \le f_v(\xi) < 1$ to be the rational number such that the stabilizer g at $(x,g) \in \mathcal{X}_v$ acts on the fiber $L_{\xi,x}$ by $\exp(2\pi\mathtt{i} f_v(\xi))$.

Lemma 2.5 ([46, Proposition 2.3]). *The flat connection ∇ and the pairing $(\cdot,\cdot)_F$ of the quantum D-module $QDM(\mathcal{X}) = (F, \nabla, (\cdot,\cdot)_F)$ are invariant under the following map given for $\xi \in H^2(\mathcal{X},\mathbb{Z})$:*

$$\begin{aligned} H^*_{\mathrm{CR}}(\mathcal{X}) \times (U \times \mathbb{C}) &\to H^*_{\mathrm{CR}}(\mathcal{X}) \times (U \times \mathbb{C}) \\ (\phi, \tau, z) &\longmapsto (dG(\xi)(\phi), G(\xi)(\tau), z). \end{aligned}$$

Here $G(\xi), dG(\xi)\colon H^*_{\mathrm{CR}}(\mathcal{X}) \to H^*_{\mathrm{CR}}(\mathcal{X})$ is defined by

$$G(\xi)(\tau_0 \oplus \bigoplus_{v \in \mathsf{T}'} \tau_v) = (\tau_0 - 2\pi\mathtt{i}\xi_0) \oplus \bigoplus_{v\in\mathsf{T}'} e^{2\pi\mathtt{i} f_v(\xi)} \tau_v$$

$$dG(\xi)(\tau_0 \oplus \bigoplus_{v \in \mathsf{T}'} \tau_v) = \tau_0 \oplus \bigoplus_{v\in\mathsf{T}'} e^{2\pi\mathtt{i} f_v(\xi)} \tau_v$$

where we used the decomposition $H^*_{\mathrm{CR}}(\mathcal{X}) = H^*(\mathcal{X}_0) \oplus \bigoplus_{v\in\mathsf{T}'} H^*(\mathcal{X}_v)$ and $\tau_v \in H^*(\mathcal{X}_v)$. (Here we implicitly assume that U is invariant under the map $G(\xi)$, but we can assume this without loss of generality.)

We refer to the symmetry in Lemma 2.5 as a *Galois action* of $H^2(\mathcal{X},\mathbb{Z})$ or a *local monodromy at the large radius limit.* When $\xi \in H^2(X,\mathbb{Z})$, this is the same as the aforementioned one. Note that the new symmetry can act non-trivially on the fiber of the quantum D-module. The quantum D-module descends to a flat connection on $F/H^2(\mathcal{X},\mathbb{Z}) \to (U/H^2(\mathcal{X},\mathbb{Z})) \times \mathbb{C}$. We call this flat connection on the quotient space also the *quantum D-module.*

We can construct a partial compactification $\overline{V}$ of the quotient $V = U/H^2(\mathcal{X},\mathbb{Z})$ such that $\overline{V}$ contains the large radius limit point and that the quantum D-module on V extends to a D-module on $\overline{V}$ with a logarithmic singularity along the normal crossing divisor[3] $\overline{V} \setminus V$. Choose a $\mathbb{Z}$-basis $p_1, \dots, p_r$ of $H^2(X,\mathbb{Z})/\text{torsion}$ such that p_a intersects every effective curve class $d \in \mathrm{Eff}_{\mathcal{X}}$ non-negatively (*i.e.* p_a is nef). Then we

[3]More precisely, $\overline{V} \setminus V$ is a normal crossing divisor on an étale cover of $\overline{V}$.

have the embedding

$$U/H^2(X,\mathbb{Z}) \hookrightarrow \mathbb{C}^r \times W, \qquad \left[\sum_{a=1}^r t^a p_a + \tau'\right] \mapsto (e^{t_1}, \dots, e^{t_r}, \tau')$$

where $W = \bigoplus_{p\neq 2} H^p(\mathcal{X}_0) \oplus \bigoplus_{v\in\mathsf{T}'} H^*(\mathcal{X}_v)$ and $\tau' \in W$. By Assumption 2.1 and the choice of p_a, the image of this embedding contains the open set $((\mathbb{C}^*)^r \times W) \cap \Delta_M$ for a sufficiently big $M > 0$, where

$$\Delta_M = \{(q^1, \dots, q^r, \tau') \in \mathbb{C}^r \times W \ ; \ |q^a| < e^{-M}, \|\tau'\| < e^{-M}\}.$$

We set $\overline{U/H^2(X,\mathbb{Z})} := (U/H^2(X,\mathbb{Z})) \cup \Delta_M \subset \mathbb{C}^r \times W$. For $\tau_{0,2} = \sum_{a=1}^r t^a p_a$, we have $e^{\langle \tau_{0,2}, d\rangle} = (e^{t^1})^{\langle p_1, d\rangle} \cdots (e^{t^r})^{\langle p_r, d\rangle}$. Therefore by the formula (3), since p_a is nef, the quantum product $\circ_\tau$ on $U/H^2(X,\mathbb{Z})$ extends to $\overline{U/H^2(X,\mathbb{Z})}$. The Dubrovin connection on Δ_M in the direction of $q^a = e^{t^a}$ can be written as

$$\nabla_{\frac{\partial}{\partial t^a}} = q^a \frac{\partial}{\partial q^a} + \frac{1}{z} p_a \circ_\tau .$$

Hence it has a logarithmic pole along $q^1 \cdots q^r = 0$. We can now define $\overline{V}$ as the quotient space (or stack):

$$\overline{V} := \overline{U/H^2(X,\mathbb{Z})}/(H^2(\mathcal{X},\mathbb{Z})/H^2(X,\mathbb{Z})).$$

This contains both $U/H^2(\mathcal{X},\mathbb{Z})$ and the large radius limit point $q = \tau' = 0$.

Remark 2.6. The partial compactification $\overline{V}$ depends on the choice of a nef basis p_a. A canonical partial compactification is given by the possibly singular quotient stack $[(\operatorname{Spec}\mathbb{C}[\operatorname{Eff}_{\mathcal{X}}] \times W)/(H^2(\mathcal{X},\mathbb{Z})/H^2(X,\mathbb{Z}))]$. Note that there exists a natural map from $\overline{V}$ to this stack.

Remark 2.7. Due to the new discrete symmetries, the large radius limit point in $\overline{V}$ can have an orbifold singularity when $\mathcal{X}$ is an orbifold. Also, the quantum D-module $F/H^2(X,\mathbb{Z})$ on the quotient space may not be trivialized in the standard way. In other words, an element of $H^*_{\mathrm{CR}}(\mathcal{X})$ gives a possibly multi-valued section of $F/H^2(\mathcal{X},\mathbb{Z})$.

2.3. Fundamental solution $L(\tau, z)$ and the space $\mathcal{S}(\mathcal{X})$ of flat sections

We introduce a *fundamental solution* for ∇-flat sections of the quantum D-module (F, ∇). Orbifold Gromov–Witten theory also has (gravitational) descendant invariants (as opposed to the primary invariants

(1)) of the form

$$\left\langle \alpha_1\psi_1^{k_1}, \dots, \alpha_m\psi_m^{k_m} \right\rangle_{0,m,d}$$

where $\alpha_i \in H^*_{\mathrm{CR}}(\mathcal{X})$, $d \in \mathrm{Eff}_{\mathcal{X}}$ and k_i is a non-negative integer. The symbol ψ_i represents the first Chern class of the line bundle on the moduli space of stable maps formed by the cotangent lines at the i-th marked point of the coarse domain curve. As is well-known in manifold Gromov–Witten theory (see *e.g.* [58, Proposition 2]), we can write the fundamental solution to the equation $\nabla s = 0$ by using descendant invariants. Let $\mathrm{pr}\colon I\mathcal{X} \to \mathcal{X}$ be the natural projection. For $\tau_0 \in H^*(\mathcal{X}_0)$, we define the action of τ_0 on $H^*_{\mathrm{CR}}(\mathcal{X})$ as

$$\tau_0 \cdot \alpha = \mathrm{pr}^*(\tau_0) \cup \alpha$$

where the right-hand side is the cup product on $H^*(I\mathcal{X})$. (This is known to be the same as the orbifold cup product $\tau_0 \cup_{\mathrm{CR}} \alpha$.) Let $\{\phi_k\}_{k=1}^N$ and $\{\phi^k\}_{k=1}^N$ be bases of $H^*_{\mathrm{CR}}(\mathcal{X})$ dual with respect to the orbifold Poincaré pairing, *i.e.* $(\phi_i, \phi^j)_{\mathrm{orb}} = \delta_i^j$.

Proposition 2.8 (See *e.g.* [46, Proposition 2.4]). *Let $L(\tau, z)$ be the following $\mathrm{End}(H^*_{\mathrm{CR}}(\mathcal{X}))$-valued function on $U \times \mathbb{C}^*$:*

$$\begin{aligned} L(\tau,z)\phi &= e^{-\tau_{0,2}/z}\phi \\ &\quad - \sum_{\substack{(d,m)\neq(0,0),\\ d\in\mathrm{Eff}_{\mathcal{X}},1\le k\le N}} \frac{e^{\langle \tau_{0,2}, d\rangle}}{m!}\phi_k \left\langle \phi^k, \tau', \dots, \tau', \frac{e^{-\tau_{0,2}/z}\phi}{z+\psi_{m+2}} \right\rangle_{0,m+2,d}, \end{aligned} \tag{9}$$

where $\tau = \tau_{0,2} + \tau'$ is the decomposition in (4) and $1/(z+\psi_{m+2})$ in the correlator should be expanded in the z^{-1}-series $\sum_{k\ge 0}(-1)^k z^{-k-1}\psi_{m+2}^k$. Set $\rho := c_1(\mathcal{X}) \in H^2(\mathcal{X}_0)$ and

$$z^{-\mu}z^{\rho} := \exp(-\mu\log z)\exp(\rho\log z), \quad \mu \text{ is given in (8).}$$

Then we have

$$\begin{aligned} &\nabla_i(L(\tau,z)z^{-\mu}z^{\rho}\phi) = 0, \quad \nabla_{z\partial_z}(L(\tau,z)z^{-\mu}z^{\rho}\phi) = 0, \\ &(L(\tau,-z)\phi_i, L(\tau,z)\phi_j)_{\mathrm{orb}} = (\phi_i,\phi_j)_{\mathrm{orb}}. \end{aligned}$$

In particular, $s_i(\tau,z) = L(\tau,z)z^{-\mu}z^{\rho}\phi_i$, $1 \le i \le N$, form a basis of multi-valued ∇-flat sections of F satisfying the asymptotic initial condition at the large radius limit (5):

$$s_i(\tau,z) \sim z^{-\mu}z^{\rho}e^{-\tau_{0,2}}\phi_i.$$

Remark 2.9. The convergence of the fundamental solution $L(\tau, z)$ is not a priori clear. Under the Assumption 2.1, however, we know that $L(\tau, z)$ also converges on $U \times \mathbb{C}^*$ because this is a solution to the linear partial differential equations $\nabla s = 0$.

Definition 2.10. Define $\mathcal{S}(\mathcal{X})$ to be the space of multi-valued ∇-flat sections of the quantum D-module $QDM(\mathcal{X}) = (F, \nabla, (\cdot, \cdot)_F)$:

$$\mathcal{S}(\mathcal{X}) := \{s(\tau, z) \in \Gamma(U \times \widetilde{\mathbb{C}^*}, \mathcal{O}(F)) \ ; \ \nabla s = 0\}.$$

This is a $\mathbb{C}$-vector space with $\dim_{\mathbb{C}} \mathcal{S}(\mathcal{X}) = \dim_{\mathbb{C}} H^*_{\mathrm{CR}}(\mathcal{X})$. $\mathcal{S}(\mathcal{X})$ is endowed with the pairing $(\cdot, \cdot)_{\mathcal{S}}$:

$$(s_1, s_2)_{\mathcal{S}} := (s_1(\tau, e^{\pi \mathtt{i}} z), s_2(\tau, z))_{\mathrm{orb}} \in \mathbb{C},$$

where $s_1(\tau, e^{\pi \mathtt{i}} z)$ denotes the parallel translate of $s_1(\tau, z)$ along the counterclockwise path $[0, 1] \ni \theta \mapsto e^{\mathtt{i}\pi\theta} z$. Because s_1, s_2 are flat sections, the right-hand side is a complex number independent of (τ, z). $\mathcal{S}(\mathcal{X})$ is also equipped with the automorphism $G^{\mathcal{S}}(\xi)$ for $\xi \in H^2(\mathcal{X}, \mathbb{Z})$ induced from the Galois action in Lemma 2.5:

$$G^{\mathcal{S}}(\xi) \colon \mathcal{S}(\mathcal{X}) \to \mathcal{S}(\mathcal{X}), \quad s(\tau, z) \mapsto dG(\xi)(s(G(\xi)^{-1}\tau, z)).$$

In general, $(\cdot, \cdot)_{\mathcal{S}}$ is neither symmetric nor anti-symmetric. When $\mathcal{X}$ is Calabi–Yau, *i.e.* $\rho = c_1(\mathcal{X}) = 0$, $(\cdot, \cdot)_{\mathcal{S}}$ is symmetric when $n = \dim_{\mathbb{C}} \mathcal{X}$ is even and is anti-symmetric when n is odd.

The fundamental solution in Proposition 2.8 gives the *cohomology framing* $\mathcal{Z}_{\mathrm{coh}}$ of $\mathcal{S}(\mathcal{X})$:

$$\mathcal{Z}_{\mathrm{coh}} \colon H^*_{\mathrm{CR}}(\mathcal{X}) \xrightarrow{\cong} \mathcal{S}(\mathcal{X}), \quad \phi \mapsto L(\tau, z) z^{-\mu} z^{\rho} \phi. \tag{10}$$

In terms of this cohomology framing $\mathcal{Z}_{\mathrm{coh}}$, it is easy to check that the pairing and Galois actions on $\mathcal{S}(\mathcal{X})$ can be written as follows:

$$\begin{aligned} &(\mathcal{Z}_{\mathrm{coh}}(\alpha), \mathcal{Z}_{\mathrm{coh}}(\beta))_{\mathcal{S}} = (e^{\pi \mathtt{i} \rho}\alpha, e^{\pi \mathtt{i} \mu}\beta)_{\mathrm{orb}} \\ &G^{\mathcal{S}}(\xi)(\mathcal{Z}_{\mathrm{coh}}(\alpha)) = \mathcal{Z}_{\mathrm{coh}}((\bigoplus_{v \in \mathsf{T}} e^{-2\pi \mathtt{i} \xi_0} e^{2\pi \mathtt{i} f_v(\xi)})\alpha) \end{aligned} \tag{11}$$

where we used the decomposition $H^*_{\mathrm{CR}}(\mathcal{X}) = \bigoplus_{v \in \mathsf{T}} H^*(\mathcal{X}_v)$ in the second line. (See the paragraph before Lemma 2.5 for $\xi_0 \in H^2(\mathcal{X}_0)$ and $f_v(\xi) \in [0, 1)$.)

2.4. K-theory integral lattice of flat sections

We will introduce an integral lattice in the space $\mathcal{S}(\mathcal{X})$ of flat sections using the K-group and the characteristic class called the $\widehat{\Gamma}$-class. Let $K(\mathcal{X})$ be the Grothendieck group of topological orbifold vector bundles over $\mathcal{X}$ (see *e.g.* [2] for orbifold vector bundles and orbifold K-theory). For simplicity, we assume that $\mathcal{X}$ is isomorphic to a quotient orbifold $[M/G]$ as a topological orbifold, where M is a compact manifold and G is a compact Lie group acting on M with at most finite stabilizers. Under this assumption, $K(\mathcal{X})$ is isomorphic to the G-equivariant K-theory $K_G^0(M)$ and is a finitely generated abelian group [2]. For an orbifold vector bundle V on $I\mathcal{X}$ and a component $\mathcal{X}_v$ of $I\mathcal{X}$, we denote the eigenbundle decomposition of $V|_{\mathcal{X}_v}$ with respect to the stabilizer action as follows:

$$V|_{\mathcal{X}_v} = \bigoplus_{0\le f<1} V_{v,f},$$

where the stabilizer of $\mathcal{X}_v$ acts on $V_{v,f}$ by $\exp(2\pi \mathtt{i} f)$. The Chern character $\widetilde{\mathrm{ch}}\colon K(\mathcal{X}) \to H^*(I\mathcal{X})$ for orbifold vector bundles is defined as follows:

$$\widetilde{\mathrm{ch}}(V) := \bigoplus_{v\in\mathsf{T}} \sum_{0\le f<1} e^{2\pi\mathtt{i}f}\,\mathrm{ch}((\mathrm{pr}^* V)_{v,f}),$$

where $\mathrm{pr}\colon I\mathcal{X} \to \mathcal{X}$ is the natural projection. For an orbifold vector bundle V on $\mathcal{X}$, let $\delta_{v,f,i}$, $i=1,\dots,l_{v,f}$ be the Chern roots of the vector bundle $(\mathrm{pr}^* V)_{v,f}$ on $\mathcal{X}_v$ (where $l_{v,f} = \mathrm{rank}(\mathrm{pr}^* V)_{v,f}$). The Todd class $\widetilde{\mathrm{Td}}(V)$ is defined by

$$\widetilde{\mathrm{Td}}(V) := \bigoplus_{v\in\mathsf{T}} \prod_{0<f<1,1\le i\le l_{v,f}} \frac{1}{1-e^{-2\pi\mathtt{i}f}e^{-\delta_{v,f,i}}} \prod_{f=0,1\le i\le l_{v,0}} \frac{\delta_{v,0,i}}{1-e^{-\delta_{v,0,i}}}.$$

When the orbifold vector bundle V admits the structure of a holomorphic orbifold vector bundle, the holomorphic Euler characteristic $\chi(V) := \sum_{i=1}^{n}(-1)^i \dim_{\mathbb{C}} H^i(\mathcal{X},V)$ is given by the Kawasaki–Riemann–Roch formula [52]:

$$\chi(V) = \int_{I\mathcal{X}} \widetilde{\mathrm{ch}}(V)\cup\widetilde{\mathrm{Td}}(T\mathcal{X}). \tag{12}$$

Note that $\chi(V)$ is an integer by definition. For a (not necessarily holomorphic) topological orbifold vector bundle V on $\mathcal{X}$, we *define* $\chi(V)$ to be the right-hand side of the above formula (12). It follows from Kawasaki's index theorem [53] for elliptic operators on orbifolds that $\chi(V)$ is an integer for any V. In fact, the right-hand side of (12) equals the index of an elliptic operator $\overline{\partial} + \overline{\partial}^* : V \otimes \mathcal{A}_{\mathcal{X}}^{0,\mathrm{even}} \to V \otimes \mathcal{A}_{\mathcal{X}}^{0,\mathrm{odd}}$, where $\overline{\partial}$ is a not necessarily integrable $(0,1)$-connection on V and $\overline{\partial}^*$ is its adjoint with respect to a hermitian metric on V.

Define a multiplicative characteristic class $\widehat{\Gamma}\colon K(\mathcal{X}) \to H^*(I\mathcal{X})$ as follows:

$$\widehat{\Gamma}(V) := \bigoplus_{v\in\mathsf{T}} \prod_{0\le f<1} \prod_{i=1}^{l_{v,f}} \Gamma(1 - f + \delta_{v,f,i}).$$

Here $\delta_{v,f,i}$ is the same as above. The Gamma function in the right-hand side should be expanded in Taylor series at $1 - f > 0$. The $\widehat{\Gamma}$-class can be viewed as a "square root" of the Todd class (more precisely, $\widehat{A}$-class). In fact, using the Gamma function equality $\Gamma(z)\Gamma(1-z) = \pi/\sin(\pi z)$, we find

$$\begin{aligned}
&[(e^{\pi\mathtt{i}\deg/2}\widehat{\Gamma}(V))\cup \mathrm{inv}^*\,\widehat{\Gamma}(V)]_v \cup e^{\pi\mathtt{i}(c_1(\mathrm{pr}^* V)|_{\mathcal{X}_v} + \mathrm{age}_v(V))} \\
&\qquad = (2\pi\mathtt{i})^{\sum_{f\neq 0} l_{v,f}} [(2\pi\mathtt{i})^{\deg/2}\widetilde{\mathrm{Td}}(V)]_v,
\end{aligned}$$

where $\deg\colon H^*(I\mathcal{X}) \to H^*(I\mathcal{X})$ is the ordinary grading operator defined by $\deg = p$ on $H^p(I\mathcal{X})$, $\mathrm{age}_v(V) = \sum_{0<f<1} f l_{v,f}$ is the age of V along $\mathcal{X}_v$, and $[\cdots]_v$ is the $H^*(\mathcal{X}_v)$-component. In this sense[4], the K-group framing $\mathcal{Z}_K\colon K(\mathcal{X}) \to \mathcal{S}(\mathcal{X})$ below can be considered as a "Mukai vector" in quantum cohomology.

Definition 2.11. We define the *K-group framing* $\mathcal{Z}_K\colon K(\mathcal{X}) \to \mathcal{S}(\mathcal{X})$ of the space $\mathcal{S}(\mathcal{X})$ of flat sections by the formula:

$$\begin{aligned}
&\mathcal{Z}_K(V) := \mathcal{Z}_{\mathrm{coh}}(\Psi(V)) = L(\tau, z) z^{-\mu} z^{\rho} \Psi(V), \\
&\text{where } \Psi(V) := (2\pi)^{-\frac{n}{2}} \widehat{\Gamma}(T\mathcal{X}) \cup (2\pi\mathtt{i})^{\deg/2}\, \mathrm{inv}^*\, \widetilde{\mathrm{ch}}(V).
\end{aligned} \tag{13}$$

Here $\mathcal{Z}_{\mathrm{coh}}$ is the cohomology framing (10), $L(\tau,z)z^{-\mu}z^{\rho}$ is the fundamental solution in Proposition 2.8, $(2\pi\mathtt{i})^{\deg/2} \in \mathrm{End}(H^*(I\mathcal{X}))$ is defined by $(2\pi\mathtt{i})^{\deg/2}|_{H^{2p}(I\mathcal{X})} = (2\pi\mathtt{i})^p$ and $\widehat{\Gamma}(T\mathcal{X})\cup$ is the cup product in $H^*(I\mathcal{X})$. The image $\mathcal{S}(\mathcal{X})_{\mathbb{Z}} := \mathcal{Z}_K(K(\mathcal{X})) \subset \mathcal{S}(\mathcal{X})$ of the K-group framing is called *the K-theory integral structure* on the quantum cohomology.

The notation $\mathcal{Z}_K$ for the K-group framing is motivated by the *central charge* in physics. Conjecturally, the integral

$$Z(V) := c(z) \int_{\mathcal{X}} \mathcal{Z}_K(V)(\tau, z) = c(z)(\mathbf{1}, \mathcal{Z}_K(V)(\tau,z))_{\mathrm{orb}} \tag{14}$$

[4] A Mukai vector is given by $\mathrm{ch}(V)\sqrt{\mathrm{Td}(TX)}$ for a vector bundle V on X.

with $c(z) = (2\pi z)^{\frac{n}{2}}/(2\pi\mathtt{i})^n$, $n = \dim \mathcal{X}$ gives the central charge of a B-type D-brane in the class V at the point τ of the (extended) Kähler moduli space. This plays a central role in stability conditions on the derived category $D^b_{\mathrm{coh}}(\mathcal{X})$ [30, 8]. It would be very interesting to find an intrinsic explanation for the formula (13) from this point of view. In the language of quantum D-modules, $Z(V)$ is a coefficient of the unit section $\mathbf{1}$ expressed in a ∇-flat frame.

Proposition 2.12 ([46, Proposition 2.10]). (i) *The image $\mathcal{S}(\mathcal{X})_{\mathbb{Z}}$ of the K-group framing $\mathcal{Z}_K$ is a lattice in $\mathcal{S}(\mathcal{X})$:*

$$\mathcal{S}(\mathcal{X})_{\mathbb{Z}} \otimes_{\mathbb{Z}} \mathbb{C} = \mathcal{S}(\mathcal{X}).$$

(ii) *The pairing $(\cdot,\cdot)_{\mathcal{S}}$ on $\mathcal{S}(\mathcal{X})$ corresponds to the Mukai pairing on $K(\mathcal{X})$ through the K-group framing $\mathcal{Z}_K$:*

$$(\mathcal{Z}_K(V_1), \mathcal{Z}_K(V_2))_{\mathcal{S}} = \chi(V_1 \otimes V_2^\vee).$$

Therefore, we have a $\mathbb{Z}$-valued pairing $\mathcal{S}(\mathcal{X})_{\mathbb{Z}} \times \mathcal{S}(\mathcal{X})_{\mathbb{Z}} \to \mathbb{Z}$.

(iii) *For $\xi \in H^2(\mathcal{X},\mathbb{Z})$, the Galois action $G^{\mathcal{S}}(\xi)$ on $\mathcal{S}(\mathcal{X})$ corresponds to the tensor by the orbifold line bundle $L_\xi^\vee$ (corresponding to $-\xi$) on $K(\mathcal{X})$:*

$$\mathcal{Z}_K(L_\xi^\vee \otimes V) = G^{\mathcal{S}}(\xi)(\mathcal{Z}_K(V)).$$

In particular, the lattice $\mathcal{S}(\mathcal{X})_{\mathbb{Z}}$ is invariant under the Galois action.

The statement (i) follows from the Adem–Ruan decomposition theorem [2, Theorem 5.1], which implies that $\widetilde{\mathrm{ch}}\colon K(\mathcal{X}) \to H^*(I\mathcal{X})$ is an isomorphism when tensored with $\mathbb{C}$. The statements (ii) and (iii) follow from straightforward calculations. It is somewhat surprising that many complicated terms finally give the Mukai pairing in (ii) via the Kawasaki–Riemann–Roch formula (12).

Remark 2.13. The formula (13) arose in [45, 46] from the study of mirror symmetry for toric orbifolds. The mirror Landau–Ginzburg model has the natural integral structure and we can shift it to the quantum cohomology. Katzarkov–Kontsevich–Pantev [49] also proposed essentially the same definition (for a rational structure) when $\mathcal{X}$ is a manifold. Closely related results have been observed in the context of mirror symmetry. Calculations and conjectures of Hosono [42], [43, Conjecture 6.3] are compatible with the integral structure above; the works of Horja [40, 41] and Borisov–Horja [7] strongly suggest a relation between K-group and quantum D-module.

Example 2.14. (i) $\mathcal{X} = \mathbb{P}^1$. Let $\omega \in H^2(\mathbb{P}^1, \mathbb{Z})$ be the integral Kähler class. We take $1, \omega$ as a basis of $H^*(\mathbb{P}^1)$. In terms of the cohomology framing $\mathcal{Z}_{\mathrm{coh}}\colon H^*(\mathbb{P}^1) \cong \mathcal{S}(\mathbb{P}^1)$ in (10), the Galois action and the pairing on $\mathcal{S}(\mathbb{P}^1)$ is represented by the matrices:

$$G^{\mathcal{S}}(\omega) = \begin{bmatrix} 1 & 0 \\ -2\pi\mathtt{i} & 1 \end{bmatrix}, \quad (\cdot,\cdot)_{\mathcal{S}} = \begin{bmatrix} 2\pi & \mathtt{i} \\ -\mathtt{i} & 0 \end{bmatrix}.$$

If an integral lattice L in $H^*(\mathbb{P}^1) \cong \mathcal{S}(\mathbb{P}^1)$ is invariant under $G^{\mathcal{S}}(\omega)$ and if the restriction of $(\cdot,\cdot)_{\mathcal{S}}$ to L gives a perfect pairing $L \times L \to \mathbb{Z}$, then L must take the following form:

$$L = \mathbb{Z}\sqrt{\frac{n}{2\pi}}(1 + c\omega) \oplus \mathbb{Z}\mathtt{i}\sqrt{\frac{2\pi}{n}}\omega$$

for some $n \in \mathbb{Z} \setminus \{0\}$ and $c \in \mathbb{C}$. The K-theory integral structure corresponds to the choice $n = 1$ and $c = -2\gamma$, where $\gamma = 0.57721...$ is Euler's constant (coming from the $\widehat{\Gamma}$-class $\widehat{\Gamma}(T\mathbb{P}^1) = 1 - 2\gamma\omega$).

(ii) When $\mathcal{X} = X$ is a Calabi–Yau threefold, the $\widehat{\Gamma}$ class is given by

$$\widehat{\Gamma}(TX) = 1 - \frac{\pi^2}{6}c_2(X) - \zeta(3)c_3(X)$$

where $\zeta(3)$ is the special value of Riemann's zeta function. From this, it follows that the central charges (14) of $\mathcal{O}_{\mathrm{pt}}$, $\mathcal{O}_C$, $\mathcal{O}_S$ and $\mathcal{O}$ (for any smooth curve C and surface S) restricted to $H^2(X)$ are

$$\begin{aligned}
Z(\mathcal{O}_{\mathrm{pt}}) &= 1, \\
Z(\mathcal{O}_C) &= ((1 - g(C)) - \frac{\tau}{2\pi\mathtt{i}} \cap [C], \\
Z(\mathcal{O}_S) &= \frac{[S]^3}{8} + \frac{\chi(S)}{24} + \frac{\tau}{2\pi\mathtt{i}} \cap \frac{[S]^2}{2} + \frac{d_{[S]}F_0(\tau)}{(2\pi\mathtt{i})^2}, \\
Z(\mathcal{O}) &= -\frac{\zeta(3)}{(2\pi\mathtt{i})^3}\chi(X) - \frac{\tau}{2\pi\mathtt{i}} \cdot \frac{c_2(X)}{24} + \frac{H(\tau)}{(2\pi\mathtt{i})^3},
\end{aligned}$$

where $\tau = \tau_{0,2} \in H^2(X)$, $g(C)$ is the genus of C, and $\chi(X)$ and $\chi(S)$ are the Euler numbers of X and S. $F_0(\tau)$ is the genus zero potential of X

$$F_0(\tau) := \frac{1}{6}\int_X \tau^3 + \sum_{d \in \mathrm{Eff}_X \setminus \{0\}} \langle\ \rangle_{0,0,d}\, e^{\langle \tau, d\rangle},$$

$d_{[S]}F_0$ is its derivative in the direction of the Poincaré dual of $[S]$ and $H(\tau) := 2F_0(\tau) - \sum_i t^i \partial_i F_0(\tau)$. The zeta value $\zeta(3)$ also appeared in the quintic mirror calculation of Candelas–de la Ossa–Green–Parkes [16].

(iii) When $\mathcal{X}$ is a weak Fano compact toric orbifold, it is shown in [45], [46, Theorem 4.13] that the central charge of the structure sheaf can be written as an oscillating integral of the mirror Landau–Ginzburg model $W_\tau \colon (\mathbb{C}^*)^n \to \mathbb{C}$:

$$Z(\mathcal{O}_{\mathcal{X}})(\tau, z) = \frac{1}{(2\pi\mathtt{i})^n} \int_{\Gamma_{\mathbb{R}} \subset (\mathbb{C}^*)^n} e^{-W_\tau(y)/z} \frac{dy}{y}, \quad n = \dim_{\mathbb{C}} \mathcal{X}.$$

Here dy/y is an invariant holomorphic n-form on $(\mathbb{C}^*)^n$ and $\Gamma_{\mathbb{R}}$ is a non-compact cycle (Lefschetz thimble) in $(\mathbb{C}^*)^n$. (Strictly speaking, we need a "mirror map" between $\tau \in H^2_{\mathrm{CR}}(\mathcal{X})$ in the left-hand side and the parameter τ in the Landau–Ginzburg potential W_τ.) Under mirror symmetry, the unit section $\mathbf{1}$ of the quantum D-module corresponds to the oscillating form $e^{-W_\tau/z}(dy/y)$ in the Landau–Ginzburg model. Thus the formula above shows that the structure sheaf $\mathcal{O}_{\mathcal{X}}$ corresponds to the Lefschetz thimble $\Gamma_{\mathbb{R}}$. Moreover, it turns out that the K-theory integral structure in Definition 2.11 corresponds to the lattice of Lefschetz thimbles under mirror symmetry. See [45, 46] for more details.

(iv) The $\widehat{\Gamma}$-class contains odd zeta values $\zeta(3), \zeta(5), \dots$ and products of Gamma values. When $\mathcal{X}$ is holomorphic symplectic, however, the $\widehat{\Gamma}$-class is defined over $\mathbb{Q}(\zeta)[\pi]$ for some root of unity ζ. This could be related to the fact that there is no quantum correction[5].

Remark 2.15. We can consider the Grothendieck group of algebraic vector bundles or coherent sheaves on $\mathcal{X}$ instead of topological K-groups. In this case, the K-theory integral structure is defined on the algebraic part of the orbifold cohomology $H^*_{\mathrm{CR}}(\mathcal{X})$, *i.e.* cohomology classes on $I\mathcal{X}$ which can be written as linear combinations of Poincaré duals of algebraic cycles with complex coefficients. The algebraic part of orbifold quantum cohomology makes sense due to the algebraic construction of orbifold Gromov–Witten theory [1]. A theoretical difficulty is that we do not know if the orbifold Poincaré pairing is non-degenerate when restricted to the algebraic part of $H^*_{\mathrm{CR}}(\mathcal{X})$: This would be a consequence of the famous Hodge conjecture/Grothendieck standard conjecture. Apart from this point, many discussions in this paper can be equally applied to *algebraic* K-theory integral structures.

2.5. Remark on non-compact case

Even when the space $\mathcal{X}$ is non-compact, we can sometimes define the (orbifold) quantum cohomology. Non-compact local cases are important in the study of Ruan's conjecture. One standard way is to use

[5]In toric examples [27, 23], the zeta-values appear in the analytic continuation of the J-function, a generating function of Gromov–Witten invariants.

the *torus-equivariant* Gromov–Witten theory. If $\mathcal{X}$ admits a torus action and the fixed point set is compact, we can define torus-equivariant orbifold Gromov–Witten invariants using the Atiyah–Bott style localization on the moduli space of stable maps [35]. In good cases, we can take the non-equivariant limit and obtain the non-equivariant quantum cohomology. In general, we can define Gromov–Witten invariants if the moduli space of stable maps to $\mathcal{X}$ is compact[6]. More generally, even when the moduli space may not be compact, if the evaluation map from the moduli space to the inertia stack $I\mathcal{X}$ is proper, we can define the quantum product by the push-forward by the evaluation map at the "last" marked point. As suggested in [14], this happens for example when X is semi-projective, *i.e.* projective over an affine scheme. In this section, assuming the existence of a well-defined orbifold quantum cohomology for a non-compact space, we describe a possible framework for K-theory integral structures in this case.

Assume that the (non-equivariant) quantum cohomology of $\mathcal{X}$ is well-defined. Quantum cohomology defines the Dubrovin connection and the quantum D-module in the same fashion as in Definition 2.3. The discrete Galois symmetry in Lemma 2.5 is also well-defined. A problem in the non-compact case is that the orbifold Poincaré pairing on $H^*_{\mathrm{CR}}(\mathcal{X})$ is degenerate. However, we have a non-degenerate pairing between $H^*_{\mathrm{CR}}(\mathcal{X})$ and the *compactly supported* orbifold cohomology $H^*_{\mathrm{CR,c}}(\mathcal{X})$, which is defined to be the direct sum of compactly supported cohomology groups of the inertia components $\mathcal{X}_v$ (with the same grading shift as before):

$$(\cdot,\cdot)_{\mathrm{orb}}\colon H^*_{\mathrm{CR,c}}(\mathcal{X}) \times H^*_{\mathrm{CR}}(\mathcal{X}) \to \mathbb{C}.$$

This pairing defines the *dual Dubrovin connection* on the $H^*_{\mathrm{CR,c}}(\mathcal{X})$-bundle $F_{\mathrm{c}} := H^*_{\mathrm{CR,c}}(\mathcal{X}) \times (U \times \mathbb{C}) \to U \times \mathbb{C}$:

$$\begin{aligned}
\nabla_i &= \frac{\partial}{\partial t^i} + \frac{1}{z}(\phi_i \circ_\tau)^\dagger, \\
\nabla_{z\partial_z} &= z\frac{\partial}{\partial z} - \frac{1}{z}(E\circ_\tau)^\dagger + \mu
\end{aligned}$$

where $(\phi_i\circ_\tau)^\dagger, (E\circ_\tau)^\dagger \in \mathrm{End}(H^*_{\mathrm{CR,c}}(\mathcal{X}))$ are the adjoint operators with respect to $(\cdot,\cdot)_{\mathrm{orb}}$. We call (F_{c}, ∇) the *compactly supported quantum*

[6]However, the degree zero moduli space always has a non-compact component, so we indeed need that the evaluation map is proper as stated. This is particularly relevant to the orbifold case where degree zero moduli spaces give a lot of non-trivial Gromov–Witten invariants.

D-module. Note that the dual product $(\phi_i \circ_\tau)^\dagger$ is defined by essentially the same formula as the original product. In fact, $(\alpha \circ_\tau \beta, \gamma)_{\rm orb} = (\alpha, (\beta \circ_\tau)^\dagger \gamma)_{\rm orb}$ may be defined by the right-hand side of (2) with $\alpha, \beta \in H^*_{\rm CR}(\mathcal{X})$ and $\gamma \in H^*_{\rm CR,c}(\mathcal{X})$ (under the assumption that the evaluation map is proper). Tautologically, one has a ∇-flat pairing:

$$(-)^*\mathcal{O}(F_{\rm c}) \otimes \mathcal{O}(F) \to \mathcal{O}_{U \times \mathbb{C}}$$

induced from the orbifold Poincaré pairing, where we recall that $(-): U \times \mathbb{C} \to U \times \mathbb{C}$ is the map sending (τ, z) to $(\tau, -z)$. One has a natural map

$$(F_{\rm c}, \nabla) \to (F, \nabla)$$

induced from $H^*_{\rm CR,c}(\mathcal{X}) \to H^*_{\rm CR}(\mathcal{X})$. The fundamental solution in Proposition 2.8 also makes sense. We have two fundamental solutions $\tilde{L}(\tau, z)$, $L(\tau, z)$ taking values in $\mathrm{End}(H^*_{\rm CR,c}(\mathcal{X}))$ and $\mathrm{End}(H^*_{\rm CR}(\mathcal{X}))$ respectively such that

$$\nabla(\tilde{L}(\tau, z) z^{-\mu} z^{\rho} \varphi) = 0, \quad \nabla(L(\tau, z) z^{-\mu} z^{\rho} \phi) = 0,$$
$$(\tilde{L}(\tau, -z)\varphi, L(\tau, z)\phi)_{\rm orb} = (\varphi, \phi)_{\rm orb},$$

where $\varphi \in H^*_{\rm CR,c}(\mathcal{X})$ and $\phi \in H^*_{\rm CR}(\mathcal{X})$. Here again, $\tilde{L}(\tau, z)$ and $L(\tau, z)$ can be defined by the same formula (9), with different domains of definitions[7]. The spaces $\mathcal{S}(\mathcal{X})$, $\mathcal{S}_{\rm c}(\mathcal{X})$ of multi-valued flat sections of F, $F_{\rm c}$ are defined in the same way as in Definition 2.10. The symmetries in Lemma 2.5 act on these spaces as automorphisms preserving the pairing:

$$(\cdot, \cdot)_{\mathcal{S}} \colon \mathcal{S}_{\rm c}(\mathcal{X}) \times \mathcal{S}(\mathcal{X}) \to \mathbb{C}, \quad (s_1, s_2) \mapsto (s_1(\tau, e^{\pi \mathtt{i}} z), s_2(\tau, z))_{\rm orb}.$$

Likewise, the formula (13) defines K-group framings

$$\mathcal{Z}_K \colon K(\mathcal{X}) \to \mathcal{S}(\mathcal{X}), \quad \mathcal{Z}_{K,\rm c} \colon K_{\rm c}(\mathcal{X}) \to \mathcal{S}_{\rm c}(\mathcal{X})$$

where $K_{\rm c}(\mathcal{X})$ is the *compactly supported* K-group. (We need to use $\tilde{L}(\tau, z)$ instead of $L(\tau, z)$ in (13) for the compact support version.) For example, when $\mathcal{X}$ is of the form M/G, one can define $K_{\rm c}(\mathcal{X})$ as the G-equivariant reduced K-group $\widetilde{K}^0_G(M^+)$ of the one-point compactification M^+ of M (as in [69]). One can also use the Grothendieck group $K_Z(\mathcal{X})$ of coherent sheaves on $\mathcal{X}$ supported on a compact set Z. In non-compact

[7] Here one of the dual pairs $\{\phi_k\}, \{\phi^k\}$ in (9) should be taken from $H^*_{\rm CR,c}(\mathcal{X})$ and the other from $H^*_{\rm CR}(\mathcal{X})$. We take $\phi^k \in H^*_{\rm CR,c}(\mathcal{X})$ when defining $L(\tau, z)$ and $\phi_k \in H^*_{\rm CR,c}(\mathcal{X})$ when defining $\tilde{L}(\tau, z)$.

case, the definition of $K(\mathcal{X})$ may be subject to change *e.g.* we may need to include perfect complexes or infinite dimensional bundles *cf.* [71]. We will not pursue a more precise formulation here. Note that we have a well-defined central charge $Z(V) := c(z)\int_{\mathcal{X}} \mathcal{Z}_{K,c}(V)$ for $V \in K_c(\mathcal{X})$.

Example 2.16 (*cf.* [45, Example 6.5], [46, Section 5.4]). (i) $\mathcal{X} = [\mathbb{C}^2/G]$ where G is a finite subgroup of $SL(2,\mathbb{C})$. The inertia stack $I\mathcal{X}$ is given by

$$I\mathcal{X} = \mathcal{X} \sqcup \bigsqcup_{(g)\neq 1} \mathcal{X}_{(g)}, \quad \mathcal{X}_{(g)} = [\{0\}/C(g)] \quad (g \neq 1),$$

where (g) is a conjugacy class of $g \in G$, and $C(g)$ is the centralizer of g in G. Let $\mathbf{1}$ be the unit class supported on $\mathcal{X}$ and $\mathbf{1}_{(g)} \in H^*_{\mathrm{CR}}(\mathcal{X})$ be the unit class supported on $\mathcal{X}_{(g)}$. The grading is given by

$$\deg \mathbf{1} = 0, \quad \deg \mathbf{1}_{(g)} = 2 \quad (g \neq 1).$$

Since $\mathcal{X}$ is holomorphic symplectic, there is no quantum deformation and $\circ_\tau$ is trivial: $\mathbf{1} \circ_\tau \mathbf{1}_{(g)} = \mathbf{1}_{(g)}$ and all other products are zero. (We can have non-trivial quantum cohomology by considering the equivariant version.) The $\widehat{\Gamma}$-class is given by

$$\widehat{\Gamma}(T\mathcal{X}) = \mathbf{1} \oplus \bigoplus_{(g)\neq(1)} \frac{\pi}{\sin(\pi f_g)} \mathbf{1}_{(g)} \in H^0(I\mathcal{X})$$

where $0 \le f_g \le 1/2$ is the rational number such that the eigenvalues of $g \in SL(2,\mathbb{C})$ are $\exp(\pm 2\pi\mathtt{i} f_g)$. Let $\beta, \mathbf{1}_{(g)}$ $(g \neq 1)$ be compactly supported cohomology classes on $\mathcal{X}$, $\mathcal{X}_{(g)}$ such that

$$(\beta, \mathbf{1})_{\mathrm{orb}} = \frac{1}{|G|}, \quad (\mathbf{1}_{(g)}, \mathbf{1}_{(g^{-1})})_{\mathrm{orb}} = \frac{1}{|C(g)|} \quad (g \neq 1).$$

Here $\deg \beta = 4$. We consider the Grothendieck group $K_0^G(\mathbb{C}^2)$ of G-equivariant coherent sheaves on $\mathbb{C}^2$ supported at the origin. A finite dimensional representation ϱ of G defines a G-equivariant sheaf $\mathcal{O}_0 \otimes \varrho$ on $\mathbb{C}^2$. These sheaves generate $K_0^G(\mathbb{C}^2)$ and the Galois action corresponds to the tensor product by a one-dimensional representation. By the equivariant Koszul resolution:

$$0 \to \mathcal{O}_{\mathbb{C}^2} \otimes \varrho \to \mathcal{O}_{\mathbb{C}^2} \otimes \varrho \otimes Q^\vee \to \mathcal{O}_{\mathbb{C}^2} \otimes \varrho \to \mathcal{O}_0 \otimes \varrho \to 0,$$

where $Q = \mathbb{C}^2$ is the standard G-representation defined by the inclusion $G \subset SL(2,\mathbb{C})$, we compute the Chern character as

$$\widetilde{\mathrm{ch}}(\mathcal{O}_0 \otimes \varrho) = (\dim \varrho)\beta \oplus \bigoplus_{(g)\neq(1)} \mathrm{Tr}(g|\varrho \otimes (\mathbb{C}^2 - Q))\, \mathbf{1}_{(g)} \in H^*_c(I\mathcal{X}).$$

Here $\mathrm{Tr}(g|\varrho \otimes (\mathbb{C}^2 - Q))$ is the trace of g on the virtual representation $\varrho \otimes (\mathbb{C}^2 - Q)$. Therefore, using $\tilde{L}(\tau, z) = \exp(-(\tau \circ_\tau)^\dagger / z)$, we find

$$(15) \qquad Z(\mathcal{O}_0 \otimes \varrho) = e^{-t^0/z} \left(\frac{\dim \varrho}{|G|} + \sum_{(g) \neq 1} \frac{\mathrm{Tr}(g|\varrho) \sin(\pi f_g)}{|C(g)| \pi} t^{(g)} \right),$$

where we put $\tau = t^0 \mathbf{1} + \sum_{(g)\neq 1} t^{(g)} \mathbf{1}_{(g)}$. The simplest central charge is given by the regular representation ϱ_{reg}:

$$Z(\mathcal{O}_0 \otimes \varrho_{\mathrm{reg}}) = e^{-t^0/z}.$$

The vector $[\mathcal{O}_0 \otimes \varrho_{\mathrm{reg}}] \in K_0^G(\mathbb{C}^2)$ is invariant under every Galois action.

(ii) $\mathcal{X} = \mathbb{C}^3/G$ where G is a finite subgroup of $SL(3, \mathbb{C})$. Unlike the previous case (i), $\mathcal{X}$ can have a non-trivial (non-equivariant) quantum cohomology. The inertia stack $I\mathcal{X}$ is given by

$$I\mathcal{X} = \mathcal{X} \sqcup \bigsqcup_{(g)\neq(1)} \mathcal{X}_{(g)}, \quad \mathcal{X}_{(g)} = [(\mathbb{C}^3)^g / C(g)],$$

where $(\mathbb{C}^3)^g \subset \mathbb{C}^3$ is the subspace fixed by g. The ordinary and compactly supported orbifold cohomology are

$$H^*_{\mathrm{CR}}(I\mathcal{X}) = \mathbb{C}\mathbf{1} \oplus \bigoplus_{(g)\neq 1} \mathbb{C}\mathbf{1}_{(g)},$$

$$H^*_{\mathrm{CR,c}}(I\mathcal{X}) = \mathbb{C}\alpha \oplus \bigoplus_{(g): n_g = 1} \mathbb{C}\beta_{(g)} \oplus \bigoplus_{(g): n_g = 0} \mathbb{C}\mathbf{1}_{(g)},$$

where $n_g = \dim \mathcal{X}_{(g)}$. Here $\mathbf{1}_{(g)}$ is the unit class supported on $\mathcal{X}_{(g)}$ and α, $\beta_{(g)}$ are top classes on $\mathcal{X}$, $\mathcal{X}_{(g)}$ respectively (with $n_g = 1$) such that

$$(\alpha, \mathbf{1})_{\mathrm{orb}} = \frac{1}{|G|}, \quad (\beta_{(g)}, \mathbf{1}_{(g^{-1})})_{\mathrm{orb}} = (\mathbf{1}_{(g)}, \mathbf{1}_{(g^{-1})})_{\mathrm{orb}} = \frac{1}{|C(g)|}.$$

Note that $\deg \mathbf{1}_{(g)} = 2\iota_{(g)}$, $\iota_{(g)} = 1$ if $n_g = 1$, $\deg \alpha = 6$ and $\deg \beta_{(g)} = 4$. When $n_g = 1$, let $0 < f_g \le 1/2$ be a rational number such that $1, e^{\pm 2\pi \mathtt{i} f_g}$ are the eigenvalues of $g \in SL(3, \mathbb{C})$. When $n_g = 0$, let $0 < f_{g,1} \le f_{g,2} \le f_{g,3} < 1$ be rational numbers such that $e^{2\pi \mathtt{i} f_{g,j}}$, $j = 1, 2, 3$, are the eigenvalues of g. Consider again the Grothendieck group $K_0^G(\mathbb{C}^3)$ of G-equivariant coherent sheaves supported at the origin. A finite dimensional representation ϱ of G gives a class $[\mathcal{O}_0 \otimes \varrho] \in K_0^G(\mathbb{C}^3)$. This yields a dual flat section $\mathcal{Z}_{K,\mathrm{c}}(\mathcal{O}_0 \otimes \varrho) = \tilde{L}(\tau, z) z^{-\mu} \Psi(\mathcal{O}_0 \otimes \varrho)$ with $\Psi(\mathcal{O}_0 \otimes \varrho)$ given by

$$(\dim \varrho)\alpha \oplus \bigoplus_{(g): n_g = 1} (-1) A^{\varrho}_{g^{-1}} \beta_{(g)} \oplus \bigoplus_{(g): n_g = 0} (-1)^{1+\iota_{(g)}} B^{\varrho}_{g^{-1}} \mathbf{1}_{(g)}.$$

Here

$$A_g^{\varrho} = \mathrm{Tr}(g|\varrho)\frac{\sin(\pi f_g)}{\pi}, \quad B_g^{\varrho} = \frac{\mathrm{Tr}(g|\varrho)}{\prod_{j=1}^{3}\Gamma(1-f_{g,j})}.$$

The corresponding central charge restricted to $H^2_{\mathrm{CR}}(\mathcal{X})$ is

$$(16)\quad Z(\mathcal{O}_0\otimes\varrho) = \frac{\dim\varrho}{|G|} + \sum_{(g):n_g=1}\frac{A_g^{\varrho}}{|C(g)|}t^{(g)} + \sum_{(g):n_g=0} B_g^{\varrho}F_{0,(g^{-1})}(\tau),$$

where $\tau = \sum_{\iota_{(g)}=1} t^{(g)}\,\mathbf{1}_{(g)} \in H^2_{\mathrm{CR}}(\mathcal{X})$ and

$$(17)\quad F_{0,(g^{-1})}(\tau) = \begin{cases} t^{(g)}/|C(g)|, & \iota_{(g)}=1,\\ \sum_{m\geq 2}\frac{1}{m!}\left\langle \mathbf{1}_{(g^{-1})},\tau,\dots,\tau\right\rangle_{0,m+1,0}, & \iota_{(g)}=2.\end{cases}$$

This follows from $Z(\mathcal{O}_0\otimes\varrho) = (\tilde{L}(\tau,z)^{\dagger}\,\mathbf{1}, z^{-\mu}\Psi(\mathcal{O}_0\otimes\varrho))_{\mathrm{orb}}$ and the formula for the *J-function* $J(\tau,-z) = \tilde{L}(\tau,z)^{\dagger}\,\mathbf{1}$:

$$J(\tau,-z) = \mathbf{1} - \frac{\tau}{z} + \sum_{\iota_{(g)}=2} F_{0,(g^{-1})}(\tau)|C(g)|\frac{\mathbf{1}_{(g)}}{z^2}.$$

Again the regular representation ρ_{reg} gives the simplest charge 1. The Γ-product $\prod_{j=1}^{3}\Gamma(1-f_{g,j})$ in the central charge may have something to do with the Chowla–Selberg formula [22].

§3. Ruan's conjecture

We incorporate our K-theory integral structure into the Ruan's conjecture [62, 63] and discuss what follows from this. We propose the conjecture that an isomorphism between K-theory induces an isomorphism of quantum D-modules via the K-group framing (13).

Ruan's conjecture can be discussed in many situations. It basically asserts that two birational spaces $\mathcal{X}_1$, $\mathcal{X}_2$ in a "crepant" relationship have isomorphic (orbifold) quantum cohomology under a suitable identification of quantum parameters. One such relationship is a *crepant resolution.* Let $\mathcal{X}$ be a Gorenstein orbifold without generic stabilizers, *i.e.* the automorphism group at every point x is contained in $SL(T_x\mathcal{X})$. Then the canonical line bundle $K_{\mathcal{X}}$ of $\mathcal{X}$ becomes the pull-back of K_X of the coarse moduli space X. A resolution of singularity $\pi\colon Y\to X$ is called *crepant* if $\pi^*K_X\cong K_Y$. We can regard Y and $\mathcal{X}$ as two different crepant resolutions of the same space X:

$$\mathcal{X} \longrightarrow X \longleftarrow Y.$$

In this case, Ruan's conjecture for a pair $(\mathcal{X}, Y)$ is called the *crepant resolution conjecture* and has been studied in many works [15, 59, 14, 27, 11, 5, 13, 23]. Ruan's conjecture have been discussed also for *flops*. Li–Ruan [55] showed that the quantum cohomology is invariant under flops between Calabi–Yau 3-folds. Recently, this was generalized to the case of simple $\mathbb{P}^r$-flops and Mukai flops [54] in any dimension. The case of certain singular flops between orbifolds are also studied in [19, 20].

More generally, Ruan's conjecture may hold for *K-equivalences*. We say that two smooth Deligne–Mumford stacks $\mathcal{X}_1$, $\mathcal{X}_2$ are K-equivalent if there exist a smooth Deligne–Mumford stack $\mathcal{X}$ and a diagram of projective birational morphisms

$$\mathcal{X}_1 \xleftarrow{p_1} \mathcal{X} \xrightarrow{p_2} \mathcal{X}_2 \tag{18}$$

such that $p_1^* K_{\mathcal{X}_1} \cong p_2^* K_{\mathcal{X}_2}$. The most general form of Ruan's conjecture would be the invariance of quantum cohomology under *D-equivalences*, *i.e.* the equivalence of derived categories of coherent sheaves. It is conjectured in [50] that K-equivalence is equivalent to D-equivalence for smooth birational varieties, but D-equivalence does not imply birational equivalence in general. An interesting example is reported [61, 44] where the Gromov–Witten theories of non-birational but D-equivalent Calabi–Yau 3-folds have the same mirror family and, in particular, should be equivalent.

One striking feature in Ruan's conjecture is that we need the *analyticity* of the quantum cohomology. In the crepant resolution conjecture, the orbifold quantum cohomology is identified with the *expansion* of the manifold quantum cohomology around a point where the quantum parameter $q = e^{\tau_{0,2}}$ is a root of unity. In the flop conjecture, the two quantum cohomology algebras are identified under the transformation $q \mapsto q^{-1}$, where q is the parameter of the exceptional curve.

3.1. A picture of the global quantum D-module

Let $\mathcal{X}_1, \mathcal{X}_2$ be a pair of smooth Deligne–Mumford stacks. For a complex analytic space $\mathcal{M}$, let $\pi\colon \mathcal{M} \times \mathbb{C} \to \mathcal{M}$ be the projection to the first factor, z be the co-ordinate on the $\mathbb{C}$ factor and $(-)\colon \mathcal{M} \times \mathbb{C} \to \mathcal{M} \times \mathbb{C}$ be the map sending (τ, z) to $(\tau, -z)$ as before.

Assumption 3.1 (Global quantum D-modules: See Figure 1). *There exists a* global quantum D-module $(F, \nabla, (\cdot,\cdot)_F, F_{\mathbb{Z}})$ *over a* global Kähler moduli space $\mathcal{M}$ *given by the following data:*

—A connected complex analytic space $\mathcal{M}$;

—A holomorphic vector bundle F of rank N over $\mathcal{M} \times \mathbb{C}$;

—*A meromorphic flat connection* ∇ *on* F *(with poles along* $z=0$*):*

$$\nabla\colon \mathcal{O}(F) \to \mathcal{O}(F)(\mathcal{M}\times\{0\}) \otimes_{\mathcal{O}_{\mathcal{M}\times\mathbb{C}}} (\pi^*\Omega^1_{\mathcal{M}} \oplus \mathcal{O}_{\mathcal{M}\times\mathbb{C}}\frac{dz}{z})$$

where $\mathcal{O}(F)(\mathcal{M}\times\{0\})$ *is the sheaf of meromorphic sections of* F *with poles of order less than or equal to one only along the divisor* $\mathcal{M}\times\{0\}$*;*

—*A non-degenerate,* ∇*-flat pairing* $(\cdot,\cdot)_F$*:*

$$(\cdot,\cdot)_F\colon (-)^*\mathcal{O}(F)\otimes\mathcal{O}(F)\to\mathcal{O}_{\mathcal{M}\times\mathbb{C}};$$

—*An integral local system (*$\mathbb{Z}^N$*-subbundle)* $F_{\mathbb{Z}}\to\mathcal{M}\times\mathbb{C}^*$ *underlying the flat vector bundle* $F|_{\mathcal{M}\times\mathbb{C}^*}$ *such that*

$$F_{\mathbb{Z}}\subset\operatorname{Ker}(\nabla),\quad F|_{\mathcal{M}\times\mathbb{C}^*}=F_{\mathbb{Z}}\otimes\mathbb{C},\quad ((-)^*F_{\mathbb{Z}},F_{\mathbb{Z}})_F\subset\mathbb{Z}.$$

We postulate that the 4-tuple $(F,\nabla,(\cdot,\cdot)_F,F_{\mathbb{Z}})$ *satisfies the following.*

(i) *There exist open subsets* $V_i\subset\mathcal{M}$, $i=1,2$, *such that* V_i *is identified with the base space of the quantum D-module* $QDM(\mathcal{X}_i)$*:*

$$V_i\cong U_i/H^2(\mathcal{X}_i,\mathbb{Z}),$$

and that the restriction of $(F,\nabla,(\cdot,\cdot)_F)$ *to* $V_i\times\mathbb{C}$ *is isomorphic to* $QDM(\mathcal{X}_i)$*:*

$$(F,\nabla,(\cdot,\cdot)_F)|_{V_i\times\mathbb{C}}\cong QDM(\mathcal{X}_i),\quad i=1,2.$$

Here $U_i\subset H^*_{\rm CR}(\mathcal{X}_i)$ *is the convergence domain of the quantum product in Assumption 2.1 and* $U_i/H^2(\mathcal{X},\mathbb{Z})$ *is the quotient by the Galois action. Moreover, this isomorphism matches the integral local system* $F_{\mathbb{Z}}$ *with the K-theory integral structure of* $QDM(\mathcal{X}_i)$ *in Definition 2.11.*

(ii) Assume that $\mathcal{X}_1$ *and* $\mathcal{X}_2$ *are K-equivalent (18) and also related by a birational correspondence*

$$\mathcal{X}_1 \xrightarrow{\pi_1} Z \xleftarrow{\pi_2} \mathcal{X}_2 \tag{19}$$

such that $\pi_1\circ p_1=\pi_2\circ p_2$*. Take base points* $x_i\in V_i$*. For a line bundle* L *on* Z*, denote by* $l_i(L)\in\pi_1(V_i,x_i)$ *the homotopy class of a loop given by the class* $[\pi_i^*(L)]\in H^2(\mathcal{X}_i,\mathbb{Z})$*. (Recall that* $V_i\cong U_i/H^2(\mathcal{X}_i,\mathbb{Z})$*.) There exists a path* $\gamma\colon[0,1]\to\mathcal{M}$ *from* $\gamma(0)=x_1$ *to* $\gamma(1)=x_2$ *such that* $\gamma_*(l_1(L))=l_2(L)$ *for any line bundle* L *on* Z*. Here* γ *is independent of* L*.*

As far as the author knows, all the concrete examples of global quantum D-modules arise from mirror symmetry. For example, in the case of toric flops or toric crepant resolutions (and complete intersections in them), we can construct a global quantum D-module using the mirror Landau–Ginzburg model and $\mathcal{M}$ is identified with the complex moduli space of the mirror [27, 24, 23]. The space of stability conditions on the derived category $D^b_{\mathrm{coh}}(\mathcal{X}_i)$ due to Douglas and Bridgeland [30, 8] gives a candidate for the universal cover of $\mathcal{M}$. Another candidate (though being only infinitesimal at present) is the space of A_∞-deformations of the derived Fukaya category of $\mathcal{X}_i$.

We assume the existence of a global quantum D-module F connecting $QDM(\mathcal{X}_1)$ and $QDM(\mathcal{X}_2)$. Choosing a path $\gamma\colon [0,1] \to \mathcal{M}$ from a point $x_1 \in V_1$ to a point $x_2 \in V_2$, we have an analytic continuation map P_γ of flat sections

$$P_\gamma\colon \mathcal{S}(\mathcal{X}_1) \to \mathcal{S}(\mathcal{X}_2) \tag{20}$$

along the path $\hat{\gamma} = (\gamma, 1)\colon [0,1] \to \mathcal{M} \times \mathbb{C}^*$. Here by (i), we identified the space of flat sections of F over $V_i \times \mathbb{C}^*$ with $\mathcal{S}(\mathcal{X}_i)$. This preserves the K-theory integral structures $P_\gamma(\mathcal{S}(\mathcal{X}_1)_\mathbb{Z}) = \mathcal{S}(\mathcal{X}_2)_\mathbb{Z}$ and the pairing $(\cdot,\cdot)_\mathcal{S}$. Then it would be natural to conjecture the following.

Conjecture 3.2. *For each path γ, there exists an isomorphism of K-groups*

$$\mathbb{U}_{K,\gamma}\colon K(\mathcal{X}_1) \to K(\mathcal{X}_2) \tag{21}$$

which induces the analytic continuation map P_γ in (20) through the K-group framing (13). $\mathbb{U}_{K,\gamma}$ preserves the Mukai pairing $\chi(\mathbb{U}_{K,\gamma}(V_1) \otimes \mathbb{U}_{K,\gamma}(V_2)^\vee) = \chi(V_1 \otimes V_2^\vee)$. Here $\mathbb{U}_{K,\gamma}$ gives the full relationships between $QDM(\mathcal{X}_1)$ and $QDM(\mathcal{X}_2)$, i.e. we know the quantum cohomology of $\mathcal{X}_2$ once we know the isomorphism $\mathbb{U}_{K,\gamma}$ and the complete information on the analytic continuation of the quantum cohomology of $\mathcal{X}_1$.

We expect that the K-group isomorphisms $\mathbb{U}_{K,\gamma}$ are given by geometric correspondences such as Fourier–Mukai transformations [9, 51]. This conjecture is compatible with Borisov–Horja's result [7], where they identified the K-group of toric Calabi–Yau orbifold with the space of solutions to the GKZ system and also identified the analytic continuation of GKZ solutions with the Fourier–Mukai transformations between K-groups. If the path γ is the same as what appeared in (ii) of Assumption 3.1, we also expect that $\mathbb{U}_{K,\gamma}$ commutes with the actions of line bundles pulled back from Z, *i.e.* $\mathbb{U}_{K,\gamma}(\pi_1^*(L) \otimes V) = \pi_2^*(L) \otimes \mathbb{U}_{K,\gamma}(V)$ for a line bundle L on Z. This is compatible with (ii) in Assumption 3.1 and the fact that tensoring by the $\pi_i^* L$ on $K(\mathcal{X}_i)$ corresponds to the monodromy (Galois) action on $\mathcal{S}(\mathcal{X}_i)$ along the loop $l_i(L)$.

Remark 3.3. (i) Unlike the original quantum D-module, the *global* quantum D-module F is not a priori trivialized in the standard way. This is an important point in this formulation. In fact, for the crepant resolution of $\mathbb{C}^3/\mathbb{Z}_3$ (or its compactification $\mathbb{P}(1,1,1,3)$), F has different trivializations over V_1 and V_2 [3, 27]. Here different trivializations correspond to different Frobenius/flat structures on the base $\mathcal{M}$.

(ii) The flat connection can have poles along $z=0$. For a local section s of F around $z=0$, $\nabla_X s$ has a pole of order ≤ 1 along $z=0$ for $X \in T\mathcal{M}$ and $\nabla_{\partial_z} s$ has a pole of order ≤ 2 along $z=0$.

(iii) The K-theory isomorphism (21) depends on the choice of a path γ. It would be very interesting to study the global monodromy of $(F, \nabla, (\cdot,\cdot)_F, F_{\mathbb{Z}})$.

Remark 3.4. In the context of Ruan's conjecture, the picture of the global quantum D-module has been proposed in [27], [28] in terms of the Givental formalism. An integral structure was incorporated in this picture in [45, 46]. The structure analogous to the global quantum D-module $(F, \nabla, (\cdot,\cdot)_F, F_{\mathbb{Z}})$ first emerged in singularity theory [66] and has been studied under various names: *Frobenius manifolds* [31]; *semi-infinite Hodge structures* [4]; *TE(R)P structures* [37, 38]; *twistor structures* [70, 65]; *non-commutative Hodge structures* [49] etc.

3.2. Family of algebras: isomorphism of F-manifolds

We explain that Assumption 3.1 implies the deformation equivalence of quantum cohomology. Choosing a local trivialization of F, we can write the connection operator ∇_X with $X \in T\mathcal{M}$ in the form

$$\nabla_X = X + \frac{1}{z}\mathcal{A}_X(\tau, z).$$

The residual part $\mathcal{A}_X(\tau,0) = [z\nabla_X]|_{z=0}$ defines a well-defined endomorphism of $F|_{\mathcal{M}\times\{0\}}$. The flatness of the connection ∇ implies the commutativity of these operators $[\mathcal{A}_X(\tau,0), \mathcal{A}_Y(\tau,0)] = 0$. Note that on $V_i \subset \mathcal{M}$, $\mathcal{A}_X(\tau,0)$ is identified with the quantum product $X\circ_\tau$. (Here we identify the tangent vector X with an element of $H^*_{\mathrm{CR}}(\mathcal{X}_i)$.) We say that (F,∇) is *miniversal* at a point $\tau \in \mathcal{M}$ if there exists a vector $v \in F_{(\tau,0)}$ such that the map

$$T_\tau\mathcal{M} \to F_{(\tau,0)}, \quad X \mapsto \mathcal{A}_X(\tau,0)v \tag{22}$$

is an isomorphism. This property clearly holds at $\tau \in V_i$ since we can choose v to be the unit $\mathbf{1} \in H^*_{\mathrm{CR}}(\mathcal{X})$. The miniversality may fail along a complex analytic subvariety of $\mathcal{M}$. In the sequel, by deleting such loci if necessary, we assume that (F, ∇) is miniversal everywhere on $\mathcal{M}$. Then we can define the product $\circ_\tau$ on the tangent space $T_\tau\mathcal{M}$ by the formula:

$$\mathcal{A}_{X\circ_\tau Y}(\tau,0)v = \mathcal{A}_X(\tau,0)(\mathcal{A}_Y(\tau,0)v),$$

where $v \in F_{(\tau,0)}$ is a vector which makes the map (22) an isomorphism. The unit vector $e \in T_\tau\mathcal{M}$ is defined by

$$\mathcal{A}_e(\tau,0)v = v.$$

Then $(T_\tau M, \circ_\tau, e)$ becomes an associative commutative ring by the commutativity of $\mathcal{A}_X(\tau,0)$. This definition does not depend on the choice of v. In fact, the inclusion

$$T_\tau\mathcal{M} \hookrightarrow \mathrm{End}(F_{(\tau,0)}), \quad X \mapsto \mathcal{A}_X(\tau,0)$$

becomes a homomorphism of rings. This product $\circ_\tau$ endows the base space $\mathcal{M}$ with the structure of an F-manifold [36]. Here an F-manifold is a complex manifold $\mathcal{M}$ endowed with an associative and commutative multiplication $T\mathcal{M}\otimes_{\mathcal{O}_\mathcal{M}} T\mathcal{M} \to T\mathcal{M}$ satisfying a certain axiom (see [36]). This is a weak version of the Frobenius manifold structure explained below in Section 3.4.

The F-manifold $\mathcal{M}$ here admits an *Euler vector field*. In a local frame of F, we can write the connection in the z-direction as

$$\nabla_{z\partial_z} = z\partial_z - \frac{1}{z}\mathcal{U}(\tau) + \mathcal{V}(\tau,z), \quad \mathcal{V}(\tau,z) \text{ is regular at } z=0. \tag{23}$$

The residual part $\mathcal{U}(\tau) = [z^2\nabla_{\partial_z}]|_{z=0}$ again defines a well-defined endomorphism of the bundle $F|_{\mathcal{M}\times\{0\}}$. The flatness of ∇ implies that the endomorphism $\mathcal{U}(\tau)$ commutes with $\mathcal{A}_X(\tau,0)$ for every $X \in T\mathcal{M}$. From this (and miniversality) it follows that there exists a unique vector field $E \in \Gamma(\mathcal{M}, T\mathcal{M})$ such that

$$\mathcal{U}(\tau) = \mathcal{A}_E(\tau,0).$$

This satisfies the axiom of the Euler vector field:

$$[E, X\circ_\tau Y] = [E,X]\circ_\tau Y + X\circ_\tau [E,Y] + X\circ_\tau Y. \tag{24}$$

Proposition 3.5. *Under Assumption 3.1, the quantum cohomologies of $\mathcal{X}_1$ and $\mathcal{X}_2$ underlie the same F-manifold $\mathcal{M}$ with the Euler vector field E.*

3.3. Semi-infinite variation of Hodge structures

The deformation equivalence explained in the previous section is a rather weak relationship. The global quantum D-module F has much more information than just a family of algebras. We consider the *semi-infinite variation of Hodge structures* or $\frac{\infty}{2}$VHS associated to F. This notion was introduced by Barannikov [4]. The information of $\frac{\infty}{2}$VHS is in fact equivalent to that of the meromorphic flat connection $(F, \nabla, (\cdot,\cdot)_F)$, but the analogy with the ordinary Hodge theory is clearer in this language.

We will work over the universal cover $\widetilde{\mathcal{M}}$ of $\mathcal{M}$. Let $\mathcal{H}$ be the space of flat sections of F over $\widetilde{\mathcal{M}} \times \mathbb{C}^*$:

$$\mathcal{H} := \{s \in \Gamma(\widetilde{\mathcal{M}} \times \mathbb{C}^*, \mathcal{O}(F)) \ ; \ \nabla_X s = 0, \ \forall X \in T\mathcal{M}\}.$$

Note that $s \in \mathcal{H}$ is flat only in the direction of $\mathcal{M}$ and not necessarily flat in the z direction. This is infinite dimensional over $\mathbb{C}$. For $\tau \in \widetilde{\mathcal{M}}$, every section $s(\tau, \cdot) \in \Gamma(\{\tau\} \times \mathbb{C}^*, F)$ can be uniquely extended to a section s over $\widetilde{\mathcal{M}} \times \mathbb{C}^*$ which is flat in the $\mathcal{M}$-direction. Therefore $\mathcal{H}$ is isomorphic to $\Gamma(\{\tau\} \times \mathbb{C}^*, F)$ and is a free $\mathcal{O}(\mathbb{C}^*)$-module of rank N, where $\mathcal{O}(\mathbb{C}^*)$ is the space of holomorphic functions on $\mathbb{C}^*$ and N is the rank of F. The pairing on $\mathcal{H}$ is defined by

$$(s_1, s_2)_{\mathcal{H}} := (s_1(\tau, -z), s_2(\tau, z))_F \in \mathcal{O}(\mathbb{C}^*).$$

Note that the right-hand side does not depend on τ since s_1, s_2 are flat in the $\mathcal{M}$-direction. This pairing satisfies $(s_2, s_1)_{\mathcal{H}} = (-)^*(s_1, s_2)_{\mathcal{H}}$. For $\tau \in \widetilde{\mathcal{M}}$, the space of sections of F over $\{\tau\} \times \mathbb{C}$ is naturally embedded into $\mathcal{H}$ (via the $\mathcal{M}$-flat extension as above)

$$\Gamma(\{\tau\} \times \mathbb{C}, F) \hookrightarrow \mathcal{H}.$$

We denote by $\mathbb{F}_\tau$ the image of this embedding. Recall that the image of $\Gamma(\{\tau\} \times \mathbb{C}^*, F)$ gives the whole space $\mathcal{H}$. $\mathbb{F}_\tau$ consists of flat sections $s \in \mathcal{H}$ such that $s(\tau, \cdot)$ is regular at $z = 0$. We call $\mathbb{F}_\tau$ the *semi-infinite Hodge structure.* $\mathbb{F}_\tau$ is a free $\mathcal{O}(\mathbb{C})$-submodule of $\mathcal{H}$ and can be regarded as a point on the Segal–Wilson Grassmannian [60] of $\mathcal{H}$ as follows: Fix an $\mathcal{O}(\mathbb{C}^*)$-basis $e_1, \dots, e_N$ of $\mathcal{H}$. An $\mathcal{O}(\mathbb{C})$-basis $s_1, \dots, s_N$ of $\mathbb{F}_\tau$ can be written as $s_j = \sum_{i=1}^N e_i c_{ij}(\tau, z)$. By restricting z to lie on S^1, the $N \times N$ matrix $(c_{ij}(\tau, z))$ defines an element of the loop group $LGL(N, \mathbb{C})$. A change of the basis s_j changes the matrix (c_{ij}) by the left multiplication by an element of the positive loop group $LGL^+(N, \mathbb{C})$ (whose entries are holomorphic functions on $\mathbb{C}$). Thus the subspace $\mathbb{F}_\tau$ is identified with an element $[(c_{ij}(\tau,z))]$ of $LGL(N,\mathbb{C})/LGL^{+}(N,\mathbb{C}) =: \mathrm{Gr}_{\frac{\infty}{2}}(\mathcal{H})$. We call the map

$$\widetilde{\mathcal{M}} \ni \tau \longmapsto \mathbb{F}_\tau \in \mathrm{Gr}_{\frac{\infty}{2}}(\mathcal{H})$$

the *semi-infinite period map.*

Proposition 3.6 ([27, Proposition 2.9]). *The semi-infinite period map $\tau \mapsto \mathbb{F}_\tau$ satisfies the following:*

$$\begin{aligned} &X\mathbb{F}_\tau \subset z^{-1}\mathbb{F}_\tau, \quad X \in T_\tau\mathcal{M},\\ &(\mathbb{F}_\tau, \mathbb{F}_\tau)_{\mathcal{H}} \subset \mathcal{O}(\mathbb{C}),\\ &(\nabla_{z\partial_z} + E)\mathbb{F}_\tau \subset \mathbb{F}_\tau, \end{aligned}$$

where we used the fact that $\nabla_{z\partial_z}$ acts on $\mathcal{H}$ as a $\mathbb{C}$-endomorphism. The first property is an analogue of Griffiths transversality and the second is the Hodge–Riemann bilinear relation.

3.4. Opposite subspace and Frobenius manifolds

As we remarked, the global quantum D-module is not a priori trivialized. A good trivialization is given by the choice of an *opposite subspace* to the $\frac{\infty}{2}$VHS. The choice of an opposite subspace and a dilaton shift defines a Frobenius structure on the universal cover of $\mathcal{M}$. The Frobenius/flat structure was discovered by K. Saito [66] as a structure on a miniversal deformation of isolated hypersurface singularities and the use of opposite subspaces goes back to M. Saito's work [67] in that context. Let $\mathcal{O}(\mathbb{P}^1 \setminus \{0\})$ be the space of holomorphic functions on $\mathbb{P}^1 \setminus \{0\}$. This is contained in $\mathcal{O}(\mathbb{C}^*)$.

Definition 3.7. An *opposite subspace* $\mathcal{H}_-$ at $\tau \in \widetilde{\mathcal{M}}$ is a free $\mathcal{O}(\mathbb{P}^1 \setminus \{0\})$-submodule of $\mathcal{H}$ such that the natural map

$$\mathcal{H}_- \oplus \mathbb{F}_\tau \to \mathcal{H} \tag{25}$$

is an isomorphism. $\mathcal{H}_-$ is said to be *homogeneous* if

$$\nabla_{z\partial_z}\mathcal{H}_- \subset \mathcal{H}_-$$

and *isotropic* if

$$(\mathcal{H}_-, \mathcal{H}_-)_{\mathcal{H}} \subset z^{-2}\mathcal{O}(\mathbb{P}^1 \setminus \{0\}).$$

In terms of the loop Grassmannian $LGL(N,\mathbb{C})/LGL^{+}(N,\mathbb{C})$, $\mathcal{H}_-$ is opposite at τ if $\mathbb{F}_\tau$ lies on the "big cell": an open orbit of $LGL^{-}(N,\mathbb{C})$. Therefore, the opposite property ((25) is an isomorphism) is an open condition: If $\mathcal{H}_-$ is opposite at τ, then it is opposite in a neighborhood of

τ. Given an opposite subspace $\mathcal{H}_-$ at some point, the opposite property may fail along a complex analytic subvariety of $\widetilde{\mathcal{M}}$.

We explain in the lemma below that a homogeneous opposite subspace corresponds to an extension of (F, ∇) across $z = \infty$ such that the connection ∇ has a logarithmic singularity along $z = \infty$.

Lemma 3.8. *For a point $\tau \in \widetilde{\mathcal{M}}$, the following are equivalent:*
(i) *$\mathcal{H}_-$ is a homogeneous opposite subspace at τ.*
(ii) *$\mathcal{H}_-$ is homogeneous and at least one of the following projections*

$$z\mathcal{H}_-/\mathcal{H}_- \longleftarrow z\mathcal{H}_- \cap \mathbb{F}_\tau \longrightarrow \mathbb{F}_\tau/z\mathbb{F}_\tau$$

is an isomorphism of finite dimensional $\mathbb{C}$-vector spaces.

(iii) *Define an extension $\widehat{F}_\tau \to \{\tau\} \times \mathbb{P}^1$ of the vector bundle $F|_{\{\tau\}\times\mathbb{C}}$ to $\{\tau\} \times \mathbb{P}^1$ as follows: A section $s \in \Gamma(\{\tau\} \times \mathbb{C}^*, F)$ extends to a regular section of $\widehat{F}_\tau$ over $\{\tau\} \times (\mathbb{P}^1 \setminus \{0\})$ if the image of s in $\mathcal{H}$ lies in $z\mathcal{H}_-$. The extension $(\widehat{F}_\tau, \nabla)$ is a trivial vector bundle over $\mathbb{P}^1$ and ∇ has a logarithmic singularity at $z = \infty$.*

Proof. (i) $\Rightarrow$ (ii). The injectivity of the maps in (ii) is obvious. For $[v] \in z\mathcal{H}_-/\mathcal{H}_-$ with $v \in z\mathcal{H}_-$, write $v = v_0 + v_-$ where $v_0 \in \mathbb{F}_\tau$ and $v_0 \in \mathcal{H}_-$. Then $v_0 = v - v_- \in z\mathcal{H}_- \cap \mathbb{F}_\tau$ and $[v] = [v_0]$. This shows the surjectivity of $z\mathcal{H}_- \cap \mathbb{F}_\tau \to z\mathcal{H}_-/\mathcal{H}_-$. For $[v] \in \mathbb{F}_\tau/z\mathbb{F}_\tau$ with $v \in \mathbb{F}_\tau$, write $z^{-1}v = v_- + v_0$, where $v_- \in \mathcal{H}_-$ and $v_0 \in \mathbb{F}_\tau$. Then $zv_- = v - zv_0 \in \mathbb{F}_\tau \cap z\mathcal{H}_-$ and $[v] = [zv_-]$. This shows the surjectivity of $z\mathcal{H}_- \cap \mathbb{F}_\tau \to \mathbb{F}_\tau/z\mathbb{F}_\tau$.

(ii) $\Rightarrow$ (iii). Consider the extension $\widehat{F}_\tau \to \{\tau\} \times \mathbb{P}^1$ in (iii). We can identify $z\mathcal{H}_-/\mathcal{H}_-$ with the fiber $\widehat{F}_{(\tau,\infty)}$, $z\mathcal{H}_-\cap\mathbb{F}_\tau$ with the space of global sections $\Gamma(\mathbb{P}^1, \widehat{F}_\tau)$ and $\mathbb{F}_\tau/z\mathbb{F}_\tau$ with the fiber $\widehat{F}_{(\tau,0)}$. Since the maps in (ii) are induced from the restrictions, that one of them is an isomorphism implies that $\widehat{F}_\tau$ is a trivial holomorphic vector bundle. For a local coordinate $w = z^{-1}$ around $z = \infty$, we have $\nabla_{w\partial_w} = -\nabla_{z\partial_z}$. Hence the homogeneity implies $\nabla_{w\partial_w}(z\mathcal{H}_-) \subset (z\mathcal{H}_-)$, so ∇ has a logarithmic singularity at $w = 0$.

(iii) $\Rightarrow$ (i). Note that $\mathcal{H}$ is identified with the space of sections of $\widehat{F}_\tau$ over $\{\tau\} \times \mathbb{C}^*$. Because $\widehat{F}_\tau$ is trivial, that (25) is an isomorphism follows from the decomposition

$$\mathcal{O}(\mathbb{C}^*) = z^{-1}\mathcal{O}(\mathbb{P}^1 \setminus \{0\}) \oplus \mathcal{O}(\mathbb{C}).$$

The logarithmic singularity of ∇ implies the homogeneity of $\mathcal{H}_-$. Q.E.D.

By the isomorphism in (ii) of Lemma 3.8, a homogeneous opposite subspace $\mathcal{H}_-$ gives a local trivialization of F. In fact, since $F|_{\{\tau\}\times\mathbb{C}}$

extends to a trivial vector bundle $\widehat{F}_\tau$ over $\{\tau\} \times \mathbb{P}^1$, we have

$$F_{(\tau,z)} \cong \Gamma(\{\tau\} \times \mathbb{P}^1, \widehat{F}_\tau) \cong z\mathcal{H}_- \cap \mathbb{F}_\tau \cong z\mathcal{H}_-/\mathcal{H}_-. \tag{26}$$

The finite dimensional vector space $z\mathcal{H}_-/\mathcal{H}_-$ does not depend on τ, so this defines a trivialization of F over an open subset of $\widetilde{\mathcal{M}}$. Under this trivialization, the flat connection ∇ can be written as follows (see *e.g.* [27, Proposition 2.11])

$$\begin{aligned} \nabla_X &= X + \frac{1}{z}\mathcal{A}_X(\tau), \quad X \in T\mathcal{M}, \\ \nabla_{z\partial_z} &= z\partial_z - \frac{1}{z}\mathcal{U}(\tau) + \mathcal{V}, \end{aligned} \tag{27}$$

where $\mathcal{A}(\tau)$ is an $\mathrm{End}(z\mathcal{H}_-/\mathcal{H}_-)$-valued 1-form, $\mathcal{U}(\tau)$ is an $\mathrm{End}(z\mathcal{H}_-/\mathcal{H}_-)$-valued function, and $\mathcal{V}$ is a constant operator in $\mathrm{End}(z\mathcal{H}_-/\mathcal{H}_-)$. Here $\mathcal{A}(\tau), \mathcal{U}(\tau)$ are independent of z and defined on an open subset of $\widetilde{\mathcal{M}}$. Note that $\mathcal{U}(\tau) = \mathcal{A}_E(\tau)$ by the definition of the Euler vector field E.

In order to have a Frobenius structure on $\widetilde{\mathcal{M}}$, in addition to $\mathcal{H}_-$, we need to choose[8] an eigenvector $v_0 \in z\mathcal{H}_-/\mathcal{H}_-$ of $\mathcal{V}$ satisfying the miniversality condition:

$$T_\tau\widetilde{\mathcal{M}} \to z\mathcal{H}_-/\mathcal{H}_-, \quad X \mapsto \mathcal{A}_X(\tau)v_0 \quad \text{is an isomorphism.} \tag{28}$$

We call v_0 a *dilaton shift*. The isomorphism $T_\tau\widetilde{\mathcal{M}} \cong z\mathcal{H}_-/\mathcal{H}_-$ above defines an affine flat structure[9] on $\widetilde{\mathcal{M}}$. A vector field X is defined to be flat if $\mathcal{A}_X(\tau)v_0$ is a constant element in $z\mathcal{H}_-/\mathcal{H}_-$. A flat co-ordinate system on $\widetilde{\mathcal{M}}$ is constructed as follows. Let $\hat{v}_0 + \psi(\tau)$ be the unique intersection point of $\mathbb{F}_\tau$ and the affine subspace $\hat{v}_0 + \mathcal{H}_-$, where $\hat{v}_0 \in z\mathcal{H}_-$ is an (arbitrarily fixed) lift of v_0 and $\psi(\tau) \in \mathcal{H}_-$. See Figure 2. Then the map

$$\widetilde{\mathcal{M}} \ni \tau \mapsto [\psi(\tau)] \in \mathcal{H}_-/z^{-1}\mathcal{H}_-$$

is a local isomorphism and gives a flat co-ordinate system. In fact, the differential of this map is identified with (28). Varying τ, the intersection point $\hat{v}_0 + \psi(\tau) \in \mathbb{F}_\tau$ gives a section s_0 of F which corresponds to $v_0 \in z\mathcal{H}_-/\mathcal{H}_-$ in the trivialization (26). (Note that $\hat{v}_0 + \psi(\tau) \in z\mathcal{H}_- \cap \mathbb{F}_\tau$.) This section s_0 is called a *primitive section.* In Gromov–Witten theory, the corresponding vector $\hat{v}_0 + \psi(\tau) \in \mathcal{H}$ is called the *J-function.*

[8]The author does not claim that v_0 always exists. The action of $\mathcal{V}$ on $z\mathcal{H}_-/\mathcal{H}_-$ is induced from that of $\nabla_{z\partial_z}$ on $z\mathcal{H}_-$.

[9]An affine flat manifold is a manifold with a torsion free flat connection on the tangent bundle.

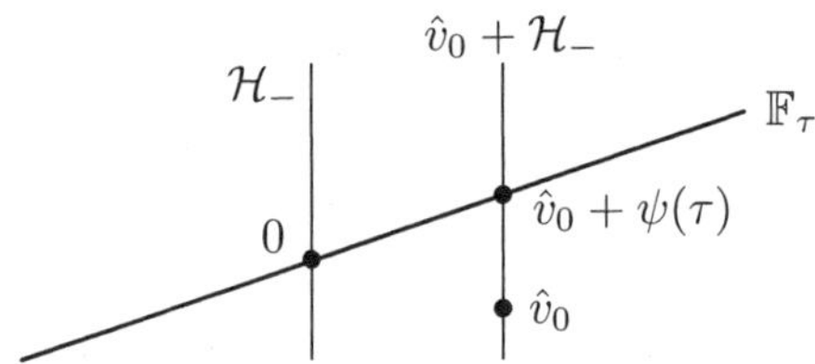

Fig. 2. J-function $\hat{v}_0 + \psi(\tau)$ and flat co-ordinates $[\psi(\tau)] \in \mathcal{H}_-/z^{-1}\mathcal{H}_-$.

For a flat vector field X, we have $\mathcal{V}(\mathcal{A}_X v_0) = \mathcal{A}_{(\alpha+1)X-[X,E]}v_0$ where α is the eigenvalue of v_0 with respect to $\mathcal{V}$.

When $\mathcal{H}_-$ is isotropic, the pairing $(\cdot,\cdot)_\mathcal{H}$ on $\mathcal{H}$ restricts to a symmetric bilinear $\mathbb{C}$-valued pairing on $z\mathcal{H}_- \cap \mathbb{F}_\tau \cong z\mathcal{H}_-/\mathcal{H}_-$. By pulling back this pairing on $z\mathcal{H}_-/\mathcal{H}_-$ to $T_\tau\widetilde{\mathcal{M}}$ by the map (28), we obtain a $\mathbb{C}$-bilinear metric $g\colon T_\tau\widetilde{\mathcal{M}} \times T_\tau\widetilde{\mathcal{M}} \to \mathbb{C}$. The metric tensor of g is constant in the flat co-ordinates above, so the metric g is flat.

Proposition 3.9 ([27, Proposition 2.12]). *Take an isotropic homogeneous opposite subspace $\mathcal{H}_-$ and a dilaton shift $v_0 \in z\mathcal{H}_-/\mathcal{H}_-$ satisfying (28) at some point τ. Then the F-manifold structure $(\circ_\tau, e, E)$ in Proposition 3.5 can be lifted to the Frobenius manifold structure $(\circ_\tau, e, E, g)$ on the complement of a complex analytic subvariety in $\widetilde{\mathcal{M}}$. These data satisfy:*

(i) *the Levi-Civita connection ∇^{LC} of g is flat;*
(ii) *$(T_\tau\mathcal{M}, \circ_\tau, g)$ is a commutative Frobenius algebra;*
(iii) *the pencil of flat connections $\nabla^\lambda_X = \nabla^{\mathrm{LC}}_X + \lambda X \circ_\tau$ is flat;*
(iv) *the unit vector field e is flat;*
(v) *the Euler vector field E satisfies (24), $(\nabla^{\mathrm{LC}})^2 E = 0$ and*

$$Eg(X,Y) = g([E,X],Y) + g(X,[E,Y]) + 2(\alpha+1)g(X,Y),$$

where $\alpha \in \mathbb{C}$ is the eigenvalue of v_0: $\mathcal{V}v_0 = \alpha v_0$.

3.5. Opposite subspaces at cusps

We call the large radius limit point of $\mathcal{X}_i$ a *cusp* of the global Kähler moduli space $\mathcal{M}$ and V_i a neighborhood of a cusp. Since the base space of $QDM(\mathcal{X}_i)$ is a quotient of a vector space, V_i is equipped with the standard Frobenius/flat structure as described in [57, 31]. We will show that, under certain conditions, the Frobenius structure (or the corresponding opposite subspace) of V_i can be uniquely characterized by the monodromy invariance and the compatibility with the Deligne extension. This means that *there is a canonical choice of the Frobenius manifold structure at each cusp from a purely D-module theoretic viewpoint.* The characterization here was shown in the case $\mathcal{X} = \mathbb{P}(1,1,1,3)$ in [27].

Henceforth we study the global quantum D-module restricted to V_i *i.e.* $QDM(\mathcal{X}_i)$. We omit the subscript i and write $V, \mathcal{X}$ for $V_i, \mathcal{X}_i$ etc. The open set $U \subset H^*_{\mathrm{CR}}(\mathcal{X})$ in Assumption 2.1 is identified with the universal cover of $V \cong U/H^2(\mathcal{X},\mathbb{Z})$.

Definition 3.10 (Givental space [26, 33])**.** The *Givental symplectic space* $\mathcal{H}^{\mathcal{X}}$ is defined to be the free $\mathcal{O}(\mathbb{C}^*)$-module

$$\mathcal{H}^{\mathcal{X}} := H^*_{\mathrm{CR}}(\mathcal{X}) \otimes \mathcal{O}(\mathbb{C}^*),$$

endowed with an $\mathcal{O}(\mathbb{C}^*)$-valued pairing $(\cdot,\cdot)_{\mathcal{H}}$:

$$(f(z), g(z))_{\mathcal{H}} = (f(-z), g(z))_{\mathrm{orb}}.$$

As an infinite dimensional vector space over $\mathbb{C}$, $\mathcal{H}^{\mathcal{X}}$ has the following symplectic form:

$$\Omega(f,g) = \mathrm{Res}_{z=0}(f(-z), g(z))_{\mathrm{orb}} dz. \tag{29}$$

We identify the Givental space $\mathcal{H}^{\mathcal{X}}$ with the space $\mathcal{H}$ of flat sections of $QDM(\mathcal{X})$ over U through the fundamental solution in Proposition 2.8:

$$\mathcal{H}^{\mathcal{X}} \cong \mathcal{H}, \quad \phi(z) \mapsto L(\tau,z)\phi(z).$$

This identification preserves the pairing.

In terms of the Givental space, the semi-infinite Hodge structure $\mathbb{F}_\tau$ is identified with the Lagrangian subspace:

$$\mathbb{F}_\tau = L(\tau,z)^{-1}(H^*_{\mathrm{CR}}(\mathcal{X}) \otimes \mathcal{O}(\mathbb{C})) \subset \mathcal{H}^{\mathcal{X}}, \quad \tau \in U. \tag{30}$$

The Givental space has a standard opposite subspace $\mathcal{H}^{\mathcal{X}}_-$:

$$\mathcal{H}^{\mathcal{X}}_- := z^{-1} H^*_{\mathrm{CR}}(\mathcal{X}) \otimes \mathcal{O}(\mathbb{P}^1 \setminus \{0\}) \subset \mathcal{H}^{\mathcal{X}}.$$

In fact, this is opposite to $\mathbb{F}_\tau$ (*i.e.* $\mathcal{H}^{\mathcal{X}}_- \oplus \mathbb{F}_\tau = \mathcal{H}^{\mathcal{X}}$) for every $\tau \in U$ because $L(\tau,z)$ is regular at $z=\infty$ and $L(\tau,z) = \mathrm{id} + O(z^{-1})$.

Proposition 3.11. *The standard opposite subspace $\mathcal{H}^{\mathcal{X}}_-$ is homogeneous and isotropic. This $\mathcal{H}^{\mathcal{X}}_-$ and the standard dilaton shift $v_0 = \mathbf{1} \in z\mathcal{H}^{\mathcal{X}}_-/\mathcal{H}^{\mathcal{X}}_-$ endow the base space $V \cong U/H^2(\mathcal{X},\mathbb{Z})$ of the quantum D-module with the standard Frobenius manifold structure coming from the linear structure on $U \subset H^*_{\mathrm{CR}}(\mathcal{X})$ and the orbifold Poincaré pairing on $T_\tau U \cong H^*_{\mathrm{CR}}(\mathcal{X})$. (See Proposition 3.9 for the construction of Frobenius manifolds.)*

Proof. It follows from Proposition 2.8 that $L(\tau, z)$ satisfies the differential equation $\nabla_{z\partial_z} L(\tau, z)\phi = L(\tau, z)(\mu - \rho/z)\phi$ for $\phi \in H^*_{\mathrm{CR}}(\mathcal{X})$. This shows that the action of $\nabla_{z\partial_z}$ on the Givental space is given by

$$\nabla_{z\partial_z} = z\partial_z + \mu - \frac{\rho}{z} \quad \text{on } \mathcal{H}^{\mathcal{X}}. \tag{31}$$

Therefore the standard opposite subspace is homogeneous $\nabla_{z\partial_z}\mathcal{H}^{\mathcal{X}}_- \subset \mathcal{H}^{\mathcal{X}}_-$. It is obvious that $\mathcal{H}^{\mathcal{X}}_-$ is isotropic. Because $L(\tau, z)^{-1}\phi = \phi + O(z^{-1})$ for $\phi \in H^*_{\mathrm{CR}}(\mathcal{X})$, we have $L(\tau, z)^{-1}\phi \in z\mathcal{H}^{\mathcal{X}}_- \cap \mathbb{F}_\tau$. Therefore, the constant section ϕ of $QDM(\mathcal{X})$ corresponds to (again) the constant element $\phi \in z\mathcal{H}^{\mathcal{X}}_-/\mathcal{H}^{\mathcal{X}}_-$ under the trivialization (26). This means that $\mathcal{H}^{\mathcal{X}}_-$ yields exactly the *given* trivialization of $QDM(\mathcal{X})$. In particular, the connection operators $\mathcal{A}_X$, $\mathcal{U}$, $\mathcal{V}$ in (27) are identified with $X\circ_\tau$, $E\circ_\tau$, μ and $\mathbf{1} \in H^*_{\mathrm{CR}}(\mathcal{X})$ is the eigenvector of $\mathcal{V} = \mu$ of eigenvalue $-\dim_{\mathbb{C}} \mathcal{X}/2$. Now we only need to check that the corresponding flat metric g is the orbifold Poincaré pairing. But this is obvious from $(L(\tau, -z)^{-1}\phi_1, L(\tau, z)^{-1}\phi_2)_{\mathrm{orb}} = (\phi_1, \phi_2)_{\mathrm{orb}}$. Q.E.D.

Here we describe the two characteristic properties of $\mathcal{H}^{\mathcal{X}}_-$: the *monodromy invariance* and the *compatibility with the Deligne extension.*

The monodromy invariance of $\mathcal{H}^{\mathcal{X}}_-$ —We see that $\mathcal{H}^{\mathcal{X}}_-$ is invariant under the local monodromy (or Galois actions) around the large radius limit. The Galois action in Lemma 2.5 acts on the Givental space $\mathcal{H}^{\mathcal{X}}$ by $G^{\mathcal{H}}(\xi)$:

$$G^{\mathcal{H}}(\xi) = \bigoplus_{v \in \mathsf{T}} e^{-2\pi \mathtt{i} \xi_0/z} e^{2\pi \mathtt{i} f_v(\xi)}, \quad \xi \in H^2(\mathcal{X}, \mathbb{Z}),$$

where we used the decomposition $\mathcal{H}^{\mathcal{X}} = \bigoplus_{v \in \mathsf{T}} H^*(\mathcal{X}_v) \otimes \mathcal{O}(\mathbb{C}^*)$. Since $G^{\mathcal{H}}(\xi)$ contains only negative powers in z, we have

$$G^{\mathcal{H}}(\xi)\mathcal{H}^{\mathcal{X}}_- \subset \mathcal{H}^{\mathcal{X}}_-. \tag{32}$$

The semi-infinite Hodge structures are monodromy-equivariant: $G^{\mathcal{H}}(\xi)\mathbb{F}_\tau = \mathbb{F}_{G(\xi)\tau}$. The monodromy-invariance of $\mathcal{H}^{\mathcal{X}}_-$ corresponds to the fact that the corresponding Frobenius manifold structure is well-defined[10] on the quotient $V \cong U/H^2(\mathcal{X}, \mathbb{Z})$. The induced action of $G^{\mathcal{H}}(\xi)$ on $z\mathcal{H}^{\mathcal{X}}_-/\mathcal{H}^{\mathcal{X}}_-$ is given by $\bigoplus_{v \in \mathsf{T}} e^{2\pi \mathtt{i} f_v(\xi)}$. Because $f_v(\xi)$ is a rational number, there exists a positive integer $k_0 > 0$ such that

$$(G^{\mathcal{H}}(\xi))^{k_0} = \mathrm{id} \quad \text{on } z\mathcal{H}^{\mathcal{X}}_-/\mathcal{H}^{\mathcal{X}}_-. \tag{33}$$

[10]More precisely, we also need the fact that the vector $\mathbf{1} \in z\mathcal{H}^{\mathcal{X}}_-/\mathcal{H}^{\mathcal{X}}_-$ is invariant under the Galois action.

This corresponds to the fact that the monodromy of the Levi-Civita connection ∇^{LC} of the flat metric g (or the monodromy of the trivialization (26)) becomes trivial on a k_0-fold cover of V. In fact, one can see that the monodromy of ∇^{LC} is trivial on the cover $U/H^2(X,\mathbb{Z}) \to U/H^2(\mathcal{X},\mathbb{Z}) \cong V$, where X is the coarse moduli space of $\mathcal{X}$.

Compatibility with Deligne's canonical extension —As we did at the end of Section 2.2, we can extend the quantum D-module on the cover $U/H^2(X,\mathbb{Z})$ to a connection on $\overline{U/H^2(X,\mathbb{Z})}$ with a logarithmic pole along $q^1 \cdots q^r = 0$ by choosing a nef basis $p_1, \dots, p_r$ of $H^2(X,\mathbb{Z})/\mathrm{tors}$. This is a Deligne's canonical extension [29] of ∇ for a fixed $z \in \mathbb{C}^*$. In order to have a Deligne's extension, we need to choose a logarithm of the monodromy $M_a := G^{\mathcal{H}}(p_a) = e^{-2\pi \mathtt{i} p_a/z}$ around the axis $q^a = 0$. In our case, M_a has the standard logarithm $\mathrm{Log}(M_a) = -2\pi \mathtt{i} p_a/z$ since M_a is unipotent. We can define the Deligne extension here as follows. A section $s(\tau, z)$ of F over $(U/H^2(X,\mathbb{Z})) \times \mathbb{C}^*$ is defined to be extendible to $\overline{U/H^2(X,\mathbb{Z})} \times \mathbb{C}^*$ if the image $\iota_\tau(s) \in \mathcal{H}^{\mathcal{X}}$ of $s(\tau,\cdot) \in \Gamma(\{\tau\} \times \mathbb{C}^*, F)$ satisfies the following: *the family of elements in* $\mathcal{H}^{\mathcal{X}}$

$$U/H^2(X,\mathbb{Z}) \ni [\tau] \mapsto \tilde{s}_\tau := \exp\left(\sum_{a=1}^{r} \frac{\log q^a}{2\pi \mathtt{i}} \mathrm{Log}(M_a)\right) \iota_\tau(s) \in \mathcal{H}^{\mathcal{X}}$$

extends holomorphically to $\overline{U/H^2(X,\mathbb{Z})}$, where we put $\tau = \tau_{0,2} + \tau'$ as in (4) and $\tau_{0,2} = \sum_{a=1}^r p_a \log q^a$. Note that $\tilde{s}_\tau$ is single-valued on $U/H^2(X,\mathbb{Z})$ since the exponential factor offsets the monodromy. Moreover, the limit of $s(\tau,z)$ at $q = \tau' = 0$ is regular at $z = 0$ if $\tilde{s}_\tau|_{q=\tau'=0}$ lies in the *limiting Hodge structure* $\mathbb{F}_{\mathrm{lim}}$:

$$\mathbb{F}_{\mathrm{lim}} := \lim_{\substack{q \to 0 \\ \tau' \to 0}} \exp\left(\sum_{a=1}^{r} \frac{\log q^a}{2\pi \mathtt{i}} \mathrm{Log}(M_a)\right) \mathbb{F}_\tau,$$

where we put $\tau = \sum_{a=1}^r p_a \log q^a + \tau'$ as in (4). By using (30) and the definition (9) of $L(\tau, z)$, one can check that $\mathbb{F}_{\mathrm{lim}}$ exists and

$$\mathbb{F}_{\mathrm{lim}} = H^*_{\mathrm{CR}}(\mathcal{X}) \otimes \mathcal{O}(\mathbb{C}) \subset \mathcal{H}^{\mathcal{X}}. \tag{34}$$

The existence of $\mathbb{F}_{\mathrm{lim}}$ is an analogue of the *nilpotent orbit theorem* [68] in quantum cohomology. This means that the semi-infinite Hodge structure $\mathbb{F}_\tau$ is approximated by the nilpotent orbit $e^{-\sum_{a=1}^r \log q^a \mathrm{Log}(M_a)/(2\pi \mathtt{i})}\mathbb{F}_{\mathrm{lim}}$ as $q, \tau' \to 0$. The standard opposite subspace is opposite to $\mathbb{F}_{\mathrm{lim}}$:

$$\mathcal{H}^{\mathcal{X}}_- \oplus \mathbb{F}_{\mathrm{lim}} = \mathcal{H}^{\mathcal{X}}. \tag{35}$$

This corresponds to the fact that the trivialization induced from $\mathcal{H}^{\mathcal{X}}_-$ is compatible with the Deligne extension at $q = 0$, *i.e.* a section which is constant in the trivialization (26) is extendible across $q = 0$ in the Deligne extension. Note that this is a stronger condition than that $\mathcal{H}_-$ is opposite to $\mathbb{F}_\tau$ for every $\tau \in U$.

For a multiplicative character $\alpha\colon H^2(\mathcal{X},\mathbb{Z}) \to \mathbb{C}^*$, we put

$$\mathsf{T}_\alpha := \{v \in \mathsf{T}\ ;\ \exp(2\pi \mathtt{i} f_v(\xi)) = \alpha(\xi), \forall \xi \in H^2(\mathcal{X},\mathbb{Z})\}.$$

Because $e^{2\pi \mathtt{i}\iota_v} = \alpha([-K_{\mathcal{X}}])$ for $v \in \mathsf{T}_\alpha$, the age ι_v for $v \in \mathsf{T}_\alpha$ have the common fractional part for each α. Consider the following two conditions.

$$\forall\alpha,\ \exists n_\alpha \in \mathbb{Q} \text{ such that } \forall v \in \mathsf{T}_\alpha\ (n_v + 2\iota_v = n_\alpha \text{ or } n_\alpha + 1). \tag{36}$$

$$\iota_v = 0,\ v \neq 0 \implies \exists \xi \in H^2(\mathcal{X},\mathbb{Z}) \text{ such that } f_v(\xi) > 0. \tag{37}$$

Here $n_v := \dim_{\mathbb{C}} \mathcal{X}_v$. The first condition is a weaker version of the Hard Lefschetz condition we will see later[11]. (There we have $n_v + 2\iota_v = \dim_{\mathbb{C}} \mathcal{X}$ for all v.) When (36) is satisfied, we put

$$\mathsf{T}_{\alpha,j} = \{v \in \mathsf{T}_\alpha\ ;\ n_v + 2\iota_v = n_\alpha + j\}, \quad \mathsf{T}_\alpha = \mathsf{T}_{\alpha,0} \sqcup \mathsf{T}_{\alpha,1}. \tag{38}$$

Example 3.12. If $\mathcal{X}$ is isomorphic to a quotient $[M/G]$ of a manifold M by an *abelian* Lie group G as a topological orbifold, the conditions (36), (37) are satisfied since every T_α consists of one element. In fact, there are sufficiently many line bundles on $[M/G]$ arising from characters of G which "separate" different inertia components. In particular, these hold for toric orbifolds.

Theorem 3.13. *Assume that the coarse moduli space X of $\mathcal{X}$ is projective and that the quantum cohomology of $\mathcal{X}$ is convergent (Assumption 2.1). The standard opposite subspace $\mathcal{H}_- = \mathcal{H}_-^{\mathcal{X}}$ and the standard dilaton shift $v_0 = \mathbf{1}$ of the quantum D-module $QDM(\mathcal{X})$ are characterized as follows.*

(i) Under the condition (36), there exists a unique homogeneous opposite subspace satisfying the monodromy invariance (32), (33) and the compatibility with the Deligne extension (35).

(ii) Under the condition (37), there exists a unique vector $v_0 \in z\mathcal{H}_-^{\mathcal{X}}/\mathcal{H}_-^{\mathcal{X}}$ (up to a scalar multiple) such that v_0 is an eigenvector of $\mu = \mathcal{V} = [\nabla_{z\partial_z}]$ of the smallest eigenvalue $-\dim_{\mathbb{C}} \mathcal{X}/2$ and invariant under every Galois action on $z\mathcal{H}_-^{\mathcal{X}}/\mathcal{H}_-^{\mathcal{X}}$.

Thus under (36) and (37), the above conditions determine a canonical Frobenius structure at the cusp up to a constant multiple of the flat metric.

[11] The condition (36) says that $V_\alpha = \bigoplus_{v \in \mathsf{T}_\alpha} H^{*-2\iota_v}(\mathcal{X}_v)$ is bicentric HL in the sense of Definition 3.20. See also Remark 3.21.

Proof. Let $\mathcal{H}_- \subset \mathcal{H}^{\mathcal{X}}$ be any homogeneous opposite subspace satisfying (32), (33) and (35). We decompose the Galois action as

$$G^{\mathcal{H}}(\xi) = e^{-2\pi\mathtt{i}\xi_0/z} \circ G_0^{\mathcal{H}}(\xi), \quad G_0^{\mathcal{H}}(\xi) = \bigoplus_{v\in\mathsf{T}} e^{2\pi\mathtt{i}f_v(\xi)}.$$

Claim: $\mathcal{H}_-$ satisfies the following:

$$\xi_0 \cdot \mathcal{H}_- \subset \mathcal{H}_-, \quad G_0^{\mathcal{H}}(\xi)\mathcal{H}_- \subset \mathcal{H}_-, \quad (z\partial_z + \mu)\mathcal{H}_- \subset \mathcal{H}_-.$$

Take a sufficiently big $k_0 > 0$ such that $(G_0^{\mathcal{H}}(\xi))^{k_0} = \mathrm{id}$ and (33) hold. Then $(G^{\mathcal{H}}(\xi))^{k_0} = e^{-k_0 2\pi\mathtt{i}\xi_0/z}$ preserves $\mathcal{H}_-$ and acts trivially on $z\mathcal{H}_-/\mathcal{H}_-$. Then $\mathrm{Log}((G^{\mathcal{H}}(\xi))^{k_0}) = -k_0 2\pi\mathtt{i}\xi_0/z$ sends $z\mathcal{H}_-$ to $\mathcal{H}_-$. This implies the first equation. The second equation follows from $G_0^{\mathcal{H}}(\xi) = e^{2\pi\mathtt{i}\xi_0/z} \circ G^{\mathcal{H}}(\xi)$ and (32). The third equation follows from $\nabla_{z\partial_z}\mathcal{H}_- \subset \mathcal{H}_-$, the formula (31) for $\nabla_{z\partial_z}$ and $(\rho/z)\mathcal{H}_- \subset \rho\mathcal{H}_- \subset \mathcal{H}_-$.

The third equation in the claim means that $\mathcal{H}_-$ is homogeneous with respect to the usual grading on $H^*_{\mathrm{CR}}(\mathcal{X})$ together with $\deg z = 2$. The opposite property (35) and the formula (34) for $\mathbb{F}_{\mathrm{lim}}$ imply that

$$z\mathcal{H}_- \cap \mathbb{F}_{\mathrm{lim}} \cong \mathbb{F}_{\mathrm{lim}}/z\mathbb{F}_{\mathrm{lim}} = H^*_{\mathrm{CR}}(\mathcal{X}). \tag{39}$$

Since $z\partial_z + \mu$ preserves $z\mathcal{H}_- \cap \mathbb{F}_{\mathrm{lim}}$, this is an isomorphism of graded vector spaces. Also $G_0^{\mathcal{H}}(\xi)$ preserves $z\mathcal{H}_- \cap \mathbb{F}_{\mathrm{lim}}$ and (39) is equivariant with respect to the action of $G_0^{\mathcal{H}}(\xi)$. Therefore (39) is decomposed into the sum of simultaneous eigenspaces of the commuting operators $G_0^{\mathcal{H}}(\xi)$. Recall that the condition (36) gives the decomposition (38). Take a multiplicative character $\alpha\colon H^2(\mathcal{X},\mathbb{Z}) \to \mathbb{C}^*$ and set

$$V_{\alpha,j} = \bigoplus_{v\in\mathsf{T}_{\alpha,j}} H^{*-2\iota_v}(\mathcal{X}_v), \quad j=0,1, \quad V_\alpha = V_{\alpha,0}\oplus V_{\alpha,1}.$$

Then V_α is the simultaneous eigenspace of $G_0^{\mathcal{H}}(\xi)$ of eigenvalue α. By (39), for a homogeneous element $\phi \in V_{\alpha,j}$, there exists a unique lift $\hat{\phi} \in z\mathcal{H}_- \cap \mathbb{F}_{\mathrm{lim}}$ such that

$$\hat{\phi} = \phi + O(z), \quad \deg\hat{\phi} = \deg\phi, \quad \hat{\phi} \in V_\alpha \otimes \mathcal{O}(\mathbb{C}^*).$$

By the Claim above, the $H^2(\mathcal{X})$-action also preserves $z\mathcal{H}_-\cap\mathbb{F}_{\mathrm{lim}}$. Therefore we have $\widehat{\omega\cdot\phi} = \omega\cdot\hat{\phi}$ for a Kähler class ω. Because X is Kähler, the cohomology ring $H^*(\mathcal{X}_v)$ of every inertia component has the Hard Lefschetz property. Hence under the condition (36), the following holds with respect to the grading of the Chen–Ruan cohomology $H^*_{\mathrm{CR}}(\mathcal{X})$.

$$\omega^i\colon V_{\alpha,j}^{n_\alpha+j-i} \to V_{\alpha,j}^{n_\alpha+j+i} \quad \text{is an isomorphism} \quad j=0,1. \tag{40}$$

We also have the Lefschetz decomposition of $V_{\alpha,j}$:

$$V_{\alpha,j} = \bigoplus_{k\geq 0} \bigoplus_{i=0}^{k} \omega^i PV_{\alpha,j}^{n_\alpha+j-k}$$

where $PV_{\alpha,j}^{n_\alpha+j-k} = \mathrm{Ker}(\omega^{k+1} \colon V_{\alpha,j}^{n_\alpha+j-k} \to V_{\alpha,j}^{n_\alpha+j+k+2})$ is the primitive part. By the property $\widehat{\omega\cdot\phi} = \omega\cdot\hat{\phi}$, we only need to know $\hat{\phi}$ for $\phi \in PV_{\alpha,j}^{n_\alpha+j-k}$. For $\phi \in PV_{\alpha,j}^{n_\alpha+j-k}$, we can put

$$\hat{\phi} = \phi + z\phi_1 + z^2\phi_2 + \cdots .$$

where $\phi_i \in V_\alpha^{n_\alpha+j-k-2i}$. Then $0 = \widehat{\omega^{k+1}\phi} = \sum_{i\geq 1} z^i\omega^{k+1}\phi_i$. This implies $\omega^{k+1}\phi_i = 0$. Note that $\phi_i \in V_{\alpha,0}^{n_\alpha-(k+2i-j)} \oplus V_{\alpha,1}^{n_\alpha+1-(k+2i+1-j)}$. Then the Hard Lefschetz (40) for $V_{\alpha,*}$ implies $\phi_i = 0$ and so $\hat{\phi} = \phi$. By the Lefschetz decomposition, we have $\hat{\phi} = \phi$ for every $\phi \in V_{\alpha,j}$. Therefore $z\mathcal{H}_- \cap \mathbb{F}_{\mathrm{lim}} = H^*_{\mathrm{CR}}(\mathcal{X})$ and $z\mathcal{H}_- = H^*_{\mathrm{CR}}(\mathcal{X}) \otimes \mathcal{O}(\mathbb{P}^1 \setminus \{0\})$.

It is easy to show the characterization of v_0. When v_0 is replaced with λv_0 for some $\lambda \in \mathbb{C}$, the flat metric g is multiplied by λ^2. Q.E.D.

Remark 3.14. The limiting Hodge structure $\mathbb{F}_{\mathrm{lim}}$ depends on the choice of co-ordinates $q^1, \dots, q^r$ on $\overline{U/H^2(X,\mathbb{Z})}$. Another co-ordinate system $\hat{q}^a := c^a q^a \exp(F_a(q))$ with $F_a(0) = 0$ changes $\mathbb{F}_{\mathrm{lim}}$ by the multiplication by $\exp(\sum_a \log c^a \,\mathrm{Log}(M_a)/(2\pi\mathtt{i}))$. Under the monodromy invariance (32) for $\mathcal{H}_-$, $\mathcal{H}_-$ being opposite to $\mathbb{F}_{\mathrm{lim}}$ (35) is independent of the choice of a co-ordinate system since $\mathrm{Log}(M_a)$ preserves $\mathcal{H}_-$.

Remark 3.15. We can normalize the dilaton shift $v_0 \in z\mathcal{H}_-/\mathcal{H}_-$ using the integral structure $F_{\mathbb{Z}}$. The dilaton shift v_0 defines a primitive section s_0 of the quantum D-module via the trivialization (26). Under the condition (37), there exists a one-dimensional subspace $\mathbb{C}A_0$ of the space $\mathcal{S}(\mathcal{X})$ of flat sections which is invariant under every Galois action and contained in the image of $(\mathrm{id} - G^{\mathcal{S}}(\xi))^n$ for some unipotent operator $G^{\mathcal{S}}(\xi)$ with the maximum unipotency $n = \dim_{\mathbb{C}} \mathcal{X}$. (This can be seen from the cohomology framing. See (11).) An integral generator A_0 of this subspace is determined up to sign: In fact, this is given by the structure sheaf of a non-stacky point $A_0 = \pm\mathcal{Z}_K(\mathcal{O}_{\mathrm{pt}})$. The choice $v_0 = \pm\mathbf{1}$ corresponds to the normalization $(s_0, A_0)_F \sim (2\pi\mathtt{i})^n/(2\pi z)^{\frac{n}{2}}$ in the large radius limit.

3.6. Symplectic transformation between Givental spaces

Here we see that Assumption 3.1 gives rise to a symplectic transformation $\mathbb{U}$ between the Givental spaces $\mathcal{H}^{\mathcal{X}_1}$ and $\mathcal{H}^{\mathcal{X}_2}$. The transformation $\mathbb{U}$ was introduced in [27] to describe relationships between the genus zero Gromov–Witten theories of $\mathcal{X}_1$ and $\mathcal{X}_2$. As we have seen, the genus zero theory defines a semi-infinite variation of Hodge structures $\mathbb{F}_\tau^{\mathcal{X}_i} \subset \mathcal{H}^{\mathcal{X}_i}$ in the Givental spaces. We shall see in (48) that they match under $\mathbb{U}$: $\mathbb{U}\mathbb{F}_\tau^{\mathcal{X}_1} = \mathbb{F}_\tau^{\mathcal{X}_2}$. This implies that Givental's Lagrangian cones $\mathcal{L}_i \subset \mathcal{H}^{\mathcal{X}_i}$ [26] swept by the semi-infinite subspaces $z\mathbb{F}_\tau^{\mathcal{X}_i}$ are mapped to each other under $\mathbb{U}$:

$$\mathbb{U}\mathcal{L}_1 = \mathcal{L}_2, \quad \text{where} \quad \mathcal{L}_i := \bigcup_\tau z\mathbb{F}_\tau^{\mathcal{X}_i} \subset \mathcal{H}^{\mathcal{X}_i}.$$

The Lagrangian cone $\mathcal{L}_i \subset \mathcal{H}^{\mathcal{X}_i}$ can be also described as the graph of the genus zero descendant potential of $\mathcal{X}_i$ [26] and encodes all the information on genus zero Gromov–Witten theory. In the literature [27, 28, 23], the crepant resolution conjecture was formulated in this way and verified in several examples. See these references for more details and examples of $\mathbb{U}$.

Take a path $\gamma\colon [0,1] \to \mathcal{M}$ connecting two cusp neighborhoods V_1, V_2. Then we have the analytic continuation map (20) $P_\gamma\colon \mathcal{S}(\mathcal{X}_1) \to \mathcal{S}(\mathcal{X}_2)$ along the path $\hat{\gamma} = (\gamma, 1)\colon [0,1] \to \mathcal{M} \times \mathbb{C}^*$. Through the cohomology framing $\mathcal{Z}_{\mathrm{coh}}$ (10), P_γ induces the following isomorphism:

$$\mathbb{U}_{\mathrm{coh}}\colon H^*_{\mathrm{CR}}(\mathcal{X}_1) \to H^*_{\mathrm{CR}}(\mathcal{X}_2), \quad \mathbb{U}_{\mathrm{coh}} = \mathcal{Z}_{\mathrm{coh}}^{-1} P_\gamma \mathcal{Z}_{\mathrm{coh}}. \tag{41}$$

Recall that the Givental space $\mathcal{H}^{\mathcal{X}_i}$ is identified with the space of (multi-valued) sections of F over $V_i \times \mathbb{C}^*$ which are flat in the V_i direction. Therefore, the analytic continuation along $\hat{\gamma}$ also induces a map between the Givental spaces:

$$\mathbb{U}\colon \mathcal{H}^{\mathcal{X}_1} \to \mathcal{H}^{\mathcal{X}_2}. \tag{42}$$

The map $\mathbb{U}$ is an $\mathcal{O}(\mathbb{C}^*)$-linear isomorphism preserving the pairing $(\cdot,\cdot)_{\mathcal{H}}$ on the Givental spaces. In particular, $\mathbb{U}$ is a symplectic transformation with respect to the symplectic form (29). Recall that the cohomology framing identifies $\phi \in H^*_{\mathrm{CR}}(\mathcal{X}_i)$ with a flat section $L(\tau, z) z^{-\mu_i} z^{\rho_i} \phi$ of $QDM(\mathcal{X}_i)$. Also recall that $\phi(z)$ in the Givental space $\mathcal{H}^{\mathcal{X}}$ corresponds to the flat section $L(\tau, z)\phi(z)$. Therefore, one has the commutative diagram involving "multi-valued" Givental spaces:

$$\begin{array}{ccc} H^*_{\mathrm{CR}}(\mathcal{X}_1) & \xrightarrow{\mathbb{U}_{\mathrm{coh}}} & H^*_{\mathrm{CR}}(\mathcal{X}_2) \\ {\scriptstyle z^{-\mu_1} z^{\rho_1}} \downarrow & & \downarrow {\scriptstyle z^{-\mu_2} z^{\rho_2}} \\ \mathcal{H}^{\mathcal{X}_1} \otimes_{\mathcal{O}(\mathbb{C}^*)} \mathcal{O}(\widetilde{\mathbb{C}^*}) & \xrightarrow{\mathbb{U}} & \mathcal{H}^{\mathcal{X}_2} \otimes_{\mathcal{O}(\mathbb{C}^*)} \mathcal{O}(\widetilde{\mathbb{C}^*}) \end{array} \tag{43}$$

where $\rho_i = c_1(\mathcal{X}_i)$ and μ_i is the Hodge grading operator of $\mathcal{X}_i$.

For a rational number $f \in [0,1)$, we set

$$H^*_{\mathrm{CR}}(\mathcal{X})_f := \bigoplus_{\langle \iota_v \rangle = f} H^{*-2\iota_v}(\mathcal{X}_v) =^{12} \bigoplus_{\langle p/2 \rangle = f} H^p_{\mathrm{CR}}(\mathcal{X}). \tag{44}$$

Here $\langle \iota_v \rangle$ is the fractional part of ι_v. Correspondingly, we set

$$\mathcal{H}^{\mathcal{X}}_f := H^*_{\mathrm{CR}}(\mathcal{X})_f \otimes \mathcal{O}(\mathbb{C}^*) \subset \mathcal{H}^{\mathcal{X}}.$$

We list basic properties of $\mathbb{U}_{\mathrm{coh}}$ and $\mathbb{U}$, some of which already appeared in [27, 28]. We will use these later.

Lemma 3.16. *Under Assumption 3.1, the analytic continuation maps $\mathbb{U}_{\mathrm{coh}}$ and $\mathbb{U}$ given in (41), (42) satisfy the following:*

$$\mathbb{U}_{\mathrm{coh}}\rho_1 = \rho_2\mathbb{U}_{\mathrm{coh}}, \quad \mathbb{U}\rho_1 = \rho_2\mathbb{U}, \tag{45}$$

$$\mathbb{U}_{\mathrm{coh}}H^*_{\mathrm{CR}}(\mathcal{X}_1)_f = H^*_{\mathrm{CR}}(\mathcal{X}_2)_f, \quad \mathbb{U}\mathcal{H}^{\mathcal{X}_1}_f = \mathcal{H}^{\mathcal{X}_2}_f, \tag{46}$$

$$\mathbb{U} = z^{-\mu_2}\mathbb{U}_{\mathrm{coh}}z^{\mu_1}, \tag{47}$$

$$\mathbb{U}\mathbb{F}^{\mathcal{X}_1}_\tau = \mathbb{F}^{\mathcal{X}_2}_\tau, \quad \tau \in \mathcal{M}. \tag{48}$$

Here $\mathbb{F}^{\mathcal{X}_i}_\tau \subset \mathcal{H}^{\mathcal{X}_i} \cong \mathcal{H}$ is the semi-infinite Hodge structure (30) at $\tau \in \mathcal{M}$ considered as a subspace of the Givental space. The equation (47) shows that $\mathbb{U}$ is degree-preserving, where the grading on $\mathcal{H}^{\mathcal{X}}$ is given by the usual grading on $H^*_{\mathrm{CR}}(\mathcal{X})$ and $\deg z = 2$.

Assume that $\mathcal{X}_1$ and $\mathcal{X}_2$ are K-equivalent and related by the diagrams (18), (19) such that $\pi_1 \circ p_1 = \pi_2 \circ p_2$. Let γ be the path in (ii) of Assumption 3.1. Then for a class $\alpha \in H^2(Z,\mathbb{C})$,

$$\mathbb{U}_{\mathrm{coh}}(\pi_1^*\alpha) = (\pi_2^*\alpha)\mathbb{U}_{\mathrm{coh}}, \quad \mathbb{U}(\pi_1^*\alpha) = (\pi_2^*\alpha)\mathbb{U}. \tag{49}$$

Proof. The analytic continuation along $\hat{\gamma} = (\gamma, 1)$ must be equivariant under the monodromy in $z \in \mathbb{C}^*$. A simple calculation shows that the monodromy in z acts on $\mathcal{S}(\mathcal{X}_i) \cong H^*_{\mathrm{CR}}(\mathcal{X}_i)$ by

$$M_i = (-1)^n e^{-2\pi\mathtt{i}\rho_i} \bigoplus_{v \in \mathsf{T}_i} e^{2\pi\mathtt{i}\iota_v}, \quad n = \dim \mathcal{X}_i, \tag{50}$$

where T_i is the index set of the inertia component of $\mathcal{X}_i$. Then $M_2\mathbb{U}_{\mathrm{coh}} = \mathbb{U}_{\mathrm{coh}}M_1$. Taking sufficiently high powers of M_i, we have $e^{-k_0 2\pi\mathtt{i}\rho_2}\mathbb{U}_{\mathrm{coh}} = \mathbb{U}_{\mathrm{coh}}e^{-k_0 2\pi\mathtt{i}\rho_1}$. This shows the first equation of (45). Therefore we also have $\mathbb{U}_{\mathrm{coh}}\bigoplus_{v\in\mathsf{T}_1} e^{2\pi\mathtt{i}\iota_v} = \bigoplus_{v\in\mathsf{T}_2} e^{2\pi\mathtt{i}\iota_v}\mathbb{U}_{\mathrm{coh}}$. This shows the first

[12] This equality holds since we ignore cohomology classes of odd parity.

equation of (46). Since $\mathbb{U}_{\rm coh}$ commutes with ρ_i, z^{ρ_i}'s in the commutative diagram (43) cancel each other. This shows (47) and in turn shows the second equations of (45), (46). The equation (48) is a tautological relation since $\mathbb{F}_\tau^{\mathcal{X}_1}$ and $\mathbb{F}_\tau^{\mathcal{X}_2}$ arise from the same subspace $\mathbb{F}_\tau$ of $\mathcal{H}$.

When $\mathcal{X}_1$ and $\mathcal{X}_2$ are related by the birational correspondences (18), (19), the analytic continuation P_γ is equivariant under the monodromy (Galois) action coming from a line bundle L on Z. By the formula (11) of the Galois action in terms of $\mathcal{Z}_{\rm coh}$, we have $\mathbb{U}_{\rm coh} e^{-2\pi\mathtt{i}\pi_1^* c_1(L)} = e^{-2\pi\mathtt{i}\pi_2^* c_1(L)}\mathbb{U}_{\rm coh}$ and (49) follows. Q.E.D.

3.7. Hard Lefschetz condition

We have seen under Assumption 3.1 that quantum cohomologies of $\mathcal{X}_1$ and $\mathcal{X}_2$ underlie the same F-manifold $\mathcal{M}$ (Proposition 3.5) and that the F-manifold structure can be (canonically) lifted over V_i to a Frobenius manifold structure by the opposite subspace $\mathcal{H}_-^{\mathcal{X}_i}$ (Propositions 3.9 and Theorem 3.13). Since a Frobenius structure is well-defined over the complement of an analytic subvariety of $\widetilde{\mathcal{M}}$, we can compare the two Frobenius structures arising from different cusps V_1, V_2. However, we have many examples (*e.g.* a crepant resolution of $\mathbb{P}(1,1,1,3)$) where they do not coincide [3, 27]. The *Hard Lefschetz condition* introduced by [27] (and adopted by [14]) is a criterion for the two Frobenius structures to match. The point is that the monodromy action coming from line bundles on Z uniquely fixes opposite subspaces under this condition.

In this section, we consider the case where $\mathcal{X}_1$ and $\mathcal{X}_2$ are K-equivalent (18) and related by the birational correspondence:

$$\mathcal{X}_1 \xrightarrow{\pi_1} Z \xleftarrow{\pi_2} \mathcal{X}_2$$

such that $\pi_1 \circ p_1 = \pi_2 \circ p_2$.

Definition 3.17. Assume that $H^*_{\rm CR}(\mathcal{X}_i)$ is graded by integers. We say that $\pi_i \colon \mathcal{X}_i \to Z$ satisfies the *Hard Lefschetz condition* if the map

$$(\pi_i^*\omega_Z)^k \colon H^{n-k}_{\rm CR}(\mathcal{X}_i) \to H^{n+k}_{\rm CR}(\mathcal{X}_i)$$

is an isomorphism for a class ω_Z of an ample line bundle on Z.

Remark 3.18. In the context of crepant resolution conjecture, one can take $\mathcal{X}_1 = \mathcal{X}$, Z to be the coarse moduli space X of $\mathcal{X}$ and $\mathcal{X}_2$ to be a crepant resolution Y of X. The Hard Lefschetz condition was originally discussed in [27, 14] for the natural map $\mathcal{X} \to X$. As was observed in [32], the Hard Lefschetz condition for $\mathcal{X} \to X$ is equivalent to

$$\iota_v = \iota_{{\rm inv}(v)}, \ \forall v \in \mathsf{T}. \tag{51}$$

For a non-compact $\mathcal{X}$, the Hard Lefschetz condition for $\mathcal{X} \to X$ can be defined by this condition (51). It is important to consider non-compact cases, but unfortunately, the discussion in this section does not apply to a non-compact $\mathcal{X}$.

Remark 3.19. Cataldo–Migliorini [17] showed that when $\mathcal{X}_i = Y$ is a smooth projective variety, $\pi\colon Y \to Z$ satisfies the Hard Lefschetz condition if and only if π is semismall. Here a proper morphism $\pi\colon Y \to Z$ is said to be *semismall* if $\dim Z^k + 2k \le \dim Y$, where $Z^k = \{z \in Z\ ;\ \dim \pi^{-1}(z) = k\}$.

We will consider a generalization of the Hard Lefschetz condition, where we do not assume the integer grading and also include the "bicentric" case.

Definition 3.20. (i) We say that a pair (V, ω) of a $\mathbb{Q}$-graded complex vector space V and a nilpotent endomorphism $\omega \in \mathrm{End}(V)$ of degree 2 is *bicentric HL* if there exists a rational number $n \in \mathbb{Q}$ and a graded decomposition $V = V_0 \oplus V_1$ such that $V^p = 0$ unless $p \in n + \mathbb{Z}$ and

$$\omega^k\colon V_j^{n+j-k} \to V_j^{n+j+k} \text{ is an isomorphism for } j = 0, 1 \text{ and all } k \ge 0.$$

We call the set $\{n, n+1\}$ the *bicenter.* Note that this definition contains the "mono-centric" case where V_0 or V_1 vanishes.

(ii) We say that a proper morphism $\pi\colon \mathcal{X} \to Z$ satisfies the *generalized Hard Lefschetz condition* if for every rational number $f \in [0, 1)$, the pair $(H^*_{\mathrm{CR}}(\mathcal{X})_f, \pi^*\omega_Z)$ is bicentric HL, where $H^*_{\mathrm{CR}}(\mathcal{X})_f$ is the graded subspace of $H^*_{\mathrm{CR}}(\mathcal{X})$ defined in (44) and ω_Z is a class of an ample line bundle on Z.

Remark 3.21. When π is the natural map $\mathcal{X} \to X$ to the coarse moduli, the generalized Hard Lefschetz condition for π reads as follows: For every rational number $f \in [0, 1)$, there exists $n_f \in \mathbb{Q}$ such that

$$\langle \iota_v \rangle = f \implies \dim_{\mathbb{C}} \mathcal{X}_v + 2\iota_v = n_f \text{ or } n_f + 1.$$

Here $\{n_f, n_f + 1\}$ is the bicenter of $(H^*_{\mathrm{CR}}(\mathcal{X})_f, \omega_X)$.

Theorem 3.22. *Let $\mathcal{X}_1, \mathcal{X}_2$ be K-equivalent smooth Deligne–Mumford stacks related by the diagrams (18), (19) such that $p_1^* K_{\mathcal{X}_1} = p_2^* K_{\mathcal{X}_2}$ and $\pi_1 \circ p_1 = \pi_2 \circ p_2$. Assume that $\pi_1\colon \mathcal{X}_1 \to Z$ satisfies the (generalized) Hard Lefschetz condition. Under Assumption 3.1, the standard opposite subspaces $\mathcal{H}_-^{\mathcal{X}_1}$, $\mathcal{H}_-^{\mathcal{X}_2}$ coincide under the analytic continuation along the path γ in (ii) of Assumption 3.1, i.e. $\mathbb{U}(\mathcal{H}_-^{\mathcal{X}_1}) = \mathcal{H}_-^{\mathcal{X}_2}$. Moreover,*

(i) If $\mathcal{X}_1$ or $\mathcal{X}_2$ does not have generic stabilizers, the Frobenius manifold structures on $\mathcal{M}$ coming from the quantum cohomology of $\mathcal{X}_1$ and

$\mathcal{X}_2$ coincide up to a scalar multiple of the flat metric though the analytic continuation along γ.

(ii) There is a graded isomorphism $(H^*_{\mathrm{CR}}(\mathcal{X}_1), \pi_1^*\omega_Z) \cong (H^*_{\mathrm{CR}}(\mathcal{X}_2), \pi_2^*\omega_Z)$ preserving the actions of ω_Z. In particular, $\pi_2 \colon \mathcal{X}_2 \to Z$ also satisfies the (generalized) Hard Lefschetz condition.

This theorem is a generalization of a result in [27]. We use the following lemma in the proof.

Lemma 3.23. *Let V_i, $i = 1, 2$ be $\mathbb{Q}$-graded vector spaces and $\omega_i \in \mathrm{End}(V_i)$ be nilpotent endomorphisms of degree two. Assume that V_1 and V_2 are isomorphic as graded vector spaces and that there exists a (not necessarily graded) linear isomorphism $\mathbb{U} \colon V_1 \to V_2$ such that $\mathbb{U}\omega_1 = \omega_2\mathbb{U}$. If (V_1, ω_1) is bicentric HL, then there exists a (not canonical) graded isomorphism $\varphi \colon V_1 \to V_2$ such that $\varphi\omega_1 = \omega_2\varphi$. In particular, (V_2, ω_2) is also bicentric HL.*

Proof. Let V be a $\mathbb{Q}$-graded vector space and ω be a nilpotent operator on V of degree 2. Let $a_1 \ge a_2 \ge \cdots \ge a_l$ be lengths of the Jordan cells appearing in the Jordan normal form of ω. Then we can take a basis of V of the form

$$\{\omega^k \phi_j \; ; \; 1 \le j \le l, \; 0 \le k \le a_j\}, \quad a_1 \ge a_2 \ge \cdots \ge a_l \tag{52}$$

such that $\omega^{a_j+1}\phi_j = 0$. Here we can assume that ϕ_j is homogeneous. Set $\deg \phi_j = -a_j + \lambda_j$ for some $\lambda_j \in \mathbb{Q}$. By rearranging the basis, we can assume that $\lambda_j \ge \lambda_{j+1}$ if $a_j = a_{j+1}$. The sequence $\{(a_j, \lambda_j)\}_{j \ge 1}$ is uniquely determined by (V, ω) and we call it the *type* of (V, ω). It suffices to show that (V_i, ω_i), $i = 1, 2$ have the same type. Let $\{(a_j^{(i)}, \lambda_j^{(i)})\}_{j \ge 1}$ be the type of (V_i, ω_i). Since ω_1 and ω_2 are conjugate, we have $a_j := a_j^{(1)} = a_j^{(2)}$. Because (V_1, ω_1) is bicentric HL, there exists $n \in \mathbb{Q}$ such that $\lambda_j^{(1)} = n$ or $n+1$ for all j. Then the degree spectrum of V_1 is contained in $[-a_1 + n, a_1 + n + 1]$. Since V_1 and V_2 are isomorphic as graded vector spaces, we know that $[-a_j + \lambda_j^{(2)}, a_j + \lambda_j^{(2)}] \subset [-a_1 + n, a_1 + n + 1]$. Therefore, $\lambda_j^{(2)} = n$ or $n+1$ if $a_j = a_1$. Take $k > 0$ such that $a_1 = \cdots = a_k > a_{k+1}$. We calculate

$$\begin{aligned}
\dim V_1^{a_1+n+1} + \dim V_1^{-a_1+n} &= k \\
\dim V_2^{a_1+n+1} + \dim V_2^{-a_1+n} &= k + \sharp\{j > k \; ; \; -a_j + \lambda_j^{(2)} = -a_1 + n\} \\
&\quad + \sharp\{j > k \; ; \; a_j + \lambda_j^{(2)} = a_1 + n + 1\}.
\end{aligned}$$

Since these are equal, we have $[-a_j+\lambda_j^{(2)}, a_j+\lambda_j^{(2)}] \subset (-a_1+n, a_1+n+1)$ if $j > k$. Therefore,

$$\begin{aligned}\sharp\{j \le k\ ;\ \lambda_j^{(1)} = n+1\} &= \dim V_1^{a_1+n+1} = \dim V_2^{a_1+n+1} \\ &= \sharp\{j \le k\ ;\ \lambda_j^{(2)} = n+1\}.\end{aligned}$$

Hence $\lambda_j^{(1)} = \lambda_j^{(2)}$ for $j \le k$. This shows that (V_1, ω_1) and (V_2, ω_2) contains an isomorphic graded subspace (V', ω') of the type $\{(a_j, \lambda_j^{(1)})\}_{1\le j\le k}$. By taking the quotient by this subspace, one can proceed by induction on dimensions. Q.E.D.

Proof of Theorem 3.22. Take a path $\gamma\colon [0,1] \to \mathcal{M}$ satisfying the condition (ii) of Assumption 3.1. The analytic continuation map P_γ (20) along the path $\hat{\gamma} = (\gamma, 1)$ induces maps $\mathbb{U}_{\rm coh}$ (41) and $\mathbb{U}$ (42). Recall that $\mathbb{U}_{\rm coh}$ splits into isomorphisms $\mathbb{U}_{{\rm coh},f}\colon H^*_{\rm CR}(\mathcal{X}_1)_f \to H^*_{\rm CR}(\mathcal{X}_2)_f$ for each $f \in [0,1)$ by (46). By (49), we have

$$\mathbb{U}_{{\rm coh},f}(\pi_1^*\omega_Z) = (\pi_2^*\omega_Z)\mathbb{U}_{{\rm coh},f} \tag{53}$$

for an ample class ω_Z on Z. On the other hand, by the theorem of Lupercio–Poddar [56] and Yasuda [72, 73], $H^*_{\rm CR}(\mathcal{X}_1)$ and $H^*_{\rm CR}(\mathcal{X}_2)$ are isomorphic as graded vector spaces when $\mathcal{X}_1$ and $\mathcal{X}_2$ are K-equivalent. Thus $H^*_{\rm CR}(\mathcal{X}_1)_f$ and $H^*_{\rm CR}(\mathcal{X}_2)_f$ are also isomorphic as graded vector spaces. By Lemma 3.23 and (53), we know that there is a graded isomorphism

$$\varphi\colon (H^*_{\rm CR}(\mathcal{X}_1)_f, \pi_1^*\omega_Z) \to (H^*_{\rm CR}(\mathcal{X}_2)_f, \pi_2^*\omega_Z)$$

and $(H^*_{\rm CR}(\mathcal{X}_2)_f, \pi_2^*\omega_Z)$ is also bicentric HL.

In general, a nilpotent operator ω on a vector space V defines a unique (increasing) *weight filtration* $W_i(V)$ of V such that $\omega W_i(V) \subset W_{i-2}(V)$ and that $\omega^i\colon {\rm Gr}_i^W(V) \to {\rm Gr}_{-i}^W(V)$ is an isomorphism. Here ${\rm Gr}_i^W(V) = W_i(V)/W_{i-1}(V)$. When V is a graded vector space, ω is of degree two and (V, ω) is bicentric HL with a graded decomposition $V = V_0 \oplus V_1$ and a bicenter $\{n, n+1\}$ (as in Definition 3.20), the weight filtration of V is given by

$$W_k(V) = V_0^{\ge n-k} \oplus V_1^{\ge n+1-k}.$$

Consider the case $(V, \omega) = (H^*_{\rm CR}(\mathcal{X}_i)_f, \pi_i^*\omega_Z)$. Since the isomorphism $\mathbb{U}_{{\rm coh},f}$ preserves the weight filtration (by (53)) and $(H^*_{\rm CR}(\mathcal{X}_i)_f, \pi_i^*\omega_Z)$ is bicentric HL, we have

$$\mathbb{U}_{{\rm coh},f}(H^p_{\rm CR}(\mathcal{X}_1)_f) \subset H^{\ge p-1}_{\rm CR}(\mathcal{X}_2)_f. \tag{54}$$

When $\phi \in H^p_{\rm CR}(\mathcal{X}_1)$, this together with the formula (47) implies that $\mathbb{U}\phi$ cannot contain positive powers in z. Therefore a matrix representation $U(z)$ of $\mathbb{U}$ with respect to a basis of $H^*_{\rm CR}(\mathcal{X}_i)$ does not contain positive powers in z. Since $\mathbb{U}$ preserves the pairing $(\cdot,\cdot)_{\mathcal{H}}$, the same is true for the inverse $U(z)^{-1}$ which is the adjoint of $U(-z)$ with respect to the Poincaré pairing. Thus we have $\mathbb{U}\mathcal{H}_-^{\mathcal{X}_1} \subset \mathcal{H}_-^{\mathcal{X}_2}$ and $\mathbb{U}^{-1}\mathcal{H}_-^{\mathcal{X}_2} \subset \mathcal{H}_-^{\mathcal{X}_1}$. Hence $\mathbb{U}\mathcal{H}_-^{\mathcal{X}_1} = \mathcal{H}_-^{\mathcal{X}_2}$.

Now we assume $\mathcal{X}_i$ does not have generic stabilizers. Let $\mathcal{H}_- \subset \mathcal{H}$ be the common opposite subspace. Then the dilaton shift $v_0 \in z\mathcal{H}_-/\mathcal{H}_-$ is characterized up to a constant by the condition that v_0 is an eigenvector of $\nabla_{z\partial_z}$ on $z\mathcal{H}_-/\mathcal{H}_-$ of the smallest eigenvalue. This shows (i). The rest of the statements follows from what we already showed. Q.E.D.

Remark 3.24. We used the theorem of Lupercio–Poddar and Yasuda [56, 72, 73] in the proof. However, as [27] did, we can deduce the graded isomorphism $H^*_{\rm CR}(\mathcal{X}_1) \cong H^*_{\rm CR}(\mathcal{X}_2)$ from Assumption 3.1 and certain additional assumptions. For example, we can show $H^*_{\rm CR}(\mathcal{X}_1) \cong H^*_{\rm CR}(\mathcal{X})$ under the assumption that $\mathcal{H}_-^{\mathcal{X}_2}$ *is opposite to the limiting Hodge structure* $\mathbb{F}^{\mathcal{X}_1}_{\rm lim}$ *at the cusp of* V_1, *i.e.*

$$\mathbb{U}(\mathbb{F}^{\mathcal{X}_1}_{\rm lim}) \oplus \mathcal{H}_-^{\mathcal{X}_2} = \mathcal{H}^{\mathcal{X}_2}. \tag{55}$$

The equality (55) was conjectured to hold for a general crepant resolution $\mathcal{X}_2 = Y \to X \leftarrow \mathcal{X}_1$ in [28]. Interestingly, under the generalized Hard Lefschetz condition, the equality (55) is a consequence of the weaker Assumption 3.1 and Lupercio–Poddar–Yasuda's theorem.

By Theorem 3.22 and Cataldo–Migliorini's theorem [17] (see Remark 3.19), Assumption 3.1 has the following interesting consequences:

- Let $\mathcal{X}$ be a Gorenstein orbifold and $Y \to X$ be a crepant resolution. Then $\mathcal{X}$ satisfies the Hard Lefschetz condition if and only if $Y \to X$ is semismall.
- Let X_1 and X_2 be K-equivalent smooth projective varieties related by the diagrams (18), (19) with $\pi_1 \circ p_1 = \pi_2 \circ p_2$. Then $X_1 \to Z$ is semismall if and only if $X_2 \to Z$ is semismall.

The author learned from Tom Coates that the first statement has been conjectured by Jim Bryan [10].

3.8. Integral periods (Central charges)

Up to now, we have not used the integral structure $F_{\mathbb{Z}}$ of the global quantum D-module. In this section, we will see that the integral structure defines an integral co-ordinate—integral period—on the global Kähler moduli space. This is called a *central charge* (see (14)) in physics.

For example, using this, we can give a "reason" why the specialization value of quantum parameters should be a root of unity in the crepant resolution conjecture [45, 46]. In this section, we restrict our attention to the case of crepant resolution $\mathcal{X}_1 = \mathcal{X} \to X \leftarrow Y = \mathcal{X}_2$. Also we assume that Y and $\mathcal{X}$ are Calabi–Yau. The case where $c_1(\mathcal{X})$ is semi-positive can be discussed in a similar way by using the *conformal limit*[13] introduced in [45, 46]. See [45, 46] for semi-positive case.

Let $\mathcal{X}$ be a Calabi–Yau Gorenstein orbifold of dimension n and $\pi\colon Y \to X$ be a crepant resolution of the coarse moduli space X. Note that the Gorenstein assumption implies that $H^*_{\mathrm{CR}}(\mathcal{X})$ is graded by even integers. In the Calabi–Yau case, the base space of the quantum D-module has a distinguished locus where the Euler vector field E vanishes. By the formula (7), this is exactly the small (orbifold) quantum cohomology locus $H^2_{\mathrm{CR}}(\mathcal{X})$ or $H^2(Y)$. Recall that the Euler vector field is globally defined on $\mathcal{M}$ by Section 3.2.

Assumption 3.25. *The locus $\mathcal{M}_0 \subset \mathcal{M}$ where the Euler vector field vanishes is connected. Also the path $\gamma\colon [0,1] \to \mathcal{M}$ in (ii) of Assumption 3.1 can be chosen so that it is contained in $\mathcal{M}_0$.*

In Calabi–Yau case ($\rho = 0$), the situation is greatly simplified. The monodromy in $z \in \mathbb{C}^*$ is almost trivial and given by $(-1)^n$ by (50). Over the locus $\mathcal{M}_0$, the global quantum D-module gives rise to a *finite dimensional variation of Hodge structures (VHS).* The finite dimensional VHS arises from the filtration of flat sections by the pole/zero orders at $z = 0$. The space $\mathcal{S}$ of multi-valued ∇-flat sections of F is single-valued in $w = z^{1/2}$ since the monodromy in z is ± 1. Moreover, over the locus $\mathcal{M}_0$, the flat connection ∇ has a logarithmic pole at $z = 0$ since $\mathcal{U} = \mathcal{A}_E(\tau, 0)$ in (23) is zero. Therefore, a ∇-flat section $s(\tau, z) \in \mathcal{S}$ is at worst meromorphic at $w = z^{1/2} = 0$. This introduces the decreasing filtration $\mathcal{S} = F^0_\tau(\mathcal{S}) \supset F^1_\tau(\mathcal{S}) \supset \cdots \supset F^n_\tau(\mathcal{S}) \supset 0$ for $\tau \in \mathcal{M}_0$:

$$F^p_\tau(\mathcal{S}) = \{s \in \mathcal{S}\ ;\ z^{\frac{n}{2}-p} s(\tau, z) \text{ is regular at } z = 0\}.$$

Note that the factor $z^{\frac{n}{2}}$ kills the monodromy of $s(\tau, z)$ in z. On the neighborhoods V_1, V_2 of cusps, $\mathcal{S}$ is identified with $\mathcal{S}(\mathcal{X}), \mathcal{S}(Y)$ and $F^p(\mathcal{S})$ can be described as follows. Because $E = 0$ on $\mathcal{M}_0$, $\nabla_{z\partial_z} = z\partial_z + \mu$ for

[13]This is very close to Y. Ruan's quantum corrected cohomology ring of Y which has the quantum correction only from the exceptional locus [63]; In the abstract Hodge theory, this is also known as a graded quotient by the Sabbah filtration [64, 38].

quantum D-modules and we have

$$
\begin{aligned}
F^p_\tau(\mathcal{S}) &\cong \{s \in \mathcal{S}(\mathcal{X}) \; ; \; s(\tau,z) = z^{-\mu}\phi, \; \exists \phi \in H^{\le 2n-2p}_{\mathrm{CR}}(\mathcal{X})\} \\
&\cong \{s \in \mathcal{S}(Y) \; ; \; s(\tau,z) = z^{-\mu}\phi, \; \exists \phi \in H^{\le 2n-2p}(Y)\}
\end{aligned} \tag{56}
$$

on $V_1 \cap \mathcal{M}_0$ and $V_2 \cap \mathcal{M}_0$ respectively. The usual Griffiths transversality and Hodge–Riemann bilinear relation hold for $F^p_\tau(\mathcal{S})$:

$$
dF^p_\tau(\mathcal{S}) \subset F^{p-1}_\tau(\mathcal{S}) \otimes \Omega^1_{\mathcal{M}_0}, \quad (F^p_\tau(\mathcal{S}), F^{n-p+1}_\tau(\mathcal{S}))_{\mathcal{S}} = 0.
$$

Here the pairing $(\cdot,\cdot)_{\mathcal{S}}$ is defined in the same way as in the case of quantum D-modules (see Definition 2.10). The $\frac{\infty}{2}$VHS $\mathbb{F}_\tau$ at $\tau \in \mathcal{M}_0$ can be recovered from $F^p_\tau(\mathcal{S})$ as follows:

$$
\mathbb{F}_\tau = (z^{-\frac{n}{2}} F^n_\tau(\mathcal{S}) + z^{-\frac{n}{2}+1} F^{n-1}_\tau(\mathcal{S}) + \cdots + z^{\frac{n}{2}} F^0_\tau(\mathcal{S})) \otimes \mathcal{O}(\mathbb{C}).
$$

We introduce an integral period on $\mathcal{M}_0$ corresponding to an element of $\mathcal{S}_\mathbb{Z}$, *i.e.* a section of the integral local system $F_\mathbb{Z}$. This coincides with the central charge introduced in (14) for quantum D-modules. Recall that the analytic continuation map $\mathcal{S}(\mathcal{X}) \cong \mathcal{S} \cong \mathcal{S}(Y)$ along the path $\hat{\gamma}$ in Assumption 3.1 is equivariant under the Galois action of line bundles of the coarse moduli space X. Take an ample line bundle L on X and consider the corresponding Galois action $M = G^{\mathcal{S}}([L])$ on $\mathcal{S}$.

Lemma 3.26. (i) $F^n_\tau(\mathcal{S}) \subset \mathcal{S}$ *is a one dimensional subspace for a generic* $\tau \in \mathcal{M}_0$.

(ii) *There exists a unique (up to sign) integral vector* $A_0 \in \mathcal{S}_\mathbb{Z}$ *contained in the image of* $(\mathrm{Log}(M) - 1)^n$. *Under the* K*-group framing (13)* $\mathcal{Z}_K \colon K(\mathcal{X}) \to \mathcal{S}(\mathcal{X})$ *(or* $K(Y) \to \mathcal{S}(Y)$*),* A_0 *is identified with the structure sheaf of a non-stacky point* $A_0 = \pm \mathcal{Z}_K(\mathcal{O}_{\mathrm{pt}})$.

Proof. Since $\dim F^n_\tau$ is upper semi-continuous, (i) follows from the description (56) of $F^n_\tau(\mathcal{S})$ near the cusps. The operator M corresponds to the unipotent operator $e^{-2\pi i c_1(L)}$ on $H^*_{\mathrm{CR}}(\mathcal{X})$ through the cohomology framing (10), thus $\mathrm{Im}(\mathrm{Log}(M) - 1)^n \cong \mathrm{Im}\, c_1(L)^n = H^{2n}(\mathcal{X})$ is one-dimensional. This contains an integral vector $\mathcal{Z}_K(\mathcal{O}_{\mathrm{pt}})$. Q.E.D.

By Lemma 3.26, the following definition makes sense.

Definition 3.27. Let $\mathbb{C}^*_w \to \mathbb{C}^* = \mathbb{C}^*_z$ be the double cover of the z-plane with a co-ordinate $w = z^{1/2}$. Take a flat section $A_0 \in \mathcal{S}_\mathbb{Z}$ in Lemma 3.26. A *normalized primitive section* is a section $\tilde{s}_0 \in \Gamma(\mathcal{M}_0 \times \mathbb{C}^*_w, F)$ satisfying

- For every $\tau \in \mathcal{M}_0$, $\tilde{s}_0(\tau, z)$ is the restriction of an element of $F^n_\tau(\mathcal{S})$ to $\{\tau\} \times \mathbb{C}^*_w$.

- $(\tilde{s}_0(\tau, e^{\pi \mathtt{i}} z), A_0(\tau, z))_F = 1$.

This $\tilde{s}_0$ is unique up to sign since so is A_0. The *integral period* Π_A associated to $A \in \mathcal{S}_{\mathbb{Z}}$ is the function on $\mathcal{M}_0$ defined by

$$\Pi_A(\tau) := (\tilde{s}_0(\tau, e^{\pi \mathtt{i}} z), A(\tau, z))_F, \quad \tau \in \mathcal{M}_0. \tag{57}$$

We compute the normalized primitive section and integral periods for the quantum D-modules of $\mathcal{X}$ and Y. Using the fundamental solution $L(\tau, z)$ in Proposition 2.8, we define the *J-function* by

$$J(\tau, -z) := L(\tau, z)^{\dagger} \mathbf{1},$$

where $L(\tau, z)^{\dagger}$ is the adjoint with respect to the Poincaré pairing. The J-function has the following expression:

$$J(\tau, -z) = e^{-\tau_{0,2}/z} \Bigg(1 - \frac{\tau'}{z} + \\ + \sum_{\substack{d \in \mathrm{Eff}_{\mathcal{X}}, 1 \le k \le N \\ d=0 \Rightarrow m \ge 2}} \left\langle \tau', \dots, \tau', \frac{\phi_k}{z(z + \psi_{m+1})} \right\rangle_{0, m+1, d} \frac{e^{\langle \tau_{0,2}, d \rangle}}{m!} \phi^k \Bigg).$$

Here $\tau = \tau_{0,2} + \tau'$ is the decomposition in (4). (This can be derived from (9) and the String equation [1, Theorem 8.3.1].) When $\mathcal{X}$ is Calabi–Yau and $\tau \in H^2_{\mathrm{CR}}(\mathcal{X})$, the J-function is homogeneous of degree zero and is of the form

$$J(\tau, -z) = 1 - \frac{\tau}{z} + \sum_{k \ge 2} \frac{\alpha_k(\tau)}{z^k}, \quad \alpha_k(\tau) \in H^{2k}_{\mathrm{CR}}(\mathcal{X}). \tag{58}$$

Proposition 3.28. *(This proposition applies to the resolution Y as well.) The normalized primitive section of the quantum D-module of $\mathcal{X}$ is given by*

$$\tilde{s}_0(\tau, z) = \frac{(2\pi z)^{\frac{n}{2}}}{(-2\pi)^n} \mathbf{1}.$$

Therefore, the integral period Π_A (57) associated to an integral flat section $A = \mathcal{Z}_K(V)$, $V \in K(\mathcal{X})$ equals the central charge $Z(V)$ (14) of V. This is a component of the J-function:

$$\Pi_A = Z(V) = (2\pi)^{-\frac{n}{2}} \mathtt{i}^{-n} (J(\tau, -1), \Psi(V))_{\mathrm{orb}}, \quad \tau \in H^2_{\mathrm{CR}}(\mathcal{X}),$$

where $\Psi(V)$ was defined in (13) and $J(\tau, -z)$ is the J-function.

Proof. By (56), $\tilde{s}_0$ satisfies the first condition in Definition 3.27. From $A_0 = \mathcal{Z}_K(\mathcal{O}_{\rm pt}) = L(\tau, z)((2\pi\mathtt{i})^n/(2\pi z)^{\frac{n}{2}})[{\rm pt}]$ and the formula (58) for the J-function, the second condition follows. The rest of the statements just follows from the definition (57) of Π_A with the formulas (13), (14), (58) and $\mu^\dagger = -\mu$. Q.E.D.

Remark 3.29. The above calculation shows that the "normalized" primitive section is (up to a function in z) nothing but the primitive section $s_0 = \mathbf{1}$ associated to the standard opposite subspace and dilaton shift (see Section 3.4). The existence of a canonical (normalized) primitive section along the locus $\mathcal{M}_0$ does not mean that the Frobenius manifold structures of $\mathcal{X}$ and Y are the same. In fact, the primitive sections s_0 of $\mathcal{X}$ and Y may differ outside the locus $\mathcal{M}_0 \subset \mathcal{M}$.

Corollary 3.30. *Under the Assumption 3.1 and Conjecture 3.2, the central charges of the corresponding K-group elements define the same function (up to sign) on $\mathcal{M}_0$:*

$$Z^Y(V) = \pm Z^{\mathcal{X}}(\mathbb{U}_K^{-1}(V)), \quad V \in K(Y),$$

where $Z^{\mathcal{X}}$ and Z^Y are the central charges (14) of $\mathcal{X}$ and Y respectively and $\mathbb{U}_K = \mathbb{U}_{K,\gamma}\colon K(\mathcal{X}) \cong K(Y)$ is the isomorphism in Conjecture 3.2. The sign $\pm$ depends on the sign of $\mathbb{U}_K(\mathcal{O}_{\rm pt}) = \pm\mathcal{O}_{\rm pt}$[14].

It is interesting to study what integral periods are *affine linear* functions on $H^2_{\rm CR}(\mathcal{X})$ or $H^2(Y)$. For example, there exists an affine co-ordinate system on $H^2(\mathcal{X}) \oplus \bigoplus_{{\rm codim}\,\mathcal{X}_v=2} H^0(\mathcal{X}_v) \subset H^2_{\rm CR}(\mathcal{X})$ or on $H^2(Y)$ consisting of integral periods [45, Proposition 6.3], [46, Proposition 5.5]. If we have a stratum $\mathcal{X}_v$ of codimension ≥ 3 with $\iota_v = 1$, the corresponding linear projection $H^2_{\rm CR}(\mathcal{X}) \to H^0(\mathcal{X}_v) = \mathbb{C}$ may not be written as an affine linear combination of integral periods. Also, an affine linear integral period on $H^2(Y)$ may not correspond to an affine linear integral period on $H^2_{\rm CR}(\mathcal{X})$. In the next section, we will examine some local examples.

3.9. Local examples

We consider the crepant resolution conjecture for $\mathcal{X} = [\mathbb{C}^n/G]$ where $G \subset SL(n, \mathbb{C})$ is a finite subgroup and $n = 2$ or 3. A standard crepant resolution of $X = \mathbb{C}^n/G$ is given by the G-Hilbert scheme [9]:

$$\pi\colon Y := G\text{-Hilb}(\mathbb{C}^n) \to X = \mathbb{C}^n/G.$$

[14]The author guesses that the sign should be plus.

Moreover, an equivalence of derived categories $D(Y) \cong D(\mathcal{X}) := D^G(\mathbb{C}^n)$ is given by the Fourier–Mukai transformation $\Phi\colon D(Y) \to D(\mathcal{X})$ [9]:

$$\Phi = \boldsymbol{R}q_* \circ p^*, \quad Y \xleftarrow{\ p\ } \mathcal{Z} \xrightarrow{\ q\ } \mathbb{C}^n$$

where $\mathcal{Z} \subset Y \times \mathbb{C}^n$ is the universal subscheme and p and q are natural projections. It would be natural to conjecture that our K-group isomorphism $\mathbb{U}_K$ comes from this derived equivalence:

$$\mathbb{U}_K^{-1}\colon K_E(Y) \cong K_0^G(\mathbb{C}^n), \quad [V] \longmapsto [\boldsymbol{R}q_*(p^*V)],$$

where $E = \pi^{-1}(0) \subset Y$ is the exceptional set. Recall that we need to use compactly supported K-groups in order to get well-defined central charges. For a rational curve $\mathbb{P}^1 \cong C \subset E$ in the exceptional set, the central charge of the class $[\mathcal{O}_C(-1)] \in K_E(Y)$ is given by (*cf.* Example 2.14)

$$Z^Y(\mathcal{O}_C(-1)) = -\frac{1}{2\pi\mathtt{i}}\tau \cap [C]$$

for $\tau \in H^2(Y)$. Let $\tau_C := \tau \cap [C]$, $\tau \in H^2(Y)$ be the co-ordinate on $H^2(Y)$ and ϱ_C be the virtual representation of G given by the Fourier–Mukai transform $[\varrho_C \otimes \mathcal{O}_0] = [\boldsymbol{R}q_*(p^*\mathcal{O}_C(-1))]$. Corollary 3.30 suggests the following conjecture:

Conjecture 3.31. *The small quantum cohomology (or D-modules) of $\mathcal{X}$ and Y are isomorphic under the co-ordinate change*

$$\tau_C = -2\pi\mathtt{i}Z^{\mathcal{X}}(\mathcal{O}_0 \otimes \varrho_C) \tag{59}$$

where the right-hand side is the central charge function on $H^2_{\mathrm{CR}}(\mathcal{X})$. See (15) and (16) for formulas of $Z^{\mathcal{X}}(\mathcal{O}_0 \otimes \varrho_C)$. In particular, the quantum variable $q_C = \exp(\tau_C)$ specializes to $\exp(-2\pi\mathtt{i}(\dim \varrho_C)/|G|)$ *at the large radius limit point of $\mathcal{X}$.*

Remark 3.32. (i) Because $\mathcal{X}$ is not compact, the characterization of the vector A_0 in Lemma 3.26 does not hold. However, we can expect that the conclusion of Corollary 3.30 still holds because the K-group class[15] $[\mathcal{O}_{\mathrm{pt}}]$ of a non-stacky point should correspond to each other (*i.e.* $\mathbb{U}_K([\mathcal{O}_{\mathrm{pt}}]) = [\mathcal{O}_{\mathrm{pt}}]$) under a birational transformation.

(ii) Since the H^2-variables do not have the degree, we can expect that the co-ordinate change above is also correct for the $\mathbb{C}^*$-equivariant quantum cohomology. Here $\mathbb{C}^*$ acts on $\mathbb{C}^n$ diagonally. In dimension two, the non-equivariant quantum product is constant in τ, so it is interesting to study the equivariant version.

[15]This corresponds to $[\mathcal{O}_0 \otimes \varrho_{\mathrm{reg}}]$ in $K_0^G(\mathbb{C}^n)$.

(iii) The specialization of q_C to a root of unity comes from the fact that the central charges (15), (16) of $[\mathcal{O}_0 \otimes \varrho_C] = \mathbb{U}_K^{-1}[\mathcal{O}_C(-1)]$ take rational values at the orbifold large radius limit point $\tau = 0$. In [45, 46], the rationality of the central charge of $\mathbb{U}_K^{-1}[\mathcal{O}_C(-1)]$ at the large radius limit was also discussed without assuming the precise form of the K-group framing. When the coarse moduli space X is projective, under the assumption that $H^*(\mathcal{X})$ is generated by $H^2(\mathcal{X})$ and the condition (37), the rationality here is forced only by the monodromy consideration [45, 46].

We have two cases.
(Case 1) *When the Hard Lefschetz condition holds for $\mathcal{X} \to X$.* Then we have [13, Lemma 3.4.1]

- $n = 2$ *or*
- $n = 3$ and G is conjugate to a subgroup of $SL(2, \mathbb{C})$ *or*
- $n = 3$ and G is conjugate to a subgroup of $SO(3, \mathbb{R})$.

In these cases, every inertia component has age $\iota_v = 1$ and the small quantum cohomology is already "big" (ignoring the unit direction), so the above conjecture determines the full relationships of quantum cohomology algebras. Because all the central charges $Z^{\mathcal{X}}(\mathcal{O}_0 \otimes \varrho)$ are affine linear on $H^2_{\mathrm{CR}}(\mathcal{X})$ (the third term in (16) does not exist), the co-ordinate change (59) preserves the flat structure on the base and the Frobenius structures match. Each irreducible component C of the exceptional set E is a rational curve and corresponds to a non-trivial irreducible representation ϱ_C under the Fourier–Mukai transformation[16] (see [48, 34, 6]). The formula (59) agrees with the conjecture of Bryan–Gholampour [11, 13, 14]. The conjecture has been proved for A_n surface singularities $\mathcal{X} = [\mathbb{C}^2/\mathbb{Z}_n]$ [24] and for $\mathcal{X} = [\mathbb{C}^3/\mathbb{Z}_2 \times \mathbb{Z}_2]$ and $[\mathbb{C}^3/A_4]$ [12] (where $G = A_4$ is the alternating group; this is the only case where the non-abelian crepant resolution conjecture has been proved).

(Case 2) *When the Hard Lefschetz condition fails for $\mathcal{X} \to X$.* This happens only when $n = 3$. In this case, since we have the component with age ≥ 2, the above conjecture does not give a full co-ordinate change between Frobenius manifolds (see Remark 3.33 below). As we can see from (16), integral periods can be non-linear functions on $H^2_{\mathrm{CR}}(\mathcal{X})$, so the co-ordinate change (59) can be also non-linear. Consider the case $\mathcal{X} = \mathbb{C}^3/\mathbb{Z}_3$, where $\mathbb{Z}_3$ acts on $\mathbb{C}^3$ by the weight $\frac{1}{3}(1,1,1)$. Then Y is the total space of the canonical bundle of $\mathbb{P}^2$ with the exceptional set

[16]The author thanks Samuel Boissiere for explaining this for $G \subset SO(3, \mathbb{R})$.

$E = \mathbb{P}^2$. The Fourier–Mukai transformation is given by the diagram

$$Y = \mathcal{O}_{\mathbb{P}^2}(-3) \xleftarrow{p} \mathcal{Z} = \mathcal{O}_{\mathbb{P}^2}(-1) \xrightarrow{q} \mathbb{C}^3.$$

Let ϱ_1, ϱ_2 be the representations of $\mathbb{Z}_3$ such that $\varrho_k(1 \mod 3) = e^{2\pi \mathtt{i} k/3}$. For a degree one rational curve $\mathbb{P}^1 \cong C \subset E$, the Fourier–Mukai transform of $\mathcal{O}_C(-1)$ gives the representation $\varrho_C = 2\varrho_1 \oplus \varrho_2$. Thus the predicted co-ordinate change is

$$\tau_C = -2\pi\mathtt{i} - \frac{2\pi\sqrt{3}}{3\Gamma(\frac{2}{3})^3}\alpha^2 t + \frac{2\pi\sqrt{3}}{\Gamma(\frac{1}{3})^3}\alpha \frac{\partial F_0^{\mathcal{X}}}{\partial t}, \tag{60}$$

where t is a co-ordinate on the twisted sector $H^2_{\mathrm{CR}}(\mathcal{X})$ dual to $\mathbf{1}_{\frac{1}{3}}$, $\alpha = e^{2\pi\mathtt{i}/3}$ and $F_0^{\mathcal{X}}$ is the genus zero potential of $\mathcal{X}$ (see (17)). Since we have [27, 24]:

$$F_0^{\mathcal{X}}(t) = \frac{1}{3\cdot 3!}t^3 - \frac{1}{3^3\cdot 6!}t^6 + \frac{1}{3^2\cdot 9!}t^9 - \frac{1093}{3^5\cdot 12!}t^{12} + \cdots,$$

the co-ordinate change (60) is quite non-linear. This (60) agrees with the computation in [27, 23] up to the Galois actions $\tau_C \mapsto \tau_C + 2\pi\mathtt{i}, t \mapsto \alpha^2 t$.

Coates [23] studied other non-Hard Lefschetz examples $[\mathbb{C}^2/\mathbb{Z}_4]$ with weight $\frac{1}{4}(1,1,2)$ and $[\mathbb{C}^3/\mathbb{Z}_5]$ with weight $\frac{1}{5}(1,1,3)$. It would be an interesting exercise to compare Conjecture 3.31 with Coates' calculations.

Remark 3.33. In the second case, we can predict the full relationships between the small quantum cohomology by considering the central charges of $[\mathcal{O}_S] \in K_E(Y)$ associated to surfaces $S \subset E$ in Corollary 3.30. Note that $Z^Y(\mathcal{O}_S)$ contains the information of the derivative of the potential F_0^Y (see Example 2.14, (ii)). The co-ordinate change of big quantum cohomology can be also determined by $\mathbb{U}_K$ in principle, but the formula could be very complicated.

References

[1] D. Abramovich, T. Graber and A. Vistoli, Gromov–Witten theory of Deligne–Mumford stacks, Amer. J. Math., **130** (2008), 1337–1398; arXiv:math/0603151[math.AG].

[2] A. Adem and Y. Ruan, Twisted orbifold K-theory, Comm. Math. Phys., **237** (2003), 533–556.

[3] M. Aganagic, V. Bouchard and A. Klemm, Topological strings and (almost) modular forms, Comm. Math. Phys., **277** (2008), 771–819; arXiv:hep-th/0607100.

[4] S. Barannikov, Quantum periods. I. Semi-infinite variations of Hodge structures, Internat. Math. Res. Notices, **2001** (2001), 1243–1264.

[5] S. Boissière, E. Mann and F. Perroni, The cohomological crepant resolution conjecture for $P(1,3,4,4)$, Internat. J. Math., **20** (2009), 791–801; arXiv:0712.3248.

[6] S. Boissière and A. Sarti, Contraction of excess fibres between the McKay correspondences in dimensions two and three, Ann. Inst. Fourier (Grenoble), **57** (2007), 1839–1861.

[7] L. A. Borisov and R. P. Horja, Mellin–Barnes integrals as Fourier–Mukai transforms, Adv. Math., **207** (2006), 876–927.

[8] T. Bridgeland, Stability conditions on triangulated categories, Ann. of Math. (2), **166** (2007), 317–345.

[9] T. Bridgeland, A. King and M. Reid, The McKay correspondence as an equivalence of derived categories, J. Amer. Math. Soc., **14** (2001), 535–554.

[10] J. Bryan, An e-mail to T. Coates, December, 2006.

[11] J. Bryan and A. Gholampour, Root systems and the quantum cohomology of ADE resolutions, Algebra Number Theory, **2** (2008), 369–390; arXiv:0707.1337.

[12] J. Bryan and A. Gholampour, Hurwitz–Hodge integrals, the E_6, D_4 root systems, and the crepant resolution conjecture, Adv. Math., **221** (2009), 1047–1068; arXiv:0708.4244.

[13] J. Bryan and A. Gholampour, The quantum McKay correspondence for polyhedral singularities, Invent. Math., **178** (2009), 655–681; arXiv:0803.3766.

[14] J. Bryan and T. Graber, The crepant resolution conjecture, In: Algebraic Geometry—Seattle 2005. Part 1, Proc. Sympos. Pure Math., **80**, Part 1, Amer. Math. Soc., Providence, RI, 2009, pp. 23–42; arXiv:math/0610129[math.AG].

[15] J. Bryan, T. Graber and R. Pandharipande, The orbifold quantum cohomology of $\mathbb{C}_2/\mathbb{Z}_3$ and Hurwitz–Hodge integrals, J. Algebraic Geom., **17** (2008), 1–28.

[16] P. Candelas, X. C. de la Ossa, P. S. Green and L. Parkes, A pair of Calabi–Yau manifolds as an exactly soluble superconformal theory, Nuclear Phys. B, **359** (1991), 21–74.

[17] M. de Cataldo and L. Migliorini, The Hard Lefschetz Theorem and the topology of semismall maps, Ann. Sci. École Norm. Sup. (4), **35** (2002), 759–772; arXiv:math/0006187.

[18] S. Cecotti and C. Vafa, Topological—anti-topological fusion, Nuclear Phys. B, **367** (1991), 359–461.

[19] B. Chen, A.-M. Li and G. Zhao, Ruan's conjecture on singular symplectic flops, preprint, arXiv:0804.3143.

[20] B. Chen, A.-M. Li and Q. Zhang and G. Zhao, Singular symplectic flops and Ruan cohomology, Topology, **48** (2009), 1–22; arXiv:0804.3144.

[21] W. Chen and Y. Ruan, Orbifold Gromov–Witten theory, In: Orbifolds in Mathematics and Physics, Madison, WI, 2001, Contemp. Math., **310**, Amer. Math. Soc., Province, RI, 2002, pp. 25–85.

[22] S. Chowla and A. Selberg, On Epstein's zeta-function, J. Reine Angew. Math., **227** (1967), 86–110.

[23] T. Coates, On the crepant resolution conjecture in the local case, Comm. Math. Phys., **287** (2009), 1071–1108; a longer preprint version: Wall-crossing in toric Gromov–Witten theory II: local examples, arXiv:0804.2592.

[24] T. Coates, A. Corti, H. Iritani and H.-H. Tseng, Computing genus-zero twisted Gromov–Witten invariants, Duke Math. J., **147** (2009), 377–438; arXiv:math/0702234.

[25] T. Coates, A. Corti, H. Iritani and H.-H. Tseng, Gromov–Witten theory of toric stacks, in preparation.

[26] T. Coates and A. Givental, Quantum Riemann–Roch, Lefschetz and Serre, Ann. of Math. (2), **165** (2007), 15–53.

[27] T. Coates, H. Iritani and H.-H. Tseng, Wall-crossing in toric Gromov–Witten theory I: crepant examples, Geom. Topol., **13** (2009), 2675–2744; arXiv:math/0611550[math.AG].

[28] T. Coates and Y. Ruan, Quantum cohomology and crepant resolutions: A conjecture, preprint, arXiv:0710.5901.

[29] P. Deligne, Équations Différentielles à Points Singuliers Réguliers, Lecture Notes in Math., **163**, Springer-Verlag, Berlin-New York, 1970.

[30] M. R. Douglas, Dirichlet branes, homological mirror symmetry, and stability, In: Proceedings of the International Congress of Mathematicians, Vol. III, Beijing, 2002, Higher Ed. Press, Beijing, 2002, pp. 395–408.

[31] B. Dubrovin, Geometry of 2D topological field theories, In: Integrable Systems and Quantum Groups, Montecatini Terme, 1993, Lecture Notes in Math., **1620**, Springer-Verlag, Berlin, 1996, pp. 120–348.

[32] J. Fernandez, Hodge structures for orbifold cohomology, Proc. Amer. Math. Soc., **134** (2006), 2511–2520.

[33] A. Givental, Gromov–Witten invariants and quantization of quadratic Hamiltonians, Mosc. Math. J., **1** (2001), 551–568, 645.

[34] Y. Gomi, I. Nakamura and K.-I. Shinoda, Coinvariant algebras of finite subgroups of $SL(3,\mathbb{C})$, Canad. J. Math., **56** (2004), 495–528.

[35] T. Graber and R. Pandharipande, Localization of virtual classes, Invent. Math., **135** (1999), 487–518.

[36] C. Hertling and Yu. I. Manin, Weak Frobenius manifolds, Internat. Math. Res. Notices, **1999** (1999), 277–286.

[37] C. Hertling, tt^*-geometry, Frobenius manifolds, their connections, and the construction for singularities, J. Reine Angew. Math., **555** (2003), 77–161.

[38] C. Hertling and C. Sevenheck, Nilpotent orbits of a generalization of Hodge structures, J. Reine Angew. Math., **609** (2007), 23–80.

[39] K. Hori and C. Vafa, Mirror symmetry, preprint, arXiv:hep-th/0002222.

[40] R. P. Horja, Hypergeometric functions and mirror symmetry in toric varieties, preprint, arXiv:math/9912109[math.AG].
[41] R. P. Horja, Derived category automorphisms from mirror symmetry, Duke Math. J., **127** (2005), 1–34.
[42] S. Hosono, Local mirror symmetry and type IIA monodromy of Calabi–Yau manifolds, Adv. Theor. Math. Phys., **4** (2000), 335–376.
[43] S. Hosono, Central charges, symplectic forms, and hypergeometric series in local mirror symmetry, In: Mirror Symmetry. V, AMS/IP Stud. Adv. Math., **38**, Amer. Math. Soc., Providence, RI, 2006, pp. 405–439.
[44] S. Hosono and Y. Konishi, Higher genus Gromov-Witten invariants of the Grassmannian, and the Pfaffian Calabi–Yau threefolds, Adv. Theor. Math. Phys., **13** (2009), 463–495; arXiv:0704.2928.
[45] H. Iritani, Real and integral structures in quantum cohomology I: toric orbifolds, preprint, arXiv:0712.2204.
[46] H. Iritani, An integral structure in quantum cohomology and mirror symmetry for toric orbifolds, Adv. Math., **222** (2009), 1016–1079; arXiv:0903.1463.
[47] H. Iritani, tt^*-geometry in quantum cohomology, preprint, arXiv:0906.1307.
[48] Y. Ito and I. Nakamura, Hilbert schemes and simple singularities, In: New Trends in Algebraic Geometry, Warwick, 1996, London Math. Soc. Lecture Note Ser., **264**, Cambridge Univ. Press, Cambridge, 1999, pp. 151–233.
[49] L. Katzarkov, M. Kontsevich and T. Pantev, Hodge theoretic aspects of mirror symmetry, In: From Hodge Theory to Integrability and TQFT tt^*-Geometry, Proc. Sympos. Pure Math., **78**, Amer. Math. Soc., Providence, RI, 2008, pp. 87–174; arXiv:0806.0107.
[50] Y. Kawamata, D-equivalence and K-equivalence, J. Differential Geom., **61** (2002), 147–171.
[51] Y. Kawamata, Log crepant birational maps and derived categories, J. Math. Sci. Univ. Tokyo, **12** (2005), 211–231.
[52] T. Kawasaki, The Riemann–Roch theorem for complex V-manifolds, Osaka J. Math., **16** (1979), 151–159.
[53] T. Kawasaki, The index of elliptic operators over V-manifolds, Nagoya Math. J., **84** (1981), 135–157.
[54] Y.-P. Lee, H.-W. Lin and C.-L.Wang, Flops, motives and invariance of quantum rings, preprint, arXiv:math/0608370[math.AG].
[55] A.-M. Li and Y. Ruan, Symplectic surgery and Gromov–Witten invariants of Calabi–Yau 3-folds, Invent. Math., **145** (2001), 151–218.
[56] E. Lupercio and M. Poddar, The global McKay-Ruan correspondence via motivic integration, Bull. London Math. Soc., **36** (2004), 509–515.
[57] Yu. I. Manin, Frobenius Manifolds, Quantum Cohomology, and Moduli Spaces, Amer. Math. Soc. Colloq. Publ., **47**, Amer. Math. Soc., Providence, RI, 1999.

[58] R. Pandharipande, Rational curves on hypersurfaces (after Givental), Séminaire Bourbaki, **1007/98**, Astérisque, **252** (1998), Exp. No. 848, 5, 307–340.

[59] F. Perroni, Orbifold Cohomology of ADE-singularities, Ph.D. thesis at SISSA, Trieste, arXiv:math/0510528; short published version: Chen–Ruan cohomology of ADE singularities, Internat. J. Math., **18** (2007), 1009–1059.

[60] A. Pressley and G. Segal, Loop Groups, Oxford Math. Monogr., The Clarendon Press, Oxford Univ. Press, 1986.

[61] Einer Andreas Rødland, The Pfaffian Calabi–Yau, its Mirror and their link to the Grassmannian $G(2,7)$, Compositio Math., **122** (2000), 135–149.

[62] Y. Ruan, Stringy geometry and topology of orbifolds, In: Symposium in Honor of C. H. Clemens, Salt Lake City, UT, 2000, Contemp. Math., **312**, Amer. Math. Soc., Providence, RI, 2002, pp. 187–233.

[63] Y. Ruan, Cohomology ring of crepant resolution of orbifolds, In: Gromov–Witten Theory of Spin Curves and Orbifolds, Contemp. Math., **403**, Amer. Math. Soc., Providence, RI, 2006, pp. 117–126.

[64] C. Sabbah, Hypergeometric period for a tame polynomial, preprint, arXiv:math/9805077[math.AG]; a short version published in: C. R. Acad. Sci. Paris Sér. I Math., **328** (1999), 603–608.

[65] C. Sabbah, Polarizable Twistor D-Modules, Astérisque, **300**, Soc. Math. France, 2005.

[66] K. Saito, Period mapping associated to a primitive form, Publ. Res. Inst. Math. Sci., **19** (1983), 1231–1264.

[67] M. Saito, On the structure of Brieskorn lattice, Ann. Inst. Fourier (Grenoble), **39** (1989), 27–72.

[68] W. Schmid, Variation of Hodge structure: the singularities of the period mapping, Invent. Math., **22** (1973), 211–319.

[69] G. Segal, Equivariant K-theory, Inst. Hautes Études Sci. Publ. Math., **34** (1968), 129–151.

[70] C. T. Simpson, Mixed twistor structures, preprint, arXiv:math/9705006 [math.AG].

[71] B. Totaro, The resolution property of schemes and stacks, J. Reine Angew. Math., **577** (2004), 1–22.

[72] T. Yasuda, Twisted jets, motivic measure and orbifold cohomology, Compos. Math., **140** (2004), 396–422.

[73] T. Yasuda, Motivic integration over Deligne–Mumford stacks, Adv. Math., **207** (2006), 707–761.

Department of Mathematics
Graduate School of Science
Kyoto University
Kitashirakawa Oiwake-cho
Sakyo-ku, Kyoto 606-8502
Japan
E-mail address: iritani@math.kyoto-u.ac.jp

Advanced Studies in Pure Mathematics 59, 2010
New Developments in Algebraic Geometry,
Integrable Systems and Mirror Symmetry (Kyoto, 2008)
pp. 167–200

Logarithmic stable maps

Bumsig Kim

Abstract.

We introduce the notion of a logarithmic stable map from a minimal log prestable curve to a log twisted semi-stable variety of form $xy = 0$. We study the compactification of the moduli spaces of such maps and provide a perfect obstruction theory, applicable to the moduli spaces of (un)ramified stable maps and stable relative maps. As an application, we obtain a modular desingularization of the main component of Kontsevich's moduli space of elliptic stable maps to a projective space.

§1. Introduction

1.1.

In papers [11, 21], the admissibility, or equivalently, the predeformability of J. Li was introduced. It is a condition on maps from curves to semi-stable varieties which are étale locally of form $xy = 0$. The condition is natural. It is a necessary and sufficient condition to deform, étale locally at the domain, such a map to a map from a smooth domain curve to a smooth target. It is, however, not a condition friendly to moduli problems ([21, 22, 10]). If suitable log structures are given both on the domain curve and the target space, the admissibility amounts to the requirement that the map is a log morphism which is simple at the inverse image of the singular locus. For the separatedness of moduli spaces of such maps, it will be imposed that the log structures on the domain prestable curves are minimal (see 3.7); the log structures on the targets are extended log twisted ones (see 4.3); and the automorphism groups of the log morphisms are finite. Those maps, satisfying the above conditions, will be called log stable maps. Once suitable log structures

Received September 12, 2008.
2000 *Mathematics Subject Classification.* Primary 14H10; Secondary 14N35.
Partly supported by JSPS Grant-in-Aid for Scientific Research (S-19104002).

are introduced, due to the construction of log cotangent complexes by Olsson in [29] among other things, it is straightforward to show that the moduli spaces of log stable maps carry the relative perfect obstruction theory identical to that of usual stable maps case if the tangent sheaves of the targets are replaced by the log tangent sheaves (see 7.1). In log sense, both sources and targets of the maps we consider are smooth, which is one of reasons why log geometry works well in the study of moduli spaces of such stable maps. The advantage of the usage of log geometry is manifest in 7.2: The log admissibility suffices to deform, étale locally at the target, such a map to a map from a smooth domain curve to a smooth target.

We explain the results of this paper, more precisely. Let $\mathbf{k}$ be a fixed algebraically closed field of characteristic zero. Let $\mathcal{B}$ be an algebraic stack whose objects form a collection of $(W/S, W \to X)$, where S, X are schemes over a base algebraic $\mathbf{k}$ scheme Λ; W is an algebraic space over S; W/S is a proper flat family of semi-stable varieties of form $xy = 0$; and $W \to X$ is a map. Assume that $\mathcal{B}$ is smooth over Λ and has a universal family $\mathcal{U}$ over $\mathcal{B}$. Main examples we have in mind are these: the stack $\mathfrak{X}$ of FM spaces of a smooth projective variety X ([15]); the stack of expanded degeneration spaces of a smooth projective variety with respect to a smooth divisor ([21, 19]); and the stack of expanded degeneration spaces of a semi-stable, projective, degeneration space of form $xy = 0$ ([21]).

Main Theorem A. *The moduli stack $\overline{M}^{\log}_{g,n}(\mathcal{U}/\mathcal{B})$ of (g, n) log stable maps to $W/S \in \mathcal{B}$ is an algebraic stack with a perfect obstruction theory over Λ. Furthermore if the moduli stack $\overline{M}_{g,n}(\mathcal{U}/\mathcal{B})$ of its underlying stable maps is a proper DM stack over Λ, so is $\overline{M}^{\log}_{g,n}(\mathcal{U}/\mathcal{B})$ over Λ.*

We refer to subsection 6.3 for the precise meaning of the statement of Main Theorem A. The theorem yields an explicit description of a perfect obstruction theory for the log versions of the stack of stable (un)ramified maps ([15]) and the stack of stable relative maps of J. Li ([22]). In fact, the moduli stack $\overline{M}^{\log}_{g,n}(\mathcal{U}/\mathcal{B})$ is equipped with a log structure, and hence it is a log algebraic stack. When $\mathcal{B}$ is the stack $\mathfrak{X}$ of FM spaces of a projective nonsingular variety X over $\mathbf{k}$ ([15]), after (un)ramified condition and the strong stability being imposed, the stack $\overline{M}^{\log}_{g,n}(\mathfrak{X}^+/\mathfrak{X})$ is a proper DM log stack over $\mathbf{k}$. Here $\mathfrak{X}^+$ is the universal family of $\mathfrak{X}$.

When the genus is 1, we will consider a variant of stable (un)ramified maps, namely, elliptic log stable maps to chain type FM spaces of X. The key condition on these, possibly non-finite, maps is that either the genus 1 components or the loops of the rational components are nonconstantly mapped under the maps. See 8.1 for the precise definition.

Main Theorem B. *The moduli stack* $\overline{M}_1^{\log,\,\mathrm{ch}}(\mathfrak{X}^+/\mathfrak{X})$ *of elliptic log stable maps to chain type FM spaces of* X *is a proper DM-stack over* $\mathbf{k}$*, carrying a perfect obstruction theory. When* X *is a projective space* $\mathbb{P}^r_{\mathbf{k}}$*, the stack is smooth over* $\mathbf{k}$*.*

Here the smoothness means the usual smoothness as a DM stack. Hence it provides a moduli-theoretic desingularization of the main component of stable maps. It would be interesting to find an explicit relationship with Vakil and Zinger's desingularization in [34]. The space $\overline{M}_1^{\log,\,\mathrm{ch}}(\mathfrak{X}^+/\mathfrak{X})$ can be perhaps used to algebro-geometrically establish quantum Lefschetz hyperplane section principle for elliptic case and to prove elliptic Mirror Conjecture for Calabi–Yau hypersurfaces in a projective space. The hyperplane section principle for reduced genus 1 Gromov–Witten invariants and the comparisons between reduced and standard genus 1 invariants are accomplished in [24, 36, 37].

1.2. Remarks

The minimality condition on log prestable curves, defined here, is a condition stronger than the minimality introduced by Wewers in [35]. The idea that log structures should be useful in (relative) Gromov–Witten theory has been around many years ([22, 33]). After finishing the paper, the author learned of Siebert's long unfinished project [33] on log GW invariants that includes a construction of the virtual fundamental class without using log cotangent complexes. After posting the paper, the author was informed that there is another approach by Abramovich and Fantechi using twisted stacky structures along the singular loci ([2]).

1.3. Conventions

Throughout the paper, all schemes are locally noetherian schemes over Λ unless otherwise stated. The readers who are familiar with log geometry can skip Section 2, keeping in mind the following conventions. Log structures considered are always fine. Log schemes will be denoted by (X, M) (or $X^\dagger$) when X is an underlying scheme and M is a sheaf of monoid on the étale site $X_{\text{ét}}$ of X, equipped with a homomorphism $\alpha : M \to \mathcal{O}_X$ such that $\alpha^{-1}(\mathcal{O}_X^\times) \cong \mathcal{O}_X^\times$ under α. Sometimes X alone will denote the log scheme. For a monoid P, P_X or simply P will mean a constant sheaf of the monoid on $X_{\text{ét}}$. The relative log differential sheaf of $X^\dagger \to Y^\dagger$ will be written by $\Omega^\dagger_{X/Y}$. In order to avoid confusion, we often use separated notations f and $\underline{f}$ for a log morphism and its underlying morphism, respectively. The symbol e_i is the i-th element of the standard basis of $\mathbb{N}^m$. For maps $X \to Y$ and $Z \to Y$, the fiber product $X \times_Z Y$ will be written by $X_{|Z}$; and by X_z if Z is a point z. We will consider log structures also on algebraic spaces, extending the definition obviously.

1.4. Acknowledgements

The author thanks A. Bertram, I. Ciocan-Fontanine, A. Givental, T. Graber, A. Kresch, J. Li, Y.-G. Oh, R. Pandharipande, B. Siebert, R. Vakil, and A. Zinger for helpful discussions; M. Olsson for delightful suggestions; C. Cadman, Y.-P. Lee, F. Sato and Referee for useful comments on the paper; and Department of Mathematics, UC-Berkeley for providing excellent environments while the paper is written there. This work is partially supported by NRF grant 2009-0063179.

§2. Basics on Logarithmic Geometry

Closely following [14, 26], we recall the basic terminology on log geometry which we use in this paper.

2.1. Monoids

A *monoid* is a set P with an associative commutative binary operation $+$ with a unity, zero. We assume that a homomorphism between monoids preserves the unities. An equivalence relation on a monoid P is called a *congruence relation* if $a \sim b$ iff $a + p \sim b + p$ for all $p \in P$. Note that when $\sim$ is a congruence relation there is a unique monoid structure on $P/\sim$ such that the projection $P \to P/\sim$ is a monoid homomorphism. A monoid P is called *integral* if $p + p_1 = p + p_2$ implies that $p_1 = p_2$. A homomorphism $h : P \to Q$ between integral monoids is called an *integral homomorphism* if for any $h(p_1) + q_1 = h(p_2) + q_2$, there are $p_1', p_2' \in P, q \in Q$ such that $q_1 = h(p_1') + q$, $q_2 = h(p_2') + q$, and $p_1 + p_1' = p_2 + p_2'$. The *cokernel* of a homomorphism $h : P \to Q$ is defined to be the induced monoid on Q/P where the coset is given as: $q \sim q'$ iff $q + h(p) = q' + h(p')$ for some $p, p' \in P$. The *group* P^{gp} associated to a monoid P is defined to be the induced monoid on $P \times P/\sim$, where $(p, q) \sim (p', q')$ iff $p + q' + r = p' + q + r$ for some $r \in P$. A monoid is called *sharp* if there is no nonzero unit (here units are, by definition, invertible elements). A nonzero element u in a sharp monoid is called *irreducible* if $u = p + q$ implies that p or q is zero.

2.2. Log structures

A *pre-logarithmic* structure on a scheme X is a pair (M, α) where M is a sheaf of unital commutative monoids on the étale site $X_{\text{ét}}$ of X

and α is a monoidal sheaf homomorphism from M to $\mathcal{O}_X$. Here $\mathcal{O}_X$ is taken to be a sheaf of monoid with respect to the multiplications. When $\alpha^{-1}(\mathcal{O}_X^\times) \cong \mathcal{O}_X^\times$ via α, it is called *logarithmic* structure. For a given pre-log structure (M, α), the *associated* log structure M^a is defined to be the amalgamated sum (i.e., the pushout)

$$M \oplus_{\alpha^{-1}(\mathcal{O}_X)} \mathcal{O}_X^\times := M \oplus \mathcal{O}_X^\times / \sim,$$

where $(m, u) \sim (m', u')$ if there are $a, a' \in \alpha^{-1}(\mathcal{O}_X^\times)$ for which $a + m = a' + m'$ and $\alpha(a')u = \alpha(a)u'$.

If $f : X \to Y$ is a map between schemes and N is a log structure on Y, then define the *pullback* f^*N to be the associated log structure of the prelog structure of $f^{-1}N \to f^{-1}\mathcal{O}_Y \to \mathcal{O}_X$.

A *log morphism* $f : (X, M) \to (Y, N)$ between log schemes consists of a scheme morphism $f : X \to Y$ and a monoidal sheaf homomorphism $f^b : f^*N \to M$ making the diagram

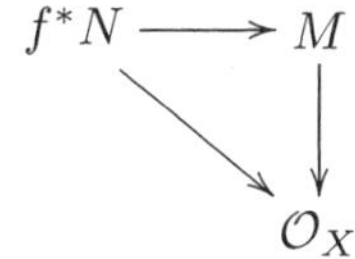

commute. Later, for simplicity, we often say that $f^b : f^*N \to M$ is a homomorphism *over log structure maps* if the diagram commutes.

A *chart* P_X of a log structure M on X is $\theta : P_X \to M$, where P_X is the constant sheaf of a finitely generated integral monoid P making its associated log structure isomorphic to the log structure M under θ. When a chart exits étale locally everywhere on X, we say that the log structure is *fine*. From now on, we consider only fine log structures. We will denote the separable closure of $p \in X$ by $\bar{p}$. The quotient sheaf $M/\alpha^{-1}(\mathcal{O}_X^\times)$, denoted by $\overline{M}$, is called the *characteristic*. It is useful to note that $\overline{f^*N} = f^{-1}\overline{N}$. A fine log structure is called *locally free* if for every point $p \in X$, $\overline{M}_{\bar{p}}$ is finitely generated and free, i.e. $\overline{M}_{\bar{p}} \cong \mathbb{N}^r$ for some integer r which possibly depends on p. If M is locally free, then for every $p \in X$, there is a chart $\theta : \mathbb{N}^r_X \to M$ by which $\mathbb{N}^r \to \overline{M}_{\bar{p}}$ is an isomorphism and $M_{\bar{p}} \cong \overline{M}_{\bar{p}} \oplus \mathcal{O}_{\bar{p}}^\times$ (Lemma 3.3.1).

Let M and M' be log structures on X with $\overline{M}_{\bar{p}}$ and $\overline{M}_{\bar{p}'}$ sharp, where $p \in X$. A homomorphism h from M to M' will be called *simple* at $p \in X$ if: $\bar{h} : \overline{M}_{\bar{p}} \to \overline{M'}_{\bar{p}}$ is injective, and for any irreducible element $b \in \overline{M'}_{\bar{p}}$, there is an irreducible element a in $\overline{M}_{\bar{p}}$ such that $\bar{h}(a) = mb$ for some positive integer m, where $\bar{h}$ is the induced map from h. When M and M' are locally free, it means that $\overline{M}_{\bar{p}} \to \overline{M'}_{\bar{p}}$ is form of a diagonal map $d = (d_1, ..., d_r) : \mathbb{N}^r \to \mathbb{N}^r$, where the standard basis e_i maps to $d_i e_i$ and $d_i \neq 0$ for all i.

A *chart of a log morphism* f is a triple $(P_X \to M, Q_Y \to N, Q \to P)$ of the chart P for M, the chart Q for N, and a homomorphism $Q \to P$ making the diagram

$$\begin{array}{ccc} P_X & \longrightarrow & M \\ \uparrow & & \uparrow \\ Q_X & \longrightarrow & f^*N \end{array}$$

commutative. For any log morphism $f : (X, M) \to (Y, N)$ and any étale local chart $Q \to N$, there is a chart $(P_X \to M, Q_Y \to N, Q \to P)$ of the map f étale locally by Lemma 2.10 in [14].

A log morphism $f : (X, M) \to (Y, N)$ is called:

- *log smooth* if étale locally, there is a chart $(P_X \to M, Q_Y \to N, Q \to P)$ of f such that:
  - $\mathrm{Ker}(Q^{gp} \to P^{gp})$ and the torsion part of $\mathrm{Coker}(Q^{gp} \to P^{gp})$ are finite groups.
  - The induced map $X \to Y \times_{\mathrm{Spec}(\mathbb{Z}[Q])} \mathrm{Spec}\mathbb{Z}[P]$ is smooth in the usual sense.
- *integral* if for every $p \in X$, the induced map $\overline{N}_{f(\bar{p})} \to \overline{M}_{\bar{p}}$ is integral.
- *vertical* if M/f^*N is a sheaf of groups under the induced monoidal operation.
- *strict* if $f^b : f^*N \to M$ is an isomorphism.

All above notions are preserved under base changes.

The amalgamation of integral monoids

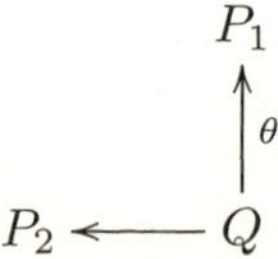

is not longer integral in general. However, when θ is integral, the amalgamation is always integral. Hence in that case, the base change, in other words, the fiber product of an integral log morphism f_1 is defined to be

$$(X_1, M_1) \times_{(Y,N)} (X_2, M_2) := (X_1 \times_Y X_2, (P_1 \oplus_Q P_2)^a)$$

where P_1, P_2, Q are charts of the log morphisms $f_i : (X_i, M_i) \to (Y, N)$. See Proposition 4.1 in [14] for various equivalent definitions of integral morphisms. For instance, we see that the underlying map of a smooth and integral morphism is flat (Corollary 4.5 [14]).

§3. Log Curves

This section deals with the definition and some properties of the allowed domains of log stable maps. The domains will be minimal log prestable curves defined in Subsection 3.7. We start with fixing a notion of log prestable curves basically following [25, 35, 13, 27] for our purpose.

3.1. 1st definition

A log morphism $\pi : (C, M) \to (S, N)$ is called a *log prestable curve* over a fine log scheme (S, N) if every geometric fiber of $\underline{\pi}$ is a prestable curve and π is a proper, log smooth, integral, and vertical morphism.

According to [13] and [27], this definition is equivalent to the following second definition.

3.2. 2nd definition

A log morphism $\pi : (C, M) \to (S, N)$ is called a *log prestable curve* over (S, N) if:

(1) The underlying map $\underline{\pi} : C \to S$ is a S-family of prestable curves.

(2)
- The restriction of $\pi^*N \to M$ to the $\underline{\pi}$-smooth locus $(C/S)^{\mathrm{sm}}$ is an isomorphism.
- If p is a singular point with respect to $\underline{\pi}$, then there are an étale neighborhood U of $\bar{p}$, an affine étale neighborhood $\mathrm{Spec}A$ of $\pi(\bar{p})$ in S, and a chart $((\mathbb{N}^2 \oplus_{\mathbb{N}} Q)_C \to M, Q_S \to N, Q \to \mathbb{N}^2 \oplus_{\mathbb{N}} Q)$ of π such that the induced map

$$U \to \mathrm{Spec}(A \otimes_{\mathbb{Z}[Q]} \mathbb{Z}[\mathbb{N}^2 \oplus_{\mathbb{N}} Q])$$

  is étale, where $\mathbb{N} \to \mathbb{N}^2$ is the diagonal Δ, sending $e_1 \mapsto e_1 + e_2$.

(3) The log structure of N is fine.

3.3. Remarks

Here are some remarks on the definitions above.

3.3.1. Since the homomorphism Δ and the monoid Q are integral, the monoid $\mathbb{N}^2 \oplus_{\mathbb{N}} Q$ and the homomorphism $\mathbb{N}^2 \oplus_{\mathbb{N}} Q \to Q$ are also integral by Proposition 4.1 in [14].

3.3.2. Let $Q \to N$ be a chart at $s := \pi(\bar{p})$, which induces an isomorphism $Q \to \overline{N}_{\bar{s}}$. Such a chart is called good at s in the past literature. Then at any node $\bar{p}$ of $C_{\bar{s}}$, étale locally f has a chart $((\mathbb{N}^2 \oplus_{\mathbb{N}} Q)_C \to \mathcal{O}_C, Q_S \to \mathcal{O}_S, Q \to \mathbb{N}^2 \oplus_{\mathbb{N}} Q)$ satisfying: the induced map $\mathcal{O}_S \otimes_{\mathbb{Z}[Q]} \mathbb{Z}[\mathbb{N}^2 \oplus_{\mathbb{N}} Q] \to \mathcal{O}_C$ is étale, and the induced map $\mathbb{N}^2 \oplus_{\mathbb{N}} Q \to \overline{M}_{\bar{p}}$ is an isomorphism. We can see this by working on the characteristics $Q = \overline{N}_{\bar{s}}$ and $P = \overline{M}_{\bar{p}}$, as following. By the second definition of log prestable curves and the statement 1 in Lemma 3.3.1 below, we know that P is the amalgamation sum of

$$\begin{array}{ccc} \mathbb{N} & \xrightarrow[e_1 \mapsto e_1 + e_2]{} & \mathbb{N}^2 \\ \downarrow & & \\ Q. & & \end{array}$$

Now we lift P and Q to make charts for M and N, using 2 and 3 in Lemma 3.3.1.

Lemma 3.3.1. *Let (M, α) be a fine log structure on a scheme X and let $x \in X$.*

1. A homomorphism $\theta : P \to M_{\bar{x}}$ induces a chart étale locally at $\bar{x}$ if and only if the induced homomorphism $P/(\alpha \circ \theta)^{-1}(\mathcal{O}_{\bar{x}}^{\times}) \to \overline{M}_{\bar{x}}$ is an isomorphism.

2. Suppose that $\theta : \overline{M}_{\bar{x}} \to M_{\bar{x}}$ is a lift of $M_{\bar{x}} \to \overline{M}_{\bar{x}}$, then θ provides a local chart of M at $\bar{x}$.

3. (Proposition 2.1 [28]) *Let (X, M) be a fine log scheme, and let $x \in X$ such that the characteristic of the residue field of x is zero. There is a lift of $M_{\bar{x}} \to \overline{M}_{\bar{x}}$.*

Proof. 1. This is alluded in [14]. The "only if" part is clear. We prove the "if" part. It amounts to showing the natural homomorphism

$$P \oplus_{(\alpha \circ \theta)^{-1}(\mathcal{O}_{\bar{x}}^{\times})} \mathcal{O}_{\bar{x}}^{\times} \to M_{\bar{x}}$$

is an isomorphism. It is clearly surjective. For 1-1, let $\theta(p) + u = \theta(p') + u'$, where $(p, u), (p', u') \in M_{\bar{x}} \times \mathcal{O}_{\bar{x}}^{\times}$. Then $\overline{\theta(p)} = \overline{\theta(p')}$ in $\overline{M}_{\bar{x}}$ and so by assumption

$$p + v = p' + v' \tag{3.3.1}$$

for some $v, v' \in (\alpha \circ \theta)^{-1}(\mathcal{O}_{\bar{x}}^{\times})$. Here we abuse notation. Thus, $\theta(p) + v = \theta(p') + v'$. This together with $\theta(p) + u = \theta(p') + u'$ implies that $\theta(p) + v + u' = \theta(p') + v' + u' = \theta(p) + u + v'$, which in turn shows that

$$u' + v = u + v'. \tag{3.3.2}$$

By (3.3.1) and (3.3.2), $(p, u) \sim (p', u')$.

2. This follows from 1 since $(\alpha \circ \theta)^{-1}(\mathcal{O}_{\bar{x}}^{\times}) = \{0\}$. Q.E.D.

3.4.

For a prestable curve C/S, there is a canonical log structure $M^{C/S}$ on C (resp. $N^{C/S}$ on S) induced from the log structure on the stacks $\mathfrak{M}_{g,1}$ (resp. $\mathfrak{M}_g$) of genus g, 1-pointed (resp. no-pointed) prestable curves ([25, 13]). Here the log structures on a smooth cover is given by the boundary divisors of singular curves and the log structures are defined on smooth topology ([28]). Thus, the natural log morphism $(C, M^{C/S}) \to (S, N^{C/S})$ is a log prestable curve. More generally, for a given homomorphism $N^{C/S} \to N$ from the canonical log structure $N^{C/S}$ to a fine log structure N on S, the amalgamated sum M defined to be the log structure associated to the prelog structure

$$M^{C/S} \oplus_{\pi^{-1}N^{C/S}} \pi^{-1}N$$

yields a log prestable curve over (S, N). All log prestable curves are obtained in this manner according to Proposition 2.3 (and Remark 2.4) in [13] and Theorem 2.7 in [27].

3.5. 3rd definition

A pair $(C/S, N^{C/S} \to N)$ is called a *log prestable curve* over (S, N) if C/S is a prestable curve over S, and $N^{C/S} \to N$ is a homomorphism from the canonical log structure $N^{C/S}$ on S, induced from the family C/S, to a fine log structure N on S.

3.6. Special coordinates

Let C/S be a prestable curve; let p be a nodal point; $A := \mathcal{O}_{\pi(\bar{p})}$; and $R := \mathcal{O}_{\bar{p}}$ be the strict henselianization of $A[u,v]/(uv - t)$ at the ideal $(u, v, \mathfrak{m}_A)$, where $t \in \mathfrak{m}_A$.

Then we call (u, v) a *special coordinate pair* at the node p defined by the ideal $(u, v, \mathfrak{m}_A)$. Based on the coordinates, in fact, the canonical log structure is constructed. Locally define prelog structures $\mathbb{N}^2 \to \mathcal{O}_C$ by sending $e_1 \mapsto u$, $e_2 \mapsto v$ and define $\mathbb{N} \to \mathcal{O}_S$ by sending $e_1 \mapsto uv$. Since by Lemma 3.6.1 special coordinates are unique up to multiplications by unique elements in $\mathcal{O}_C^{\times}$ whose product is in $\mathcal{O}_S^{\times}$, such prelog structures are isomorphic up to unique isomorphism. They can be glued together. See [13] for the detail of the construction of the canonical log structure on a (pre)stable curve C/S.

Lemma 3.6.1. ([13]) *Provided with the notation as in the beginning of 3.6, we have:*

1. Let u', v' be elements in R. Suppose that the ideals $(u', v', \mathfrak{m}_A)$ and $(u, v, \mathfrak{m}_A)$ coincide, and the product $u'v'$ is an element in A. Then there are $a, b \in R^\times$ such that $u' = au$, $v' = bv$ (or $u' = av$, $v' = bu$) and $ab \in A^\times$.

2. Suppose that $u^l = au^l$ and $v^l = bv^l$, where $l \in \mathbb{N}_{\geq 1}$, $a, b \in R^\times$. If $ab \in A^\times$, then $a = b = 1$.

Proof. The first statement and the second statement with $l = 1$ are proven in [13]. The proof of Lemma 2.2 in [13] for $l = 1$ works also for the general l since $R \subset \hat{R}$, where $\hat{R}$ is the t-adic completion of R. Q.E.D.

For example, in the localization of $\Lambda[u, v]/(uv)$ at the ideal (u, v), let $u' = (1 + u)u$, $v' = (1 + v)v$. Then we have $u' = (1 + u - \frac{v}{1+v})u$ and $v' = (1 + v - \frac{u}{1+u})v$, where the product $(1 + u - \frac{v}{1+v})(1 + v - \frac{u}{1+u}) = 1$.

Corollary 3.6.2. *Let l be a positive integer and let $\pi : (C, M) \to (S, N)$ be a log prestable curve. Then with the notation as in the beginning of Subsection 3.6, in $M_{\bar{p}}$ there is a unique pair γ_u, γ_v—which will be denoted by $l \log u$ and $l \log v$, respectively—such that $\gamma_u + \gamma_v \in N_{\pi(\bar{p})}$ and $\alpha(\gamma_u) = u^l$, $\alpha(\gamma_v) = v^l$, where α is the structure map $M_{\bar{p}} \to \mathcal{O}_{\bar{p}}$.*

Proof. The existence follows from the second definition of log prestable curves. If (γ'_u, γ'_v) is another such pair. Then it is clear that $\gamma'_u = \gamma_u$ and $\gamma'_v = \gamma_v$ in $\overline{M}_{\bar{p}}$. Hence, $\gamma'_u = \gamma_u + c_u$, $\gamma'_v = \gamma_v + c_v$, where $c_u, c_v \in \mathcal{O}^\times_{\bar{p}}$. Since $\gamma'_u + \gamma'_v, \gamma_u + \gamma_v \in N_{\pi(\bar{p})}$ and $\gamma'_u + \gamma'_v = \gamma_u + \gamma_v$ in $\overline{N}_{\pi(\bar{p})}$, we see that $c_u c_v \in \mathcal{O}^\times_{\pi(\bar{p})}$. Now we apply Lemma 3.6.1, to conclude the proof. Q.E.D.

3.7. A minimal log prestable curve

We call a log prestable curve

$$(C/S, N^{C/S} \to N)$$

minimal in weak sense ([35]) when N is locally free and there is no proper *locally free* submonoid of N containing the image of $N^{C/S}$. This amounts that for all $s \in S$, $\overline{N}_{\bar{s}}$ is a finitely generated free monoid and there is no proper free submonoid of $\overline{N}_{\bar{s}}$ containing

$$\mathrm{Im}(\overline{N}^{C/S}_{\bar{s}} \to \overline{N}_{\bar{s}}).$$

It is called *minimal* in strong sense furthermore if for every irreducible $b \in \overline{N}_{\bar{s}}$, there is an irreducible element $a \in \overline{N}^{C/S}_{\bar{s}}$ such that $a = lb$ for some positive integer l. Since we will use only 'minimal in strong sense' in this paper, by 'minimal', we will mean 'minimal in strong sense' unless otherwise stated.

3.8. Remark

It is straightforward to check that the fibered categories $\mathfrak{M}_g^{\log}$ of log prestable curves in either definitions, 1st one and 3rd one, are equivalent as following. Let us choose pullbacks once for all and then identify them, for example $(f \circ g)^* M = g^*(f^* M)$, if there exist canonical isomorphisms. A morphism, i.e., arrow, $(C', M')/(S', N') \to (C, M)/(S, N)$ is a pair (h, ϕ), where $h : (C', M') \to (C, M)$ and $\phi : (S', M') \to (S, M)$ are morphisms between log schemes such that:

- The underlying scheme morphisms provide a cartesian square diagram

$$\begin{array}{ccc} C' & \xrightarrow[\underline{h}]{} & C \\ \downarrow{\scriptstyle \underline{\pi}'} & & \downarrow{\scriptstyle \underline{\pi}} \\ S' & \xrightarrow[\underline{\phi}]{} & S. \end{array} \tag{3.8.1}$$

- $h^b : h^* M \to M'$ and $\phi^b : \phi^* N \to N'$ are isomorphisms making the diagram

$$\begin{array}{ccc} M' & \longleftarrow & \underline{h}^* M \\ \uparrow & & \uparrow \\ \underline{\pi}^* N' & \longleftarrow & \underline{h}^* \underline{\pi}^* N = (\underline{\pi}')^* \underline{\phi}^* N \end{array}$$

commute over log structure maps.

In view point of the third definition, a morphism $(C'/S', N^{C'/S'} \to N') \to (C/S, N^{C/S} \to N)$ is a pair $(\underline{h} : C' \to C, \phi : (S', N') \to (S, N))$ such that $\underline{h}$ and $\underline{\phi}$ give rise to a cartesian square diagram of underlying schemes as in (3.8.1) and the diagram

$$\begin{array}{ccc} \underline{\phi}^* N & \xrightarrow[\phi^b]{} & N' \\ \uparrow & & \uparrow \\ \underline{\phi}^* N^{C/S} & \xrightarrow[=]{} & N^{C'/S'} \end{array}$$

which commutes over log structure maps. Here the equality means the second part in the pair of the canonical isomorphisms $(\underline{h}^* M^{C/S}$,

$\underline{\phi}^* N^{C/S}) \to (M^{C'/S'}, N^{C'/S'})$ between two pairs of canonical log structures on C'/S'. We will see later that the fibered category of log prestable curves is an algebraic stack over the category (Sch/Λ) of schemes over Λ.

§4. Log twisted FM type spaces

This section deals with the allowed targets of log stable maps, called log twisted FM type spaces.

4.1. (log) FM type spaces

An algebraic space W over a scheme S is called a *FM type space* if $W \to S$ is a projective, flat map whose geometric fibers are semistable varieties of form $xy = 0$, i.e., for every $s \in S$, étale locally, there is an étale map $W_{\bar{s}} \to \mathrm{Spec} k(\bar{s})[x, y, z_1, ..., z_{r-1}]/(xy)$ where x, y, z_i are independent variables with only one relation $xy = 0$.

Furthermore, if W/S allows a special log morphism, then we call it a *log FM type space.*

Here we say that for a FM type space W/S, a log smooth morphism

$$\pi : (W, M^{W/S}) \to (S, N^{W/S})$$

is *special* ([27]) if:

- $M^{W/S}$ and $N^{W/S}$ are locally free.
- For any $w \in W$, the induced map $\overline{\pi^* N_{\bar{w}}}^{W/S} \to \overline{M}_{\bar{w}}^{W/S}$ is either an isomorphism or the part of the cocatesian diagram

$$\begin{array}{ccc} \mathbb{N} & \xrightarrow[e_1 \mapsto e_1 + e_2]{\Delta} & \mathbb{N}^2 \\ {\scriptstyle h}\downarrow & & \downarrow \\ \overline{N}_{\pi(\bar{w})}^{W/S} & \longrightarrow & \overline{M}_{\bar{w}}^{W/S}, \end{array}$$

  where $h(e_1)$ is an irreducible element.
- The natural map $W_{\bar{s}}^{\mathrm{sing}} \to \mathrm{Irr}\overline{N}_{\bar{s}}$ induced by the above diagram gives rise to a bijection

$$\mathrm{Irr} W_{\bar{s}}^{\mathrm{sing}} \to \mathrm{Irr}\overline{N}_{\bar{s}},$$

  where $\mathrm{Irr} W_{\bar{s}}^{\mathrm{sing}}$ is the set of irreducible components of the singular locus $W_{\bar{s}}^{\mathrm{sing}}$ of $W_{\bar{s}}$ and $\mathrm{Irr}\overline{N}_{\bar{s}}$ is the set of irreducible elements of $\overline{N}_{\bar{s}}$.

We will also call the special log structure the *canonical* one (since it is unique up to a unique isomorphism by Theorem 2.7 in [27]) and denote it by $M^{W/S}$ and $N^{W/S}$, respectively as in the definition above.

4.2.

As a generalization of log twisted curves ([31]) to a FM type space W/S we define the following.

A *log twisted FM type space* is a pair $(W/S, N^{W/S} \to N)$, where W/S is a log FM type space and $N^{W/S} \to N$ is a *simple* map from the canonical log structure $N^{W/S}$ to a locally free log structure N of S. By Lemma 3.3.1, this amounts that étale locally at $s \in S$ there is a commuting diagram of charts:

(4.2.1)
$$\begin{array}{ccc} N^{W/S} & \longrightarrow & N \\ \uparrow{\scriptstyle \theta^{W/S}} & & \uparrow{\scriptstyle \theta} \\ \mathbb{N}^m & \longrightarrow & \mathbb{N}^m, \end{array}$$

where the bottom map is a diagonal map between the constant sheaves $\mathbb{N}^m$ and $\theta^{W/S}$ (resp. θ) induces an isomorphism from $\mathbb{N}^m$ to $\overline{N}_{\bar{s}}^{W/S}$ (resp. $\overline{N}_{\bar{s}}$).

4.3.

An *extended log twisted FM type space* is a pair $(W/S, N^{W/S} \to N)$, where W/S is a log FM type space and $N^{W/S} \to N$ is an extended simple map from the canonical log structure $N^{W/S}$ to a log structure N of S. This means, as the definition, that étale locally there is a commuting diagram of charts:

(4.3.1)
$$\begin{array}{ccccc} N^{W/S} & & \longrightarrow & & N \\ \uparrow & & & & \uparrow \\ \mathbb{N}^m & \longrightarrow & \mathbb{N}^m & \longrightarrow & \mathbb{N}^m \oplus \mathbb{N}^{m'}, \end{array}$$

where the first bottom map is a diagonal map, the second bottom map is the natural monomorphism $(\mathrm{id}_{\mathbb{N}^m}, 0)$, and vertical maps induce isomorphisms between $\mathbb{N}^m$ (resp. $\mathbb{N}^m \oplus \mathbb{N}^{m'}$) and $\overline{N}_{\bar{s}}^{W/S}$ (resp. $\overline{N}_{\bar{s}}$). In particular, N is a locally free log structure on S. We often write simply $(d, 0) : N^{W/S} \to N$ for (4.3.1) with the diagonal map being $d = (d_1, ..., d_m)$.

We endow a log structure M on W by the amalgamated sum

$$(M^{W/S} \oplus_{\pi^{-1}N^{W/S}} \pi^{-1}N)^a.$$

Conversely, from Theorem 2.7 of [27] we arrive at another equivalent definition.

4.4.

An *extended log twisted FM type space* is a proper, log smooth, integral, vertical morphism $\pi : (W, M) \to (S, N)$ such that the underlying map $\pi : W \to S$ is a FM space; the log structure N is locally free; π^b is an isomorphism on the smooth locus of W/S; and at a singular point of W/S, étale locally π has a chart:

$$\begin{array}{ccc} \pi^{-1}N & \longrightarrow & M \\ \uparrow & & \uparrow \\ Q := \mathbb{N}^{r-1} \oplus \mathbb{N} & \xrightarrow[(\mathrm{id},\phi)]{} & P := \mathbb{N}^{r-1} \oplus B \end{array}$$

where the monoid B is the amalgamated sum in

$$\begin{array}{ccc} \mathbb{N} & \xrightarrow[\Delta]{e \mapsto e_1 + e_2} & \mathbb{N}^2 \\ {\scriptstyle \times d}\downarrow & & \downarrow \\ \mathbb{N} & \xrightarrow{\phi} & B \end{array} ;$$

the induced map $\mathcal{O}_S \otimes_{\mathbb{Z}[Q]} \mathbb{Z}[P] \to \mathcal{O}_W$ is smooth; and for every $s \in S$, the natural map $W_{\bar{s}}^{\mathrm{sing}} \to \mathrm{Irr}\overline{N}_{\bar{s}}$ induced by the above diagram gives rise to an injection

$$\mathrm{Irr} W_{\bar{s}}^{\mathrm{sing}} \to \mathrm{Irr}\overline{N}_{\bar{s}}.$$

4.5. Log twisted FM spaces

For a nonsingular projective variety X over $\mathbf{k}$, let $X[n]$ be the Fulton–MacPherson configuration space of n labeled distinct points in X and let $X[n]^+$ be its universal space ([8]).

In [15], a Fulton–MacPherson degeneration space or in short, a *FM space* W of X over S is defined to be a pair $(W \to S, \pi_X : W \to X)$ of maps, where: W is an algebraic space over a scheme S; and for every $s \in S$, there are an étale neighborhood T of $\bar{s}$, and a cartesian diagram

$$\begin{array}{ccc} W_{|T} & \longrightarrow & X[n]^+ \\ \downarrow & & \downarrow \\ T & \longrightarrow & X[n] \end{array}$$

for some n such that, through the diagram, π_X is compatible with the composite

$$X[n]^+ \longrightarrow X^{n+1} \xrightarrow[\mathrm{pr}_{n+1}]{} X.$$

By the blowup construction of the universal family $X[n]^+$ from $X[n] \times X$, it is clear that $X[n]^+ \to X[n]$ is a log FM type space, and thus FM spaces are log FM type spaces.

There are two more examples constructed in [19]: $(X_D^{[n]})^+/X_D^{[n]}$ (the configuration space of n labeled points away from a smooth closed subvariety D in X) and $X_D[n]^+/X_D[n]$ (the configuration space of n labeled distinct points away from a smooth closed subvariety D in X). The constructions in [19] are valid over any closed field $\mathbf{k}$ instead of $\mathbb{C}$, without any changes.

§5. Log stable maps

In this section, we define log stable maps and study their properties. We first need to explain what the underlying maps of the log stable maps are.

5.1. Admissible maps and stable unramified maps

We recall some definitions in [21, 15].

5.1.1. A triple

$$(\star) \qquad ((C/S, \mathbf{p}), W/S, f : C \to W)$$

is called a n-pointed, genus g, *admissible map* to a FM type space W/S if:

(1) $(C/S, \mathbf{p} = (p_1, ..., p_n))$ is a n-pointed, genus g, prestable curve over S.
(2) W/S is a FM type space.
(3) $f : C \to W$ is a map over S.
(4) (*Admissibility*) If a point $p \in C$ is mapped into the relatively singular locus $(W/S)^{\mathrm{sing}}$ of W/S, then étale locally at $\bar{p}$, f is factorized as

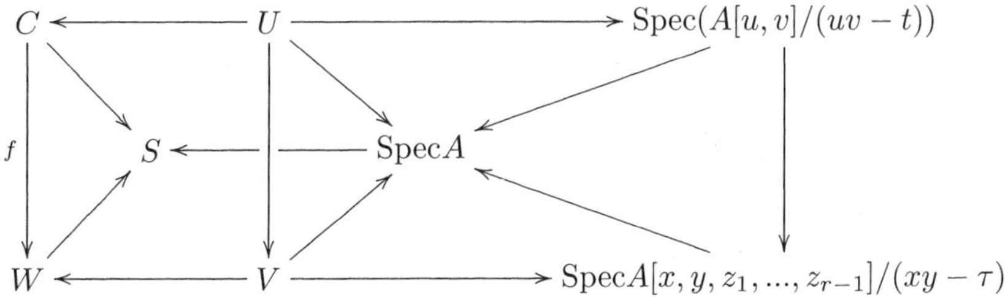

where all 5 horizontal maps are formally étale; u, v, x, y, z_i are indeterminates; $x = u^l$, $y = v^l$ under the far right vertical map for some positive integer l; t, τ are elements in the maximal ideal $\mathfrak{m}_A$ of the local ring A; and $\bar{p}$ is mapped to the point defined by the ideal $(u, v, \mathfrak{m}_A)$.

We call a node $\bar{p}$ of $C_{\bar{s}}$ *distinguished* (resp. *nondistinguished*) if $f(\bar{p}) \in W^{\mathrm{sing}}_{\pi(\bar{p})}$ (resp. $f(\bar{p}) \in W^{\mathrm{sm}}_{\pi(\bar{p})}$), where π is the map $C \to S$.

5.1.2. *Remark* The admissibility can be stated as below, too. If $p \in f^{-1}((W/S)^{\mathrm{sing}})$, then under f, $x = c_1 u^l$, $y = c_2 v^l$ for some $c_i \in R^\times$ such that $c_1 c_2 \in A^\times$ where $A := \mathcal{O}_{\pi(\bar{p})}$; $R := \mathcal{O}_{\bar{p}}$ is form of the strict henselianization of $A[u,v]/(uv - t)$ at the ideal $(u, v, \mathfrak{m}_A)$; $\mathcal{O}_{f(\bar{p})}$ is form of the strict henselianization of $A[x, y, z_1, ..., z_{r-1}]/(xy - \tau)$ at the ideal $(x, y, z_1, ..., z_{r-1}, \mathfrak{m}_A)$; and t, τ are in the maximal ideal $\mathfrak{m}_A$ of A.

Two definitions are equivalent: One direction is clear. We prove the other direction. Consider polynomials $w^l - c_i$ in $R[w]$ where w is an indeterminate. Each polynomial has a linear coprime factorization over $R/\mathfrak{m}_R$. Since R is a henselian ring, the linear factorization can be lifted over R, providing a solution $c_i^{1/l} \in R^\times$ to $w^l = c_i$. Now we have $x = (u')^l$ and $y = (v')^l$, where $u' = c_1^{1/l} u$, $v' = c_2^{1/l} v$. On the other hand, note that $u'v' \in A$ since $(u'v')^l = xy = \tau \in A$. Therefore, we can apply Lemma 3.6.1 to replace $c_i^{1/l}$ by elements $a_i \in R^\times$ such that $a_1 a_2 \in A^\times$. Finally we conclude that $u'v' \in \mathfrak{m}_A$.

5.1.3. Let a FM type space W/S be equipped with a map $\pi_X : W \to X$ from W to a scheme X. An admissible map ($\star$) is called a *stable admissible map* to a FM type space W/S of X if the *stability with respect to* π_X holds, namely: The automorphism group $\mathrm{Aut}_X(f_{\bar{s}})$ is finite for all $s \in S$, where $\mathrm{Aut}_X(f_{\bar{s}})$ consists of all pairs (h, φ) of automorphisms h of $C_{\bar{s}}$ preserving n-labeled points and automorphisms φ of $W_{\bar{s}}$ with respect X such that they are compatible with $f_{\bar{s}}$, i.e., $h(\mathbf{p}_{\bar{s}}) = \mathbf{p}_{\bar{s}}$, $\pi_X \circ \varphi = \pi_X$, and $f_{\bar{s}} \circ h = \varphi \circ f_{\bar{s}}$.

5.1.4. Let $\mu = (\mu_1, ..., \mu_n) \in \mathbb{N}^n_{\geq 1}$. A stable admissible map ($\star$) is called a *stable μ-ramified map* to a FM space of a smooth projective scheme X over $\mathbf{k}$ if:

- W/S is a FM space of X.
- For all $s \in S$, $f(p_i)_{\bar{s}}$ are pairwise distinct for all i.
- (*Strong Stability*) For all $s \in S$, $\mathrm{Aut}_X(f_{\bar{s}})$ is a finite group. Furthermore, every end component of $W_{\bar{s}}$ contains either a non-line image of an irreducible component of $C_{\bar{s}}$ or the images of at least two labeled points.

- $f((C/S)^{\text{sing}}) \subset (W/S)^{\text{sing}}$ and f is unramified everywhere on the relatively smooth locus $(C/S)^{\text{sm}}$, possibly except at the labeled points.
- The ramification order at p_i is exactly μ_i.

Here the end components of $W_{\bar{s}}$ are the screen components which correspond the end nodes of the dual graph of $W_{\bar{s}}$. See [15] for the precise definition.

When $\mu = (1, ..., 1)$, we call the map a *stable unramified map.* Here we follow the ramification order convention that ramification order 2 means the simple ramification.

5.2. Log stable maps

Combining all previous notions, we introduce a series of definitions.

5.2.1. A pair $((C, M)/(S, N), \mathbf{p})$ is called a n-pointed, genus g, (resp. minimal) log prestable curve over (S, N) if $(C, M)/(S, N)$ is a genus g (resp. minimal) log prestable curve and $(C/S, \mathbf{p})$ is a n-pointed prestable curve over S.

5.2.2. A log morphism

$$(\star\star) \qquad (f : (C, M_C, \mathbf{p}) \to (W, M_W)) \,/(S, N)$$

is called a (g, n) *log prestable map over* (S, N) if:

(1) $((C, M)/(S, N), \mathbf{p})$ is a n-pointed, genus g, minimal log prestable curve.
(2) $(W, M_W)/(S, N)$ is an extended log twisted FM type space.
(3) (*Corank = # Nondistinguished Nodes Condition*) For every $s \in S$, the rank of $\mathrm{Coker}(N_{\bar{s}}^{W/S} \to N_{\bar{s}})$ coincides with the number of nondistinguished nodes on $C_{\bar{s}}$.
(4) $f : (C, M_C) \to (W, M_W)$ is a log morphism over (S, N).
(5) (*Log Admissibility*) either of the following conditions, equivalent under the above four conditions, holds:
   - $\underline{f}$ is admissible.
   - $f^b : f^* M_W \to M_C$ is simple at every distinguished node.

5.2.3. *Log Admissibility* We want to see the explicit meaning of the last condition (5) in 5.2.2 under the rest conditions imposed. Provided with the notation in the admissible condition in 5.1.1, there exist $\log x$ and $\log y$ in $(M_W)_{f(\bar{p})}$ such that $\log x + \log y \in N_{\bar{s}}$, $\alpha_W(\log x) = x$, and $\alpha_W(\log y) = y$, where $\bar{p} \mapsto \bar{s}$ under $C \to S$, and $\alpha_W : M_W \to \mathcal{O}_W$ is the log structure map. Then $f^b(\log x)$ and $f^b(\log y)$ must be the $l \log u$ and the $l \log v$, respectively, since the pair $(f^b(\log x), f^b(\log y))$ satisfies the assumption in Corollary 3.6.2. Therefore, at a distinguished node p, the log prestable map is described as: at chart levels there is a diagram

$$\left(\begin{array}{ccc} \mathbb{N}^2 \oplus_{\mathbb{N}} \mathbb{N} & \xrightarrow{\log x,\ \log y\ \mapsto\ l \log u,\ l \log v} & \mathbb{N}^2 \oplus_{\mathbb{N}} \mathbb{N} \\ & {\scriptstyle e_1 \mapsto (0,e_1)} \nwarrow \quad \nearrow {\scriptstyle e_1 \mapsto (0,e_1)} & \\ & \mathbb{N} & \end{array} \right) \oplus \mathbb{N}^{m+m'-1}$$

for some positive integer l, where LHS and RHS amalgamated sums are given by

$$\begin{array}{ccc} \mathbb{N} \ni e_1 & \longmapsto & de_1 \in \mathbb{N} \\ \downarrow & & \\ \mathbb{N}^2 \ni\ e_1 + e_2 & & \end{array} \quad ; \quad \begin{array}{ccc} \mathbb{N} \ni e_1 & \longmapsto & \Gamma e_1 \in \mathbb{N} \\ \downarrow & & \\ \mathbb{N}^2 \ni\ e_1 + e_2 & & \end{array}$$

for some positive integers d, Γ, respectively.

Note also that there is a natural map $\mathrm{Irr}\overline{N}_{\bar{s}}^{C/S} \to \mathrm{Irr}\overline{N}_{\bar{s}}$ by sending a to b if $\Gamma b = a$ under the map $\overline{N}_{\bar{s}}^{C/S} \to \overline{N}_{\bar{s}}$ for some positive integer Γ.

Summing these observations together, we claim that if f is a log prestable map then, étale locally at every geometric point $\bar{s} \to S$, there are commuting charts

(5.2.1)

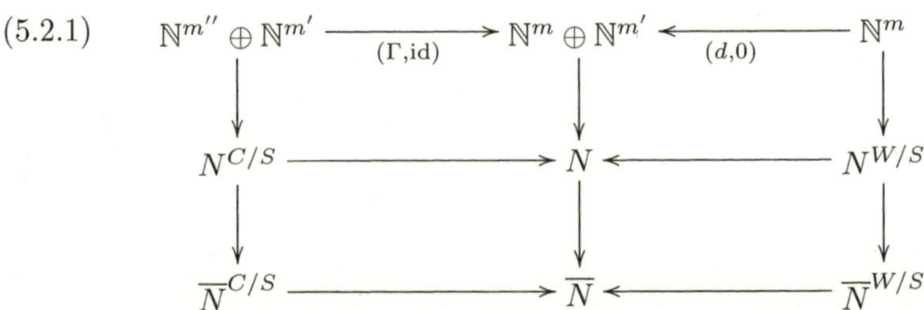

$$\begin{array}{ccccc} \mathbb{N}^{m''} \oplus \mathbb{N}^{m'} & \xrightarrow[(\Gamma,\mathrm{id})]{} & \mathbb{N}^{m} \oplus \mathbb{N}^{m'} & \xleftarrow[(d,0)]{} & \mathbb{N}^{m} \\ \downarrow & & \downarrow & & \downarrow \\ N^{C/S} & \longrightarrow & N & \longleftarrow & N^{W/S} \\ \downarrow & & \downarrow & & \downarrow \\ \overline{N}^{C/S} & \longrightarrow & \overline{N} & \longleftarrow & \overline{N}^{W/S} \end{array}$$

such that the vertical maps induce isomorphisms between chart monoids and the characteristics at $\bar{s}$, where m'' = the number of distinguished nodes in $C_{\bar{s}}$; m' = the number of nondistinguished nodes in $C_{\bar{s}}$; m = the number of irreducible components of $W_{\bar{s}}^{\mathrm{sing}}$; Γ is a 'generalized diagonal' matrix of size $m \times m''$

$$\Gamma = \left(\begin{array}{c|c|c|c} \Gamma_{1,1} \cdots \Gamma_{1,k_1} & 0 \cdots 0 & \cdots & 0 \cdots 0 \\ \hline 0 \cdots 0 & \Gamma_{2,1} \cdots \Gamma_{2,k_2} & 0 \cdots & 0 \cdots 0 \\ \hline \vdots & & \ddots & \vdots \\ \hline 0 \cdots 0 & \cdots & \cdots 0 & \Gamma_{m,1} \cdots \Gamma_{m,k_m} \end{array} \right);$$

and d is a diagonal $(d_1, ..., d_m)$. For the existence of the diagram, first note that we have such maps (Γ, id) and d at the characteristic level due to: the simplicity of $\overline{N} \leftarrow \overline{N}^{W/S}$; the simplicity of f at distinguished nodes; and the minimality of the log prestable curve at nondistinguished nodes. Now to lift such maps at chart levels, start with a chart in the middle and then build the rest of the charts.

The minimality at distinguished nodes means that there is no nontrivial common divisors of positive integers $\Gamma_{i,k_1}, ..., \Gamma_{i,k_i}$ for all $i = 1, ..., m$. Let $\{p_{i,j}\}$ be the set of distinguished nodes on $C_{\bar{s}}$, which are mapped into the i-th component of $W_{\bar{s}}^{\mathrm{sing}}$. Since $d_i = l_{i,j}\Gamma_{i,j}$, d_i must be the least common multiple of $l_{ij}, \forall j$, where $l_{i,j} = \overline{f^b}_{p_{i,j}}$ comes from the positive multiplication map

$$\overline{f^b}_{p_{i,j}} : \mathbb{Z} \cong \overline{f^*(M_W/\pi^*N)}_{p_{ij}} \to \mathbb{Z} \cong \overline{M_C/\pi^*N}_{p_{ij}}$$

induced by f^b.

5.2.4. Let a FM type space W/S be equipped with a map $\pi_X : W \to X$ from W to a scheme X. A log prestable map $(\star\star)$ is called a *log stable map* if the stability condition holds, i.e., for each $s \in S$, the group of automorphisms (h, φ) is finite where:

- h is an automorphism of $((C, M_C)/(S, N))_{\bar{s}}$ preserving n-labeled points $\mathbf{p}_{\bar{s}}$.
- φ is an automorphism of $((W, M_W)/(S, N))_{\bar{s}}$ preserving $W_{\bar{s}} \to X$.
- $\varphi \circ f_{\bar{s}} = f_{\bar{s}} \circ h$.

We will see in 6.2.4 that the above stability holds if and only if the underlying automorphism stability holds. That is, $\underline{f}$ is stable if and only f is stable.

5.2.5. Let $\beta \in A_1(X)/\sim^{\mathrm{alg}}$. A log prestable map $(\star\star)$ is called a (g, n, β) *log stable map to a FM space* W/S *of a smooth projective* **k** *variety* X if:

- W/S is a FM space of X over S.
- Stability Condition holds.
- $(\pi_X \circ f)_*[C_{\bar{s}}] = \beta$ for every $s \in S$.

5.2.6. Fix $\mu = (\mu_1, ..., \mu_n)$. A log stable map $(\star\star)$ to a FM space W/S of X is called a *log stable μ-ramified map* if $\underline{f}$ is stable μ-ramified map over S as in 5.1.4.

When $\mu = (1, ..., 1)$, we call it a *log stable unramified map.* Note that in this case, indeed, f is *log unramified*, that is, $f^*\Omega^{\dagger}_{W/S} \to \Omega^{\dagger}_{C/S}$ is surjective at every $p \in C$.

5.3. Remarks

5.3.1. Every usual stable map to a smooth projective $\mathbf{k}$ variety X from a prestable curve can be an underlying stable map of a unique log stable map.

5.3.2. Suppose that a log prestable curve $(C, M_C)/(S, N)$; an extended twisted FM space $(W, M_W)/(S, N)$; and an admissible map $f : C/S \to W/S$ are given. Then there is a unique locally closed subscheme S' of S where f is a log prestable map over $(S', N_{|_{S'}})$ satisfying the universal property: If $Z \to S$ is a scheme morphism such that $f_{|_Z} : (C_{|_Z}, (M_C)_{|_Z}) \to (W_{|_Z}, (M_W)_{|_Z})$ is a log prestable map over $(Z, N_{|_Z})$ if and only if $Z \to S$ uniquely factors as $Z \to S' \to S$. Here, for example, $N_{|_Z}$ is the log structure of the pullback of N under $Z \to S$.

First of all the above statement makes sense because of 5.2.3 where we have seen that f^b is determined by $\underline{f}$ and log structures on C/S and W/S. The minimality and the corank = # nondistinguished node condition are open conditions. Let V be the open subscheme of S where these conditions are satisfied. Then we need to consider only the condition that f becomes a log morphism. Since $f^b(\log x) = l \log u$ and $f^b(\log y) = l \log v$, the condition becomes the equality $\log x + \log y = l \log u + l \log v$ in N, which defines a locally closed subscheme Z in V by the following reason.

Let M be a fine log structure on a scheme Y_2 and φ be a scheme morphism from a scheme Y_1 to Y_2. Then $\overline{\varphi^* M} = \varphi^{-1}(\overline{M})$. This together with $M_{\bar{y}} \cong \overline{M}_{\bar{y}} \oplus \mathcal{O}_{\bar{y}}^{\times}$, $y \in Y_2$, implies that the requirement of the 'equality' of two sections of M defines a closed subscheme of an open subscheme in Y_2.

§6. Stacks

In this section, we show that the moduli stacks of log stable maps are log algebraic (resp. proper DM) stacks over Λ whenever the corresponding stacks of underlying stable maps are algebraic (resp. proper DM) over Λ. Since some stacks considered here are not separated, we use terminology algebraic stacks.

6.1. Log stacks

Following [13], we introduce log stacks and examples.

6.1.1. Let (Sch/Λ) be the category of locally noetherian schemes over Λ. A pair $(\mathcal{F}, L)$ is called a *log stack* if: $\mathcal{F}$ is a stack over (Sch/Λ), and L is a functor from $\mathcal{F}$ to the category LOG of fine log schemes over

Λ, whose morphisms are strict log morphisms, making the diagram

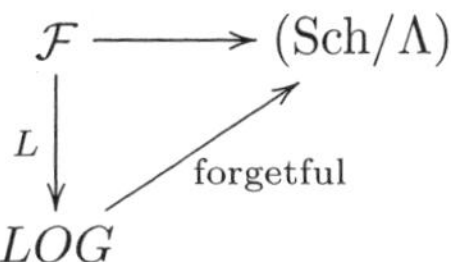

commute. Furthermore when $\mathcal{F}$ is an algebraic stack, it is called a log algebraic stack. Here a stack $\mathcal{F}$ over (Sch/Λ) is said to be *algebraic* if the diagonal $\mathcal{F} \to \mathcal{F} \times_{(\mathrm{Sch}/\Lambda)} \mathcal{F}$ is representable and of finite presentation (see §4 in [7] for the definition), and it allows a smooth cover by a scheme.

A log scheme (X, M) can be considered as a log stack (h_X, L_M), where $h_X(Y) = \mathrm{Hom}_{(\mathrm{Sch}/\Lambda)}(Y, X)$ and $L_M(f : Y \to X) = (Y, f^*M)$.

6.1.2. For a fine log scheme Y, $\mathcal{L}_Y$ will denote the stack of fine log schemes over Y: The objects are fine log schemes over the log scheme Y, and morphisms from Z/Y to Z'/Y are strict log morphisms $h : Z \to Z'$ over Y. The log stack $\mathcal{L}_Y$ was denoted by $\mathcal{L}og_Y$ in [28]. To make notation easy, we use the symbol $\mathcal{L}_Y$.

For a log stack $(\mathcal{F}, L)$, $\mathcal{L}_{\mathcal{F}}$ is defined as: An object is a pair $(x \in \mathcal{F}(X), (X, M) \leftarrow L(x))$, where $\mathcal{F}(X)$ is the collection of objects of $\mathcal{F}$ over X, and the arrow is a (possibly non-strict) log morphism. A morphism from $(x \in \mathcal{F}(X), (X, M) \leftarrow L(x))$ to $(y \in \mathcal{F}(Y), (Y, N) \leftarrow L(y))$ is a pair $(x \to y, (X, M) \to (Y, N))$, where the first arrow is a morphism in the stack $\mathcal{F}$, and the second arrow is a strict log morphism for which the diagram

$$\begin{array}{ccc} (X, M) & \longrightarrow & (Y, N) \\ \uparrow & & \uparrow \\ L(x) & \xrightarrow[L(x \to y)]{} & L(y) \end{array}$$

commutes.

Note that for a log scheme (X, M), $\mathcal{L}_{(h_X, L_M)}$ is equivalent to $\mathcal{L}_{(X,M)}$ as log stacks.

Lemma 6.1.1. *If a log stack $\mathcal{F}$ is algebraic, then $\mathcal{L}_{\mathcal{F}}$ is algebraic.*

Proof. The following argument is due to Olsson. Take a smooth cover of the algebraic stack $\mathcal{F}$ by a scheme Z. Then we obtain the diagram

$$\begin{array}{ccc} \mathcal{L}_{\mathcal{F}} & \longrightarrow & \mathcal{F} \\ \uparrow & & \uparrow \\ \mathcal{L}_{\mathcal{F}} \times_{\mathcal{F}} Z & \longrightarrow & Z. \end{array}$$

Note that the fibered category $\mathcal{L}_{\mathcal{F}} \times_{\mathcal{F}} Z$ is equivalent to $\mathcal{L}_Z$, where Z is endowed with the log structure induced from the cover. Since $\mathcal{L}_Z$ is an algebraic stack due to [28], we can apply Lemma 6.1.2 to the diagram to conclude that $\mathcal{L}_{\mathcal{F}}$ is an algebraic stack. Q.E.D.

Lemma 6.1.2. ([3]) *Let $\mathcal{F}$ be a stack. Suppose that there are: a morphism from $\mathcal{F}$ to an algebraic stack $\mathcal{G}$ and a smooth surjective map from a scheme U to $\mathcal{G}$, for which $\mathcal{F} \times_{\mathcal{G}} U$ is algebraic. Then $\mathcal{F}$ is algebraic.*

6.2. Stacks of log prestable curves and extended log twisting FM type spaces

6.2.1. As usual, we can define the category $\mathfrak{M}_g^{\log}$ of genus g, log prestable curves over (Sch/Λ). Then using the inverse image of log structures, we see that the category is fibered in groupoids over (Sch/Λ). Furthermore, using the gluing of sheaves, it is a stack. In fact, the stack $\mathfrak{M}_g^{\log}$ is equivalent to $\mathcal{L}_{\mathfrak{M}_g}$. Hence it is a log algebraic stack.

Let $\mathfrak{M}_{g,n}^{\log}$ be the algebraic stack $\mathfrak{M}_g^{\log} \times_{\mathfrak{M}_g} \mathfrak{M}_{g,n}$ so that there is no interesting log structures on markings.

6.2.2. We consider a stack $\mathcal{B}$ of certain $(W/S, W \to X)$, where W/S are log FM type spaces, and $W \to X$ are maps from W to a fixed X. A morphism between them is a cartesian diagram preserving X. By definition in 4.1, any object W/S in $\mathcal{B}$ can be realized as an underlying space of an extended log twisted FM type space. Suppose that $\mathcal{B}$ is an algebraic stack.

Since $\mathcal{B}$ is a log stack by the canonical log structures, we can consider $\mathcal{L}_{\mathcal{B}}$. Let $\mathcal{B}^{\mathrm{tw}}$ (resp. $\mathcal{B}^{\mathrm{etw}}$) be the full substack of $\mathcal{L}_{\mathcal{B}}$ whose objects are (resp. extended) log twisted FM type spaces whose underlying spaces are in $\mathcal{B}$. By Lemma 3.3.1, they are open substacks of the algebraic stack $\mathcal{L}_{\mathcal{B}}$, and hence they are also algebraic stacks.

Assume that there is a smooth scheme B over Λ and a smooth morphism $B \to \mathcal{B}$, defined by a 'universal' family U/B. Then, $\mathcal{B}^{\mathrm{tw}}$ and $\mathcal{B}^{\mathrm{etw}}$ are smooth over Λ. Indeed, we can formulate a smooth versal space as following. Let $W/k(\bar{p})$ be the pullback of U at a point $p \in B$. Then, a formal versal space of $\mathcal{B}^{\mathrm{etw}}$ at $W/k(\bar{p})$ with an extended log twisting (4.3.1) is

$$\mathrm{Spf}\hat{\mathcal{O}}_{\bar{p}}[[x_1, ..., x_m, y_1, ..., y_{m'}]]/(x_1^{d_1} - \tau_1, ..., x_m^{d_m} - \tau_m),$$

where x_i, y_j are indeterminates;

$$(d_1, ..., d_m, \underbrace{0, ..., 0}_{m'}) : \overline{N}_{\bar{p}}^{U/B} \to \overline{N}_{\bar{p}};$$

$\hat{\mathcal{O}}_{\bar{p}} \ni \tau_i$ are $\alpha^{U/B}(e_i)$ for a chart $\mathbb{N}^m \to N^{U/B}$ at $\bar{p}$; and $m =$ the number of $\mathrm{Irr}U_{\bar{p}}^{\mathrm{sing}}$.

6.2.3. Note that $\mathfrak{M}_{g,n}^{\log} \times_{LOG} \mathcal{B}^{\mathrm{etw}}$ is equivalent to the stack whose objects are pairs of log prestable curves $(C, M_C)/(S, N)$ and extended log twisting FM type spaces $(W, M_W)/(S, N)$. Since $\mathfrak{M}_{g,n}^{\log} \times_{LOG} \mathcal{B}^{\mathrm{etw}}$ is a fiber product of algebraic stacks over an algebraic stack, it is algebraic. It is formally smooth over Λ, because the projection $\mathfrak{M}_{g,n}^{\log} \times_{LOG} \mathcal{B}^{\mathrm{etw}} \to \mathcal{B}^{\mathrm{etw}}$ is smooth by Proposition 3.14 in [14], and $\mathcal{B}^{\mathrm{etw}}$ is formally smooth over Λ.

6.2.4. Consider a log twisting (4.2.1); assume that the charts are global charts over S. Then the automorphism functor

$$\mathrm{Aut}_{W/S}(W/S, N \stackrel{d}{\leftarrow} N^{W/S})$$

over W/S, i.e., fixing W/S, is representable by

$$\mathbf{Spec}\mathcal{O}_S[z_1^{\pm 1}, ..., z_m^{\pm 1}]/(z_i^{d_i} - 1, z_i\alpha(e_i) - \alpha(e_i))$$

over S, where $\alpha : N \to \mathcal{O}_S$ is the structure map of the log structure. Similarly,

$$\mathrm{Aut}_{(C/S, W/S)}(C/S, W/S, N^{C/S} \stackrel{(\Gamma, \mathrm{id})}{\to} N \stackrel{(d,0)}{\leftarrow} N^{W/S}),$$

fixing all underlying scheme structures, is representable by

$$\mathbf{Spec}\mathcal{O}_S[z_1^{\pm 1}, ..., z_m^{\pm 1}]/(z_i^{G_i} - 1, z_i\alpha(e_i) - \alpha(e_i))$$

over S, where log twistings are given as in (5.2.1) with the global charts over S, and $G_i = \mathrm{GCD}(\Gamma_{i,j}, \forall j)$.

6.3. The stack of log stable maps

Denote by $\mathcal{U}$ be the stack of all pairs $(W/S, \sigma)$, where W/S is an object in $\mathcal{B}$ and σ is a section of $W \to S$. The arrows are morphisms in $\mathcal{B}$ preserving sections. Define the category $\overline{M}_{g,n}^{\log}(\mathcal{U}/\mathcal{B})$ of (g, n) log stable maps to FM type spaces in the stack $\mathcal{B}$. A morphism from $((f : (C', M_{C'}, \mathbf{p}') \to (W', M_{W'}))/(S', N_{S'}))$ to $((f : (C, M_C, \mathbf{p}) \to$

$(W, M_W))/(S, N_S))$ is a commutative diagram

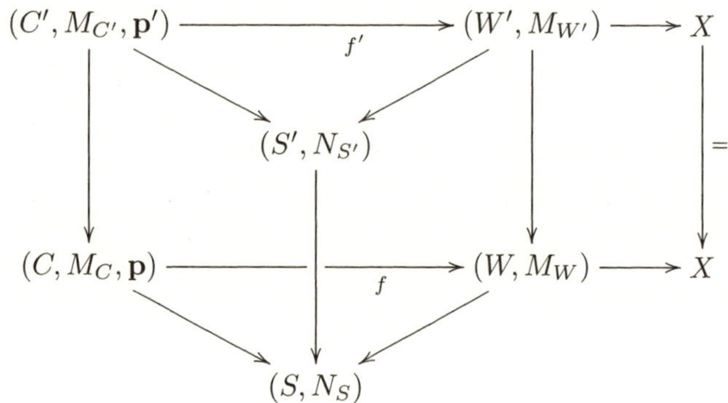

where two side squares with the edge $(S', N_{S'}) \to (S, N_S)$ are fiber products in log sense preserving labeled points, and the log structure $N_{S'}$ of S' is naturally isomorphic to the pullback of the log structure N_S of S. Then it is a log stack over (Sch/Λ); there is a natural map

$$\overline{M}^{\log}_{g,n}(\mathcal{U}/\mathcal{B}) \to \mathfrak{M}^{\log}_{g,n} \times_{LOG} \mathcal{B}^{\mathrm{etw}}.$$

Similarly define the stack $\overline{M}_{g,n}(\mathcal{U}/\mathcal{B})$ by taking off all log structures. There is a natural map

$$\overline{M}_{g,n}(\mathcal{U}/\mathcal{B}) \to \mathfrak{M}_{g,n} \times_{(\mathrm{Sch}/\Lambda)} \mathcal{B}.$$

Note that since f^b is uniquely determined by $\underline{f}$ and log structures on its source and target, as seen in 5.2.3, $\overline{M}^{\log}_{g,n}(\mathcal{U}/\mathcal{B})$ is a full substack of

$$\overline{M}_{g,n}(\mathcal{U}/\mathcal{B}) \times_{(\mathfrak{M}_{g,n} \times \mathcal{B})} (\mathfrak{M}^{\log}_{g,n} \times_{LOG} \mathcal{B}^{\mathrm{etw}}).$$

In what follows, a DM stack over Λ means an algebraic stack over Λ for which the diagonal $\mathcal{F} \to \mathcal{F} \times_{(\mathrm{Sch}/\Lambda)} \mathcal{F}$ is separated and unramified.

Theorem 6.3.1. *If $\overline{M}_{g,n}(\mathcal{U}/\mathcal{B})$ is an algebraic stack (resp. a proper DM stack) over Λ, then so is $\overline{M}^{\log}_{g,n}(\mathcal{U}/\mathcal{B})$.*

Proof. Since $\overline{M}_{g,n}(\mathcal{U}/\mathcal{B})$ is a full substack of an algebraic stack, the isomorphism functors are representable and of finite presentation. Now, the algebraic stack part of the statement follows from Remark 5.3.2. The properness follows from that of $\overline{M}_{g,n}(\mathcal{U}/\mathcal{B})$ and Lemma 6.3.2 since $\overline{M}^{\log}_{g,n}(\mathcal{U}/\mathcal{B})$ is of finite presentation over Λ. Due to the finiteness of automorphisms, the diagonal is unramified. Q.E.D.

Lemma 6.3.2. *Let R be a DVR with an algebraically closed residue field; let $(W/S, W \to X)$ be a log FM type space, where $S = \mathrm{Spec}R$; and let C/S be a prestable curve over S. Suppose that a log stable map $f_\xi : (C_\xi, M_{C_\xi}) \to (W_\xi, M_{W_\xi})$ over (ξ, N_ξ) is given, where ξ is the generic point of S. Assume that there is a stable admissible map $\underline{f} : C \to W$ over S, extending the underlying map $\underline{f_\xi}$. Then there exists a unique pair of a minimal log prestable curve $(\overline{C}, M_C)/(S, N)$ and an extended log twisted FM type space $(W, M_W)/(S, N)$, extending $(C_\xi, M_{C_\xi})/(\xi, N_\xi)$ and $(W_\xi, M_{W_\xi})/(\xi, N_\xi)$, respectively, such that the stable admissible map $\underline{f}$ becomes a log morphism over (S, N).*

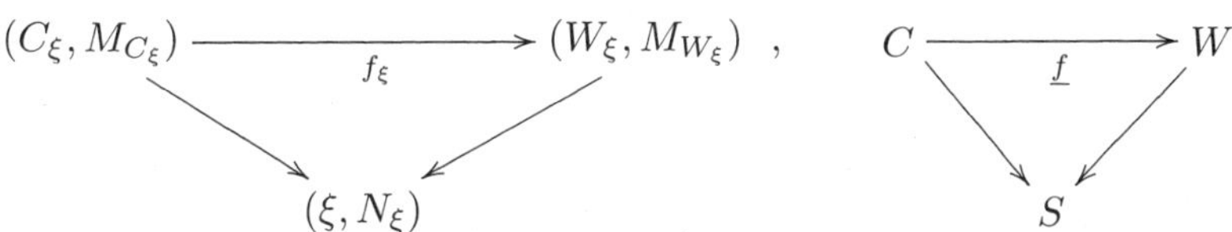

Proof. Let p be the closed point of S. We use the following index sets:

I_1 (resp. $I_1 \coprod I_2$)	$=$	an index set for the components of the singular locus of W at $\bar{\xi}$ (resp. p).
I'_1 (resp. $I'_1 \coprod I'_2$)	$=$	an index set for the nondistinguished nodes of C at $\bar{\xi}$ (resp. p).
I''_1 (resp. $I''_1 \coprod I''_2$)	$=$	an index set for the distinguished nodes of C at $\bar{\xi}$ (resp. p).

Then by 5.2.3 we may assume that the following compatible charts are given:

$$\begin{array}{ccccc} N_{\bar{\xi}}^{C/S} & \longrightarrow & N_{\bar{\xi}} & \longleftarrow & N_{\bar{\xi}}^{W/S} \\ \uparrow & & \uparrow & & \uparrow \\ \mathbb{N}^{I''_1 + I'_1} & \xrightarrow[(\Gamma, \mathrm{id})]{} & \mathbb{N}^{I_1 + I'_2} & \xleftarrow[(d^{(1)}, 0)]{} & \mathbb{N}^{I_1}, \end{array}$$

where $d^{(1)}$ is a monoid homomorphism from $\mathbb{N}^{I''_1}$ to $\mathbb{N}^{I_1}$, and the sums of index sets is used for the disjoint unions of them, for simplicity.

For $i \in I''_2$, let (u_i, v_i) be a special coordinate pair at the node p_i corresponding to i, and let (x_i, y_i) be the part of coordinates at $f(p_i)$ as in the admissible condition so that $x_i = u_i^{l_i}$ and $y_i = v_i^{l_i}$ under f. For $j \in I_2$, define $d_j^{(2)} = \mathrm{LCM}(l_i, \ \forall\, i \mapsto j)$, where $i \mapsto j$ means that $f(p_i)$ is in the component of W_p^{sing} corresponding to $j \in I_2$.

First define a prelog structure

$$\alpha : P := \mathbb{N}^{I_1+I_1'+I_2+I_2'} \to \mathcal{O}_S,$$

where α is determined by

$$\alpha(e_j) = \begin{cases} 0 & j \in I_1 + I_1' \\ (\alpha^{W/S}(e_j))^{1/d_j^{(2)}} & j \in I_2 \\ \alpha^{C/S}(e_j) & j \in I_2' \end{cases}$$

where $\alpha^{W/S}$ (resp. $\alpha^{C/S}$) is the structure map from $N^{W/S}$ (resp. $N^{C/S}$) to $\mathcal{O}_S$, and $(\alpha^{W/S}(e_j))^{1/d_j^{(2)}}$ is a root (we choose a choice of the roots) whose $d_j^{(2)}$-th power is $\alpha^{W/S}(e_j)$. Set $N = P^a$, the log structure associated to P. Now we construct a log prestable curve structure on C/S and an extended log twisting on W/S. It amounts to establishing certain log morphisms $N^{C/S} \to N \leftarrow N^{W/S}$. We define them by homomorphisms between their charts as in diagram

$$\begin{array}{ccccc} N^{C/S} & \longrightarrow & N & \longleftarrow & N^{W/S} \\ \uparrow & & \uparrow & & \uparrow \\ \mathbb{N}^{I_1''+I_2''+I_1'+I_2'} & \xrightarrow[(\Gamma_1,\Gamma_2,\mathrm{id},\mathrm{id})]{} & \mathbb{N}^{I_1+I_2+I_1'+I_2'} & \xleftarrow[(d^{(1)},d^{(2)},0,0)]{} & \mathbb{N}^{I_1+I_2} \end{array}$$

where $\Gamma_2(e_i) = \frac{d_j^{(2)}}{l_i} e_j$ for $i \in I_2''$, $i \mapsto j$, and $d^{(2)} = (d_j^{(2)})_{j \in I_2}$. The existence part of Lemma is verified.

When N is defined, the ambiguity occurs only in the choices of roots $\alpha^{W/S}(e_j)^{1/d_j^{(2)}}$. Let N' be a constructed one, using another choice. Then there is a unique isomorphism $N \to N'$ commuting with maps from $N^{C/S}$ and $N^{W/S}$: The isomorphism is determined by

$$\begin{array}{ccc} N & \to & N' \\ e_j & \mapsto & e_j + \alpha(e_j)/\alpha'(e_j), \end{array}$$

where α' is the log structure map of N', and $j \in I_2$ so that $\alpha(e_j)/\alpha'(e_j) \in \mathcal{O}_S^\times$. Q.E.D.

6.3.1. *Remarks* 1. Assume that the stability condition of admissible maps to U/B, with respect to π_X (i.e., fixing X), is an open condition. Then we claim that $\overline{M}_{g,n}(\mathcal{U}/\mathcal{B})$ is a DM stack over Λ by the following argument. Consider the stack $\overline{M}_{g,n}(U/B; X)$ of admissible maps to the rigid target U/B, *stable with respect to* π_X, which is a full substack of the

Kontsevich moduli space $\overline{M}_{g,n}(U)$ of (g,n) stable maps to U. Then the stack is a DM stack over Λ by Theorem 2.11 in [21]. This in turn shows that $\overline{M}_{g,n}(\mathcal{U}/\mathcal{B})$ is a DM stack over Λ. Here the quasi-separatedness can be directly shown.

2. The explicit description in 5.2.3 shows that when S is a geometric point Spec**k**, then for an underlying admissible map $\underline{f}$, the number of isomorphism classes of log prestable maps realizing $\underline{f}$ is finite. The numerical data l, d, Γ are uniquely determined. What remain indetermined are maps from $N^{W/S}$ and $N^{C/S}$ to N. The choices, however, are finite since f is a log morphism. We illustrate the reason by an example, for simplicity. Suppose that the ranks of $N^{W/S}$ and N are 1 so that we may write a map from $N^{W/S} = \mathbb{N} \oplus \mathbf{k}^\times$ to $N = \mathbb{N} \oplus \mathbf{k}^\times$ by $e_1 \mapsto de_1$. We express a map $N^{C/S} = \mathbb{N}^m \oplus \mathbf{k}^\times$ to N by $e_i \mapsto \Gamma_i e_i + \rho_i$. Note that ρ_i must satisfy constraint $\rho_i^{l_i} = 1$, since f is a log morphism. This shows the finiteness. Some of them could be isomorphic under maps $N \to N$ sending $e_1 \mapsto e_1 + \rho$, where $\rho \in \mathbf{k}^\times$ must satisfy the condition $\rho^d = 1$. Hence, in this example, there are $|(\Pi_i \mathbb{Z}/l_i\mathbb{Z})/(\mathbb{Z}/d\mathbb{Z})|$ many realizations of log prestable maps up to isomorphisms, where the action of $\mathbb{Z}/d\mathbb{Z}$ on $\mathbb{Z}/l_i\mathbb{Z}$ is given by $a \cdot (a_1, ..., a_m) = (\Gamma_1 a + a_1, ..., \Gamma_m a + a_m)$, $a \in \mathbb{Z}/d\mathbb{Z}$ and $a_i \in \mathbb{Z}/l_i\mathbb{Z}$.

3. Using the deformation and obstruction theory at the long exact sequence in Section 7, one may try directly to prove the first part of Theorem 6.3.1, using Artin's theorem in [4]. We do not pursue this approach here.

§7. Perfect Obstruction Theory

7.1.

In this subsection, we show that there is a natural perfect obstruction theory on the log algebraic stack $\overline{M}^{\log}_{g,n}(\mathcal{U}/\mathcal{B})$. The method parallel to [6, 5] will work if the cotangent complexes are replaced by the logarithmic cotangent complexes, as following. We first consider the diagram

$$\begin{array}{ccc} \mathcal{C} & \xrightarrow[f]{} & \mathcal{U}^{\mathrm{etw}} \\ \downarrow{\scriptstyle \pi} & & \\ \mathcal{M} & & \end{array}$$

where $\mathcal{C}$ (resp. $\mathcal{U}^{\mathrm{etw}}$) is the universal family of the moduli log stack $\mathcal{M} := \overline{M}^{\log}_{g,n}(\mathcal{U}/\mathcal{B})$ (resp. $\mathcal{B}^{\mathrm{etw}}$) and f is the evaluation map. We regard f and π as log morphisms between log stacks. Let $\mathfrak{C}$ be the universal family of $\mathfrak{M}_{g,n}^{\log}$ and let $\mathfrak{MB}$ be $\mathfrak{M}_{g,n}^{\log} \times_{LOG} \mathcal{B}^{\mathrm{etw}}$. There is the composite of natural maps

$$f^* L^\bullet_{\mathcal{U}^{\mathrm{etw}}/\mathcal{B}^{\mathrm{etw}}} \to L^\bullet_{\mathcal{C}/\mathcal{B}^{\mathrm{etw}}} \to L^\bullet_{\mathcal{C}/\mathfrak{C}\times_{LOG}\mathcal{B}^{\mathrm{etw}}} \cong \pi^* L^\bullet_{\mathcal{M}/\mathfrak{MB}}$$

between logarithmic cotangent complexes. See [29] for the definition of logarithmic complexes and functorial properties. Taking the tensor product of the composite and the relative dualizing sheaf ω_π of π, we obtain an element in

$$\mathrm{Hom}_{D^b(\mathcal{C})}(f^* L^\bullet_{\mathcal{U}^{\mathrm{etw}}/\mathcal{B}^{\mathrm{etw}}} \otimes \omega_\pi[1], \pi^* L^\bullet_{\mathcal{M}/\mathfrak{MB}} \otimes \omega_\pi[1]).$$

This in turn yields an element in

$$\mathrm{Hom}_{D^b(\mathcal{M})}(R\pi_* f^* L^\bullet_{\mathcal{U}^{\mathrm{etw}}/\mathcal{B}^{\mathrm{etw}}} \otimes \omega_\pi[1], L^\bullet_{\mathcal{M}/\mathfrak{MB}})$$

since

$$\pi^* L^\bullet_{\mathcal{M}/\mathfrak{MB}} \otimes \omega_\pi[1] \cong \pi^! L^\bullet_{\mathcal{M}/\mathfrak{MB}}$$

and $R\pi_!$ is left adjoint to $\pi^!$. Finally by Grothendieck–Verdier duality we have a natural homomorphism

$$E^\bullet \to L^{\bullet \geq -1}_{\mathcal{M}/\mathfrak{MB}} \tag{7.1.1}$$

where $E^\bullet := (R\pi_* f^* T^\dagger_{\mathcal{U}^{\mathrm{etw}}/\mathcal{B}^{\mathrm{etw}}})^\vee$ and $L^{\bullet \geq -1}_{\mathcal{M}/\mathfrak{MB}}$ is two-term $[-1, 0]$ truncation of the logarithmic relative cotangent complex. Note that the latter complex is isomorphic to the usual relative cotangent complex since the map $\mathcal{M} \to \mathfrak{MB}$ is strict. The homomorphism (7.1.1) of the complexes is a perfect obstruction theory since the relative deformation/obstruction theory for log morphisms is as expected as in Theorem 5.9 of [29], and $E^\bullet$ can be realized as a two-term complex of locally free coherent sheaves. This defines a virtual fundamental class of $\overline{M}_{g,n}^{\log}(\mathcal{U}/\mathcal{B})$ by [23, 6, 5].

The absolute obstruction theory $F^\bullet \to L^\bullet_{\mathcal{M}}$ can be obtained as in [9, 16] so that there is a distinguished triangle:

$$\tau^* L^\bullet_{\mathfrak{MB}} \to F^\bullet \to E^\bullet, \tag{7.1.2}$$

where τ is the natural map $\mathcal{M} \to \mathfrak{MB}$. Consider a deformation situation of the extensions over $\mathrm{Spec}A[I]$ of a given $(f, C^\dagger, W^\dagger, S^\dagger)$ over $S := \mathrm{Spec}A$, where A is a reduced noetherian Λ-algebra; I is a finite A-module; and $A[I]$ is the trivial ring extension of A by I. Let $\mathrm{Aut}(C^\dagger \times_{S^\dagger} W^\dagger)$ be the set of automorphisms of the trivial extension of $C^\dagger \times_{S^\dagger} W^\dagger$ over $\mathrm{Spec}A[I]^\dagger$, whose restriction to $S^\dagger$ is the identity. Here the log structure on $\mathrm{Spec}A[I]^\dagger$ is given by the pullback of $\mathrm{Spec}A[I] \to \mathrm{Spec}A$. Also, let $\mathrm{Def}_I(C^\dagger \times_{S^\dagger} W^\dagger)$ be the set of isomorphism classes of the extensions over $S^\dagger$. Then combining Proposition 3.14 in [14] and Theorem 5.9 in [29], we obtain an A-module exact sequence

$$\begin{aligned} 0 &\to \mathrm{Aut}_I(C^\dagger \times_{S^\dagger} W^\dagger) \to \mathrm{RelDef}(f) = R^0\pi_* f^* T_{W^\dagger/S^\dagger} \otimes_{\mathcal{O}_S} I \\ \to &\ \mathrm{Def}(f) \to \mathrm{Def}_I(C^\dagger \times_{S^\dagger} W^\dagger) \xrightarrow{\varphi} \mathrm{RelOb}(f) = R^1\pi_* f^* T_{W^\dagger/S^\dagger} \otimes_{\mathcal{O}_S} I \\ \to &\ \mathrm{Ob}(f) \to 0 \end{aligned}$$

where $\mathrm{Ob}(f)$ is defined to be the cokernel of φ. When $I \cong A$, we may also use (7.1.2) and Proposition III, 12.2 in [12] to derive the above long exact sequence.

7.2. Log admissible covers

Let A be an Artinian local ring over $\mathbf{k}$, with the maximal ideal $\mathfrak{m}_A$. Consider a small extension R of A by I and an admissible map f, locally at a distinguished node described by

$$\begin{array}{ccc} (A[x, y, z_1, ..., z_{r-1}]/(xy - \tau))^{\mathrm{sh}} & \xrightarrow{f^*} & (A[u, v]/(uv - t))^{\mathrm{sh}} \\ x,\ y & \mapsto & u^l,\ v^l, \end{array}$$

where superscript sh means the strict henselianization at the 'origin'. We want to compute a local obstruction of extending f to an admissible map over a given extended domain and a given extended target near the node. The obstruction is $\tilde{\tau} - \tilde{t}^l \in I$ as explained in [10] if the extensions are given by $xy = \tilde{\tau} \in \mathfrak{m}_R$ and $uv = \tilde{t} \in \mathfrak{m}_R$ (locally at singular points), respectively. Here $\tilde{\tau}$ and $\tilde{t}$ are extensions of τ and t, respectively. This vanishes if f is a log morphism and the extended source and target are logarithmic ones. Indeed, the equation $[\log \tilde{\tau}] = [l \log \tilde{t}]$ in $\overline{N}_R$ implies that $\tilde{\tau} = (1 + c)\tilde{t}^l$ where $c \in I$. Since $\tilde{t} \in \mathfrak{m}_R$, we see that $\tilde{\tau} = \tilde{t}^l$.

Provided that f is a log morphism, for a log extension on the target with $xy = \tilde{\tau}$, there is a unique log extension with $uv = \tilde{t}$ such that $\tilde{\tau} = \tilde{t}^l$. This is due to the existence of element $\tilde{s} \in \mathfrak{m}_R$ such that $\tilde{s}^d = \tilde{\tau}$ and $s^\Gamma = t$, where $\tilde{s}$ is an extension of s. Hence, an infinitesimal deformation of $(W, M_W)/(S, N)$ uniquely determines an infinitesimal deformation of $(C, M_C)/(S, N)$ at distinguished nodes. In particular, this shows that when the target is a projective smooth curve X, the natural map

$$\begin{array}{ccc} \overline{M}^{\log}_{g,n,\mu}(\mathfrak{X}^+/\mathfrak{X}, \beta) & \xrightarrow{\Psi} & X[n]^{\mathrm{tw}} \\ (f : (C, M_C, \mathbf{p}) \to (W, M_W))/(S, N) & \mapsto & (W/S, N \leftarrow N^{W/S}, f(\mathbf{p})) \end{array}$$

is étale, where $\overline{M}^{\log}_{g,n,\mu}(\mathfrak{X}^+/\mathfrak{X}, \beta)$ is the moduli stack of (g, n, β) log stable μ-ramified maps (5.2.6), and $X[n]^{\mathrm{tw}}$ is the stack of n-pointed log twisted stable FM spaces of X. Here 'stable' means that there are at least two special points on screens of FM spaces. By the following Lemma $X[n]^{\mathrm{tw}}$ is an open stack of $\mathcal{L}_{X[n]}$. (Note that $X[n]^{\mathrm{tw}}$ is not separated over $\mathbf{k}$ if $n \geq 2$.) Essentially, it is already proven in [35] that the map Ψ is étale.

Lemma 7.2.1. ([15]) *Let X be a smooth variety over $\mathbf{k}$. The stack of FM spaces with n distinct, smooth, sections is equivalent to the scheme $X[n]$ with the universal space $X[n]^+$.*

According to 6.2.2, $X[n]^{\mathrm{tw}}$ is smooth over $\mathbf{k}$. Therefore, the finite map from the smooth stack $\overline{M}^{\log}_{g,n,\mu}(\mathfrak{X}^+/\mathfrak{X},\beta)$ to $\overline{M}_{g,n,\mu}(\mathfrak{X}^+/\mathfrak{X},\beta)$ is the normalization map. The normalization is also constructed by the stack of balanced twisted covers in [1].

§8. Chain Type

8.1.

In this section, we provide a modular desingularization of the main component of the moduli space of elliptic stable maps to a projective space. The main component is, by definition, the irreducible component of the moduli space, containing all the elliptic stable maps with smooth domains. This makes sense due to the connectedness result in [18]. First, note that any prestable curve C over $\mathbf{k}$ of genus 1 has a unique subcurve C_0 which is either arithmetic $g = 1$ irreducible component or a loop of rational curves. We call C_0 the essential part of the curve C. Its dual graph looks like:

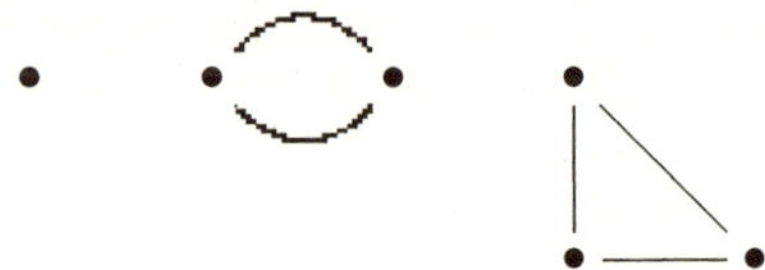

Note that the dualizing sheaf ω_{C_0} is trivial. Let $\overline{M}^{\log,\,\mathrm{ch}}_1(\mathfrak{X}^+/\mathfrak{X},\beta)$ be the moduli stack of $(g = 1, n = 0, \beta \neq 0)$ log stable maps (f, C, W) satisfying the following conditions additional to those in 5.2.5. For every $s \in S$:

- Every end component of $W_{\bar{s}}$ contains the entire image of the essential part of $C_{\bar{s}}$ under $f_{\bar{s}}$.
- The image of the essential part of $C_{\bar{s}}$ is nonconstant.

Here, it is possible that some of irreducible components in the essential part are mapped to points. Note that the dual graph of the target W_s must be a chain. Such a log stable map is called an elliptic log stable map to a chain type FM space W of the smooth projective variety X. Set $\overline{M}^{\log,\,\mathrm{ch}}_1(\mathfrak{X}^+/\mathfrak{X}) := \coprod_\beta \overline{M}^{\log,\,\mathrm{ch}}_1(\mathfrak{X}^+/\mathfrak{X},\beta)$.

8.2. Proof of Main Theorem B

By Theorem 6.3.1, it is enough to prove the properties for the moduli stack of the non-log version of elliptic log stable maps to chain type FM spaces of X. It is a DM stack of finite type over $\mathbf{k}$ by Remark 1 in 6.3.1.

Properness. We use the valuative criterion of properness. We show that the criterion holds, by adapting the properness argument in [15]. Let $f_\xi : C_\xi \to W_\xi$ be an elliptic stable map to a chain type W over a quotient field of the DVR $R = \mathbf{k}[[t]]$, and let ξ and p be the generic point and the closed point of R, respectively. For simplicity assume that $W_\xi = X \times \xi$.

— Uniqueness: Suppose that there are two extensions $f_i : C^{(i)} \to W^{(i)}$ of f over R. To show that f_1 and f_2 are equivalent over f, it is enough to verify that $W^{(1)}$ and $W^{(2)}$ are isomorphic over X and R. To prove it, we construct sections of $W^{(i)} \to \mathrm{Spec}R$. Consider two sections passing through the end component of $W_p^{(i)}$. Also, consider, for each ruled component, a section passing through the ruled component. Let N_i be two plus the number of ruled component of $W_p^{(i)}$ unless $W_p^{(i)} = X$. Let $N_i = 0$ if $W_p^{(i)} = X$. We may assume that those N_i sections are pairwise distinct at p. Let W be the FM type space of X over R, separating each set of N_i sections: W is defined to be $g^*X[N_1, N_2]^+$, where $g : \mathrm{Spec}R \to X[N_1, N_2]$ is the map associated to two sets of sections. Here $X[N_1, N_2]$ denotes the compactification of configurations of two sets of N_i distinct labeled points, $i = 1, 2$. The precise construction of $X[N_1, N_2]$ can found in [15]. The space $X[N_1, N_2]$ has the universal space $X[N_1, N_2]^+$ with two sets of N_i distinct sections. We claim that W is isomorphic to W_i over X and R. To see it, we first remove two sections associated to the end component of $W^{(2)}$. Then still, W with the remained sections is stable since the domain of the stable map limit to the new target W has genus 1 and the stable map limit agrees with f_i after the contraction of suitable screens of the target W and then stable contraction of the domain curve. Now remove the rest of sections associated ruled components of $W^{(2)}$. W with N_1 sections remains stable otherwise there will be a component of curve $C_p^{(1)}$ which is mapped into a singular locus of $W^{(1)}$. Thus, $W \cong W^{(1)}$ over X and R. The same method shows that $W \cong W^{(2)}$ over X and R, and hence $W^{(1)} \cong W^{(2)}$ over X and R.

— Existence: Consider each irreducible component C_γ of the essential subcurve of C_ξ where $f_\xi(C_\gamma)$ is nonconstant. The f_ξ restricted to C_γ is μ-ramified, for some $\mu = (\mu_1, ..., \mu_n)$. Consider the screens created by taking the limit of f_ξ restricted to C_γ as a μ-ramified stable map. See [15] for the construction. There is a natural partial order on the set of all the screens for all γ. Take the first one (which needs the smallest magnifying power to be visible). It exits since $g = 1$. Now consider two general sections passing through the first screen, and then the stable map limit with the new target separating the sections. If necessary, take a further expansion along the singular locus of the target at the special fiber so that there are no domain components which are mapped into singular locus of the new target. This process ends in a finite step as in [15] and yields a stable admissible map to a chain type FM space of X over R, which satisfies the desired requirements.

Smoothness. This is simply because the relative obstruction space in Section 7 vanishes as following. Let $f : C^\dagger \to W^\dagger$ over $\mathbf{k}^\dagger$ be an elliptic log stable map to a chain type FM space W of $X := \mathbb{P}^r_{\mathbf{k}}$. In what follows, $T^\dagger$ denotes the log tangent sheaf $T_{W^\dagger/\mathbf{k}^\dagger}$.

Decompose C to be the union $\bigcup C_i$ of the essential part C_0 and the irreducible rational components C_i, $i \geq 1$; let f_i be the map f restricted to C_i. To prove $H^1(C, f^*T^\dagger) = 0$, we claim that $H^1(C_i, f^*T^\dagger) = 0$ for all i and the evaluation map

$$\bigoplus_i H^0(C_i, f_i^*T^\dagger) \to \bigoplus_{\{i,j:i\neq j\}} f^*T^\dagger|_{C_i\cap C_j}$$

is surjective. For the latter, by the induction argument on the number of irreducible rational components C_i, $i \geq 1$, it suffices that $H^1(C_i, f_i^*T^\dagger \otimes \mathcal{O}_{C_i}(-p)) = 0$ if p is a point in C_i and $i \geq 1$. When $W = X$ case, it follows from the Euler sequence, since $H^1(C_0, f_0^*\mathcal{O}_X(1)) = H^0(C_0, f_0^*\mathcal{O}_X(-1) \otimes \omega_{C_0})^\vee = 0$ and $H^1(C_i, f_i^*\mathcal{O}_X(1) \otimes \mathcal{O}_{C_i}(-p)) = 0$ for all $i \geq 1$. Similarly, when W is singular, the claim follows from the generalized Euler sequences:

$$\begin{array}{ccccccccc}
0 & \to & \mathcal{O}_Y & \to & (\bigoplus \mathcal{O}_Y(1)) \oplus \mathcal{O}_Y & \to & T^\dagger_{|Y} & \to & 0 \\
0 & \to & \mathcal{O}_Y(-E) & \to & (\bigoplus \pi^*\mathcal{O}_{\mathbb{P}^r}(1))(-E) & \to & T^\dagger_{|Y} & \to & 0 \\
0 & \to & \mathcal{O}_Y(-E) & \to & (\bigoplus \pi^*\mathcal{O}_{\mathbb{P}^r}(1)(-E)) \oplus \mathcal{O}_Y(-E) & \to & T^\dagger_{|Y} & \to & 0
\end{array}$$

for the logarithmic tangent sheaf restricted to the end, the root, and a ruled component Y of W, respectively. The explanation of the sequences is in order. The first sequence is standard. The second one can be obtained from an isomorphism

$$\pi^*T_X(-E) \to T^\dagger_{|Y}.$$

This isomorphism can be seen by a local description of blowup $\pi : Y \to X$ at a point, with the exceptional divisor E (for example, see the proof of Lemma 1 in [17]). The third one results from the gluing of the previous two sequences. Q.E.D.

References

[1] D. Abramovich, A. Corti and A. Vistoli, Twisted bundles and admissible covers, Special issue in honor of Steven L. Kleiman, Comm. Algebra, **31** (2003), 3547–3618.

[2] D. Abramovich and B. Fantechi, Orbifold techiques in degeneration formula, in preparation.

[3] D. Abramovich, M. Olsson and A. Vistoli, Twisted stable maps to tame Artin stacks, arXiv:0801.3040.

[4] M. Artin, Versal deformations and algebraic stacks, Invent. Math., **27** (1974), 165–189.

[5] K. Behrend, Gromov–Witten invariants in algebraic geometry, Invent. Math., **127** (1997), 601–617.

[6] K. Behrend and B. Fantechi, Intrinsic normal cone, Invent. Math., **128** (1997), 45–88.

[7] P. Deligne and D. Mumford, The irreducibility of the space of curves of given genus, Inst. Hautes Études Sci. Publ. Math., **36** (1969), 75–109.

[8] W. Fulton and R. MacPherson, A compactification of configuration spaces, Ann. of Math. (2), **139** (1994), 183–225.

[9] T. Graber and R. Pandharipande, Localization of virtual classes, Invent. Math., **135** (1999), 487–518.

[10] T. Graber and R. Vakil Relative virtual localization and vanishing of tautological classes on moduli spaces of curves, Duke Math. J., **130** (2005), 1–37.

[11] J. Harris and D. Mumford, On the Kodaira dimension of the moduli space of curves, with an appendix by W. Fulton, Invent. Math., **67** (1982), 23–88.

[12] R. Hartshorne, Algebraic Geometry, Grad. Texts in Math., **52**, Springer-Verlag, New York-Heidelberg, 1977.

[13] F. Kato, Log smooth deformation and moduli of log smooth curves, Internat. J. Math., **11** (2000), 215–232.

[14] K. Kato, Logarithmic structures of Fontaine–Illusie, In: Algebraic Analysis, Geometry, and Number Theory, Baltimore, MD, 1988, Johns Hopkins Univ. Press, Baltimore, MD, 1989, pp. 191–224.

[15] B. Kim, A. Kresch and Y.-G. Oh, A compactification of the space of maps from curves, in preparation.

[16] B. Kim, A. Kresch and T. Pantev, Functoriality in intersection theory and a conjecture of Cox, Katz, and Lee, J. Pure Appl. Algebra, **179** (2003), 127–136.

[17] B. Kim, Y. Lee and K. Oh, Rational curves on blowing-ups of projective spaces, Michigan Math. J., **55** (2007), 335–345.

[18] B. Kim and R. Pandharipande, The connectedness of the moduli space of maps to homogeneous spaces, In: Symplectic Geometry and Mirror Symmetry, Seoul, 2000, World Sci. Publ., River Edge, NJ, 2001, pp. 187–201.

[19] B. Kim and F. Sato, A generalization of Fulton–MacPherson configuration spaces, Selecta Math. (N.S.), **15** (2009), 435–443.

[20] G. Laumon and L. Moret-Bailly, Champs Algébriques, Ergeb. Math. Grenzgeb. (3), **39**, Springer-Verlag, 2000.

[21] J. Li, Stable morphisms to singular schemes and relative stable morphisms, J. Differential Geom., **57** (2000), 509–578.

[22] J. Li, A degeneration formula of GW-invariants, J. Differential Geom., **60** (2002), 199–293.

[23] J. Li and G. Tian, Virtual moduli cycles and Gromov–Witten invariants of algebraic varieties, J. Amer. Math. Soc., **11** (1998), 119–174.

[24] J. Li and A. Zinger, On the genus-one Gromov–Witten invariants of complete intersections, J. Differential Geom., **82** (2009), 641–690.

[25] S. Mochizuki, The geometry of the compactification of the Hurwitz scheme, Publ. Res. Inst. Math. Sci., **31** (1995), 355–441.

[26] A. Ogus, Lectures on Logarithmic Algebraic Geometry, TeXed note in 2006.

[27] M. Olsson, Universal log structures on semi-stable varieties, Tohoku Math. J. (2), **55** (2003), 397–438.

[28] M. Olsson, Logarithmic geometry and algebraic stacks, Ann. Sci. École Norm. Sup. (4), **36** (2003), 747–791.

[29] M. Olsson, The logarithmic cotangent complex, Math. Ann., **333** (2005), 859–931.

[30] M. Olsson, Deformation theory of representable morphisms of algebraic stacks, Math. Z., **253** (2006), 25–62.

[31] M. Olsson, (Log) twisted curves, Compos. Math., **143** (2007), 476–494.

[32] E. Sernesi, Deformations of Algebraic Schemes, Grundlehren Math. Wiss., **334**, Springer-Verlag, 2006.

[33] B. Siebert, Log Gromov–Witten invariants, in preparation.

[34] R. Vakil and A. Zinger, A Desingularization of the main component of the moduli space of genus-one stable maps into $\mathbb{P}^n$, Geom. Topol., **12** (2008), 1–95.

[35] S. Wewers, Construction of Hurwitz spaces, Thesis, Universität-Gesamthochschule Essen, 1998.

[36] A. Zinger, Reduced genus-one Gromov–Witten invariants, arXiv:math/0507103.

[37] A. Zinger, Standard versus reduced genus-one Gromov–Witten invariants, Geom. Topol., **12** (2008), 1203–1241.

School of Mathematics
Korea Institute for Advanced Study
Hoegiro 87, Dongdaemun-gu, Seoul, 130-722
Korea
E-mail address: `bumsig@kias.re.kr`

Advanced Studies in Pure Mathematics 59, 2010
New Developments in Algebraic Geometry,
Integrable Systems and Mirror Symmetry (Kyoto, 2008)
pp. 201–223

Integrable structure of melting crystal model with external potentials

Toshio Nakatsu and Kanehisa Takasaki

Abstract.

This is a review of the authors' recent results on an integrable structure of the melting crystal model with external potentials. The partition function of this model is a sum over all plane partitions (3D Young diagrams). By the method of transfer matrices, this sum turns into a sum over ordinary partitions (Young diagrams), which may be thought of as a model of q-deformed random partitions. This model can be further translated to the language of a complex fermion system. A fermionic realization of the quantum torus Lie algebra is shown to underlie therein. With the aid of hidden symmetry of this Lie algebra, the partition function of the melting crystal model turns out to coincide, up to a simple factor, with a tau function of the 1D Toda hierarchy. Some related issues on 4D and 5D supersymmetric Yang–Mills theories, topological strings and the 2D Toda hierarchy are briefly discussed.

§1. Introduction

The melting crystal model is a model of statistical mechanics that describes a melting corner of a semi-infinite crystal (Figure 1). The crystal is made of unit cubes, which are initially placed at regular positions and fills the positive octant $x, y, z \geq 0$ of the three dimensional Euclidean space. As the crystal melts, a finite number of cubes are removed from the corner. The present model excludes such crystals that have "overhangs" viewed from the $(1, 1, 1)$ direction. In other words, the complement of the crystal in the positive octant is assumed to be a 3D analogue of Young diagrams (Figure 2). Since 3D Young diagrams are represented by "plane partitions", the melting crystal model is also referred to as a model of "random plane partitions".

Received July 31, 2008.
2000 *Mathematics Subject Classification.* Primary 35Q58; Secondary 17B65, 82B20.

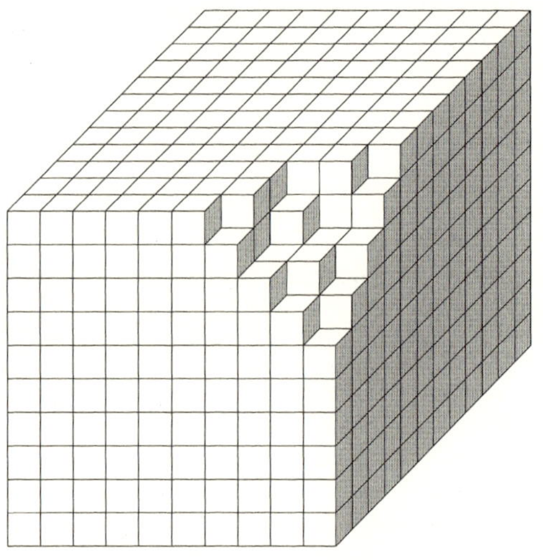

Fig. 1. Melting crystal corner

Though combinatorics of plane partitions has a rather long history [1], Okounkov and Reshetikhin [2] proposed an entirely new approach in the course of their study on a kind of stochastic process of random partitions (the Schur process). Their approach was based on "diagonal slices" of 3D Young diagrams and "transfer matrices" between those slices. As a byproduct, they could re-derive a classical result of MacMahon [1] on the generating function of the numbers of plane partitions. Actually, this generating function is nothing but the partition function of the aforementioned melting crystal model. The method of Okounkov and Reshetikhin was soon generalized [3] to deal with the topological vertex [4, 5] of A-model topological strings on toric Calabi–Yau threefolds.

The melting crystal model is also closely related to supersymmetric gauge theories. Namely, with slightest modification, the partition function can be interpreted as the instanton sum of 5D $\mathcal{N} = 1$ supersymmetric (SUSY) $U(1)$ Yang–Mills theory on partially compactified space-time $\mathbf{R}^4 \times S^1$ [6]. This instanton sum is a 5D analogue of Nekrasov's instanton sum for 4D $\mathcal{N} = 2$ SUSY gauge theories [7, 8]. The 4D instanton sum is a statistical sum over ordinary partitions (or "colored" partitions in the case of $SU(N)$ theory), hence a model of random partitions. Nekrasov and Okounkov [9] used such models of random partitions to re-derive the Seiberg–Witten solutions [10] of 4D $\mathcal{N} = 2$ SUSY gauge

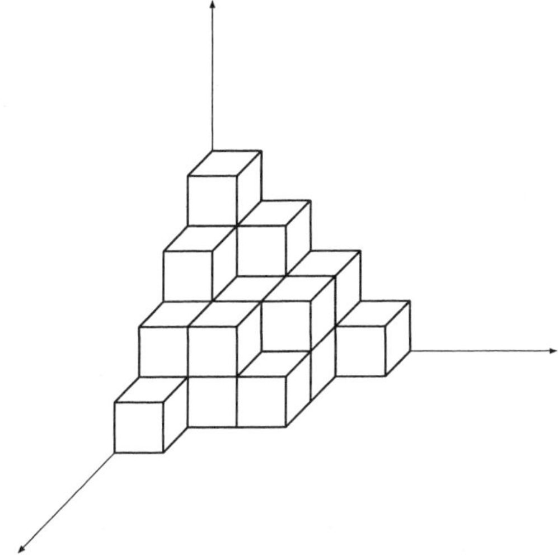

Fig. 2. 3D Young diagram as complement of crystal corner

theories. Actually, by the aforementioned method of transfer matrices, the statistical sum over plane partitions can be reorganized to a sum over partitions. This is a kind of q-deformations of 4D instanton sums. A 5D analogue of the Seiberg–Witten solution can be derived from this q-deformed instanton sum [9, 11].

In this paper, we review our recent results [12] on an integrable structure of the melting crystal model (and the 5D $U(1)$ instanton sum) with external potentials. The partition function $Z_p(t)$ of this model is a function of the coupling constants $t = (t_1, t_2, \ldots)$ of the external potentials. A main conclusion of these results is that $Z_p(t)$ is, up to a simple factor, a tau function of the 1D Toda hierarchy, in other words, a tau function $\tau_p(t, \bar{t})$ of the 2D Toda hierarchy [13] that depends only on the difference $t - \bar{t}$ of the two sets $t, \bar{t}$ of time variables. To derive this conclusion, we first rewrite $Z_p(t)$ in terms of a complex fermion system. In the case of 4D instanton sum, such a fermionic representation was proposed by Nekrasov et al. [14, 9]. In the present case, we can use the aforementioned transfer matrices [2] to construct a fermionic representation. This fermionic representation, however, does not take the form of a standard fermionic representation of the (1D or 2D) Toda hierarchy [15, 16]. To resolve this problem, we derive a set of algebraic relations (referred to as "shift symmetry") satisfied by the transfer matrices and a set of fermion bilinear forms. (Actually, these fermion bilinear forms turn out to give a realization of "quantum torus Lie algebra".) These algebraic relations enable us to rewrite the fermionic representation of $Z_p(t)$ to the standard form of Toda tau functions.

In the 4D case, a similar partition function with external potentials has been studied by Marshakov and Nekrasov [17, 18]. According to their results, the 1D Toda hierarchy is also a relevant integrable structure therein. Unfortunately, our method developed for the 5D case relies heavily on the structure of quantum torus Lie algebra, which ceases to exist in the 4D setup. We shall return to this issue, along with some other issue, in the end of this paper.

This paper is organized as follows. Section 2 is a brief review of the melting crystal model and its mathematical background. Section 3 presents the fermionic formula of the partition function. The method of transfer matrices is reviewed in detail. Section 4 deals with the quantum torus Lie algebra and its shift symmetries. In Section 5, we use this symmetry to rewrite the fermionic representation of the partition function to the standard form as a Toda tau function. Section 6 is devoted to concluding remarks.

§2. Melting crystal model

2.1. Young diagrams and partitions

Let us recall [19] that an ordinary 2D Young diagram is represented by an integer partition, namely, a sequence

$$\lambda = (\lambda_1, \lambda_2, \ldots), \quad \lambda_1 \geq \lambda_2 \geq \cdots,$$

of nonincreasing integers $\lambda_i \in \mathbf{Z}_{\geq 0}$ with only a finite number of λ_i's being nonzero. λ_i is the length of i-th row of the Young diagram viewed as a collection of unit squares. We shall always identify such a partition λ with a Young diagram. The total area of the diagram is given by the degree

$$|\lambda| = \sum_i \lambda_i$$

of the partition.

It was shown by Euler that the generating function of the number $p(N)$ of partitions λ of degree N has an infinite product formula:

$$\sum_{N=0}^{\infty} p(N) q^N = \prod_{n=1}^{n} (1 - q^n)^{-1}, \tag{2.1}$$

where q is assumed to be in the range $0 < q < 1$. One can interpret this generating function as the partition function of a model of statistical mechanics,

$$Z_{2\mathrm{D}} = \sum_{N=0}^{\infty} p(N)q^N = \sum_{\lambda} q^{|\lambda|},$$

in which each partition λ is assigned an energy proportional to $|\lambda|$, and q is related to the temperature T as $q = e^{-\mathrm{const.}/T}$

2.2. 3D Young diagrams and plane partitions

A 3D Young diagram can be represented by a "plane partition", namely, a 2D array

$$\pi = (\pi_{ij})_{i,j=1}^{\infty} = \begin{pmatrix} \pi_{11} & \pi_{12} & \cdots \\ \pi_{21} & \pi_{22} & \cdots \\ \vdots & \vdots & \ddots \end{pmatrix}$$

of nonnegative integers $\pi_{ij} \in \mathbf{Z}_{\geq 0}$ such that

$$\pi_{ij} \geq \pi_{i,j+1}, \quad \pi_{ij} \geq \pi_{i+1,j}.$$

π_{ij} is the height of the stack of cubes placed at the (i,j)-th position of the plane. We shall identify such a plane partition with the corresponding 3D Young diagram. The total volume of the 3D Young diagram is given by

$$|\pi| = \sum_{i,j=1}^{\infty} \pi_{ij}.$$

As an analogue of $p(N)$, one can consider the number $\mathrm{pp}(N)$ of plane partitions π with $|\pi| = N$. The generating function of these numbers was studied by MacMahon [1] and shown to be given, again, by an infinite product:

$$\sum_{N=0}^{\infty} \mathrm{pp}(N)q^N = \prod_{n=1}^{\infty} (1-q^n)^{-n}. \tag{2.2}$$

The right hand side is now called the MacMahon function. In statistical mechanics, this generating function becomes the partition function

$$Z_{3\mathrm{D}} = \sum_{N=0}^{\infty} \mathrm{pp}(N)q^N = \sum_{\pi} q^{|\pi|}$$

of a canonical ensemble of plane partitions, in which each plane partition π has an energy proportional to the volume $|\pi|$.

We shall deform this simplest model by external potentials. To this end, we have to introduce the notion of "diagonal slices" of a plane partition.

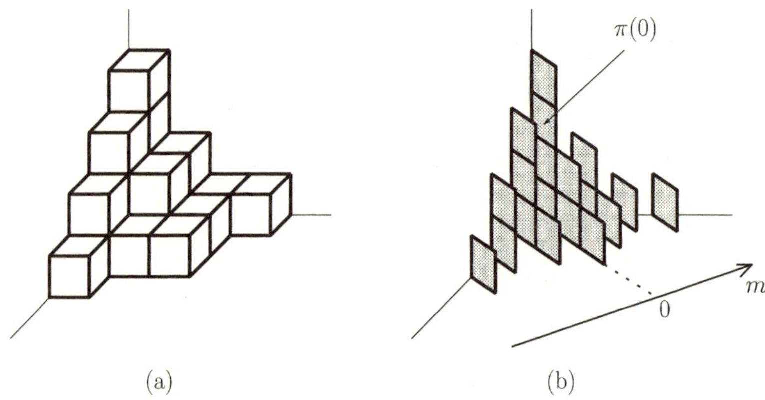

Fig. 3. Diagonal slices (b) of plane partition (a)

2.3. Diagonal slices of 3D Young diagrams

Given a plane partition $\pi = (\pi_{ij})_{i,j=1}^{\infty}$, the partition

$$\pi(m) = \begin{cases} (\pi_{i,i+m})_{i=1}^{\infty} & \text{if} \quad m \geq 0 \\ (\pi_{j-m,j})_{j=1}^{\infty} & \text{if} \quad m < 0 \end{cases}$$

is called the m-th diagonal slice of π. These partitions $\{\pi(m)\}_{m=-\infty}^{\infty}$ represent a sequence of 2D Young diagrams that are literally obtained by slicing the 3D Young diagrams (Figure 3).

The diagonal slices are not arbitrary but satisfy the condition [2, 3]

$$\cdots \prec \pi(-2) \prec \pi(-1) \prec \pi(0) \succ \pi(1) \succ \pi(2) \succ \cdots, \tag{2.3}$$

where "$\succ$" denotes *interlacing relation*, namely,

$$\lambda = (\lambda_1, \lambda_2, \ldots) \succ \mu = (\mu_1, \mu_2, \ldots) \overset{\text{def}}{\Longleftrightarrow} \lambda_1 \geq \mu_1 \geq \lambda_2 \geq \mu_2 \geq \cdots.$$

Because of these interlacing relations, a pair (T, T') of semi-standard tableaux is obtained on the main diagonal slice $\lambda = \pi(0)$ by putting "$m+1$" in boxes of the skew diagram $\pi(\pm m)/\pi(\pm(m+1))$.

By this mapping $\pi \mapsto (T, T')$, the partition function Z_{3D} of the plane partitions can be converted to a triple sum over the tableau T, T' and their shape λ:

$$Z_{\text{3D}} = \sum_{\lambda} \sum_{T,T':\text{shape } \lambda} q^T q^{T'}, \tag{2.4}$$

where

$$q^T = \prod_{m=0}^{\infty} q^{(m+1/2)|\pi(-m)/\pi(-m-1)|},$$

$$q^{T'} = \prod_{m=0}^{\infty} q^{(m+1/2)|\pi(m)/\pi(m+1)|}.$$

By the well known combinatorial definition of the Schur functions [19], the partial sum over the semi-standard tableaux turn out to be a special value of the Schur functions:

$$\sum_{T:\text{shape } \lambda} q^T = \sum_{T':\text{shape } \lambda} q^{T'} = s_\lambda(q^\rho), \tag{2.5}$$

where

$$q^\rho = (q^{1/2}, q^{3/2}, \ldots, q^{n+1/2}, \cdots).$$

Thus the partition function can be eventually rewritten as

$$Z_{\text{3D}} = \sum_{\lambda} s_\lambda(q^\rho)^2. \tag{2.6}$$

Let us note that the special value of the Schur functions has the so called Hook formula [19]

$$s_\lambda(q^\rho) = q^{n(\lambda)+|\lambda|/2} \prod_{(i,j)\in\lambda} (1 - q^{h(i,j)})^{-1}, \tag{2.7}$$

where (i, j) stands for the (i, j)-th box in the Young diagram, and $n(\lambda)$ is given by

$$n(\lambda) = \sum_{i=1}^{\infty} (i-1)\lambda_i.$$

2.4. Melting crystal model with external potentials

We now deform the foregoing melting crystal model by introducing the external potentials

$$\Phi_k(\lambda, p) = \sum_{i=1}^{\infty} q^{k(p+\lambda_i-i+1)} - \sum_{i=1}^{\infty} q^{k(-i+1)}$$

with coupling constants t_k, $k = 1, 2, 3, \ldots$, on the main diagonal slice $\lambda = \pi(0)$. The right hand side of the definition of $\Phi_k(\lambda, p)$ is understood to be a finite sum (hence a rational function of q) by cancellation of terms between the two sums:

$$\Phi_k(\lambda, p) = \sum_{i=1}^{\infty} (q^{k(p+\lambda_i-i+1)} - q^{k(p-i+1)}) + q^k \frac{1-q^{pk}}{1-q^k}.$$

The partition function of the deformed model reads

$$Z_p(t) = \sum_{\pi} q^{|\pi|} e^{\Phi(t,\pi(0),p)}, \tag{2.8}$$

where

$$\Phi(t, \lambda, p) = \sum_{k=1}^{\infty} t_k \Phi_k(\lambda, p).$$

We can repeat the previous calculations in this setting to rewrite the new partition function $Z_p(t)$ as

$$Z_p(t) = \sum_{\lambda} s_\lambda(q^\rho)^2 q^{\Phi(t,\lambda,p)}. \tag{2.9}$$

Modifying this partition function slightly, we obtain the instanton sum of 5D $\mathcal{N} = 1$ SUSY $U(1)$ Yang–Mills theory [6]:

$$Z_p(t) = \sum_{\pi} q^{|\pi|} Q^{\pi(0)} e^{\Phi(t,\pi(0),p)} = \sum_{\lambda} s_\lambda(q^\rho)^2 Q^{|\lambda|} e^{\Phi(t,\lambda,p)}. \tag{2.10}$$

q and Q are related to physical parameters $R, \Lambda, \hbar$ of 5D Yang–Mills theory as

$$q = e^{-R\hbar}, \quad Q = (R\Lambda)^2.$$

The external potentials represent the contribution of Wilson loops along the fifth dimension [20]. In this sense, $Z_p(t)/Z_p(0)$ is a generating function of correlation functions of those Wilson loop operators.

Our goal is to show that the partition function $Z_p(t)$ is, up to a simple factor, the tau function of the (1D) Toda hierarchy. To this end, we now consider a fermionic representation of this partition function.

§3. Fermionic formula of partition function

3.1. Complex fermion system

Let $\psi(z)$ and $\psi^*(z)$ denote complex 2D fermion fields

$$\psi(z) = \sum_{m=-\infty}^{\infty} \psi_m z^{-m-1}, \quad \psi^*(z) = \sum_{m=-\infty}^{\infty} \psi_m^* z^{-m}.$$

The Fourier modes ψ_m and ψ^*_m of $\psi(z)$ and $\psi^*(z)$ satisfy the anti-commutation relations

$$\{\psi_m, \psi^*_n\} = \delta_{m+n,0}, \quad \{\psi_m, \psi_n\} = \{\psi^*_m, \psi^*_n\} = 0.$$

The Fock space F splits into charge p subspaces F_p:

$$F = \bigoplus_{p=-\infty}^{\infty} F_p.$$

The charge p subspace F_p has a unique normalized ground state (charge p vacuum) $|p\rangle$ and an orthonormal basis $|\lambda; p\rangle$ labeled by partitions λ. $|p\rangle$ is characterized by the vacuum condition

$$\psi_m|p\rangle = 0 \quad \text{for } m \geq -p, \quad \psi^*_m|p\rangle = 0 \quad \text{for } m \geq p+1.$$

If the partition is of the form $\lambda = (\lambda_1, \ldots, \lambda_n, 0, 0, \ldots)$, the associated element $|\lambda; p\rangle$ of the basis is obtained from $|p\rangle$ by the action of fermion operators as

$$|\lambda; p\rangle = \psi_{-(p+\lambda_1-1)-1} \cdots \psi_{-(p+\lambda_n-n)-1} \psi^*_{(p-n)+1} \cdots \psi^*_{(p-1)+1}|p\rangle.$$

They are orthonormal in the sense that their inner products have the normalized values

$$\langle\lambda; p|\mu; q\rangle = \delta_{pq}\delta_{\lambda\mu}.$$

3.2. $U(1)$ current and fermionic representation of tau function

The $U(1)$ current $J(z)$ of the complex fermion system is defined as

$$J(z) = {:}\psi(z)\psi^*(z){:} = \sum_{k=-\infty}^{\infty} J_m z^{-m-1},$$

where $:\ :$ denotes the normal ordering with respect to the vacuum $0\rangle$:

$${:}\psi_m\psi^*_n{:} = \psi_m\psi^*_n - \langle 0|\psi_m\psi^*_n|0\rangle.$$

The Fourier modes

$$J_m = \sum_{n=-\infty}^{\infty} {:}\psi_{m-n}\psi^*_n{:}$$

of $J(z)$ satisfy the commutation relations

$$[J_m, J_n] = m\delta_{m+n,0} \tag{3.1}$$

of the A_∞ Heisenberg algebra, and play the role of "Hamiltonians" in the usual fermionic formula of the KP and 2D Toda hierarchies [15, 16]. For the case of tau functions $\tau(t,\bar{t})$, $t = (t_1, t_2, \ldots)$, $\bar{t} = (\bar{t}_1, \bar{t}_2, \ldots)$, of the 2D Toda hierarchy, the fermionic formula reads

$$\tau_p(t,\bar{t}) = \Big\langle p \,\Big|\, \exp\Big(\sum_{m=1}^{\infty} t_m J_m\Big) g \exp\Big(-\sum_{m=1}^{\infty} \bar{t}_m J_{-m}\Big) \,\Big|\, p \Big\rangle \tag{3.2}$$

where g is an element of the infinite dimensional Clifford group $GL(\infty)$.

3.3. Fermionic representation of $Z_p(t)$

The partition function $Z_p(t)$ of the deformed melting crystal has a fermionic representation of the form

$$Z_p(t) = \langle p|G_+ e^{H(t)} G_-|p\rangle. \tag{3.3}$$

Let us explain the constituents of this formula along with an outline of the derivation of this formula.

$H(t)$ is the linear combination

$$H(t) = \sum_{k=1}^{\infty} t_k H_k$$

of the "Hamiltonians"

$$H_k = \sum_{n=-\infty}^{\infty} q^{kn} {:}\psi_{-n}\psi_n^*{:}.$$

The aforementioned basis elements $|\lambda; p\rangle$ of the Fermion Fock space turn out to be eigenvectors of these Hamiltonians. The eigenvalues are nothing but the the potential functions $\Phi_k(\lambda, p)$:

$$H_k|\lambda; p\rangle = \Phi_k(\lambda, p)|\lambda; p\rangle. \tag{3.4}$$

$G_\pm$ are $GL(\infty)$ elements of the special form

$$G_\pm = \exp\Big(\sum_{k=1}^{\infty} \frac{q^{k/2}}{k(1-q^k)} J_{\pm k}\Big).$$

Since the numerical factors $q^{k/2}/(1-q^k)$ in this definition can be expanded as

$$\frac{q^{k/2}}{1-q^k} = \sum_{m=-\infty}^{-1} q^{-k(m+1/2)} = \sum_{m=0}^{\infty} q^{k(m+1/2)},$$

one can factorize these operators as

$$G_+ = \prod_{m=-\infty}^{-1} \Gamma_+(m), \quad G_- = \prod_{m=0}^{\infty} \Gamma_-(m), \tag{3.5}$$

where

$$\Gamma_\pm(m) = \exp\Bigl(\sum_{k=1}^{\infty} \frac{1}{k} q^{\mp k(m+1/2)} J_{\pm k}\Bigr).$$

These $\Gamma_\pm(m)$'s are a specialization of the so called vertex operators

$$V_\pm(z) = \exp\Bigl(\sum_{k=1}^{\infty} \frac{z^k}{k} J_{\pm k}\Bigr)$$

for bosonization of the complex fermions.

Following the idea of Okounkov and Reshetikhin [2], we now consider m as a fictitious "time" variable. A plane partition then may be thought of as the "path" (or "world volume") of discrete time evolutions of a partition λ that starts from the empty partition $\emptyset = (0, 0, \ldots)$ at infinite past and ends again in $\emptyset$ at infinite future (see Figure 3). The vertex operators $\Gamma_\pm(m)$ play the role of transfer matrices between neighboring diagonal slices.

The vertex operators $\Gamma_\pm(m)$ act on the aforementioned orthonormal bases $|\lambda; p\rangle$ and $\langle\lambda; p|$ as

$$\langle\lambda; p|\Gamma_+(m) = \sum_{\mu\succ\lambda} \langle\mu; p| q^{-(m+1/2)(|\mu|-|\lambda|)} \tag{3.6}$$

for $m = -1, -2, \ldots$ and

$$\Gamma_-(m)|\lambda; p\rangle = \sum_{\mu\succ\lambda} q^{(m+1/2)(|\mu|-|\lambda|)} |\mu; p\rangle \tag{3.7}$$

for $m = 0, 1, \ldots$ [2, 3]. The right hand side of these formulas give a linear combination of all possible time evolutions of the m-th slice $\lambda = \pi(m)$ at the next time. The weight $q^{\mp(m+1/2)(|\mu|-|\lambda|)}$ of each state on the right hand side are exactly the factors assigned to the boxes of $\pi(m)/\pi(m\mp 1)$ in the definition of the weights $q^T, q^{T'}$ that appear in the combinatorial formula (2.4).

Since $G_\pm$ are products of these slice-to-slice "transfer matrices", $\langle p|G_+$ and $G_-|p\rangle$ become linear combinations of the states $\langle\lambda; p|$ and $|\lambda; p\rangle$ that evolve from the ground states $\langle p|$ and $|p\rangle$ at $m = \mp\infty$. By what we have seen above, the weights of $\langle\lambda; p|$ and $|\lambda; p\rangle$ in these linear combinations are given by the partial sums of q^T and $q^{T'}$ over all semi-standard tableaux T and T' of shape λ, namely, the special value $s_\lambda(q^\rho)$ of the Schur function. Thus $\langle p|G_+$ and $G_-|p\rangle$ can be expressed as

$$\langle p|G_+ = \sum_\lambda \sum_{T:\text{shape}\,\lambda} q^T \langle\lambda;p| = \sum_\lambda s_\lambda(q^\rho)\langle\lambda;p|, \tag{3.8}$$

$$G_-|p\rangle = \sum_\lambda \sum_{T':\text{shape}\,\lambda} q^{T'} |\lambda;p\rangle = \sum_\lambda s_\lambda(q^\rho)|\lambda;p\rangle. \tag{3.9}$$

The expectation value of $e^{H(t)}$ with respect to these states yields the fermionic representation (3.3) of the partition function $Z_p(t)$.

The fermionic representation (3.3) is apparently different from the fermionic formula (3.2) of tau functions of the 2D Toda hierarchy. To show that $Z_p(t)$ is indeed a tau function, we have to rewrite (3.3) to the form of (3.2). This is the place where the quantum torus Lie algebra joins the game.

§4. Quantum torus Lie algebra

4.1. Fermionic realization of quantum torus Lie algebra

Let $V_m^{(k)}$ ($k = 0, 1, \ldots,\ m \in \mathbf{Z}$) denote the following fermion bilinear forms:

$$\begin{aligned} V_m^{(k)} &= q^{-km/2} \sum_{n=-\infty}^{\infty} q^{kn} {:}\psi_{m-n}\psi_n^*{:} \\ &= q^{k/2} \oint \frac{dz}{2\pi i} z^m {:}\psi(q^{k/2}z)\psi^*(q^{-k/2}z){:}\ . \end{aligned}$$

Note that

$$J_m = V_m^{(0)}, \quad H_k = V_0^{(k)}.$$

Actually, $V_m^{(k)}$ coincides with Okounkov and Pandharipande's operator $\mathcal{E}_m(z)$ [21, 22] specialized to $z = q^k$. As they found for $\mathcal{E}_m(z)$, our $V_m^{(k)}$'s satisfy the commutation relations

$$[V_m^{(k)}, V_n^{(l)}] = (q^{(lm-kn)/2} - q^{(kn-lm)/2})(V_{m+n}^{(k+l)} - \delta_{m+n,0}\frac{q^{k+l}}{1-q^{k+l}}). \tag{4.1}$$

This is a (central extension of) q-deformation of the Poisson algebra of functions on a 2-torus. We refer to this Lie algebra as "quantum torus Lie algebra". More precisely, a full quantum torus Lie algebra should contain elements for $k < 0$ as well; for several reasons, we shall not include those elements.

4.2. Shift symmetry among basis of quantum torus Lie algebras

The following relations, which we call "shift symmetry", play a central role in identifying $Z_p(t)$ as a tau function:
(4.2)

$$G_-G_+\Big(V_m^{(k)} - \delta_{m,0}\frac{q^k}{1-q^k}\Big)(G_-G_+)^{-1} = (-1)^k\Big(V_{m+k}^{(k)} - \delta_{m+k,0}\frac{q^k}{1-q^k}\Big).$$

These relations are derived as follows.

Let us recall that the fermion fields $\psi(z), \psi^*(z)$ transform under adjoint action by $J_{\pm k}$'s as

$$\exp\Big(\sum_{k=1}^{\infty} c_k J_{\pm k}\Big)\psi(z)\exp\Big(-\sum_{k=1}^{\infty} c_k J_{\pm k}\Big) = \exp\Big(\sum_{k=1}^{\infty} c_k z^{\pm k}\Big)\psi(z), \tag{4.3}$$

(4.4)

$$\exp\Big(\sum_{k=1}^{\infty} c_k J_{\pm k}\Big)\psi^*(z)\exp\Big(-\sum_{k=1}^{\infty} c_k J_{\pm k}\Big) = \exp\Big(-\sum_{k=1}^{\infty} c_k z^{\pm k}\Big)\psi^*(z).$$

By letting $c_k = q^{k/2}/(1-q^k)$, the exponential operators in these formulas turn into $G_\pm$, so that we have the operator identities

$$G_+\psi(z){G_+}^{-1} = (q^{1/2}z;q)_\infty^{-1}\psi(z), \tag{4.5}$$

$$G_+\psi^*(z){G_+}^{-1} = (q^{1/2}z;q)_\infty\psi^*(z), \tag{4.6}$$

$$G_-\psi(z){G_-}^{-1} = (q^{1/2}z^{-1};q)_\infty^{-1}\psi(z), \tag{4.7}$$

$$G_-\psi^*(z){G_-}^{-1} = (q^{1/2}z^{-1};q)_\infty\psi^*(z), \tag{4.8}$$

where $(z;q)_\infty$ denotes the standard q-factorial symbol

$$(z;q)_\infty = \prod_{n=0}^{\infty}(1-zq^n).$$

We use these operator identities to derive transformation of the fermion bilinear forms $:\psi(q^{k/2}z)\psi^*(q^{-k/2}z):$ under conjugation by $G_\pm$. Since

$$:\psi(q^{k/2}z)\psi^*(q^{-k/2}z): = -\psi^*(q^{-k/2}z)\psi(q^{k/2}z) + \frac{q^{k/2}}{(1-q^k)z},$$

let us first consider $\psi^*(q^{-k/2}z)\psi(q^{k/2}z)$. Under conjugation by G_+, it transforms as

$$\begin{aligned} & G_+\psi^*(q^{-k/2}z)\psi(q^{k/2}z){G_+}^{-1} \\ &= \frac{(q^{1/2}\cdot q^{-k/2}z;q)_\infty}{(q^{1/2}\cdot q^{k/2}z;q)_\infty}\psi^*(q^{-k/2}z)\psi(q^{k/2}z) \\ &= \prod_{m=1}^{k}(1-zq^{(k+1)/2-m})\psi^*(q^{-k/2}z)\psi(q^{k/2}z). \end{aligned}$$

This implies that

$$\begin{aligned} & G_+\Big(:\psi(q^{k/2}z)\psi^*(q^{-k/2}z): - \frac{q^{k/2}}{(1-q^k)z}\Big){G_+}^{-1} \\ &= \prod_{m=1}^{k}(1-q^{(k+1)/2-m}z)\Big(:\psi(q^{k/2}z)\psi^*(q^{-k/2}z): - \frac{q^{k/2}}{(1-q^k)z}\Big). \end{aligned} \tag{4.9}$$

In much the same way, we can derive a similar transformation under conjugation by G_-. In this case, it is more convenient to rewrite the result as follows:

$$\begin{aligned} & {G_-}^{-1}\Big(:\psi(q^{k/2}z)\psi^*(q^{-k/2}z): - \frac{q^{k/2}}{(1-q^k)z}\Big)G_- \\ &= \prod_{m=1}^{k}(1-q^{-(k+1)/2+m}z^{-1})\Big(:\psi(q^{k/2}z)\psi^*(q^{-k/2}z): - \frac{q^{k/2}}{(1-q^k)z}\Big). \end{aligned} \tag{4.10}$$

We note here that the prefactors on the right hand side of the last two equations are related as

$$\prod_{m=1}^{k}(1-q^{(k+1)/2-m}z) = (-z)^k\prod_{m=1}^{k}(1-q^{-(k+1)/2+m}z^{-1}).$$

Accounting for this simple, but significant relation, we can derive the identity

$$\begin{aligned} & G_-G_+\Big(:\psi(q^{k/2}z)\psi^*(q^{-k/2}z): - \frac{q^{k/2}}{(1-q^k)z}\Big)(G_+G_-)^{-1} \\ &= (-z)^k\Big(:\psi(q^{k/2}z)\psi^*(q^{-k/2}z): - \frac{q^{k/2}}{(1-q^k)z}\Big). \end{aligned} \tag{4.11}$$

The shift symmetry (4.2) follows immediately from this identity.

When $m = 0$ and $m = -k$, (4.2) takes the particular form

$$G_-G_+\Big(V_0^{(k)} - \frac{q^k}{1-q^k}\Big)(G_-G_+)^{-1} = (-1)^k V_k^{(k)}, \tag{4.12}$$

$$(G_-G_+)^{-1}\Big(V_0^{(k)} - \frac{q^k}{1-q^k}\Big)G_-G_+ = (-1)^k V_{-k}^{(k)}. \tag{4.13}$$

It is these identities that we shall use to convert the fermionic representation of $Z_p(t)$ to the standard fermionic formula of tau functions.

§5. Integrable structure of melting crystal model

5.1. Partition function as tau function of 2D Toda hierarchy

Let us split the operator $G_+e^{H(t)}G_-$ in (3.3) into three pieces as

$$\begin{aligned} G_+e^{H(t)}G_- &= G_+e^{H(t)/2}e^{H(t)/2}G_- \\ &= G_+e^{H(t)/2}{G_+}^{-1} \cdot G_+G_- \cdot {G_-}^{-1}e^{H(t)/2}G_- \end{aligned}$$

and use the special cases (4.12) and (4.13) of the shift symmetry to rewrite those pieces.

To this end, it is convenient to rewrite (4.12) and (4.13) as

$$\begin{aligned} G_+\Big(H_k - \frac{q^k}{1-q^k}\Big){G_+}^{-1} &= (-1)^k {G_-}^{-1}V_k^{(k)}G_-, \\ {G_-}^{-1}\Big(H_k - \frac{q^k}{1-q^k}\Big)G_- &= (-1)^k G_+V_{-k}^{(k)}{G_+}^{-1}. \end{aligned}$$

Though the operators $V_{\pm k}^{(k)}$ on the right hand side are unfamiliar in the theory of integrable hierarchies, we can convert them to the familiar "Hamiltonians" $J_{\pm k} = V_{\pm k}^{(0)}$ of the Toda hierarchies as

$$q^{W/2}V_k^{(k)}q^{-W/2} = V_k^{(0)} = J_k, \quad q^{-W/2}V_{-k}^{(k)}q^{W/2} = V_{-k}^{(0)} = J_{-k}, \tag{5.1}$$

where W is a special element of W_∞ algebra:

$$W = W_0^{(3)} = \sum_{n=-\infty}^{\infty} n^2{:}\psi_{-n}\psi_n^*{:}\ .$$

We thus eventually obtain the relations

$$G_+\Big(H_k - \frac{q^k}{1-q^k}\Big)G_+{}^{-1} = (-1)^k G_-{}^{-1} q^{-W/2} J_k q^{W/2} G_-, \tag{5.2}$$

$$G_-{}^{-1}\Big(H_k - \frac{q^k}{1-q^k}\Big)G_- = (-1)^k G_+ q^{W/2} J_{-k} q^{-W/2} G_+{}^{-1} \tag{5.3}$$

between H_k's and $J_{\pm k}$'s.

By these relations, $G_+ e^{H(t)/2} G_+{}^{-1}$ can be calculated as

$$\begin{aligned} &G_+ e^{H(t)/2} G_+{}^{-1} \\ &= \exp\Big(\sum_{k=1}^{\infty} \frac{t_k q^k}{2(1-q^k)}\Big) G_-{}^{-1} q^{-W/2} \exp\Big(\sum_{k=1}^{\infty} \frac{(-1)^k t_k}{2} J_k\Big) q^{W/2} G_-. \end{aligned}$$

A similar expression can be derived for $G_-{}^{-1} e^{H(t)} G_-$ as well. We can thus rewrite $G_+ e^{H(t)} G_-$ as

$$\begin{aligned} G_+ e^{H(t)} G_- = \exp\Big(\sum_{k=1}^{\infty} \frac{t_k q^k}{1-q^k}\Big) G_-{}^{-1} q^{-W/2} \exp\Big(\sum_{k=1}^{\infty} \frac{(-1)^k t_k}{2} J_k\Big) \times \\ \times g \exp\Big(\sum_{k=1}^{\infty} \frac{(-1)^k t_k}{2} J_{-k}\Big) q^{-W/2} G_+{}^{-1} \end{aligned}$$

where

$$g = q^{W/2} (G_- G_+)^2 q^{W/2}. \tag{5.4}$$

The partition function $Z_p(t)$ is given by the expectation value of this operator with respect to $\langle p|$ and $|p\rangle$. Since the action by the leftmost and rightmost pieces of g yields only a scalar multiplier to $\langle p|$, $|p\rangle$ as

$$\langle p| G_-{}^{-1} q^{-W/2} = q^{-p(p+1)(2p+1)/12} \langle p|, \tag{5.5}$$

$$q^{-W/2} G_+{}^{-1} |p\rangle = q^{-p(p+1)(2p+1)/12} |p\rangle, \tag{5.6}$$

$Z_p(t)$ can be expressed as

$$\begin{aligned} Z_p(t) &= \exp\Big(\sum_{k=1}^{\infty} \frac{t_k q^k}{1-q^k}\Big) q^{-p(p+1)(2p+1)/6} \times \\ &\times \Big\langle p \Big| \exp\Big(\sum_{k=1}^{\infty} \frac{(-1)^k t_k}{2} J_k\Big) g \exp\Big(\sum_{k=1}^{\infty} \frac{(-1)^k t_k}{2} J_{-k}\Big) \Big| p \Big\rangle. \end{aligned} \tag{5.7}$$

The expectation value $\langle p| \cdots |p\rangle$ takes exactly the form of (3.2). Thus, up to the simple prefactor, $Z_p(t)$ is essentially a tau function of the 2D

Toda hierarchy. Thus we find that an integrable structure behind the melting crystal model is the 2D Toda hierarchy. This is, however, not the end of the story.

5.2. 1D Toda hierarchy as true integrable structure

The foregoing calculation is based on the splitting

$$G_+ e^{H(t)} G_- = G_+ e^{H(t)/2} {G_+}^{-1} \cdot G_+ G_- \cdot {G_-}^{-1} e^{H(t)/2} G_-.$$

Actually, we could have started from a different splitting of $G_+ e^{H(t)} G_-$, e.g.,

$$G_+ e^{H(t)} G_- = G_+ e^{H(t)} {G_+}^{-1} \cdot G_+ G_- = G_+ G_- \cdot {G_-}^{-1} e^{H(t)} G_-.$$

This leads to another set of expressions of $Z_p(t)$ in which only the $\langle p| \cdots |p\rangle$ part is different. We thus have the following three different expressions for this part:

$$\begin{aligned} &\Big\langle p \Big| \exp\Big(\sum_{k=1}^{\infty} \frac{(-1)^k t_k}{2} J_k\Big) g \exp\Big(\sum_{k=1}^{\infty} \frac{(-1)^k t_k}{2} J_{-k}\Big) \Big| p \Big\rangle \\ &= \Big\langle p \Big| \exp\Big(\sum_{k=1}^{\infty} (-1)^k t_k J_k\Big) g \Big| p \Big\rangle \\ &= \Big\langle p \Big| g \exp\Big(\sum_{k=1}^{\infty} (-1)^k t_k J_{-k}\Big) \Big| p \Big\rangle. \end{aligned} \tag{5.8}$$

These identities of the expectation values can be directly derived from the operator identities

$$J_k g = g J_{-k}, \ k = 1, 2, 3, \ldots \tag{5.9}$$

satisfied by g. (These operator identities themselves are a consequences of the shift symmetry of $V_m^{(k)}$'s.) Generally speaking, this kind of operator identities imply symmetry constrains on the tau functions [23, 24]; in the present case, the constraints read

$$\frac{\partial}{\partial t_k} \tau_p(t, \bar{t}) + \frac{\partial}{\partial \bar{t}_k} \tau_p(t, \bar{t}) = 0, \quad k = 1, 2, 3, \ldots. \tag{5.10}$$

In other words, the tau function is a function of $t - \bar{t}$,

$$\tau_p(t, \bar{t}) = \tau_p(t - \bar{t}, 0) = \tau_p(0, \bar{t} - t),$$

and reduces to a tau function $\tau_p(t)$ of the 1D Toda hierarchy that has a single series of time variables $t = (t_1, t_2, \ldots)$ rather than the two series of the 2D Toda hierarchy. Thus the 1D Toda hierarchy eventually turns out to be an underlying integrable structure of the deformed melting crystal model.

The same conclusion can be derived for the instanton sum (2.10) of 5D SUSY $U(1)$ Yang–Mills theory. It has a fermionic representation of the form

$$Z_p(t) = \langle p|G_+ Q^{L_0} e^{H(t)} G_-|p\rangle \tag{5.11}$$

where L_0 is a special element of the Virasoro algebra:

$$L_0 = \sum_{n=-\infty}^{\infty} n{:}\psi_{-n}\psi_n^*{:}.$$

One can repeat almost the same calculations as the previous case to convert $Z_p(t)$ to the form of (5.7). The counterpart of g is given by

$$g = q^{W/2} G_- G_+ Q^{L_0} G_- G_+ q^{W/2}, \tag{5.12}$$

which, too, satisfy the reduction conditions (5.9) to the 1D Toda hierarchy. Thus a relevant integrable structure is again the 1D Toda hierarchy.

§6. Concluding remarks

6.1. Problems on 4D instanton sum

In deriving the instanton sum (2.10), 5D space-time is partially compactified in the fifth dimension as $\mathbf{R}^4 \times S^1$. The parameter R is the radius of S^1. Therefore, letting $R \to 0$ amounts to 4D limit.

Unfortunately, it is not straightforward to achieve such a 4D limit in the present setup. Firstly, the 5D instanton sum with external potentials does not have a reasonable limit as $R \to 0$. Any naive prescription letting $R \to 0$ yields a result in which t dependence disappears or becomes trivial [12]. Secondly, the shift symmetry of the quantum torus Lie algebra ceases to exist in the limit as $q = e^{-R\hbar} \to 1$. Speaking more precisely, the quantum torus Lie algebra itself turns into a W_∞ algebra in this limit, but no analogue of shift symmetry (4.2) is known for the latter case. For these reasons, the 4D case has to be studied independently.

The 4D instanton sum [7, 8], too, is a sum over partitions. Moreover, this statistical sum has a fermionic representation [14, 9]. Marshakov and Nekrasov [17, 18] further introduced external potentials therein. Actually, the 4D instanton sum for $U(1)$ gauge theory is almost identical to the generating function of Gromov–Witten invariants of $\mathbf{CP}^1$ [21, 22].

This can be most clearly seen in the fermionic representation of these generating functions, which reads

$$Z_p^{4D}(t) = \Big\langle p \,\Big|\, e^{J_1/\hbar} \exp\Big(\sum_{k=1}^{\infty} t_k \frac{\mathcal{P}_{k+1}}{k+1}\Big) e^{J_{-1}/\hbar} \,\Big|\, p \Big\rangle, \tag{6.1}$$

where $\mathcal{P}_k$'s are fermion bilinear forms introduced by Okounkov and Pandharipande for a fermionic representation of (absolute) Gromov–Witten invariants of $\mathbf{CP}^1$ [21]. As regards these Gromov–Witten invariants (in other words, correlation functions of the topological σ model), it has been known for years [25, 26, 27, 28, 29] that a relevant integrable structure is the 1D Toda hierarchy.

Thus the 1D Toda hierarchy is expected to be the integrable structure of the 4D instanton sum as well. This has been confirmed by Marshakov and Nekrasov in detail [17, 18]. What is still missing, however, is a formula like (5.7) that directly connects $Z_p^{4D}(t)$ with the standard fermionic formula (3.2) of the tau function. Finding a 4D analogue of (5.7) is thus an intriguing open problem. This issue is also closely related to the fate of shift symmetry (4.2) in the $q \to 1$ limit.

6.2. Relation to topological strings

Our results are directly or indirectly connected with some aspects of topological strings as well.

1. According to the theory of topological vertex [5], the partition function $Z_p(t)$ of the deformed melting crystal model has another interpretation as the A-model topological string amplitude for the toric Calabi–Yau threefold $\mathcal{O} \oplus \mathcal{O}(-2) \to \mathbf{CP^1}$. In this interpretation, q and Q are parametrized by the string coupling constant g_{st} and the Kähler volume a of $\mathbf{CP}^1$ as

$$q = e^{-g_{\mathrm{st}}}, \quad Q = e^{-a}.$$

Specializing the value of t leads to several interesting observations [12].

2. A generating function of the two-legged topological vertex $W_{\lambda\mu} \sim c_{\lambda\mu\bullet}$ is known to give a tau function of the 2D Toda hierarchy [30]. In the fermionic representation (3.2), this amounts to the case where

$$g = q^{W/2} G_+ G_- q^{W/2}. \tag{6.2}$$

(Actually, for complete agreement with the usual convention, we have to replace W with

$$K = \sum_{n=-\infty}^{\infty} \Big(n - \frac{1}{2}\Big)^2 {:}\psi_{-n}\psi_n^*{:},$$

but this is not a serious problem. The difference can be absorbed by rescaling t_k's.) Let us stress that this $GL(\infty)$ element does not satisfy the reduction condition (5.9) to the 1D Toda hierarchy.

3. A generating function of double Hurwitz numbers for coverings of $\mathbf{CP}^1$ gives yet another type of tau function of the 2D Toda hierarchy [31]. Actually, the $GL(\infty)$ element for the fermionic representation is given by

$$g = q^{W/2}. \tag{6.3}$$

(More precisely, as in the previous case, W has to be replaced by K, but the difference is again irrelevant.) In this case, the reduction condition (5.9) to the 1D Toda hierarchy is not satisfied, but the operator identities (5.1) imply that another set of reduction conditions are hidden behind (see below).

6.3. Constraints and quantum torus Lie algebra

As a consequence of the shift symmetry of $V_m^{(k)}$'s, the $GL(\infty)$ elements g of the aforementioned models of topological strings turn out to satisfy some algebraic relations other than (5.9). According to general results on constraints of the 2D Toda hierarchy [23, 24], such relations imply the existence of constraints on the tau functions and the Lax and Orlov–Schulman operators. Those constraints inherit the structure of the quantum torus Lie algebra. Let us illustrate this observation for the case of double Hurwitz numbers over $\mathbf{CP}^1$.

The $GL(\infty)$ element $g = q^{W/2}$ for this case satisfies the operator identities

$$J_k g = g V_k^{(k)}, \quad g J_{-k} = V_{-k}^{(k)} g \tag{6.4}$$

as a consequence of (5.1). These identities can be converted to the constraints

$$L = q^{1/2} q^{\bar{M}} \bar{L}, \quad \bar{L}^{-1} = q^{-1/2} q^{M} L^{-1} \tag{6.5}$$

on the Lax and Orlov–Schulman operators $L, M, \bar{L}, \bar{M}$ of the 2D Toda hierarchy. Emergence of the exponential operators q^M and $q^{\bar{M}}$ is a manifestation of the quantum torus Lie algebra. To see this, let us recall that the Lax and Orlov–Schulman operators satisfy the (twisted) canonical commutation relations

$$[L, M] = L, \quad [\bar{L}, \bar{M}] = \bar{L}. \tag{6.6}$$

This implies that the monomials $q^{-km/2}L^m q^{kM}$ and $q^{-km/2}\bar{L}^m q^{k\bar{M}}$ of $L, q^M, \bar{L}, q^{\bar{M}}$ give two copies of realizations of the quantum torus Lie algebra.

The constraints (6.5) are remarkably similar to the "string equations"

$$L = \bar{M}\bar{L}, \quad \bar{L}^{-1} = ML^{-1} \tag{6.7}$$

of $c = 1$ strings at self-dual radius [32, 33, 23, 24]. A relevant algebraic structure of these string equations is the W_∞ algebra. Thus (6.5) may be thought of as q-deformations of these W-algebraic constraints.

Acknowledgements. We are grateful to Nikita Nekrasov and Motohico Mulase for valuable comments and fruitful discussion. K. T is partly supported by Grant-in-Aid for Scientific Research No. 18340061, No. 19104002 and No. 19540179 from the Japan Society for the Promotion of Science.

References

[1] D. M. Bressoud, Proofs and Confirmations: The Story of the Alternating Sign Matrix Conjecture, Cambridge Univ. Press, 1999.

[2] A. Okounkov and N. Reshetikhin, Correlation function of Schur Process with application to local geometry of a random 3-Dimensional young diagram, J. Amer. Math. Soc., **16**, (2003), 581–603.

[3] A. Okounkov, N. Reshetikhin and C. Vafa, Quantum Calabi–Yau and classical crystals, In: The Unity of Mathematics, (eds. P. Etingof, V. Retakh and I. M. Singer), Progr. Math., **244**, Birkhäuser, 2006, pp. 597–618.

[4] A. Iqbal, All genus topological string amplitudes and 5-brane webs as Feynman diagrams, arXiv:hep-th/0207114.

[5] M. Aganagic, A. Klemm, M. Mariño and C. Vafa, The topological vertex, Commun. Math. Phys., **254** (2005), 425–478.

[6] T. Maeda, T. Nakatsu, K. Takasaki and T. Tamakoshi, Five-dimensional supersymmetric Yang–Mills theories and random plane partitions, J. High Energy Phys., **0503** (2005), 056.

[7] N. A. Nekrasov, Seiberg–Witten Prepotential from Instanton Counting, Adv. Theor. Math. Phys., **7** (2004), 831–864.

[8] H. Nakajima and K. Yoshioka, Instanton counting on blowup, I. 4-dimensional pure gauge theory, Invent. Math., **162** (2005), 313–355.

[9] N. Nekrasov and A. Okounkov, Seiberg–Witten theory and random partitions, In: The Unity of Mathematics, (eds. P. Etingof, V. Retakh and I. M. Singer), Progr. Math., **244**, Birkhäuser, 2006, pp. 525–296.

[10] N. Seiberg and E. Witten, Electric-magnetic duality, monopole condensation, and confinement in $N = 2$ supersymmetric Yang–Mills theory,

Nuclear Phys. B, **426** (1994), 19–52; Erratum, ibid., **430** (1994), 485; Monopoles, duality and chiral symmetry breaking in $N = 2$ Supersymmetric QCD, ibid., **431** (1994), 494–550.

[11] T. Maeda, T. Nakatsu, K. Takasaki and T. Tamakoshi, Free fermion and Seiberg–Witten differential in random plane partitions, Nuclear Phys. B, **715** (2005), 275–303.

[12] T. Nakatsu and K. Takasaki, Melting crystal, quantum torus and Toda hierarchy, Commun. Math. Phys., **285** (2009), 445–468.

[13] K. Ueno and K. Takasaki, Toda lattice hierarchy, Adv. Stud. Pure Math., **4** (1984), 1–95.

[14] A. Losev, A. Marshakov and N. Nekrasov, Small instantons, little strings and free fermions, In: From Fields to Strings: Circumnavigating Theoretical Physics, Ian Kogan memorial volume, (eds. M. Shifman, A. Vainstein and J. Wheater), World Scientific, 2005, pp. 581–621.

[15] M. Jimbo and T. Miwa, Solitons and infinite dimensional Lie algebras, Publ. Res. Inst. Math. Sci., **19** (1983), 943–1001.

[16] T. Takebe, Representation theoretical meanings of the initial value problem for the Toda lattice hierarchy I, Lett. Math. Phys., **21** (1991); ditto II, Publ. Res. Inst. Math. Sci., **27** (1991), 491–503.

[17] A. Marshakov and N. Nekrasov, Extended Seiberg–Witten theory and integrable hierarchy, J. High Energy Phys., **0701** (2007), 104.

[18] A. Marshakov, On microscopic origin of integrability in Seiberg–Witten theory, Theor. Math. Phys., **154** (2008), 362–384; Seiberg–Witten theory and extended Toda hierarchy, J. High Energy Phys., **0803** (2008), 055.

[19] I. G. Macdonald, Symmetric Functions and Hall Polynomials, Oxford Univ. Press, 1995.

[20] L. Baulieu, A. Losev and N.Nekrasov, Chern–Simons and twisted supersymmetry in various dimensions, Nuclear Phys. B, **522** (1998), 82–104.

[21] A. Okounkov and R. Pandharipande, Gromov–Witten theory, Hurwitz theory, and completed cycles, Ann. of Math. (2), **163** (2006), 517–560.

[22] A. Okounkov and R. Pandharipande, The equivariant Gromov–Witten theory of $\mathbf{P}^1$, Ann. of Math. (2), **163** (2006), 561–605.

[23] T. Nakatsu, K. Takasaki and S. Tsujimaru, Quantum and classical aspects of deformed $c = 1$ strings, Nuclear Phys. B, **443** (1995), 155–197.

[24] K. Takasaki, Toda lattice hierarchy and generalized string equations, Commun. Math. Phys., **181** (1996), 131–156.

[25] T. Eguchi, K. Hori and S.-K. Yang, Topological σ models and large-N matrix integral, Internat. J. Modern Phys. A, **10** (1995), 4203–4224.

[26] T. Eguchi, K. Hori and C.-S. Xiong, Quantum cohomology and Virasoro algebra, Phys. Lett. B, **402** (1997), 71–80.

[27] E. Getzler, The Toda conjecture, In: Symplectic geometry and mirror symmetry, KIAS, Seoul, 2000, (eds. K. Fukaya et al.), World Scientific, Singapore, 2001, pp. 51–79.

[28] R. Pandharipande, The Toda equations and Gromov–Witten theory of the Riemann sphere, Lett. Math. Phys., **53** (2000), 59–74.

[29] A. Givental, Gromov–Witten invariants and quantization of quadratic Hamiltonians, Moscow Math. J., **1** (2001), 551–568.
[30] J. Zhou, Hodge integrals and integrable hierarchies, arXiv:math/0310408[math.AG].
[31] A. Okounkov, Toda equations for Hurwitz numbers, Math. Res. Lett., **7** (2000), 447–453.
[32] T. Eguchi and H. Kanno, Toda lattice hierarchy and the topological description of $c = 1$ string theory, Phys. Lett. B, **331** (1994), 330–334.
[33] K. Takasaki, Dispersionless Toda hierarchy and two-dimensional string theory, Comm. Math. Phys., **170** (1995), 101–116.

Toshio Nakatsu
Faculty of Engineering, Mathematics and Physics
Setsunan University
Ikedanakamachi, Neyagawa, Osaka 572-8508
Japan

Kanehisa Takasaki
Graduate School of Human and Environmental Studies
Kyoto University
Yoshida, Sakyo, Kyoto 606-8501
Japan
E-mail address: takasaki@math.h.kyoto-u.ac.jp

Advanced Studies in Pure Mathematics 59, 2010
New Developments in Algebraic Geometry,
Integrable Systems and Mirror Symmetry (Kyoto, 2008)
pp. 225–237

Two dimensional topological strings and gauge theory

Nikita A. Nekrasov⋆

Abstract.

We study topological A-type string on an arbitrary two dimensional target space. Using the Virasoro constraints, proven by A. Okounkov and R. Pandharipande, we find an explicit formula for the partition function. The target space field theory reproducing this partition function is proposed. This field theory has infinite set of deformations which are overlooked by the standard definition of the topological string. We also discuss the relations to the multi-trace deformations of gauge theories, and make contact with quantum integrable systems. In addition, the target space theory can be in turn coupled to gravity, thereby realizing the topological string version of M. Green's "worldsheets for worldsheets" idea.

Topological strings are a continuous source of inspiration for gauge and string theorists. They can be studied on their own, for the purely mathematical reasons. Sometimes the amplitudes of the topological string can be viewed as the subset of the "physical" superstrings. The topological strings produce exact all-loop results [4], from which one hopes to gain some intuition about the quantum theory of gravity, perhaps even at the non-perturbative level. For example, the topological strings give a realization of the quantum space foam picture [11]. The topological strings of A and B type play a crucial rôle in describing the compactifications of II string theories on Calabi–Yau manifolds, which gives rise to the $\mathcal{N} = 2$ theories in four dimension. The partition function $\mathcal{Z}(t)$ of a topological string, of A or B type, depends on a some set of couplings t, which correspond to the cohomology of the target space of string theory, valued in some sheaf. For example, for the B model on a Calabi–Yau manifold X of complex dimension d, the coupling constants

Received September 24, 2008.
⋆ On leave of absence from ITEP, Moscow, Russia

t belong to

$$H_B(X) = \bigoplus_{p,q=0}^{d} H^p(X, \Lambda^q \mathcal{T}_X) \approx H^{d-*,*}(X) ,$$

while for A model the couplings are valued in

$$H_A(X) = \bigoplus_{p,q=0}^{d} H^p(X, \Lambda^q \mathcal{T}_X^*) \approx H^{*,*}(X) ,$$

In addition, every operator $\mathcal{O}$, describing these couplings, comes with the so-called gravitational descendents $\sigma_k(\mathcal{O})$, $k = 0, 1, 2, \ldots$. Thus the full set of couplings of the topological string is an infinite dimensional space

$$H_{A,B}(X) \otimes \mathbf{C}[[z]]$$

where we using a formal variable z to label the gravitational descendents:

$$\sigma_k(\mathcal{O}) \leftrightarrow \mathcal{O} \otimes z^k.$$

In the case $d = 3$ the gravitational descendents decouple for $k > 0$, except for the dilaton $\sigma_1(1)$, which corresponds to the string coupling constant $\hbar$. The (disconnected) partition function of the topological string

$$\mathcal{Z}_X(t; \hbar) = \exp \sum_{g=0}^{\infty} \hbar^{2g-2} \mathcal{F}_g(t) ,$$

where $t \in H_{A,B}(X)$, is a generating function of genus g topological string diagrams. For the B model these diagrams can be identified with Feynman diagrams of a certain quantum field theory on X, the so-called Kodaira–Spencer theory [4]. For the A model the analogous theory, the so-called theory of Kähler gravity [3] is expected to be non-local and is constructed only in the large volume limit where the non-local effects are exponentially suppressed.

In this note we shall construct the Kähler gravity theory for the two dimensional X and will find that it is a local theory of an infinite number of fields. The proofs and derivations will appear in a companion paper [21].

Duality CY vs. $\mathbf{R}^4$*: topological string—supersymmetric gauge theory.* A topic which keeps attracting attention of many researchers in the field, is the duality between the topological strings on local Calabi–Yau manifolds and the chiral sector in the four dimensional $\mathcal{N} = 2$ and $\mathcal{N} = 1$ supersymmetric gauge theories. The simplest example of that duality is the geometrical engineering of [12]. One starts with an ADE singularity, i.e. a quotient $\mathbf{C}^2/\Gamma_G$, fibered over a $\mathbf{CP}^1$ so that the total space is a (singular) Calabi–Yau manifold. By resolving the singularities one obtains a smooth non-compact Calabi–Yau manifold X_G. If one views the IIA string on $X_G \times \mathbf{R}^{1,3}$ as a large volume limit of a compactification on a Calabi–Yau manifold with the locus of ADE singularities over an isolated rational curve, then the effective four dimensional theory will decouple from gravity. Moreover one can be model the effective theory on the four dimensional $\mathcal{N}=2$ theory with the MacKay dual ADE gauge group G, where the resolution of singularities of X_G corresponds to fixing a particular vacuum expectation value of the adjoint scalar. Then the prepotential of the low-energy effective theory is given by the genus zero prepotential of the type A topological string on X_G (more precisely, it is the prepotential of the five-dimensional gauge theory compactified on a circle which arises in this way [19][14], in order to see the four dimensional prepotential one has to go to a certain scaling limit in the CY moduli space [12]).

Duality Σ vs. $\mathbf{R}^4$: topological string—supersymmetric gauge theory. Another remarkable duality between the chiral sector of the four dimensional $\mathcal{N}=2$ theories and the topological strings on the two dimensional manifolds was discovered in [16] and further studied in [17]. It is based on the comparison of the instanton calculus in the four dimensional gauge theory [20] and the Gromov–Witten/Hurwitz correspondence of [23]. The physics of that correspondence involves the theory on a fivebrane wrapped on a Riemann surface. One can actually stretch the duality beyond the realm of the physical superstrings and conjecture a powerful S-duality at the level of the topological strings only [22], leading to the concept of the topological string version of M-theory, or Z-theory [18] [6].

The duality of [16] (see also a paper on the mathematically related subject [15] and recent works on the duality with $\mathcal{N}=1$ four dimensional theories [7]) identified the disconnected partition function of the topological string on $\mathbf{CP}^1$ in the background with the arbitrary topological descendents of the Kähler form $\sigma_k(\omega)$ turned on. The couplings t_k^ω (up to a k-dependent factor) are identified with the couplings of the operators

$$\int \mathrm{d}^4\vartheta \ \mathrm{tr}\ \Phi^{k+2}$$

in the $\mathcal{N}=2$ gauge theory:

$$\sum_{k=0}^{\infty} \frac{t_k^\omega}{k!} \int_C \sigma_k(\omega) \leftrightarrow \sum_{k=0}^{\infty} \frac{t_k^\omega}{(k+2)!} \int_{\mathbf{R}^{4|4}} \mathrm{d}^4x\mathrm{d}^4\vartheta \ \mathrm{tr}\ \Phi^{k+2} \tag{1}$$

where in the left hand side we write the worldsheet couplings. In this paper we shall deepen the duality discovered in the original paper [16].

Duality Σ vs. Σ: topological string—two dimensional gauge theory. About fifteen years ago D. Gross has proposed to attack the problem of finding the large N gauge theory description in terms of some kind of string theory via the analysis of the two dimensional gauge theories. By carefully analyzing the 't Hooft limit of the two dimensional Yang–Mills theory on a Riemann surface Σ D. Gross and W. Taylor have identified many features of the corresponding string theory, while [5] have proposed a new kind of topological string theory. An important aspect of the construction of [5] was the realization of the fact that the topological Yang–Mills theory (which is the perturbative limit of the physical Yang–Mills theory) can be described by the Hurwitz theory. The latter counts ramified coverings of a Riemann surface Σ. In this paper we shall find a different version of the string field theory, the one corresponding to the A type topological strings on a Riemann surface Σ. It will turn out to be a kind of an infinite N gauge theory, but most likely not the ordinary 't Hooft large N limit of the gauge group with the finite dimensional gauge group like $SU(N)$ or $SO(N)$.

Worldsheets for worldsheets. In [9] M. Green has proposed to study the two dimensional string backgrounds as the theories of worldsheets for yet another string theories. With the advent of the string dualities a few interesting examples of this construction were invented. For example, M-theory fivebrane wrapped on **K3** becomes a heterotic string on $\mathbf{T}^3$. This is not exactly a realization of the [9] idea as we are using the localized soliton to generate the string. One could try to study the **CY4** or $\mathbf{Spin(7)}$ compactifications of the Type II string [10]. but this is difficult due to the lack of the detailed knowledge of the moduli spaces of these manifolds. In this paper we shall approach this problem in the context of the topological string.

Very large phase space of the topological string. The conventional formulation of the A model assigns to every cohomology class $\mathbf{e}_\alpha \in H^*(X)$ of the target space X an infinite sequence of observables $\sigma_k(\mathbf{e}_\alpha)$, $k = 0, 1, 2, \ldots$. The corresponding couplings t_k^α parametrize the so-called *large phase space.* For $k = 0$ one gets the *small phase space.* Viewed from the worldsheet, the observable $\sigma_k(\mathbf{e}_\alpha)$ is the k-th descendent of $\mathbf{e}_\alpha$. However, if we think of these observables in terms of the target space we should say that $\sigma_k(\mathbf{e}_\alpha)$ is the $\dim(X) - \deg(\mathbf{e}_\alpha)$-descendent of some local BRST invariant observable $\mathcal{O}_k$:

$$\sigma_k(\mathbf{e}_\alpha) \sim \int_X \mathbf{e}_\alpha \wedge \mathcal{O}_k^{(\dim(X)-\deg(\mathbf{e}_\alpha))}. \tag{2}$$

The gravitational descendents of the top cohomology class of X therefore correspond to the zero-observables $\mathcal{O}_k^{(0)}$ of the target space theory, and as such they are the simplest to study. This is why we shall use as the starting point the so-called *stationary sector* of the theory [23], where only the couplings of these observables are turned on. The observables which are the hardest ones to study are the descendents of the puncture, i.e. unit operator. These correspond to the dim(X)-observables constructed out of $\mathcal{O}_k$ and in the standard paradigm of the topological field theory correspond to the deformations of the space-time Lagrangian.

When the topological theory is a twisted version of the supersymmetric field theory, these deformations correspond to the F-terms of the supersymmetric theory. In two dimensions they are the superpotential deformations, in four dimensions they are the prepotential deformations. Whatever is their interpretation, the target space theory has more observables. Indeed, the product of two local observables $\mathcal{O}_k$ and $\mathcal{O}_l$ and higher order products cannot be expressed, in general, as linear combinations of $\mathcal{O}_k$. In analogy with the gauge theory which we shall make much more precise, the observables $\mathcal{O}_k$ correspond to the *single trace operators*, while the products $\mathcal{O}_{k_1}\mathcal{O}_{k_2}\ldots\mathcal{O}_{k_p}$, for $p > 1$, correspond to the *multi-trace operators*. Thus the full space of deformations of the target space theory will involve couplings $\mathbf{T}_{\vec{k}}^{\alpha,\nu}$, where α label the cohomology of X, ν label the gravitational descendents in the sense of the topological gravity on X (in the problem studied in this paper, X is a two dimensional manifold and ν is a non-negative integer), and $\vec{k} = (k_1 \geq k_2 \geq k_3 \geq \ldots \geq k_p)$ is a partition labelling the multi-trace operators. We call the space of all these couplings the *Very large phase space*. We shall write an expression for the partition function of the topological string on the *Very large phase space* in genus zero (target space). The problem of finding the special coordinates on the *Very large phase space*, which is in a sense equivalent to the problem of constructing the full quantum gravity dressed string theory partition function, is beyond the scope of the present paper. Nevertheless the formulation of the problem for the general target space X is more important then the possible solution of the problem we can anticipate from the gauge theory analogy for $X = \Sigma$, a Riemann surface.

The partition function $\mathcal{Z}_X$ In this paper we study the case where X is a Riemann surface of genus h. The partition function $\mathcal{Z}_X$ of the A-model on a Riemann surface X is a function of an infinite set of couplings, $\mathbf{t} = (t_n^\alpha)$ where $\alpha = 1, \ldots, \dim H^*(X) = 2h + 2$ and $n \in \mathbf{Z}_{\geq 0}$. We introduce some additive basis $\mathbf{e}_\alpha$ of the cohomology of X, $\mathbf{e}_\alpha \in$

$H^*(X, \mathbf{C})$. We have:

$$\mathcal{Z}_X(\mathbf{t}; \hbar, q) =$$

$$\exp \sum_{g,n;\beta=0}^{\infty} \sum_{\vec{k},\vec{\alpha}} \frac{\hbar^{2g-2} q^{\beta}}{n!} \prod_{i=1}^{n} t_{k_i}^{\alpha_i} \int_{\overline{\mathcal{M}_{g,n}(X,\beta)}} \bigwedge_{i=1}^{n} \mathrm{ev}_i^*(\mathbf{e}_{\alpha_i}) \wedge \psi_i^{k_i} \tag{3}$$

where we used the standard notations [13] for the moduli space $\overline{\mathcal{M}_{g,n}(X,\beta)}$ of degree β genus g stable maps to X with n punctures, the evaluation maps:

$$\mathrm{ev}_i : \overline{\mathcal{M}_{g,n}(X,\beta)} \longrightarrow X \tag{4}$$

defined as:

$$\mathrm{ev}_i\,(C, x_1, \ldots, x_n; \phi) = \phi(x_i) \tag{5}$$

where $(C, x_1, \ldots, x_n; \phi)$ is the stable map with the n punctures $x_1, \ldots, x_n$. Finally, in (3) we have the first Chern classes of the tangent lines $\psi_i = c_1(T_{x_i}C)$ at the i'th marked point. Following [8] it is convenient to think of the partition function $\mathcal{Z}_X$ as of the functional on the space of positive loops valued in $H^*(X)$. Thus, let us introduce the $H^*(X)$-valued function:

$$\mathbf{t}(z) = \sum_{n=0}^{\infty} \frac{1}{n!} \mathbf{t}_n z^n \ , \ \mathbf{t}_n = \sum_{\alpha=1}^{2h+2} t_n^{\alpha}\, \mathbf{e}_{\alpha} \in H^*(X) \tag{6}$$

of a formal variable z. In addition we introduce another function, related to $\mathbf{t}(z)$, the Legendre transform of the antiderivative $\partial^{-1}(z - \mathbf{t}(z))$,

$$\mathbf{F}_{\mathbf{t}}(x) = x\,\mathbf{z}(x) - \tfrac{1}{2}\mathbf{z}^2(x) + \sum_{k=0}^{\infty} \frac{1}{(k+1)!} \mathbf{t}_k\, \mathbf{z}^{k+1}(x), \tag{7}$$

where $\mathbf{z}(x) \in H^*(X)$ solves:

$$x = \mathbf{z}(x) - \mathbf{t}\,(\mathbf{z}\,(x)) \tag{8}$$

and is given by the following formal power series in $\mathbf{t}_k$'s:

$$\mathbf{z}(x) \equiv x + \sum_{n=1}^{\infty} \frac{1}{n!} \left[\mathbf{t}^n(x)\right]^{(n-1)} = x + \mathbf{t}(x) + \mathbf{t}(x) \cdot \mathbf{t}'(x) + \ldots . \tag{9}$$

Note that even though $x \in \mathbf{C} = H^0(X)$, $\mathbf{z}(x) \in H^*(X)$ is an inhomogeneous cohomology class.

We shall present our results in two forms: the mathematical and the physical. The mathematical formula is explicit but is not fully transparent. The physical formula is conceptually more appealing but it requires some preparations so we shall present it afterwards.

The mathematical formula represents $\mathcal{Z}_X(\mathbf{t})$ as a sum over partitions λ (the basic notions of the theory of partitions are recalled in the main body of the paper). Given a partition $\lambda = (\lambda_i)_{i\geq 1}$, let $f_\lambda(x)$ denote its profile:

$$f_\lambda(x) = |x| + \sum_{i=1}^{\infty}\Big(|x - \hbar\,(\lambda_i - i + 1)| - |x - \hbar\,(\lambda_i - i)| + |x - \hbar\,(-i)| - |x - \hbar\,(-i+1)|\Big). \tag{10}$$

Define $\mathcal{S}_\lambda(\mathbf{t})$ as:

$$\begin{aligned} \mathcal{S}_\lambda(\mathbf{t}) &= \frac{1}{2\hbar}\int_{\mathbf{R}} \mathrm{d}x\, f_\lambda''(x) \int_X \mathbf{F}_{\mathbf{t}}(x) + \\ &\quad \frac{1}{2\hbar}\int_{\mathbf{R}} \mathrm{d}x\, f_\lambda''(x) \int_X e(X)\cdot \Delta_{\mathbf{t}}(x) + \\ &\quad \frac{1}{8}\int\!\!\int_{\mathbf{R}^2} \mathrm{d}x_1 \mathrm{d}x_2\, f_\lambda''(x_1) f_\lambda''(x_2) \int_X e(X)\cdot \mathbf{G}_{\mathbf{t}}(x_1,x_2) \end{aligned} \tag{11}$$

where $e(X) = c_1(TX)$ is the Euler class of X,

$$\chi(X) = \int_X e(X) = 2 - 2h\ ,$$

and the functions $\Delta_{\mathbf{t}}$ and $\mathbf{G}_{\mathbf{t}}$ are the particular solutions to the finite difference equations:

(12)
$$\Delta_{\mathbf{t}}(x + \tfrac{1}{2}\hbar) - \Delta_{\mathbf{t}}(x - \tfrac{1}{2}\hbar) = (x + \mathbf{t}_0)\log\left(\frac{x + \mathbf{t}_0}{\mathbf{z}(x)}\right) - \sum_{l=2}^{\infty} \frac{1}{l!}\left\{\sum_{m=2}^{l} \frac{1}{m}\right\} \mathbf{t}_l\, \mathbf{z}^l(x)\ ,$$

(the right hand side is a formal power series in x)

(13)
$$\begin{aligned} &\mathbf{G}_{\mathbf{t}}(x_1 + \tfrac{1}{2}\hbar, x_2 + \tfrac{1}{2}\hbar) - \mathbf{G}_{\mathbf{t}}(x_1 - \tfrac{1}{2}\hbar, x_2 + \tfrac{1}{2}\hbar) \\ &\quad - \mathbf{G}_{\mathbf{t}}(x_1 - \tfrac{1}{2}\hbar, x_2 + \tfrac{1}{2}\hbar) + \mathbf{G}_{\mathbf{t}}(x_1 - \tfrac{1}{2}\hbar, x_2 - \tfrac{1}{2}\hbar) \\ &\qquad = \log\left(\frac{\mathbf{z}(x_1) - \mathbf{z}(x_2)}{\hbar}\right)\ , \end{aligned}$$

which we specify in [21]. The mathematical formula is:

$$\boxed{\mathcal{Z}_X\left(\mathbf{t};\hbar,q\right)=\sum_{\lambda}(-q)^{|\lambda|}\mathrm{exp}\left(\frac{\mathcal{S}_{\lambda}(\mathbf{t})}{\hbar}\right).} \tag{14}$$

We derive it in [21] using the Virasoro constraints proven in [24].

The physical formula: A model version identifies $\mathcal{Z}_X(\mathbf{t})$ with the partition function of a two dimensional gauge theory on X. The gauge theory in question is a twisted $\mathcal{N}=2$ super-Yang–Mills theory with the gauge group $\mathbf{G}$, to be specified momentarily, perturbed by all single-trace operators, commuting with the scalar supercharge Q. More precisely, $\mathcal{Z}_X$ is equal to the generating function of the correlators of all 2, 1, and 0-observables (we remind the relevant notions in the main body of the paper), constructed out of the single-trace operators

$$\mathcal{O}_k=\mathrm{Coeff}_{u^k}\,\mathrm{Tr}_{\mathcal{H}}\,e^{u\phi}\ , \tag{15}$$

that is:

$$\boxed{\begin{aligned}&\mathcal{Z}_X(\mathbf{t};\hbar,q)=\\&\qquad\left\langle\ \mathrm{exp}-\int_X\left[\sum_{k=0}^{\infty}\sum_{\alpha=1}^{2h+2}\hbar^{k-1+\frac{\mathrm{deg}e_{\alpha}}{2}}\,\widehat{t}_k^{\alpha}\,e_{\alpha}\wedge\mathcal{O}_{k+1}^{(2-\mathrm{deg}e_{\alpha})}\right]\right\rangle\\&\qquad\qquad\qquad\widehat{t}_0^{\,2h+2}=t_0^{2h+2}-\log(q)+\chi(X)\log(\hbar)\\&\qquad\qquad\qquad\qquad\widehat{t}_1^{1}=t_1^1-1\end{aligned}} \tag{16}$$

and the other times $\widehat{t}_k^{\alpha}=t_k^{\alpha}$. The gauge group $\mathbf{G}$ consists of certain unitary transformations of a Hilbert space $\mathcal{H}$. Its definition will be given in the main paper [21].

The physical formula: B model version represents $\mathcal{Z}_X(\mathbf{t})$ as a partition function of a Landau–Ginzburg theory with the worldsheet X. The $\mathcal{N}=2$ supersymmetric Landau–Ginzburg theory without topological gravity is determined by the following data: a target space, which is a complex manifold $\mathcal{U}$, a holomorphic function $\mathcal{W}$, and a top degree holomorphic form Ω on $\mathcal{U}$. The target space $\mathcal{U}$ is an infinite-dimensional disconnected space. Its connected components $\mathcal{U}_{\lambda}$ are labelled by partitions λ. Each component is isomorphic to $\mathbf{C}^{\infty}$, the space of finite sequences of complex numbers. The superpotential is given by the regularized infinite sum

$$\mathcal{W} = \sum_{i=1}^{\infty} \left[\left(\lambda_i - i + \tfrac{1}{2}\right) z_i - \tfrac{1}{2} z_i^2 + \sum_{k=0}^{\infty} \frac{t_k}{(k+1)!} z_i^{k+1} \right].$$

The top degree form is given by the formal product:

$$\Omega = \bigwedge_{i=1}^{\infty} \mathrm{d}\varepsilon_i \tag{17}$$

where

$$1 + \sum_{i=1}^{\infty} \varepsilon_i t^i = \prod_{i=1}^{\infty} (1 + t z_i).$$

Of course the infinite-dimensionality of various ingredients involved means that this is not the conventional B model. However the theory provides a regularization of the infinite products and sums above.

Very large phase space extension In the worldsheet formulation the Very large phase space observables are non-local. In the language of topological string, the insertion of the observable $\mathcal{O}_{0,0,\ldots,0}(x_1, \ldots, x_k)[\alpha]$, where $x_i \in C$, $[\alpha] \in H_*(X)$ corresponds to the condition that the points $x_1, \ldots, x_k$ of the worldsheet are mapped to the same point $f \in X$ sitting in a cycle representing $[\alpha]$.

Note that the non-local string theories describing multi-trace deformations of gauge theories were recently studied in the context of the AdS/CFT correspondence [1].

On the very large phase space the function $\mathcal{W}$ becomes a generic symmetric function of z_i's which is formally close to the function $\sum_{i=1}^{\infty} (\lambda_i - i + \frac{1}{2}) z_i - \frac{1}{2} z_i^2$:

$$\begin{aligned} \mathcal{W}_\lambda &= \sum_{i=1}^{\infty} (\lambda_i - i + \tfrac{1}{2}) z_i - \tfrac{1}{2} p_2 + w(p_1, p_2, \ldots) \\ w(p) &= \sum_{\vec{k}} \frac{\widetilde{T}_{k_1 k_2 \ldots k_l}}{k_1! k_2! \ldots k_l!} p_{k_1} p_{k_2} \cdots p_{k_l} \, , \end{aligned} \tag{18}$$

while the holomorphic top form is given by (17). The three point function on a sphere is given by the regularized version of Grothendieck residue:

$$C_{\alpha\beta\gamma} = \sum_{\lambda} \sum_{p_\lambda : d\mathcal{W}_\lambda(p_\lambda) = 0} \frac{\Phi_\alpha(p_\lambda) \Phi_\beta(p_\lambda) \Phi_\gamma(p_\lambda)}{\mathrm{Hess}_\Omega(\mathcal{W}_\lambda)} \tag{19}$$

where α, β, γ are partitions, $\Phi_{\alpha,\beta,\gamma}$ are some formal power seria in p_k's, generalizing Schur functions,

$$p_k = \sum_i z_i^k := k!\,\mathrm{Coeff}_{u^k} \sum_i e^{uz_i}$$

and $\mathrm{Hess}_{\Omega}(\mathcal{W}_\lambda)$ is the regularized determinant, defined as a ratio of the determinant of the derivative map $\mathrm{Det}\,(d\mathcal{W}_\lambda) : \det\mathcal{T} \longrightarrow \det\mathcal{T}^*$ and Ω^2 viewed as an element of $\det^{\otimes 2}\mathcal{T}^*$, where $\mathcal{T}$ is the tangent space $T_{p_\lambda}\mathcal{U}$ at the critical point p_λ of $\mathcal{W}_\lambda$.

The couplings $\widetilde{T}_{\vec{k}}$ in (18) are most likely not the flat (or special) coordinates on the Very large phase space. The flat coordinates $T_{\vec{k}}$ are obtained from $\widetilde{T}_{\vec{k}}$ by a formal diffeomorphism. To find them is the first step in understanding the target space quantum gravity.

So far we were discussing the standard topological string on X. The target space turns out to be a topological field theory. In two dimensions such a theory is a relatively simple construction.

All one needs to determine is a commutative associative algebra $\mathbf{A}$, and a functional $\langle\cdot\rangle : \mathbf{A} \to \mathbf{C}$.

In our case the algebra is just the algebra of symmetric functions. Indeed, we discussed so far the generators of this algebra, $\mathcal{O}_k$, $k = 0, 1, 2, \ldots$. We should be able to multiply $\mathcal{O}_k$'s. In this way we shall get arbitrary polynomials of $\mathcal{O}_k$'s. We may allow a formal power series in the generators $\mathcal{O}_k$'s with the assumption that such a series is well-defined once we substitute

$$\mathcal{O}_k(\lambda) = \mathrm{Coeff}_{u^{k+1}} \sum_{i=1}^{\infty} e^{u\,\mathbf{z}\left(\lambda_i - i + \frac{1}{2}\right)} \tag{20}$$

for an arbitrary partition λ.

We computed the functional for the large phase space:

$$\langle \mathcal{O} \rangle = \sum_\lambda e^{-\mathbf{r}(\lambda, \mathbf{t}^1)} \mathcal{O}(\lambda). \tag{21}$$

We should consider the algebra $\mathbf{A}$ together with its space of deformations. The latter is the space of the couplings $T_{\vec{k}}$.

The two dimensional topological field theory can be coupled to the topological gravity. This is the target space gravity.

We can describe its observables directly in target space. We can also discuss its worldsheet definition. The latter is potentially interesting for more realistic quantum gravity theories.

The target space definition is the following. Consider the moduli space $\mathcal{M}_X$ of complex structures on X. The topological string amplitudes are independent of the choice of the complex structure on X. However, one can generalize them, so that they would define closed differential forms on $\mathcal{M}_X$. Moreover, we can consider non-compact Riemann surfaces X, i.e. curves with punctures.

Michael B. Green has proposed in [9] to study the two dimensional string backgrounds as the theories of worldsheets for yet another string theories. Our approach gives a concrete realization of that proposal.

Acknowledgements. Research was partly supported by the European commission via the RTN network "Constituents, Fundamental Forces and Symmetries of the Universe", by the ANR grants for the programs "Structure of vacuum, topological strings and black holes", and "Geometry and integrability in Mathematical Physics" and by the grants РФФИ 06-02-17382, and НШ-3036.2008.2 (for the support of Russian scientific schools). The research was initiated while I visited Princeton University during the spring semester of 2007. I thank the Departments of Mathematics and Physics for their hospitality.

The results presented in this note were reported at the IHÉS seminars in June 2007, at RIMS, Kyoto University in January 2008, at Imperial College in April 2008. I thank these institutions for the opportunity to present this paper and for the audiences for the helpful remarks.

This work would not have been possible without numerous conversations I had with Andrei Okounkov over the last five years. It is a pleasure to acknowledge the collaboration with A. Alexandrov on the companion paper [2]. I also thank E. Getzler, A. Givental and R. Pandharipande for discussions.

References

[1] O. Aharony, M. Berkooz and E. Silverstein, Multiple-trace operators and non-local string theories, J. High Energy Phys., **0108** (2001), 006; arXiv:hep-th/0105309.

[2] A. Alexandrov and N. Nekrasov, Equivariant topological string on $\mathbf{P}^1$, gauge theory, and Toda hierarchy, to appear.

[3] M. Bershadsky and V. Sadov, Theory of Kähler gravity, Internat. J. Modern Phys. A, **11** (1996), 4689–4730; arXiv:hep-th/9410011.

[4] M. Bershadsky, S. Cecotti, H. Ooguri and C. Vafa, Kodaira–Spencer theory of gravity and exact results for quantum string amplitudes, Comm. Math. Phys., **165** (1994), 311–428; arXiv:hep-th/9309140.

[5] S. Cordes, G. Moore and S. Ramgoolam, Lectures on 2-d Yang–Mills theory, equivariant cohomology and topological field theories, Nuclear Phys. B Proc. Suppl., **41** (1995), 184–244; arXiv:hep-th/9411210.
———, Large N two dimensional Yang–Mills theory and topological string theory, Comm. Math. Phys., **185** (1997), 543–619; arXiv:hep-th/9402107.
[6] R. Dijkgraaf, S. Gukov, A. Neitzke and C. Vafa, Topological M-theory as unification of form theories of gravity, Adv. Theor. Math. Phys., **9** (2005), 603–665; arXiv:hep-th/0411073.
[7] F. Ferrari, Extended $\mathcal{N} = 1$ super-Yang–Mills theory, arXiv:hep-th/0709.0472.
H. Kanno and S. Moriyama, Instanton Calculus and Loop Operator in Supersymmetric Gauge Theory.
[8] A. Givental, Gromov–Witten invariants and quantization of quadratic hamiltonians, arXiv:math/0108100.
[9] M. B. Green, World sheets for world sheets, Nuclear Phys. B, **293** (1987), 593–611.
[10] M. Green and S. Shatashvili, 1995, unpublished.
[11] A. Iqbal, N. Nekrasov, A. Okounkov and C. Vafa, Quantum foam and topological strings, arXiv:hep-th/0312022.
[12] S. Katz, A. Klemm and C. Vafa, Geometric engineering of quantum field theories, arXiv:hep-th/9609239.
[13] M. Kontsevich and Yu. Manin, Gromov–Witten classes, quantum cohomology, and enumerative geometry, arXiv:hep-th/9402147.
[14] A. Lawrence and N. Nekrasov, Instanton sums and five-dimensional gauge theories, arXiv:hep-th/9706025.
[15] W.-P. Li, Z. Qin and W. Wang, Hilbert schemes, integrable hierarchies, and Gromov–Witten theory, arXiv:math/0302211.
[16] A. Losev, A. Marshakov and N. Nekrasov, Small instantons, little strings and free fermions, arXiv:hep-th/0302191.
[17] A. Marshakov and N. Nekrasov, Extended Seiberg–Witten theory and integrable hierarchy, arXiv:hep-th/0612019.
[18] N. Nekrasov, A la recherche de la M-theorie perdue Z theory: Chasing M / F theory, Lectures given at Annual International Conference on Strings, Theory and Applications (Strings 2004), Paris, France, 28 Jun–Jul 2, 2004, C. R. Phys., **6** (2005), 261–269; arXiv:hep-th/0412021.
[19] N. Nekrasov, Five dimensional gauge theories and relativistic integrable systems, arXiv:hep-th/9609219.
[20] N. Nekrasov, Seiberg–Witten prepotential from the instanton counting, arXiv:hep-th/0206161, arXiv:hep-th/0306211.
[21] N. Nekrasov, Two dimensional topological strings revisited, to appear.
[22] N. Nekrasov, H. Ooguri and C. Vafa, S-duality and topological strings, arXiv:hep-th/0403167.
[23] A. Okounkov and R. Pandharipande, Gromov–Witten theory, Hurwitz numbers, and matrix models, I, arXiv:math/0101147.

[24] A. Okounkov and R. Pandharipande, Virasoro constraints for target curves, arXiv:math/0308097.

Institut des Hautes Études Scientifiques
Le Bois-Marie, 35 route de Chartres
91440 Bures-sur-Yvette
France
E-mail address: nikitastring@gmail.com

Advanced Studies in Pure Mathematics 59, 2010
New Developments in Algebraic Geometry,
Integrable Systems and Mirror Symmetry (Kyoto, 2008)
pp. 239–288

Enumerative geometry of Calabi–Yau 5-folds

Rahul Pandharipande and Aleksey Zinger

Abstract.

Gromov–Witten theory is used to define an enumerative geometry of curves in Calabi–Yau 5-folds. We find recursions for meeting numbers of genus 0 curves, and we determine the contributions of moving multiple covers of genus 0 curves to the genus 1 Gromov–Witten invariants. The resulting invariants, conjectured to be integral, are analogous to the previously defined BPS counts for Calabi–Yau 3 and 4-folds. We comment on the situation in higher dimensions where new issues arise.

Two main examples are considered: the local Calabi–Yau $\mathbb{P}^2$ with normal bundle $\oplus_{i=1}^{3}\mathcal{O}(-1)$ and the compact Calabi–Yau hypersurface $X_7 \subset \mathbb{P}^6$. In the former case, a closed form for our integer invariants has been conjectured by G. Martin. In the latter case, we recover in low degrees the classical enumeration of elliptic curves by Ellingsrud and Strömme.

§0. Introduction

0.1. Overview

Let X be a nonsingular projective variety over $\mathbb{C}$. Let $\overline{\mathfrak{M}}_{g,k}(X,\beta)$ be the moduli space of genus g, k pointed stable maps to X representing the class $\beta \in H_2(X,\mathbb{Z})$. Let

$$\mathrm{ev}_i \colon \overline{\mathfrak{M}}_{g,k}(X,\beta) \longrightarrow X$$

be the evaluation morphism at the ith marking. The Gromov–Witten theory of primary fields concerns the invariants

$$N_{g,\beta}(\gamma_1,\ldots,\gamma_k) = \int_{[\overline{\mathfrak{M}}_{g,k}(X,\beta)]^{vir}} \prod_{i=1}^{k} \mathrm{ev}_i^*(\gamma_i) \ \in \mathbb{Q}, \tag{0.1}$$

Received March 29, 2008.
2000 *Mathematics Subject Classification.* Primary 14N35.

where $\gamma_i \in H^*(X, \mathbb{Z})$. The relationship between the Gromov–Witten invariants and the actual enumerative geometry of curves in X is subtle. An overview of the subject in low dimensions can be found in the introduction of [10].

For Calabi–Yau 3-folds, the Aspinwall–Morrison formula [1] is conjectured to produce integer invariants in genus 0. A full integrality conjecture for the Gromov–Witten theory of Calabi–Yau 3-folds was formulated by Gopakumar and Vafa in [5, 6] in terms of BPS states with geometric motivation partially provided by [14]. The Aspinwall–Morrison prediction has been extended to all Calabi–Yau n-folds in [10]: the numbers $n_{0,\beta}(\gamma_1, \ldots, \gamma_k)$ defined by

$$\sum_{\beta \neq 0} N_{0,\beta}(\gamma_1, \ldots, \gamma_k) q^\beta = \sum_{\beta \neq 0} n_{0,\beta}(\gamma_1, \ldots, \gamma_k) \sum_{d=1}^{\infty} \frac{1}{d^{3-k}} q^{d\beta} \tag{0.2}$$

are conjectured to be integers.

Let X be a Calabi–Yau of dimension $n \geq 4$. Since Gromov–Witten invariants of genus $g \geq 2$ of X vanish for dimensional reasons, only integrality predictions for genus 1 invariants of X remain to be considered. The analogue of the genus 1 Gopakumar–Vafa integrality prediction for Calabi–Yau 4-folds has been formulated in [10]. Here, we find complete formulas in dimension 5 and reinterpret the dimension 4 predictions. The geometry becomes significantly more complicated in each dimension. We discuss new aspects of the higher dimensional cases.

The relationship between Gromov–Witten theory and enumerative geometry in dimensions greater than 3 is simplest in the Calabi–Yau case. The Fano case, even in dimension 4, involves complicated higher genus phenomena which have not yet been understood.

0.2. Elliptic invariants

If X is Calabi–Yau, the virtual moduli cycle for $\overline{\mathfrak{M}}_1(X, \beta)$ is of dimension 0. We denote the associated Gromov–Witten invariant by $N_{1,\beta}$,

$$N_{1,\beta} = \int_{[\overline{\mathfrak{M}}_1(X,\beta)]^{vir}} 1 \in \mathbb{Q}.$$

Integrality predictions for Calabi–Yau n-folds are obtained by relating curve counts to Gromov–Witten invariants in an ideal Calabi–Yau X. All genus 1 curves in X are assumed to be nonsingular, super-rigid[1],

[1]A nonsingular curve $E \subset X$ with normal bundle $\mathcal{N}_E$ is super-rigid if, for every dominant stable map $f : C \to E$, the vanishing $H^0(C, f^*\mathcal{N}_E) = 0$ holds.

and disjoint from other curves. Each genus 1 degree β curve then contributes $\sigma(d)/d$ to $N_{1,d\beta}$ for every $d\in\mathbb{Z}^+$ via étale covers, where

$$\sigma(d) = \sum_{i|d} i.$$

The genus 1 to genus 1 multiple cover contribution is independent of dimension.

If X is an ideal Calabi–Yau 3-fold, the genus 0 curves in X are also nonsingular, super-rigid, and disjoint. The contribution of a genus 0 degree β curve to $N_{1,d\beta}$ is then the integral of an Euler class of an obstruction bundle on $\overline{\mathfrak{M}}_1(\mathbb{P}^1, d)$,

$$\int_{[\overline{\mathfrak{M}}_1(\mathbb{P}^1,d)]^{vir}} e(\mathrm{Obs}) = \frac{1}{12d},$$

calculated in [14]. Thus, if X is an ideal Calabi–Yau 3-fold,

$$\sum_{\beta\neq 0} N_{1,\beta} q^\beta = \sum_{\beta\neq 0} n_{1,\beta} \sum_{d=1}^{\infty} \frac{\sigma(d)}{d} q^{d\beta} - \frac{1}{12} \sum_{\beta\neq 0} n_{0,\beta} \log(1-q^\beta)\,, \tag{0.3}$$

where the enumerative invariant $n_{1,\beta}$ is defined by (0.3) and the genus 0 invariant $n_{0,\beta}$ is defined by the Aspinwall–Morrison formula (0.2). The invariants $n_{1,\beta}$ are then conjectured to be integers for all Calabi–Yau 3-folds.

If X is an ideal Calabi–Yau 4-fold, embedded genus 0 degree β curves in X form a nonsingular, compact, 1-dimensional family $\overline{\mathcal{M}}_\beta$. The moving multiple cover calculation of Section 2 of [10] shows that $\overline{\mathcal{M}}_\beta$ contributes $\chi(\overline{\mathcal{M}}_\beta)/24d$ to $N_{1,d\beta}$ for every $d\in\mathbb{Z}^+$. The calculation is done in two steps. First, the moving multiple cover integral is done assuming every genus 0 degree β curve is nonsingular. Second, the contribution from the nodal curves is determined for a particular, but sufficiently representative, Calabi–Yau 4-fold X by localization. For an ideal Calabi–Yau 4-fold X,

$$\sum_{\beta\neq 0} N_{1,\beta} q^\beta = \sum_{\beta\neq 0} n_{1,\beta} \sum_{d=1}^{\infty} \frac{\sigma(d)}{d} q^{d\beta} - \frac{1}{24} \sum_{\beta\neq 0} \chi(\overline{\mathcal{M}}_\beta) \log(1-q^\beta)\,. \tag{0.4}$$

The topological Euler characteristic $\chi(\overline{\mathcal{M}}_\beta)$ is determined by

$$\chi(\overline{\mathcal{M}}_\beta) = -n_{0,\beta}(c_2(X)) + \sum_{\beta_1+\beta_2=\beta} m_{\beta_1,\beta_2},$$

where m_{β_1,β_2} is the number of ordered pairs (C_1, C_2) of rational curves of classes β_1 and β_2 meeting at point, see Section 1.2 of [10].

The meeting numbers m_{β_1,β_2} can be expressed in terms of the invariants $n_{0,\beta}(\gamma)$ through a recursion on the total degree $\beta_1+\beta_2$ by computing the excess contribution to the topological Kunneth decomposition of $m_{\beta_1\beta_2}$, see Sections 0.3 and 1.2 of [10]. Along with these recursions, relations (0.2) and (0.4) effectively determine the numbers $n_{1,\beta}$ in terms of the genus 0 and genus 1 Gromov–Witten invariants of X. For arbitrary Calabi–Yau 4-folds, equation (0.4) is taken to be the definition of the numbers $n_{1,\beta}$ which are conjectured always to be integers.

If X is an ideal Calabi–Yau 5-fold, embedded genus 0 degree β curves in X form a nonsingular, compact, 2-dimensional family $\overline{\mathcal{M}}_\beta$. However, as the nodal curves are more complicated, the localization strategy of [10] does not appear possible. By viewing $N_{1,d\beta}$ as the number of solutions, counted with appropriate multiplicities, of a perturbed $\bar{\partial}$-equation as in [4, 11], we show in Section 2 that $\overline{\mathcal{M}}_\beta$ contributes

$$\frac{1}{24d}\int_{\overline{\mathcal{M}}_\beta}\big(2c_2(\overline{\mathcal{M}}_\beta)-c_1^2(\overline{\mathcal{M}}_\beta)\big)$$

to $N_{1,d\beta}$ for every $d\in\mathbb{Z}^+$. Thus, for an ideal Calabi–Yau 5-fold X,

$$\begin{aligned}\sum_{\beta\neq0}N_{1,\beta}q^\beta &= \sum_{\beta\neq0}n_{1,\beta}\sum_{d=1}^{\infty}\frac{\sigma(d)}{d}q^{d\beta}\\ &\quad-\frac{1}{24}\sum_{\beta\neq0}\int_{\overline{\mathcal{M}}_\beta}\big(2c_2(\overline{\mathcal{M}}_\beta)-c_1^2(\overline{\mathcal{M}}_\beta)\big)\cdot\log(1-q^\beta)\,.\end{aligned}\tag{0.5}$$

The last term in (0.5) may be written in terms of various meeting numbers of total degree β via a Grothendieck–Riemann–Roch computation applied to the deformation characterization of the tangent bundle $T\overline{\mathcal{M}}_\beta$. We pursue a more efficient strategy in Sections 1 and 2. Degree 1 maps from genus 0 curves to degree β curves in X are regular. Thus, equation (2.15) in [23] expresses their contribution to $N_{1,\beta}$ in terms of counts of m-tuples of 1-marked curves with cotangent ψ-classes meeting at the marked point. The ψ-classes can be easily eliminated using the topological recursion relation at the cost of introducing counts of arbitrary meeting configurations of rational curves in X. The latter can be recursively defined as in the case of $m_{\beta1,\beta2}$ in dimension 4. Relations (0.2) and (0.5) then reduce the numbers $n_{1,\beta}$ to functions of genus 0 and genus 1 Gromov–Witten invariants.

Let X be an arbitrary Calabi–Yau 5-fold. Equation (0.5) together with the rules provided in Sections 1 and 2 for the calculation of

$$\int_{\overline{\mathcal{M}}_\beta} \left(2c_2(\overline{\mathcal{M}}_\beta) - c_1^2(\overline{\mathcal{M}}_\beta)\right)$$

in terms of the Gromov–Witten invariants of X *define* the invariants $n_{1,\beta}$. We view $n_{1,\beta}$ as virtually enumerating elliptic curves in X.

Conjecture 1. *For all Calabi–Yau 5-folds X and curve classes $\beta \neq 0$, the invariants $n_{1,\beta}$ are integers.*

0.3. Examples

If the Gromov–Witten invariants of X are known, equation (0.5) provides an effective determination of the elliptic invariants $n_{1,\beta}$. We consider two representative examples.

The most basic local Calabi–Yau 5-fold is the total space of the bundle

$$\mathcal{O}(-1) \oplus \mathcal{O}(-1) \oplus \mathcal{O}(-1) \longrightarrow \mathbb{P}^2. \tag{0.6}$$

The balanced property of the bundle is analogous to the fundamental local Calabi–Yau 3-fold

$$\mathcal{O}(-1) \oplus \mathcal{O}(-1) \longrightarrow \mathbb{P}^1.$$

As in the 3-fold case, we find very simple closed forms in Section 3.1 for the genus 0 and 1 Gromov–Witten invariants of the local Calabi–Yau 5-fold (0.6).

We have computed the invariants $n_{1,d}$ via equation (0.5) up to degree 200. All are integers. Even the first 60, shown in Table 0.3, suggest intriguing patterns. For example, $n_{1,d} = 0$ for all multiples of 8. G. Martin has proposed an explicit formula for $n_{1,d}$ which holds for all the numbers we have computed. We state Martin's conjecture in Section 3.2.

The Calabi–Yau septic hypersurface $X_7 \subset \mathbb{P}^6$ is a much more complicated example. Using the closed formulas for the genus 1 and 2-pointed genus 0 Gromov–Witten invariants provided by [23] and [22] respectively, we have computed $n_{1,d}$ for $d \leq 100$. All are integers. The values of $n_{1,d}$ for $d \leq 10$ are shown in Table 0.3.

The invariants $n_{1,d}$ for $d \leq 4$ agree with known enumerative results for X_7. The invariants $n_{1,1}$ and $n_{1,2}$ vanish by geometric considerations. Since every genus 1 curve of degree 3 in $\mathbb{P}^6$ is planar, the number of elliptic cubics on a general X_7 can be computed classically via Schubert

d	$n_{1,d}$	d	$n_{1,d}$	d	$n_{1,d}$	d	$n_{1,d}$	d	$n_{1,d}$	d	$n_{1,d}$
1	0	11	-225	21	3025	31	-14400	41	-44100	51	105625
2	0	12	-19	22	3870	32	0	42	-51590	52	-7119
3	-1	13	-441	23	-4356	33	18496	43	-53361	53	-123201
4	0	14	630	24	0	34	22140	44	-3645	54	0
5	-9	15	784	25	0	35	23409	45	0	55	142884
6	20	16	0	26	7560	36	0	46	74250	56	0
7	-36	17	-1296	27	0	37	-29241	47	-76176	57	164836
8	0	18	0	28	-594	38	34560	48	0	58	187740
9	0	19	-2025	29	-11025	39	36100	49	0	59	-189225
10	162	20	-153	30	-13412	40	0	50	0	60	12628

Table 1. Invariants $n_{1,d}$ for $\mathcal{O}(-1) \oplus \mathcal{O}(-1) \oplus \mathcal{O}(-1) \longrightarrow \mathbb{P}^2$

calculus. The classical calculation agrees with $n_{1,3}$. Using the expression of non-planar genus 1 curves of degree 4 as complete intersections of quadrics, Ellingsrud and Strömme have enumerated elliptic quartics on X_7 in Theorem 1.3 of [3]. The result agrees with $n_{1,4}$. To our knowledge, the numbers $n_{1,d}$ are inaccessible by classical techniques for $d \geq 5$.

0.4. BPS states

The integer expansion (0.5) can be alternatively written as

$$\begin{aligned} \sum_{\beta \neq 0} N_{1,\beta} q^\beta &= -\sum_{\beta \neq 0} \tilde{n}_{1,\beta} \cdot \log(1-q^\beta) \\ &\quad - \frac{1}{24} \sum_{\beta \neq 0} \int_{\overline{\mathcal{M}}_\beta} \big(2c_2(\overline{\mathcal{M}}_\beta) - c_1^2(\overline{\mathcal{M}}_\beta)\big) \cdot \log(1-q^\beta)\,. \end{aligned} \tag{0.7}$$

The integrality condition for the invariants $\tilde{n}_{1,\beta}$ is equivalent to the conjectured integrality for $n_{1,\beta}$. We view the invariants $\tilde{n}_{1,\beta}$ as analogous to the BPS state counts in dimensions 3 and 4.

0.5. Higher dimensions

The family $\overline{\mathcal{M}}_\beta$ of embedded genus 0 degree β curves in X is nonsingular and compact for ideal Calabi–Yau n-folds for $n = 3, 4, 5$. The moving multiple cover results for $n = 3, 4, 5$ can be summarized by the following equation. The contribution of $\overline{\mathcal{M}}_\beta$ to the genus 1 degree $d\beta$ Gromov–Witten invariant is

$$\mathsf{c}_\beta(d\beta) = \frac{1}{24d} \int_{\overline{\mathcal{M}}_\beta} \big(2c_{n-3}(\overline{\mathcal{M}}_\beta) - c_1(\overline{\mathcal{M}}_\beta) c_{n-4}(\overline{\mathcal{M}}_\beta)\big)\,. \tag{0.8}$$

For dimension 6 and higher, the family of embedded genus 0 degree β curves in X is not compact (multiple covers can occur as limits) even in

d	$n_{1,d}$
1	0
2	0
3	26123172457235
4	81545482364153841075
5	117498479295762788677099464
6	126043741686161819224278666855602
7	117293462422824431122974865933687206294
8	100945295955344375879041227482174735213546636
9	82898589348613625712387472944689576403215969839772
10	66074146583335641807745540088333857250772567526848951526

Table 2. Invariants $n_{1,d}$ for a degree 7 hypersurface in $\mathbb{P}^6$

ideal cases. Nevertheless, we expect a contribution equation of the form of (0.8) to hold. The result should yield integrality predictions in higher dimensions.

Since the complexity of the Gromov–Witten approach increases so much in every dimension, an alternate method for dimensions 6 and higher is preferable. It is hoped a connection to newer sheaf enumeration and derived category techniques will be made [15, 16].

0.6. Acknowledgments

We thank J. Bryan, I. Coskun, A. Klemm, J. Starr, and R. Thomas for several related discussions. We are grateful to G. Martin for finding the pattern governing the invariants $n_{1,d}$ for the local Calabi–Yau 5-fold geometry.

The research was started during a visit to the Centre de Recherches Mathématique in Montréal in the summer of 2007. R. P. was partial supported by DMS-0500187. A. Z. was partially supported by the Sloan foundation and DMS-0604874.

§1. Genus 0 invariants

1.1. Configuration spaces of genus 0 curves

Let X be a Calabi–Yau 5-fold. We specify here what conditions an ideal X is to satisfy with respect to genus 0 curves. We denote by

$$H_+(X) \subset H_2(X, \mathbb{Z}) - 0$$

the cone of effective curve classes. If $\beta, \beta' \in H_+(X)$, we write $\beta' < \beta$ if $\beta - \beta'$ is an element of $H_+(X)$.

If J is a finite set and $\beta \in H_+(X)$, we denote by $\overline{\mathfrak{M}}_{0,J}(X,\beta)$ the moduli space of genus 0, J-marked stable maps to X representing the class β. For $j\in J$, let

$$L_j \longrightarrow \overline{\mathfrak{M}}_{0,J}(X,\beta)$$

be the universal tangent line bundle at the jth marked point. Denote by

$$\mathcal{D}_j \in \Gamma\big(\overline{\mathfrak{M}}_{0,J}(X,\beta), \mathrm{Hom}(L_j, \mathrm{ev}_j^* TX)\big)$$

the bundle section induced by the differential of the stable maps at the jth marked point.

If Σ is a curve, a map $u\colon \Sigma \longrightarrow X$ is called *simple* if u is injective on the complement of finitely many points and of the components of Σ on which u is constant. We will call a tuple $(u_1,\ldots,u_m)$ of maps $u_i\colon \Sigma_i \longrightarrow X$ *simple* if the map

$$\bigsqcup_{i=1}^{m} \Sigma_i \longrightarrow X, \qquad z \longrightarrow u_i(z) \text{ if } z\in\Sigma_i,$$

is simple. If J is a finite set and $\beta\in H_+(X)$, let

$$\mathfrak{M}^*_{0,J}(X,\beta) \subset \overline{\mathfrak{M}}_{0,J}(X,\beta)$$

be the open subspace of stable maps $[\Sigma,u]$ such that Σ is a $\mathbb{P}^1$ and u is a simple map.

If J_1 and J_2 are two finite sets and $\beta_1,\beta_2\in H_+(X)$, we denote by

$$\begin{aligned}&\mathfrak{M}^*_{0,(J_1,J_2)}\big(X,(\beta_1,\beta_2)\big) \subset \\ &\big\{(b_1,b_2)\in \mathfrak{M}^*_{0,\{0\}\sqcup J_1}(X,\beta_1)\times \mathfrak{M}^*_{0,\{0\}\sqcup J_2}(X,\beta_2)\colon \mathrm{ev}_0(b_1)=\mathrm{ev}_0(b_2)\big\}\end{aligned}$$

the subset of simple pairs of maps. Similarly, if $\beta_1,\beta_2,\beta_3\in H_+(X)$, let

$$\begin{aligned}&\mathfrak{M}^*_{0,\emptyset}\big(X,(\beta_1,\beta_2,\beta_3)\big) \subset \\ &\big\{(b_1,b_2,b_3)\in \mathfrak{M}^*_{0,(\emptyset,\{1\})}\big(X,(\beta_1,\beta_2)\big)\times\mathfrak{M}^*_{0,\{0\}}(X,\beta_3)\colon \mathrm{ev}_1(b_2)=\mathrm{ev}_0(b_3)\big\}\end{aligned}$$

be the subset of simple triples of maps. If X is an ideal Calabi–Yau 5-fold satisfying Conditions 1 and 2 below, there are no other configurations of simple genus 0 curves in X, see Figure 1.

Denote by $\overline{\mathfrak{M}}^*_{0,J}(X,\beta) \subset \overline{\mathfrak{M}}_{0,J}(X,\beta)$ and

$$\overline{\mathfrak{M}}^*_{0,(J_1,J_2)}\big(X,(\beta_1,\beta_2)\big) \subset \overline{\mathfrak{M}}_{0,\{0\}\sqcup J_1}(X,\beta_1)\times\overline{\mathfrak{M}}_{0,\{0\}\sqcup J_2}(X,\beta_2),$$

the closures of $\mathfrak{M}^*_{0,J}(X,\beta)$ and $\mathfrak{M}^*_{0,(J_1,J_2)}\big(X,(\beta_1,\beta_2)\big)$. Let

(1.1)
$$\pi_1,\pi_2\colon \overline{\mathfrak{M}}^*_{0,(J_1,J_2)}\big(X,(\beta_1,\beta_2)\big) \to \overline{\mathfrak{M}}_{0,\{0\}\sqcup J_1}\big(X,\beta_1\big), \overline{\mathfrak{M}}_{0,\{0\}\sqcup J_2}\big(X,\beta_2\big),$$

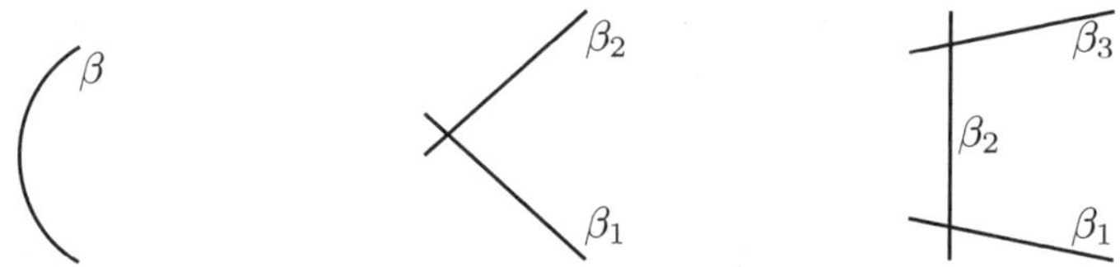

Fig. 1. The three possible configurations of rational curves in an ideal Calabi–Yau 5-fold. The label next to each component indicates the degree.

be the component projection maps.

Condition 1. *If $u : \mathbb{P}^1 \longrightarrow X$ is a simple holomorphic map, $H^1(\mathbb{P}^1, u^*TX) = 0$.*

By Condition 1, $\mathfrak{M}^*_{0,J}(X,\beta)$ is a nonsingular variety of the expected dimension $2+|J|$.

Condition 2. *For all $\beta_1,\ldots,\beta_k \in H_+(X)$, finite sets $J_1,\ldots,J_k$, and a partition of $J_1\sqcup\ldots\sqcup J_k$ into nonempty disjoint subsets $I_1,\ldots,I_m$, the restriction of the total evaluation map*[2]

$$\mathrm{ev}\colon \prod_{p=1}^{k} \mathfrak{M}^*_{0,J_p}(X,\beta_p) \longrightarrow \prod_{p=1}^{k} X^{J_p},$$
$$\mathrm{ev}\big((b_p)_{p\in[k]}\big)_{(p,j)} = \mathrm{ev}_j(b_p), \quad \forall\, p\in[k],\ j\in J_p,$$

to the open subspace of simple tuples is transverse to the diagonal

$$\big\{(x_{(p,j)})_{p\in[k],j\in J_p} : x_{(p,j)} = x_{(p',j')} \text{ if } (p,j),(p',j')\in I_q \text{ for some } q\big\}.$$

By Condition 2, $\mathfrak{M}^*_{0,(\emptyset,\emptyset)}\big(X,(\beta_1,\beta_2)\big)$ and $\mathfrak{M}^*_{0,\emptyset}\big(X,(\beta_1,\beta_2,\beta_3)\big)$ are nonsingular of dimensions 1 and 0, respectively. Furthermore, all simple genus 0 maps with reducible domains deform to curves with nonsingular domains. Furthermore, for all $\beta \in H_+(X)$, the open subspace of $\overline{\mathfrak{M}}_{0,J}(X,\beta)$ consisting of simple maps is nonsingular.

Condition 3. *For all $\beta \in H_+(X)$, the restriction of the bundle section $\mathcal{D}_1$ to $\mathfrak{M}^*_{0,1}(X,\beta)$ is transverse to the zero set. For all $\beta_1,\beta_2 \in H_+(X)$, the bundle section*

$$\pi_1^*\mathcal{D}_0 + \pi_2^*\mathcal{D}_0 \in \Gamma\big(\mathbb{P}(\pi_1^*L_0 \oplus \pi_2^*L_0)\big|_{\mathfrak{M}^*_{0,(\emptyset,\emptyset)}(X,(\beta_1,\beta_2))}, \mathrm{Hom}(\gamma, \mathrm{ev}_0^*TX)\big),$$

[2]We denote the elements of $J_1\sqcup\ldots\sqcup J_k$ by pairs (p,j), where $j \in J_p$, and let $[k] = \{1,2,\ldots,k\}$.

where $\gamma \longrightarrow \mathbb{P}(\pi_1^* L_0 \oplus \pi_2^* L_0)$ *is the tautological line bundle, is transverse to the zero set.*

By Condition 1 and the first part of Condition 3, every simple holomorphic map $u\colon \mathbb{P}^1 \longrightarrow X$ is an immersion. By Condition 2, u is injective. Thus, every irreducible genus 0 curve $C \subset X$ is nonsingular. The normal bundle to such a curve must split as

$$\mathcal{N} = \mathcal{O}(a_1) \oplus \mathcal{O}(a_2) \oplus \mathcal{O}(a_3) \oplus \mathcal{O}(a_4) \to \mathbb{P}^1, \text{with } a_i \in \mathbb{Z}, \sum_{i=1}^{i=4} a_i = -2, \ a_i \geq -1,$$

the last restriction follows from Condition 1. By the first part of Condition 4 below, $a_i \in \{0, -1\}$ for all i. The second part of Condition 3 implies that every node of a reducible genus 0 curve in X is simple.

Condition 4. *For all* $\beta \in H_+(X)$, *the bundle section*

$$\mathcal{D}\mathrm{ev}_1 \in \Gamma\big(\mathbb{P}(T\mathfrak{M}^*_{0,1}(X,\beta)), \mathrm{Hom}(\gamma, \mathrm{ev}_1^* TX)\big),$$

where $\gamma \longrightarrow \mathbb{P}(T\mathfrak{M}^*_{0,1}(X,\beta))$ *is the tautological line bundle, is transverse to the zero set. For all* $\beta_1, \beta_2 \in H_+(X)$, *the bundle section*

$$\begin{aligned} &\pi_1^* \mathcal{D}\mathrm{ev}_0 + \pi_2^* \mathcal{D}_0 \in \\ &\Gamma\big(\mathbb{P}(\pi_1^* T\mathfrak{M}^*_{0,\{0\}}(X,\beta_1) \oplus \pi_2^* L_0)\big|_{\mathfrak{M}^*_{0,(\emptyset,\emptyset)}(X,(\beta_1,\beta_2))}, \mathrm{Hom}(\gamma, \mathrm{ev}_0^* TX)\big), \end{aligned}$$

where $\gamma \longrightarrow \mathbb{P}(\pi_1^* T\mathfrak{M}^*_{0,\{0\}}(X,\beta_1) \oplus \pi_2^* L_0)$ *is the tautological line bundle, is transverse to the zero set.*

By Condition 4, neither of the two bundle sections vanishes anywhere. In the case of the first bundle section, the dimension of the base space and the rank of the vector bundle both equal 5. On the other hand, the vanishing of the bundle section here implies the differential of the evaluation map

$$\mathrm{ev}_1 \colon \mathfrak{M}^*_{0,1}(X,\beta) \longrightarrow X$$

is not injective at some simple, degree β, 1-marked map $[\mathbb{P}^1, x_1, u]$. Hence, the normal bundle must split as

$$\mathcal{N} \approx \mathcal{O}(1) \oplus \mathcal{O}(-1) \oplus \mathcal{O}(-1) \oplus \mathcal{O}(-1).$$

Therefore $\mathcal{D}\mathrm{ev}_1$ is not injective at $[\mathbb{P}^1, x, u]$ for all $x \in \mathbb{P}^1$. The zero set of the first bundle section in Condition 4 must be at least of dimension one. So by transversality, no vanishing is possible.

The non-vanishing of the second bundle section is clear from transversality since the base space is of dimension 4 and bundle is of rank 5.

Lemma 1.1. *Let X be an ideal Calabi–Yau 5-fold. If $\beta \in H_+(X)$ and J is a finite set, the space $\overline{\mathfrak{M}}^*_{0,J}(X,\beta)$ is nonsingular of dimension $2+|J|$ and consists of simple maps. Furthermore, the evaluation map*

$$\mathrm{ev}_1 : \overline{\mathfrak{M}}^*_{0,1}(X,\beta) \longrightarrow X$$

is an immersion. If $\beta_1, \beta_2 \in H_+(X)$ and J_1, J_2 are finite sets, $\overline{\mathfrak{M}}^*_{0,(J_1,J_2)}\big(X,(\beta_1,\beta_2)\big)$ is smooth of dimension $1+|J_1|+|J_2|$ and consists of simple maps.

Proof. By Condition 4, the restriction of ev_1 to the open subset

$$\mathfrak{M}^*_{0,J}(X,\beta) \subset \overline{\mathfrak{M}}^*_{0,J}(X,\beta)$$

is an immersion for every $\beta \in H_+(X)$. Therefore, by the argument given in Section 2.4, if

$$u : \Sigma \to X$$

is not simple, then no deformation of u is simple. Hence, u cannot lie in the closure of $\mathfrak{M}^*_{0,J}(X,\beta)$. We conclude $\overline{\mathfrak{M}}^*_{0,J}(X,\beta)$ consists of simple maps and therefore nonsingular of expected dimension. The proof of the claim for $\overline{\mathfrak{M}}^*_{0,(J_1,J_2)}\big(X,(\beta_1,\beta_2)\big)$ is the same. Q.E.D.

Conditions 1–4 can be extended to define an ideal Calabi–Yau n-fold for any n. However, Lemma 1.1, which depends on the dimension counting argument in the preceding paragraph, does not apply in dimensions 6 and higher. For example, if $X_8 \subset \mathbb{P}^7$ is the degree 8 Calabi–Yau hypersurface,

$$\overline{\mathfrak{M}}^*_{0,1}(X_8,1) = \overline{\mathfrak{M}}_{0,1}(X_8,1)$$

certainly consists of simple maps. However, a computation on $G(2,8)$ shows the evaluation map ev_1 is not an immersion along 133430226944 fibers of the forgetful morphism

$$\overline{\mathfrak{M}}^*_{0,1}(X_8,1) \longrightarrow \overline{\mathfrak{M}}^*_{0,0}(X_8,1).$$

A separate computation in a projective bundle over $G(3,8)$ shows the space of conics in X_8 contains 133430226944 double lines. In both cases the degenerate loci correspond to the 133430226944 lines in X_8 whose normal bundle splits as $\mathcal{O}(1)\oplus\mathcal{O}\oplus 3\mathcal{O}(-1)$, instead of the expected $3\mathcal{O}\oplus 2\mathcal{O}(-1)$. While the Calabi–Yau 6-fold X_8 is not ideal, low-degree curves in projective hypersurfaces do behave as expected. The appearance multiple covers as limits of simple maps is to be expected in dimensions 6 and higher, making a full enumerative treatment more complicated (and likely drastically so).

1.2. Genus 0 counts

We define here integer forms of the genus 0 Gromov–Witten invariants of Calabi–Yau 5-folds by considering all possible distributions of constraints and ψ-classes between the marked points. The 13 relevant types of invariants are indicated in Figure 2. We state relations motivated by ideal geometry which reduce all 13 to genus 0 Gromov–Witten invariants. These relations are taken to be the definition of 13 invariants for arbitrary Calabi–Yau 5-folds.

If J is a finite set, $J' \subset J$, and $\beta \in H_+(X)$, let

$$f_{J,J'} \colon \overline{\mathfrak{M}}_{0,J}(X,\beta) \longrightarrow \overline{\mathfrak{M}}_{0,J-J'}(X,\beta)$$

be the forgetful map dropping the marked points indexed by the set J'. If $j \in J$, let

$$\tilde{\psi}_j = f^*_{J,J-j}\psi_j \in H^2\big(\overline{\mathfrak{M}}_{0,J}(X,\beta)\big),$$

where ψ_j is the first chern class of the universal cotangent line bundle for the marked point on $\overline{\mathfrak{M}}_{0,\{j\}}(X,\beta)$.

If X is an ideal Calabi–Yau 5-fold and $\beta \in H_+(X)$, the dimension of $\overline{\mathfrak{M}}^*_{0,0}(X,\beta)$ is 2. There are 7 invariants of the form

$$n_\beta(\tilde{\psi}^a\mu_1,\mu_2,\ldots,\mu_k) = \int_{\overline{\mathfrak{M}}^*_{0,k}(X,\beta)} \tilde{\psi}^a_1 \prod_{j=1}^k \mathrm{ev}^*_j\mu_j, \qquad a \geq 0,\ \mu_j \in H^{2*}(X),$$

which we require:

(1A) $n_\beta(\mu)$ where $\mu \in H^6(X)$ counting curves through μ,
(1B) $n_\beta(\mu_1,\mu_2)$ where $\mu_1,\mu_2 \in H^4(X)$ counting curves through μ_1 and μ_2,
(1C) $n_\beta(\tilde{\psi}\mu)$ where $\mu \in H^4(X)$,
(1D) $n_\beta(\tilde{\psi}\mu_1,\mu_2)$ where $\mu_1 \in H^2(X)$ and $\mu_2 \in H^4(X)$,
(1E) $n_\beta(\tilde{\psi}^2\mu)$ where $\mu \in H^2(X)$,
(1F) $n_\beta(\tilde{\psi}^2,\mu)$ where $\mu \in H^4(X)$,
(1G) $n_\beta(\tilde{\psi}^3)$.

Let $\overline{\mathcal{M}}_\beta$ denote the unpointed space $\overline{\mathfrak{M}}^*_{0,0}(X,\beta)$. We will need the Chern number

(1H) $\gamma_1(\beta) = \int_{\overline{\mathcal{M}}_\beta} \big(c_1^2(\overline{\mathcal{M}}_\beta) - c_2(\overline{\mathcal{M}}_\beta)\big)$.

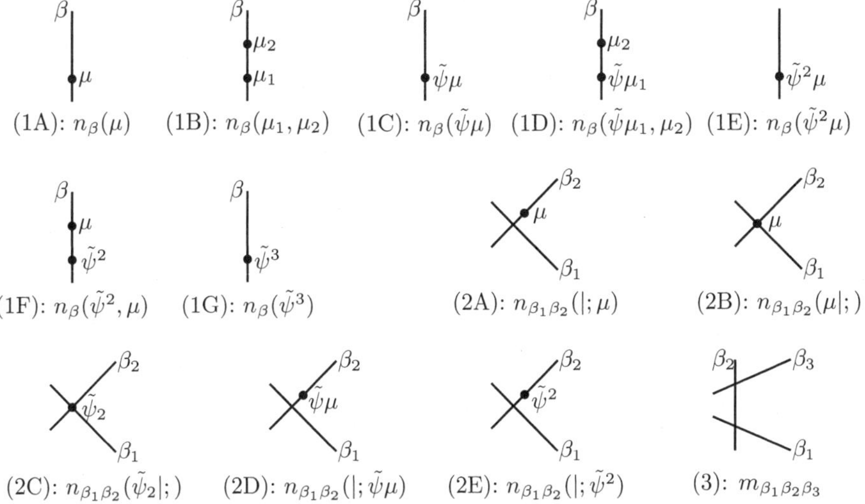

Fig. 2. Counts for Calabi–Yau 5-folds

There are 5 types of relevant counts of connected 2-component curves which we require,

$$n_{\beta_1\beta_2}(\tilde{\psi}_1^{a_1}\tilde{\psi}_2^{a_2}\mu_0|\tilde{\psi}^{b_1}\mu_{1,1}, \mu_{1,2}, \dots, \mu_{1,k_1}; \tilde{\psi}^{b_2}\mu_{2,1}, \mu_{2,2}, \dots, \mu_{2,k_2})$$

$$= \int_{\overline{\mathfrak{M}}^*_{0,([k_1],[k_2])}(X,(\beta_1,\beta_2))} \pi_1^*\left(\tilde{\psi}_0^{a_1}\tilde{\psi}_1^{b_1}\mathrm{ev}_0^*\mu_0 \prod_{j=1}^{k_1}\mathrm{ev}_j^*\mu_{1,j}\right)\pi_2^*\left(\tilde{\psi}_0^{a_2}\tilde{\psi}_1^{b_2}\prod_{j=1}^{k_2}\mathrm{ev}_j^*\mu_{2,j}\right),$$

where π_1, π_2 are the component projection maps as in (1.1), $a_i, b_i \geq 0$, and $\mu_0, \mu_{i,j} \in H^{2*}(X)$. The 5 types are represented by the following counts of (β_1, β_2)-curves:

(2A) $n_{\beta_1\beta_2}(|; \mu)$ where $\mu \in H^4(X)$,
(2B) $n_{\beta_1\beta_2}(\mu|;)$ where $\mu \in H^2(X)$,
(2C) $n_{\beta_1\beta_2}(\tilde{\psi}_2|;)$,
(2D) $n_{\beta_1\beta_2}(|; \tilde{\psi}\mu)$ where $\mu \in H^2(X)$,
(2E) $n_{\beta_1\beta_2}(|; \tilde{\psi}^2)$.

Finally, we denote the cardinality of the compact 0-dimensional space $\mathfrak{M}^*_{0,\emptyset}(\beta_1, \beta_2, \beta_3)$ for triples $\beta_1, \beta_2, \beta_3 \in H_+(X)$ by $m_{\beta_1\beta_2\beta_3}$:

(3) $m_{\beta_1\beta_2\beta_3}$ is the number of connected 3-component curves of tridegree $\beta_1, \beta_2, \beta_3$.

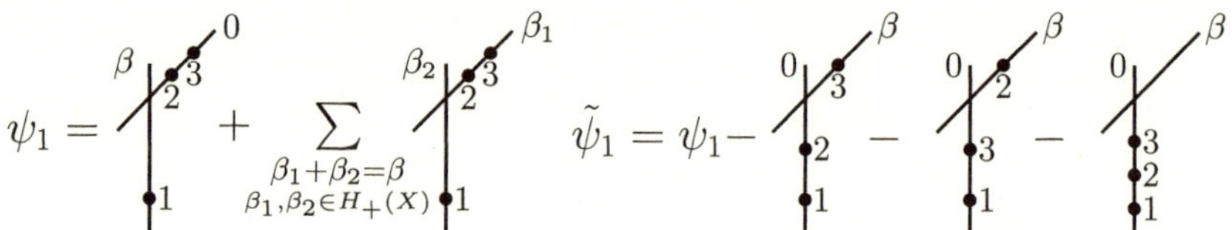

Fig. 3. Relations for ψ_1 and $\tilde{\psi}_1$ on $\overline{\mathfrak{M}}^*_{0,3}(X,\beta)$. Each curve represents the divisor in $\overline{\mathfrak{M}}^*_{0,3}(X,\beta)$ whose general element has the domain and the degree distribution specified by the curve.

The numbers (1A) and (1B) are determined from 1- and 2-pointed Gromov–Witten invariants via (0.2). The topological recursion relation for ψ_1 can be used to express $\tilde{\psi}_1$ in terms of boundary divisors on $\overline{\mathfrak{M}}^*_{0,3}(X,\beta)$, see Figure 3. The divisor relation then gives rise to the relations between the invariants (1C)–(1G) indicated in Figure 4, see also Section 3 in [13]. We now describe these relations formally. If H is a divisor on X and $H_\beta=(H,\beta)$, then

(1.2)
$$\begin{aligned}
H_\beta^2\, n_\beta(\tilde{\psi}\mu) &= n_\beta(\mu, H^2) - 2H_\beta\, n_\beta(H\mu) + \sum_{\beta_1+\beta_2=\beta} H_{\beta_1}^2 n_{\beta_1\beta_2}(|;\mu),\\
H_\beta^2\, n_\beta(\tilde{\psi}\mu_1,\mu_2) &= (\mu_1,\beta)\, n_\beta(\mu_2, H^2) - 2H_\beta\, n_\beta(H\mu_1,\mu_2)\\
&\quad + \sum_{\beta_1+\beta_2=\beta} \big((\mu_1,\beta_1)H_{\beta_2}^2 + (\mu_1,\beta_2)H_{\beta_1}^2\big) n_{\beta_1\beta_2}(|;\mu_2),\\
H_\beta^2\, n_\beta(\tilde{\psi}^2\mu) &= n_\beta(\tilde{\psi}\mu, H^2) - 2H_\beta\, n_\beta(\tilde{\psi}H\mu)\\
&\quad + \sum_{\beta_1+\beta_2=\beta} H_{\beta_1}^2 \big(n_{\beta_1\beta_2}(|;\tilde{\psi}\mu) + n_{\beta_1\beta_2}(\mu|;)\big),\\
n_\beta(\tilde{\psi}^2,\mu) &= -\sum_{\beta_1+\beta_2=\beta} n_{\beta_1\beta_2}(|;\mu),\\
H_\beta^2\, n_\beta(\tilde{\psi}^3) &= n_\beta(\tilde{\psi}^2, H^2) - 2H_\beta\, n_\beta(\tilde{\psi}^2 H)\\
&\quad + \sum_{\beta_1+\beta_2=\beta} H_{\beta_1}^2 \big(n_{\beta_1\beta_2}(|;\tilde{\psi}^2) + n_{\beta_1\beta_2}(\tilde{\psi}_2|;)\big),
\end{aligned}$$

the fourth identity above is obtained by applying the relation of Figure 4 twice. We can similarly remove ψ-classes from 2-component curves:

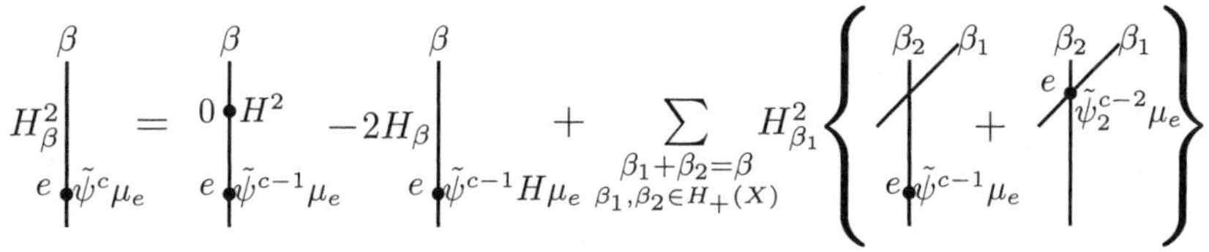

$$H_\beta^2 \;\underset{e\,\bullet\,\tilde\psi^c\mu_e}{\overset{\beta}{\Big|}} \;=\; \underset{e\,\bullet\,\tilde\psi^{c-1}\mu_e}{\overset{\beta}{0\bullet H^2}} \;-2H_\beta \underset{e\,\bullet\,\tilde\psi^{c-1}H\mu_e}{\overset{\beta}{\Big|}} \;+ \sum_{\substack{\beta_1+\beta_2=\beta\\ \beta_1,\beta_2\in H_+(X)}} H_{\beta_1}^2 \left\{ \underset{e\,\bullet\,\tilde\psi^{c-1}\mu_e}{\overset{\beta_2\;\diagup\;\beta_1}{\Big|}} \;+\; \overset{\beta_2\;\diagup\;\beta_1}{e\,\bullet\,\tilde\psi_2^{c-2}\mu_e} \right\}$$

$$H\subset X \text{ divisor},\ H_\beta = H\cdot\beta,\ H_{\beta_1}=H\cdot\beta_1$$

Fig. 4. Reducing the power of $\tilde\psi$ at marked point e in the absence of ψ-classes at other marked points.

(1.3)

$$H_{\beta_2}^2\, n_{\beta_1\beta_2}(\tilde\psi_2|;) = n_{\beta_1\beta_2}(|;H^2) - 2H_{\beta_2}\, n_{\beta_1\beta_2}(H|;) + \sum_{\beta+\beta'=\beta_2} H_\beta^2\, m_{\beta_1\beta'\beta},$$

$$H_{\beta_2}^2\, n_{\beta_1\beta_2}(|;\tilde\psi\mu) = (\mu,H)\, n_{\beta_1\beta_2}(|;H^2) - 2H_{\beta_2}\, n_{\beta_1\beta_2}(|;H\mu)$$
$$+ \sum_{\beta+\beta'=\beta_2} \big((\mu,\beta)H_{\beta'}^2 + (\mu,\beta')H_\beta^2\big) m_{\beta_1\beta'\beta},$$

$$n_{\beta_1\beta_2}(|;\tilde\psi^2) = -\sum_{\beta+\beta'=\beta_2} m_{\beta_1\beta'\beta}\,,$$

the last identity above is obtained by applying the relation of Figure 4 twice. On the other hand, by (1.15) and some manipulation,

$$\begin{aligned}\gamma_1(\beta) = \frac{1}{2}\Big(& n_\beta\big(c_3(X)\big) + n_\beta\big(\tilde\psi c_2(X)\big) + n_\beta\big(\tilde\psi^3\big)\\ &+ n_\beta\big(c_2(X),c_2(X)\big) + 4\,n_\beta\big(\tilde\psi^2, c_2(X)\big)\Big)\\ &- \sum_{\beta_1+\beta_2=\beta}\Big(2\,n_{\beta_1\beta_2}(|;\tilde\psi^2) + \frac{5}{2}\,n_{\beta_1\beta_2}(\tilde\psi_2|;)\Big).\end{aligned} \tag{1.4}$$

The meeting numbers (2A), (2B), and (3) are computed via degree reducing recursions analogous to Rules (i)–(iv) of Section 0.3 of [10] for the 4-dimensional case. Let

$$\{\omega_1,\ldots,\omega_N\}, \{\omega_1^\# \ldots, \omega_N^\#\} \subset H^4(X)\oplus H^6(X)$$

be dual bases normalized so that

$$\mathrm{PD}_{X^2}\Delta_X - \sum_{l=1}^N \omega_l\times\omega_l^\# \in \bigoplus_{k=0,1,4,5} H^{2k}(X)\otimes H^{2(5-k)}(X)\oplus H^{odd}(X)\otimes H^{odd}(X),$$

where $\Delta_X \subset X^2$ is the diagonal. Then,

$$n_{\beta_1\beta_2}(|;\mu) = \sum_{l=1}^{N} n_{\beta_1}(\omega_l)\, n_{\beta_2}(\omega_l^{\#},\mu) + \begin{cases} n_{\beta_1,\beta_2-\beta_1}(|;\mu) + n_{\beta_2-\beta_1,\beta_1}(|;\mu), & \text{if } \beta_2 > \beta_1, \\ n_{\beta_1-\beta_2,\beta_2}(|;\mu), & \text{if } \beta_2 < \beta_1, \\ n_{\beta_1}(c_2(X),\mu) + 2n_{\beta_1}(\tilde{\psi}^2,\mu), & \text{if } \beta_2 = \beta_1. \end{cases} \tag{1.5}$$

In light of the fourth identity in (1.2), the relation differs from the 4-dimensional case only by the expected adjustment for the constraint μ.

The corresponding recursions for the numbers (2B) and (3) are more complicated. For classes $\beta_1, \beta_2 \in H_+(X)$, let

$$\gamma_2(\beta_1,\beta_2) = n_{\beta_1\beta_2}(|;c_2(X)) + 2\,n_{\beta_1\beta_2}(|;\tilde{\psi}^2) + n_{\beta_1\beta_2}(\tilde{\psi}_2|;) + n_{\beta_2\beta_1}(\tilde{\psi}_2|;). \tag{1.6}$$

For $\mu \in H^2(X)$, we define

$$\mathsf{C}_{\beta_1\beta_2}(\mu) = \begin{cases} n_{\beta_2-\beta_1,\beta_1}(|;\tilde{\psi}\mu) + n_{\beta_2-\beta_1,\beta_1}(\mu|;) \\ \quad + (\mu,\beta_1)\big(\gamma_2(\beta_2-\beta_1,\beta_1) + \frac{1}{2}\sum\limits_{\beta+\beta'=\beta_2-\beta_1} m_{\beta\beta_1\beta'}\big), & \text{if } \beta_2 > \beta_1; \\ \mathsf{C}_{\beta_2\beta_1}(\mu), & \text{if } \beta_2 < \beta_1; \\ n_{\beta_1}(c_2(X)\mu) + n_{\beta_1}(\tilde{\psi}^2\mu) \\ \quad + n_{\beta_1}(c_2(X),\tilde{\psi}\mu) + (\mu,\beta_1)\gamma_1(\beta_1) & \text{if } \beta_1 = \beta_2. \\ \quad - \sum\limits_{\beta+\beta'=\beta_2} \big(2n_{\beta\beta'}(|;\tilde{\psi}\mu) + \frac{5}{2} n_{\beta\beta'}(\mu|;)\big), \end{cases} \tag{1.7}$$

For $\beta_1, \beta_2, \beta_3 \in H_+(X)$, let

$$\mathsf{C}^{(1)}_{\beta_1\beta_2\beta_3} = \begin{cases} m_{\beta_3-\beta_1,\beta_1,\beta_2}, & \text{if } \beta_3 > \beta_1; \\ m_{\beta_1-\beta_3,\beta_3,\beta_2}, & \text{if } \beta_3 < \beta_1; \\ \gamma_2(\beta_2,\beta_1), & \text{if } \beta_3 = \beta_1; \end{cases}$$

$$\mathsf{C}^{(2)}_{\beta_1\beta_2\beta_3} = -\begin{cases} m_{\beta_1,\beta_2,\beta_3-\beta_2}, & \text{if } \beta_3 > \beta_2; \\ m_{\beta_1,\beta_3,\beta_2-\beta_3} + m_{\beta_1,\beta_2-\beta_3,\beta_3}, & \text{if } \beta_3 < \beta_2; \\ n_{\beta_1\beta_2}(|;c_2(X)) + 2\,n_{\beta_1\beta_2}(|;\tilde{\psi}^2), & \text{if } \beta_3 = \beta_2; \end{cases} \tag{1.8}$$

$$\mathsf{C}^{(12)}_{\beta_1\beta_2\beta_3} = -\begin{cases} m_{\beta_3-\beta_1-\beta_2,\beta_1,\beta_2}, & \text{if } \beta_3 > \beta_1+\beta_2; \\ m_{\beta_1+\beta_2-\beta_3,\beta_3-\beta_2,\beta_2}, & \text{if } \beta_2 < \beta_3 < \beta_1+\beta_2; \\ \gamma_2(\beta_2,\beta_1), & \text{if } \beta_3 = \beta_1+\beta_2; \\ 0, & \text{otherwise.} \end{cases}$$

$n_{d_1 d_2}(H\|;)$	$d_2 = 1$	$d_2 = 2$
$d_1 = 1$	145366465734	17628837973096812
2	17628837973096812	2134616449608028257452
3	4403307962301366086458	533112594803936499402982169

Table 3. Meeting invariants $n_{d_1 d_2}(H|;)$ for a degree 7 hypersurface in $\mathbb{P}^6$ counting the virtual number of (d_1, d_2)-curves with node on a fixed hyperplane.

Then,

$$n_{\beta_1\beta_2}(\mu|;) = \sum_{l=1}^{N} n_{\beta_1}(\omega_l\mu)\, n_{\beta_2}(\omega_l^{\#}) - \sum_{\beta<\beta_1,\beta_2} (\mu,\beta)\, m_{\beta_1-\beta,\beta,\beta_2-\beta} - \mathsf{C}_{\beta_1\beta_2}(\mu), \tag{1.9}$$

$$m_{\beta_1\beta_2\beta_3} = \sum_{l=1}^{N} n_{\beta_1\beta_2}(|;\omega_l)\, n_{\beta_3}(\omega_l^{\#}) - \mathsf{C}^{(1)}_{\beta_1\beta_2\beta_3} - \mathsf{C}^{(2)}_{\beta_1\beta_2\beta_3} - \mathsf{C}^{(12)}_{\beta_1\beta_2\beta_3}. \tag{1.10}$$

A few low degree 2-component meeting numbers for a degree 7 hypersurface in $\mathbb{P}^6$ are given in Table 1.2. The number $n_{1,1}(H|;)$ can be confirmed via a Schubert computation similar to Section 3 in [9].

Configurations of rational curves in a Calabi–Yau n-fold can be studied for any n. If $n \geq 6$, such configurations include curves with non-simple nodes (several components sharing a node). While describing such curves is just notationally involved, specifying degree reducing recursions for them (following the approach of Section 1.3 below) presents new difficulties. In particular, curves with unbalanced splittings of the normal bundle will effect excess contributions via the loci of non-simple tuples of maps in the closures of simple tuples of maps, see the end of Section 1.1. Thus, separate counts must be set up for such curves, and their multiple-cover contributions to the appropriate topological intersection numbers (represented by the first terms on the right-hand side of (1.5), (1.9), and (1.10)) must be determined.

1.3. Justification of degree reducing recursions

1.3.1. *Overview* Each curve $\mathcal{C}$ of type (2A), (2B), and (3) determines a pair $(\bar{\mathcal{C}}, \mathcal{C}^*)$ of curves, where $\mathcal{C}^*$ is the last component of $\mathcal{C}$ and $\bar{\mathcal{C}}$ consists of the remaining component(s) of $\mathcal{C}$. The curve $\bar{\mathcal{C}}$ has 1 component in the first two cases and 2 components in the last case. The curves $\bar{\mathcal{C}}$ and

$\mathcal{C}^*$ carry marking $x_e \in \bar{\mathcal{C}}$ and $y_e \in \mathcal{C}^*$ satisfying $x_e = y_e$. We denote by $\overline{\mathcal{M}}$ and $\mathcal{M}^*$ the corresponding compactified spaces of curves/maps:

	$\overline{\mathcal{M}}$	$\mathcal{M}^*$
Case (2A):	$\overline{\mathfrak{M}}^*_{0,\{e\}}(X,\beta_1)$	$\{\phi \in \overline{\mathfrak{M}}^*_{0,\{e\}}(X,\beta_2)\colon (\mathrm{Im}\,\phi) \cap \mu \neq \emptyset\}$,
Case (2B):	$\{\phi \in \overline{\mathfrak{M}}^*_{0,\{e\}}(X,\beta_1)\colon \mathrm{ev}_e(\phi) \in \mu\}$	$\overline{\mathfrak{M}}^*_{0,\{e\}}(X,\beta_2)$,
Case (3):	$\overline{\mathfrak{M}}^*_{0,(\emptyset,\{e\})}(X,(\beta_1,\beta_2))$	$\overline{\mathfrak{M}}^*_{0,\{e\}}(X,\beta_3)$,

where μ above denotes a generic representative for the Poincaré dual of $\mu \in H^*(X)$. The evaluation map

$$\mathrm{ev}_{e,e}\colon \overline{\mathcal{M}} \times \mathcal{M}^* \longrightarrow X \times X, \quad \big((\bar{\mathcal{C}}, x_e), (\mathcal{C}^*, y_e)\big) \longrightarrow (x_e, y_e),$$

is then a cycle of (complex) dimension 5. The relevant meeting number is the cardinality of the subset of

$$\mathcal{Z} = \mathrm{ev}_{e,e}^{-1}(\Delta_X) = \big\{\big((\bar{\mathcal{C}}, x_e), (\mathcal{C}^*, y_e)\big) \in \overline{\mathcal{M}} \times \mathcal{M}^*\colon x_e = y_e\big\}$$

consisting of simple pairs of maps.

The homological intersection number of the cycle $\mathrm{ev}_{e,e}$ with the class of the diagonal $\Delta_X \subset X^2$ in X^2 is given by the diagonal-splitting term on the right-hand side of (1.5), (1.9), and (1.10). The homological intersection is the number of points, counted with sign, in the preimage of Δ_X under a small deformation of the map $\mathrm{ev}_{e,e}$. All such points must lie near $\mathcal{Z}$. The points of $\mathcal{Z}$ at which $\mathrm{ev}_{e,e}$ is transverse to Δ_X contribute 1 each to the homology intersection. These points include all tuples as above such that the curves $\bar{\mathcal{C}}$ and $\mathcal{C}^*$ do not have any components in common. Thus, the relevant meeting number is the diagonal-splitting term in (1.5), (1.9), and (1.10) minus the contribution to the homology intersection number of $\mathrm{ev}_{e,e}$ with Δ_X from the subset $\mathcal{Z}'$ of $\mathcal{Z}$ consisting of tuples as above such that $\bar{\mathcal{C}}$ and $\mathcal{C}^*$ have at least one component in common. In the rest of this subsection, we determine these tuples and their excess contributions.[3]

If X is an ideal Calabi–Yau 5-fold and $\beta \in H_+(X)$, the space

$$\overline{\mathcal{M}}_{\beta,1} = \overline{\mathfrak{M}}^*_{0,1}(X,\beta)$$

of simple maps to X of degree β with 1 marking is nonsingular of dimension 3, and the evaluation map

$$\mathrm{ev}\colon \overline{\mathcal{M}}_{\beta,1} \longrightarrow X$$

[3]As in the 4-dimensional case considered in [10], all contributions in case (2A) are degenerate contributions arising from loci of dimensions 1 and 2. However, in cases (2B) and (3), $\mathcal{Z}'$ includes regular points with respect to the evaluation condition which are isolated and nondegenerate.

is an immersion, see Lemma 1.1. We denote by T_β the tangent bundle of $\overline{\mathcal{M}}_{\beta,1}$ and by $\mathcal{N}_\beta$ the normal bundle to the immersion ev. Let $\mathcal{N}_\Delta \longrightarrow \Delta$ be the normal bundle to the diagonal in X^2. If $\mathcal{C}\subset X$ is a curve, let $|\mathcal{C}|$ denote the number of irreducible components of $\mathcal{C}$.

1.3.2. *Chern classes* Let X be an ideal Calabi–Yau 5-fold, and let $\beta\in H_+(X)$. We relate here the Chern classes of the normal bundle $\mathcal{N}_\beta$ to the immersion

$$\mathrm{ev}_1 : \overline{\mathcal{M}}_{\beta,1} \longrightarrow X$$

to meeting numbers. Denote by

$$f : \overline{\mathcal{M}}_{\beta,1} \longrightarrow \overline{\mathcal{M}}_\beta \tag{1.11}$$

the forgetful map to the nonsingular 2-dimensional moduli space $\overline{\mathcal{M}}_\beta = \overline{\mathfrak{M}}^*_{0,0}(X,\beta)$.

Using the bundle homomorphism $df : T\overline{\mathcal{M}}_{\beta,1} \longrightarrow f^*T\overline{\mathcal{M}}_\beta$ over $\overline{\mathcal{M}}_{\beta,1}$, we obtain

$$\begin{aligned} c_1(T_\beta) &= -\psi + f^*c_1(\overline{\mathcal{M}}_\beta), \\ c_2(T_\beta) &= \Delta - \psi\, f^*c_1(\overline{\mathcal{M}}_\beta) + f^*c_2(\overline{\mathcal{M}}_\beta), \end{aligned} \tag{1.12}$$

where ψ is the first chern class of the cotangent line bundle on $\overline{\mathcal{M}}_{\beta,1}$ viewed as a 1-pointed moduli space and $\Delta\subset\overline{\mathcal{M}}_{\beta,1}$ is the locus of singular points of f (points at which df is not surjective). On the other hand, since $c_1(X)=0$,

$$\begin{aligned} c_1(\mathcal{N}_\beta) &= -c_1(T_\beta), \\ c_2(\mathcal{N}_\beta) &= \mathrm{ev}^*c_2(X) + c_1^2(T_\beta) - c_2(T_\beta). \end{aligned} \tag{1.13}$$

Combining (1.12) and (1.13), we find

(1.14)
$$\begin{aligned} c_1(\mathcal{N}_\beta) &= \psi - f^*c_1(\overline{\mathcal{M}}_\beta), \\ c_2(\mathcal{N}_\beta) &= \mathrm{ev}^*c_2(X) + \psi^2 - \Delta - \psi\, f^*c_1(\overline{\mathcal{M}}_\beta) + f^*\big(c_1^2(\overline{\mathcal{M}}_\beta) - c_2(\overline{\mathcal{M}}_\beta)\big). \end{aligned}$$

If $\beta_1+\beta_2=\beta$ and $\beta_1\neq\beta_2$, let $D_{\beta_1,\beta_2}\subset\overline{\mathcal{M}}_\beta$ be the closure of the locus consisting of β-curves split into a β_1-curve and a β_2-curve. If $2\beta_1=\beta$, let $D_{\beta_1\beta_1}\subset\overline{\mathcal{M}}_\beta$ be twice the closure of the locus of consisting of β-curves split into two β_1-curves. In particular,

$$f_*\Delta = \frac{1}{2} \sum_{\substack{\beta_1+\beta_2=\beta \\ \beta_1,\beta_2\in H_+(X)}} D_{\beta_1,\beta_2}\,.$$

Denote by $(\psi_1+\psi_2)D_{\beta_1,\beta_2} \in H^4(\overline{\mathcal{M}}_\beta)$ the class obtained by capping Δ with the first chern class of the cotangent line bundle at the chosen node for each of the two curves. From a Grothendieck–Riemann–Roch computation applied to the deformation characterization of $T\overline{\mathcal{M}}_\beta$, we find

$$\begin{aligned} c_1(\overline{\mathcal{M}}_\beta) &= -f_*\mathrm{ev}^*c_2(X) + \sum_{\substack{\beta_1+\beta_2=\beta\\ \beta_1,\beta_2\in H_+(X)}} D_{\beta_1,\beta_2}\,, \\ 2c_2(\overline{\mathcal{M}}_\beta) - c_1^2(\overline{\mathcal{M}}_\beta) &= -f_*\big(\mathrm{ev}^*c_3(X) + \psi\,\mathrm{ev}_2^*c_2(X) + \psi^3\big) \\ &\quad + \frac{1}{2}\sum_{\substack{\beta_1+\beta_2=\beta\\ \beta_1,\beta_2\in H_+(X)}} (\psi_1+\psi_2)D_{\beta_1,\beta_2}\,. \end{aligned} \tag{1.15}$$

The 4-dimensional case of the first equation above appears in Section 1.2.4 of [10] and is also an immediate consequence of the $n=4$ analogue of (2.5) below. The second identity in (1.15) is (2.5) itself.

1.3.3. *The numbers (2A)* Suppose $\big((\bar{\mathcal{C}},x_e),(\mathcal{C}^*,x_e)\big)$ is an element of $\mathcal{Z}'$. Since the curve $\bar{\mathcal{C}}\cup\mathcal{C}^*$ passes through μ, $\bar{\mathcal{C}}\cup\mathcal{C}^*$ has at most two components. We have three possibilities for $\mathcal{Z}'$.

Case 0 *($\bar{\mathcal{C}}=\mathcal{C}^*$):* Here $\beta_1=\beta_2$ and

$$\mathcal{Z}' = \big\{\big((\mathcal{C}^*,x_e),(\mathcal{C}^*,x_e)\big) : (\mathcal{C}^*,x_e)\in\mathcal{M}^*\big\}.$$

The normal bundle of $\mathcal{Z}'$ in $\overline{\mathcal{M}}\times\mathcal{M}^*$ is isomorphic to $T_{\beta_1}\longrightarrow\mathcal{M}^*$ and the differential

$$\mathrm{dev}_{e,e} = \mathrm{dev}_e : \mathcal{N} \longrightarrow \mathrm{ev}_{e,e}^*\mathcal{N}_\Delta$$

is injective over $\mathcal{M}^*$. Thus, the contribution of $\mathcal{Z}'$ to the homology intersection number is given by

$$\big\langle e(\mathcal{N}_\Delta/\mathcal{N}),\mathcal{Z}'\big\rangle = \big\langle c_2(\mathcal{N}_{\beta_1}),\mathcal{M}^*\big\rangle.$$

Using the second equation in (1.14), the first equation in (1.15), and the fourth equation in (1.2), we obtain the $\beta_1=\beta_2$ case of (1.5).

Case 1*A ($\bar{\mathcal{C}}\subsetneq\mathcal{C}^*$):* Here $\beta_1<\beta_2$ and

$$\mathcal{Z}' = \big\{\big((\bar{\mathcal{C}},x_e),(\bar{\mathcal{C}}\vee\mathcal{C}',x_e)\big) : (\bar{\mathcal{C}},x_e)\in\overline{\mathcal{M}},\ (\bar{\mathcal{C}}\vee\mathcal{C}',x_e)\in\mathcal{Z}^*\big\},$$

where $\mathcal{Z}^*\subset\mathcal{M}^*$ is the locus consisting of 2-component curves with the marked point on the first component. Thus, $\mathcal{Z}'$ is the union of the first components of the finitely many $(\beta_1,\beta_2-\beta_1)$-curves passing through the constraint μ. The normal bundle $\mathcal{N}$ of $\mathcal{Z}'$ in $\overline{\mathcal{M}}\times\mathcal{M}^*$ contains the subbundle $\pi_1^*T_{\beta_1}$ and $\mathcal{N}/\pi_1^*T_{\beta_1}$ is isomorphic to the normal bundle $\mathcal{N}\mathcal{Z}^*$ of $\mathcal{Z}^*$ in $\mathcal{M}^*$. Since the differential

$$\mathrm{dev}_{e,e}\colon \mathcal{N} \longrightarrow \mathrm{ev}_{e,e}^*\mathcal{N}_\Delta$$

is injective over $\mathcal{Z}'$, the contribution of $\mathcal{Z}'$ to the homology intersection number is given by

$$\big\langle e(\mathcal{N}_\Delta/\mathcal{N}),\mathcal{Z}'\big\rangle = \big\langle c_1(\mathcal{N}_{\beta_1}) - c_1(\mathcal{N}\mathcal{Z}^*),\mathcal{Z}^*\big\rangle.$$

Since the degrees of the restrictions of $\mathcal{N}_{\beta_1}$ and $\mathcal{N}\mathcal{Z}^*$ to each curve $\bar{\mathcal{C}}$ are -2 and -1, respectively, we obtain the $\beta_1<\beta_2$ case of (1.5).

Case 1*B* *($\bar{\mathcal{C}}\supsetneq\mathcal{C}^*$):* Here $\beta_1>\beta_2$ and

$$\mathcal{Z}' = \big\{\big((\mathcal{C}'\vee\mathcal{C}^*,x_e),(\mathcal{C}^*,x_e)\big)\colon (\mathcal{C}'\vee\mathcal{C}^*,x_e)\in\overline{\mathcal{M}},\ (\mathcal{C}^*,x_e)\in\mathcal{Z}^*\big\},$$

where $\mathcal{Z}^*\subset\mathcal{M}^*$ is the locus of curves meeting a $(\beta_1-\beta_2)$-curve. Thus, $\mathcal{Z}'$ is the union of the second components of the finitely many $(\beta_1-\beta_2,\beta_2)$-curves whose second component passes through the constraint μ. The normal bundle $\mathcal{N}$ of $\mathcal{Z}'$ in $\overline{\mathcal{M}}\times\mathcal{M}^*$ contains the subbundle $\pi_1^*T_{\beta_1}$ and $\mathcal{N}/\pi_1^*T_{\beta_1}$ is isomorphic to the normal bundle $\mathcal{N}\mathcal{Z}^*$ of $\mathcal{Z}^*$ in $\mathcal{M}^*$. The latter is trivial. Since the differential

$$\mathrm{dev}_{e,e}\colon \mathcal{N} \longrightarrow \mathrm{ev}_{e,e}^*\mathcal{N}_\Delta$$

is injective over $\mathcal{Z}'$, the contribution of $\mathcal{Z}'$ to the homology intersection number is given by

$$\big\langle e(\mathcal{N}_\Delta/\mathcal{N}),\mathcal{Z}'\big\rangle = \big\langle c_1(\mathcal{N}_{\beta_1}) - c_1(\mathcal{N}\mathcal{Z}^*),\mathcal{Z}^*\big\rangle.$$

The $\beta_1>\beta_2$ case of (1.5) now follows from the first equation in (1.14).

1.3.4. *The numbers (2B)* Suppose $\big((\bar{\mathcal{C}},x_e),(\mathcal{C}^*,x_e)\big)$ is an element of $\mathcal{Z}'$. The curve $\bar{\mathcal{C}}\cup\mathcal{C}^*$ then has one, two, or three components and carries a marked point e lying on the divisor μ. The 6 possibilities for the connected components of $\mathcal{Z}'$ are indicated in Figure 5.

Case 0 *($\bar{\mathcal{C}}=\mathcal{C}^*$):* Here $\beta_1=\beta_2$ and $\big((\bar{\mathcal{C}},x_e),(\mathcal{C}^*,x_e)\big)$ is an element of

$$\overline{\mathcal{S}} = \big\{\big((\bar{\mathcal{C}},x_e),(\bar{\mathcal{C}},x_e)\big)\colon (\bar{\mathcal{C}},x_e)\in\overline{\mathcal{M}}\big\}\subset\mathcal{Z}'.$$

The normal bundle of $\overline{\mathcal{S}}$ in $\overline{\mathcal{M}}\times\mathcal{M}^*$ is isomorphic to $T_{\beta_2}\longrightarrow\overline{\mathcal{M}}$, and the contribution of $\overline{\mathcal{S}}$ to the homology intersection number is given by

$$\big\langle e(\mathcal{N}_\Delta/\mathcal{N}),\overline{\mathcal{S}}\big\rangle = \big\langle c_2(\mathcal{N}_{\beta_2}),\overline{\mathcal{M}}\big\rangle.$$

Using the second equation in (1.14) and the first equation in (1.15), we obtain the $\beta_1=\beta_2$ case of the last term in (1.9).

Case 1*A* *($|C^*|=2$, $\bar{C}\subsetneq C^*$):* Here $\beta_1<\beta_2$ and $((\bar{C},x_e),(C^*,x_e))$ is an element of

$$\overline{\mathcal{S}}=\big\{((\bar{C},x_e),(\bar{C}\vee C',x_e)):(\bar{C},x_e)\in\overline{\mathcal{Z}},\ (\bar{C}\vee C',x_e)\in\mathcal{M}^*\big\}\subset\mathcal{Z}',$$

where $\overline{\mathcal{Z}}\subset\overline{\mathcal{M}}$ is the locus consisting of curves meeting a $(\beta_2-\beta_1)$-curve. The normal bundle $\mathcal{N}$ of $\overline{\mathcal{S}}$ in $\overline{\mathcal{M}}\times\mathcal{M}^*$ contains the subbundle $\pi_2^*T_{\beta_2}$ and $\mathcal{N}/\pi_2^*T_{\beta_2}$ is isomorphic to the normal bundle $\mathcal{N}\overline{\mathcal{Z}}$ of $\overline{\mathcal{Z}}$ in $\overline{\mathcal{M}}$. Since the differential

$$\mathrm{dev}_{e,e}:\mathcal{N}\longrightarrow \mathrm{ev}_{e,e}^*\mathcal{N}_\Delta$$

is injective over $\overline{\mathcal{S}}$, the contribution of $\overline{\mathcal{S}}$ to the homology intersection number is given by

$$\big\langle e(\mathcal{N}_\Delta/\mathcal{N}),\overline{\mathcal{S}}\big\rangle=\big\langle c_1(\mathcal{N}_{\beta_2})-\big(c_1(\mathcal{N}_{\beta_2-\beta_1})+\tilde{\psi}_1\big),\overline{\mathcal{Z}}\big\rangle,$$

where $\tilde{\psi}_1$ is the untwisted ψ-class at the node of the $(\beta_2-\beta_1)$-component of a curve in $\overline{\mathcal{Z}}$.

Using the first equations in (1.14) and in (1.15) and the fourth equation in (1.2), we obtain the $\beta_1<\beta_2$ case of the last term in (1.9) minus the last term in (1.7). The latter arises from *Case* 2*A* below.

Case 1*B* *($|\bar{C}|=2$, $\bar{C}\supsetneq C^*$):* Here $\beta_1>\beta_2$ and $((\bar{C},x_e),(C^*,x_e))$ is an element of

$$\overline{\mathcal{S}}=\big\{((C'\vee C^*,x_e),(C^*,x_e)):(C'\vee C^*,x_e)\in\overline{\mathcal{Z}},\ (C^*,x_e)\in\mathcal{M}^*\big\}\subset\mathcal{Z}',$$

where $\overline{\mathcal{Z}}\subset\overline{\mathcal{M}}$ is the locus of $(\beta_2,\beta_1-\beta_2)$-curves with the marked point e lying on the first component. The normal bundle $\mathcal{N}$ of $\overline{\mathcal{S}}$ in $\overline{\mathcal{M}}\times\mathcal{M}^*$ contains the subbundle $\pi_2^*T_{\beta_2}$, $\mathcal{N}/\pi_2^*T_{\beta_2}$ is isomorphic to the normal bundle $\mathcal{N}\overline{\mathcal{Z}}$ of $\overline{\mathcal{Z}}$ in $\overline{\mathcal{M}}$, and the contribution of $\mathcal{Z}'$ to the homology intersection number is given by

$$\big\langle e(\mathcal{N}_\Delta/\mathcal{N}),\overline{\mathcal{S}}\big\rangle=\big\langle c_1(\mathcal{N}_{\beta_2})+(\psi_1+\psi_2),\overline{\mathcal{Z}}\big\rangle,$$

where ψ_1 and ψ_2 are the ψ-classes of the first and second components at the node of a curve in $\overline{\mathcal{Z}}$. We obtain the $\beta_1>\beta_2$ analogue of the *Case* 1*A* contribution in (1.9).

Case 2*A* *($|C^*|=3$, $\bar{C}\subsetneq C^*$):* Here $\beta_1<\beta_2$. If $|\bar{C}|=2$, $((\bar{C},x_e),(C^*,x_e))$ is an element of the space $\bar{\mathcal{S}}$ in *Case* 1*A* above. This is also the case if

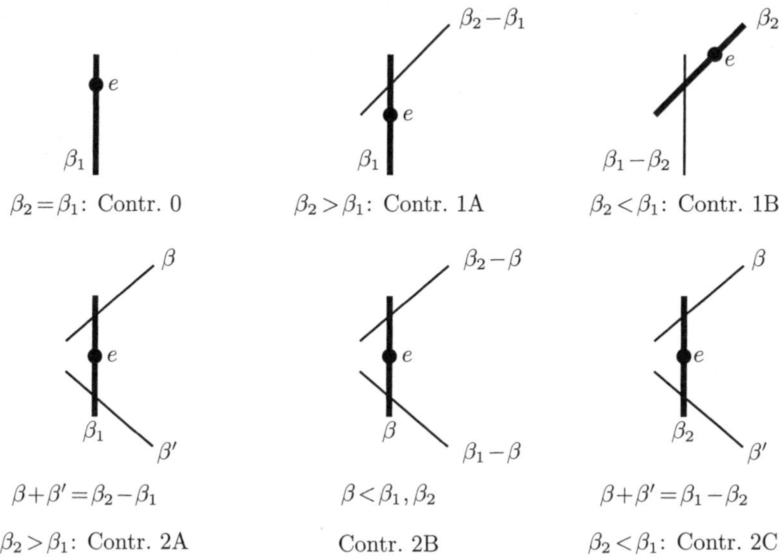

Fig. 5. Excess contributions for the meeting number $n_{\beta_1\beta_2}(\mu|;)$. The labels refer to the cases described in Section 1.3.4. The marked point e corresponds to the (former) node and lies on the divisor μ. The thicker lines indicate the multiple component. The space of curves in the first diagram in the top row is 2-dimensional. The other two spaces in the first row are 1-dimensional. All spaces in the bottom row are 0-dimensional.

$|\bar{\mathcal{C}}|=1$ and the curve $\mathcal{C}^*-\bar{\mathcal{C}}$ is connected. In the remaining case, $\bar{\mathcal{C}}$ is the middle component of the 3-component curve $\mathcal{C}^*$ and carries the marked point e, which lies on the divisor μ. Each such pair $\big((\bar{\mathcal{C}},x_e),(\mathcal{C}^*,x_e)\big)$ is a regular element of $\mathcal{Z}$ and therefore contributes 1 to the homology intersection. The contribution of such pairs is accounted for by the last term in (1.7).

Case 2B ($|\bar{\mathcal{C}}|=|\mathcal{C}^*|=2$, $\bar{\mathcal{C}}\neq\mathcal{C}^*$)*:* Here the curve $\bar{\mathcal{C}}\cup\mathcal{C}^*$ consists of three components, with the middle component meeting the hyperplane μ at the marked point e. Such pairs $\big((\bar{\mathcal{C}},x_e),(\mathcal{C}^*,x_e)\big)$ are regular elements of $\mathcal{Z}$, and their contribution is accounted for by the middle term on the right side of (1.9).

Case 2C ($|C^*|=3$, $\bar{C}\supsetneq C^*$): The analysis is the same as *Case* 2A with β_1 and β_2 interchanged.

1.3.5. *The numbers (3)* If $((\bar{C},x_e),(C^*,x_e))$ is an element of $\mathcal{Z}'$, $\bar{C}$ consists of two sets of components, $\bar{C}_1$ and $\bar{C}_2$, with the second component carrying the marked point e. Either $\bar{C}_1$ or $\bar{C}_2$ may consist of two components, while the other curve must consist of one component. The total number of components in $\bar{C}\cup C^*$ is either two or three. The 12 possibilities for the connected components of $\mathcal{Z}'$ are indicated in Figure 6.

Case 0 ($|\bar{C}\cup C^*|=2$, $\bar{C}_2\subset C^*$): If $C^*=\bar{C}_2$, then $\beta_2=\beta_3$ and $((\bar{C},x_e),(C^*,x_e))$ is an element of

$$\overline{\mathcal{S}}=\big\{\big((\bar{C}_1\vee C^*,x_e),(C^*,x_e)\big):(\bar{C}_1\vee C^*,x_e)\in\overline{\mathcal{M}}\big\}\subset\mathcal{Z}'.$$

Similarly to *Case* 0 in Sections 1.3.3 and 1.3.4, the normal bundle of $\overline{\mathcal{S}}$ in $\overline{\mathcal{M}}\times\mathcal{M}^*$ is isomorphic to $T_{\beta_2}\longrightarrow\overline{\mathcal{M}}$, and the contribution of $\overline{\mathcal{S}}$ to the homological intersection number is given by

$$\big\langle e(\mathcal{N}_\Delta/\mathcal{N}),\overline{\mathcal{S}}\big\rangle=\big\langle c_2(\mathcal{N}_{\beta_2}),\overline{\mathcal{M}}\big\rangle.$$

Using the second equation in (1.14), the first equation in (1.15), and the fourth equation in (1.2), we obtain the $\beta_3=\beta_2$ case of the term $\mathsf{C}^{(2)}_{\beta_1\beta_2\beta_3}$ in (1.10).

If $C^*=\bar{C}_1\cup\bar{C}_2$, then $\beta_1+\beta_2=\beta_3$ and $((\bar{C},x_e),(C^*,x_e))$ is an element of

$$\overline{\mathcal{S}}=\big\{\big((C^*,x_e),(C^*,x_e)\big):(C^*,x_e)\in\overline{\mathcal{M}}\big\}\subset\mathcal{Z}'.$$

The normal bundle of $\overline{\mathcal{S}}$ in $\overline{\mathcal{M}}\times\mathcal{M}^*$ is isomorphic to $T_{\beta_1+\beta_2}\longrightarrow\overline{\mathcal{M}}$, and the contribution of $\overline{\mathcal{S}}$ to the homological intersection number is given by

$$\big\langle e(\mathcal{N}_\Delta/\mathcal{N}),\overline{\mathcal{S}}\big\rangle=\big\langle c_2(\mathcal{N}_{\beta_1+\beta_2}),\overline{\mathcal{M}}\big\rangle.$$

Using the second equation in (1.14), the first equation in (1.15), and the fourth equation in (1.2), we obtain the $\beta_3=\beta_1+\beta_2$ case of the term $\mathsf{C}^{(12)}_{\beta_1\beta_2\beta_3}$ in (1.10).

Case 0′ ($|\bar{C}\cup C^*|=2$, $\bar{C}_2\not\subset C^*$): Here $\beta_1=\beta_3$ and $((\bar{C},x_e),(C^*,x_e))$ is an element of

$$\overline{\mathcal{S}}=\big\{\big((C^*\vee\bar{C}_2,x_e),(\bar{C},x_e)\big):(C^*\vee\bar{C}_2,x_e)\in\overline{\mathcal{Z}}\big\}\subset\mathcal{Z}',$$

where $\overline{\mathcal{Z}}\subset\overline{\mathcal{M}}$ consists of the pairs of 1-marked curves with the marked point at the node of the two curves. The normal bundle $\mathcal{N}$ of $\overline{\mathcal{S}}$ in

$\overline{\mathcal{M}}\times\mathcal{M}^*$ contains T_{β_1} as a subbundle, and $\mathcal{N}/T_{\beta_1}$ is isomorphic to the normal bundle of $\overline{\mathcal{Z}}$ in $\overline{\mathcal{M}}$. The latter is the universal tangent line bundle at the marked point. Since the homomorphism

$$\mathrm{dev}_{e,e}:\mathcal{N}\longrightarrow \mathrm{ev}_{e,e}^*\mathcal{N}_\Delta$$

is injective over $\mathcal{Z}'$, the contribution of $\mathcal{Z}'$ to the homological intersection number is given by

$$\langle e(\mathcal{N}_\Delta/\mathcal{N}),\overline{\mathcal{S}}\rangle=\langle c_1(\mathcal{N}_{\beta_2})+\psi_2,\overline{\mathcal{Z}}\rangle.$$

Using the first equations in (1.14) and (1.15) and the fourth equation in (1.2), we obtain the $\beta_3=\beta_1$ case of the term $\mathsf{C}^{(1)}_{\beta_1\beta_2\beta_3}$ in (1.10).

Case 1*A* *($|\bar{C}\cup C^*|=3,\ C^*\not\subset\bar{C},\ \bar{C}_2\subset C^*$):* If $\bar{C}_1\not\subset C^*$, then $\beta_2<\beta_3$ and $\big((\bar{C},x_e),(C^*,x_e)\big)$ is an element of

$$\overline{\mathcal{S}}=\big\{\big((\bar{C}_1\vee\bar{C}_2,x_e),(\bar{C}_2\vee C',x_e)\big):(\bar{C}_1\vee\bar{C}_2,x_e)\in\overline{\mathcal{Z}}\big\}\subset\mathcal{Z}',$$

where $\overline{\mathcal{Z}}\subset\overline{\mathcal{M}}$ consists of the pairs of (β_1,β_2)-curves such that the second component meets a $(\beta_3-\beta_2)$-curve. We see $\overline{\mathcal{S}}$ is the union of the middle components of $(\beta_1,\beta_2,\beta_3-\beta_2)$-curves in X, with each curve contributing -1 to the homological intersection number. The contribution accounts for the $\beta_3>\beta_2$ case of the term $\mathsf{C}^{(2)}_{\beta_1\beta_2\beta_3}$ in (1.10).

If $\bar{C}_1\subset C^*$, then $\beta_1+\beta_2<\beta_3$ and $\big((\bar{C},x_e),(C^*,x_e)\big)$ is an element of

$$\overline{\mathcal{S}}=\big\{\big((\bar{C}_1\vee\bar{C}_2,x_e),(\bar{C}_1\vee\bar{C}_2\vee C',x_e)\big):(\bar{C}_1\vee\bar{C}_2,x_e)\in\overline{\mathcal{Z}}\big\}\subset\mathcal{Z}',$$

where $\overline{\mathcal{Z}}\subset\overline{\mathcal{M}}$ consists of the pairs (β_1,β_2)-curves meeting a $(\beta_3-\beta_1-\beta_2)$-curve with the β_2-component carrying the marked point e. Here, $\bar{\mathcal{S}}$ is the union of the last components of the $(\beta_3-\beta_1-\beta_2,\beta_1,\beta_2)$-curves and the middle components of $(\beta_1,\beta_2,\beta_3-\beta_1-\beta_2)$-curves. By reasoning analogous to *Case* 1*A* in Section 1.3.4, each of the former contributes -1 to the homological intersection number, while each of the latter contributes 0. We obtain the $\beta_3>\beta_1+\beta_2$ case of the term $\mathsf{C}^{(12)}_{\beta_1\beta_2\beta_3}$ in (1.10).

Case 1*A*′ *($|\bar{C}\cup C^*|=3,\ C^*\not\subset\bar{C},\ \bar{C}_2\not\subset C^*$):* Here $\beta_3>\beta_1$ and $(\bar{C}\cup C^*,x_e)$ is a $(\beta_3-\beta_1,\beta_1,\beta_2)$-curve with the marked point lying on the node joining the last two components. Each such pair $\big((\bar{C},x_e),(C^*,x_e)\big)$ is a regular element of $\mathcal{Z}$, contributing 1 to the homology intersection number. We obtain the $\beta_3>\beta_1$ case of the term $\mathsf{C}^{(1)}_{\beta_1\beta_2\beta_3}$ in (1.10).

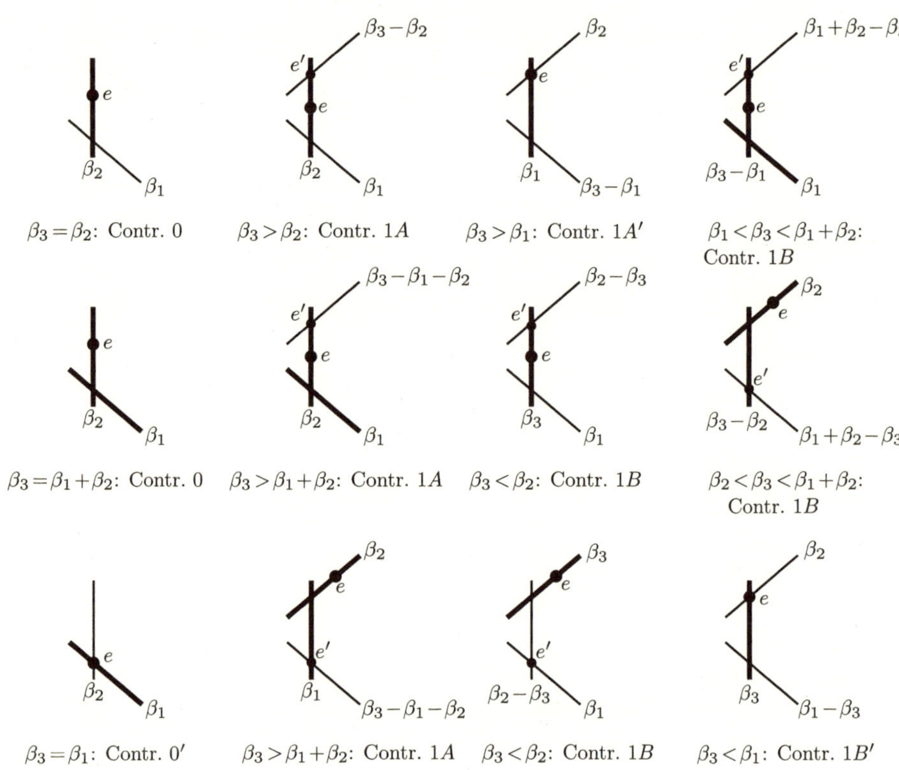

Fig. 6. Excess contributions for the meeting number $m_{\beta_1\beta_2\beta_3}$. The labels refer to the cases described in Section 1.3.5. The marked point e corresponds to the (former) node joining the β_2 and β_3 curves. For the curves of types 1A and 1B, e' indicates the new node on the (leftover) (β_1,β_2)-curve. The thicker lines represent the multiple component(s). The excess loci corresponding to *Contr.* 0 are 2-dimensional. The loci corresponding to *Contr.* $1A'$ and $1B'$ are 0-dimensional. The remaining loci are 1-dimensional.

Case $1B$ *($|\bar{\mathcal{C}}\cup\mathcal{C}^*|=3$, $\mathcal{C}^*\subset\bar{\mathcal{C}}$, $\mathcal{C}^*\not\subset\bar{\mathcal{C}}_1$):* If $\mathcal{C}^*=\bar{\mathcal{C}}_2$ or $\mathcal{C}^*=\bar{\mathcal{C}}$, $\big((\bar{\mathcal{C}},x_e),(\mathcal{C}^*,x_e)\big)$ is an element of one of the spaces $\bar{\mathcal{S}}$ defined in *Case* 0 above. Hence, we can assume that $\mathcal{C}^*\neq\bar{\mathcal{C}}_2,\bar{\mathcal{C}}$. If $\mathcal{C}^*\subset\bar{\mathcal{C}}_2$, then $\beta_2>\beta_3$ and $\big((\bar{\mathcal{C}},x_e),(\mathcal{C}^*,x_e)\big)$ is an element of

$$\overline{\mathcal{S}}=\big\{\big((\bar{\mathcal{C}}_1\vee\mathcal{C}^*\vee\mathcal{C}',x_e),(\mathcal{C}^*,x_e)\big):(\bar{\mathcal{C}}_1\vee\mathcal{C}^*\vee\mathcal{C}',x_e)\in\overline{\mathcal{Z}}\big\}\subset\mathcal{Z}',$$

where $\overline{\mathcal{Z}} \subset \overline{\mathcal{M}}$ is the locus of the pairs (β_1, β_2)-curves with the second component broken into two. As in *Case 1B* of Section 1.3.4, $\overline{\mathcal{S}}$ is the union of the middle components of $(\beta_1, \beta_3, \beta_2 - \beta_3)$-curves and the last components of $(\beta_1, \beta_2 - \beta_3, \beta_3)$, with each curve contributing -1 to the homological intersection number. The contribution accounts for the $\beta_3 < \beta_2$ case of the term $\mathsf{C}^{(2)}_{\beta_1\beta_2\beta_3}$ in (1.10).

If $\mathcal{C}^* \not\subset \bar{\mathcal{C}}_2$, then $\beta_1 + \beta_2 > \beta_3$ and $\big((\bar{\mathcal{C}}, x_e), (\mathcal{C}^*, x_e)\big)$ is an element of

$$\overline{\mathcal{S}} = \big\{\big((\bar{\mathcal{C}}_1 \vee \bar{\mathcal{C}}_2 \vee \mathcal{C}', x_e), (\bar{\mathcal{C}}_1 \vee \bar{\mathcal{C}}_2, x_e)\big) : (\bar{\mathcal{C}}_1 \vee \bar{\mathcal{C}}_2 \vee \mathcal{C}', x_e) \in \overline{\mathcal{Z}}\big\} \subset \mathcal{Z}',$$

where $\overline{\mathcal{Z}} \subset \overline{\mathcal{M}}$ is the locus of the pairs (β_1, β_2)-curves with one of the components broken into two. Here, $\overline{\mathcal{S}}$ is the union of the middle components of $(\beta_1, \beta_3 - \beta_1, \beta_1 + \beta_2 - \beta_3)$-curves, if $\beta_3 > \beta_1$, and the last components of $(\beta_1 + \beta_2 - \beta_3, \beta_3 - \beta_2, \beta_2)$-curves, if $\beta_3 > \beta_2$. Each of the latter curves contributes -1 to the homological intersection number, while each of the former contributes 0. We obtain the $\beta_3 < \beta_2$ case of the term $\mathsf{C}^{(12)}_{\beta_1\beta_2\beta_3}$ in (1.10).

Case $1B'$ *(*$|\bar{\mathcal{C}} \cup \mathcal{C}^*| = 3$, $\mathcal{C}^* \subset \bar{\mathcal{C}}_1$*)*: If $\mathcal{C}^* = \bar{\mathcal{C}}_1$, then $\big((\bar{\mathcal{C}}, x_e), (\mathcal{C}^*, x_e)\big)$ is an element of the space $\bar{\mathcal{S}}$ defined in *Case* $0'$ above. Hence, we can assume that $\mathcal{C}^* \neq \bar{\mathcal{C}}_1$. Then, $\beta_1 > \beta_3$ and $\big((\bar{\mathcal{C}}, x_e), (\mathcal{C}^*, x_e)\big)$ is an element of

$$\overline{\mathcal{S}} = \big\{\big((\mathcal{C}' \vee \mathcal{C}^* \vee \bar{\mathcal{C}}_2, x_e), (\mathcal{C}^*, x_e)\big) : (\mathcal{C}' \vee \mathcal{C}^* \vee \bar{\mathcal{C}}_2, x_e) \in \overline{\mathcal{Z}}\big\} \subset \mathcal{Z}',$$

where $\overline{\mathcal{Z}} \subset \overline{\mathcal{M}}$ is the locus of the pairs of 1-marked (β_1, β_2)-curves represented by a $(\beta_1 - \beta_3, \beta_3, \beta_2)$-curve in X with the marked point on the node of the last two components. Each such pair $\big((\bar{\mathcal{C}}, x_e), (\mathcal{C}^*, x_e)\big)$ is a regular element of $\mathcal{Z}$, contributing 1 to the homological intersection number and accounting for the $\beta_3 < \beta_1$ case of the term $\mathsf{C}^{(1)}_{\beta_1\beta_2\beta_3}$ in (1.10).

§2. Genus 1 counts

2.1. Overview

For each $\beta \in H_+(X)$, $N_{1,\beta}$ is the number of automorphism-weighted stable C^∞-maps

$$u : \Sigma \longrightarrow X$$

from prestable curve of genus 1 to X of degree β solving a perturbed Cauchy–Riemann equation,

$$\bar{\partial} u + \nu(u) = 0, \tag{2.1}$$

for a small generic multi-valued perturbation ν, see Section 1.3 of [21] for more details. If X is an ideal Calabi–Yau n-fold, $\overline{\mathfrak{M}}_1(X,\beta)$ decomposes into strata $\mathcal{Z}_\mathcal{T}$ which each have well-defined contribution to $N_{1,\beta}$ in following sense:

> *For every stratum $\mathcal{Z}_\mathcal{T}$, there exist $\mathsf{C}_\mathcal{T}(\beta)\in\mathbb{Q}$, $\epsilon_\nu\in\mathbb{R}^+$, and a compact subset K_ν of $\mathcal{Z}_\mathcal{T}$ with the following property. For every compact subset K of $\mathcal{Z}_\mathcal{T}$ and an open neighborhood U of K in the space of stable C^∞-maps, there exist an open neighborhood $U_\nu(K)$ of K and $\epsilon_\nu(U)\in(0,\epsilon_\nu)$, respectively,[4] such that*

$$^{\pm}\big|\{\bar\partial+t\nu\}^{-1}(0)\cap U\big| = \mathsf{C}_\mathcal{T}(\beta) \quad \textit{if } t\in(0,\epsilon_\nu(U)),\ K_\nu\subset K\subset U\subset U_\nu(K).$$

While there are many different strata, it turns out that $\mathsf{C}_\mathcal{T}(\beta)\neq 0$ only for strata of the three simplest types.

If X is an ideal Calabi–Yau n-fold, there are finitely many genus 1 curves in each homology class of X. Furthermore, every genus 1 curve $\mathcal{C}$ in X is embedded and super-rigid: if $\mathcal{N}$ is the normal bundle of $\mathcal{C}$ and

$$u\colon \Sigma\longrightarrow\mathcal{C}$$

is an unramified cover, then $H^0(\Sigma,u^*\mathcal{N})=0$. Hence, $H^1(\Sigma,u^*\mathcal{N})=0$ and for every $d\in\mathbb{Z}^+$

$$\mathcal{Z}_{(1,\beta/d)} = \bigcup_{[\mathcal{C}]=\beta/d}\overline{\mathfrak{M}}_1(\mathcal{C},d)$$

is a finite set of isolated regular points of $\overline{\mathfrak{M}}_1(X,\beta)$. Each such point u contributes $|1/\mathrm{Aut}(u)|$ to $N_{1,\beta}$. If $n_{1,\beta}$ is the number of genus 1 curves in the homology class β, then

$$\mathsf{C}_{(1,\beta/d)}(\beta) = \frac{\sigma(d)}{d}n_{1,\beta/d}, \tag{2.2}$$

where $\sigma(d)$ is the number of degree d unbranched covers of a genus 1 curve by connected genus 1 curves. The integral number $n_{1,\beta/d}$ is zero unless $d|\beta$, or equivalently, β/d is an integral homology class.

The remaining elements $u\colon\Sigma\longrightarrow X$ of $\overline{\mathfrak{M}}_1(X,\beta)$ are maps to genus 0 curves in X. They split into strata $\mathcal{Z}_\mathcal{T}$ indexed by combinatorial data described in Section 2.2. We will call a stratum $\mathcal{Z}_\mathcal{T}$ *basic* if either of the following conditions holds:

---

[4] $U_\nu(K)$ depends on K, while $\epsilon_\nu(U)$ depends on U.

(B1) the domain Σ of every element $[\Sigma, u]$ of $\mathcal{Z}_\mathcal{T}$ is a nonsingular genus 1 curve, or

(B2) the domain Σ of every element $[\Sigma, u]$ of $\mathcal{Z}_\mathcal{T}$ is a union of a nonsingular genus 1 curve Σ_P and a $\mathbb{P}^1$ and u is constant on Σ_P.

In both cases, the restriction of u to the non-contracted component must be a $d\!:\!1$ cover of a curve in the homology class β/d, for some $d \in \mathbb{Z}^+$. We will write $\mathcal{T}_{\text{eff}}(\beta/d, d)$ and $\mathcal{T}_{\text{gh}}(\beta/d, d)$ for the corresponding types of strata (B1) and (B2), with *eff* and *gh* standing for *effective* and *ghost* (principal component).

Theorem 2.1. *Suppose X is an ideal Calabi–Yau 5-fold.*

(i) If $\mathcal{Z}_\mathcal{T}$ is a stratum of $\overline{\mathfrak{M}}_1(X,\beta)$ consisting of maps to rational curves in X and is not basic, $\mathrm{C}_\mathcal{T}(\beta) = 0$.

(ii) For $\beta \in H_+(X)$ and $d \in \mathbb{Z}^+$,

(2.3)
$$\mathrm{C}_{\mathcal{T}_{\text{eff}}(\beta,d)}(d\beta) = \frac{d-1}{d^2}\mathrm{C}_{\mathcal{T}_{\text{gh}}(\beta,1)}(\beta), \qquad \mathrm{C}_{\mathcal{T}_{\text{gh}}(\beta,d)}(d\beta) = \frac{1}{d^2}\mathrm{C}_{\mathcal{T}_{\text{gh}}(\beta,1)}(\beta).$$

In Section 2.3, we will prove

$$\mathrm{C}_{\mathcal{T}_{\text{gh}}(\beta,1)}(\beta) = \frac{1}{24}\int_{\overline{\mathcal{M}}_\beta} \big(2c_2(\overline{\mathcal{M}}_\beta) - c_1^2(\overline{\mathcal{M}}_\beta)\big). \tag{2.4}$$

On the other hand, the space $\overline{\mathcal{M}}_\beta = \overline{\mathfrak{M}}_0^*(X,\beta)$ consists of regular maps to X. Thus, the contribution to $N_{1,\beta}$ is given by the right side of equation (2.15) in [23]:

$$\mathrm{C}_{\mathcal{T}_{\text{gh}}(\beta,1)}(\beta) = \frac{1}{24}\bigg(- n_\beta\Big(\frac{c(X)}{1-\psi}\Big) + \frac{1}{2}\sum_{\substack{\beta_1+\beta_2=\beta \\ \beta_1,\beta_2 \in H_+(X)}} n_{\beta_1\beta_2}(\psi_1 + \psi_2|;)\bigg).$$

Comparing the above identity with (2.4), we find that

(2.5)
$$\int_{\overline{\mathcal{M}}_\beta} \big(2c_2(\overline{\mathcal{M}}_\beta) - c_1^2(\overline{\mathcal{M}}_\beta)\big) = -n_\beta\Big(\frac{c(X)}{1-\psi}\Big) + \frac{1}{2}\sum_{\substack{\beta_1+\beta_2=\beta \\ \beta_1,\beta_2 \in H_+(X)}} n_{\beta_1\beta_2}(\psi_1 + \psi_2|;).$$

We calculate the left side in terms of the Gromov–Witten invariants of X by expanding the right side via the equations of Section 1.

Our proof of Theorem 2.1 applies also in dimensions 3 and 4. In particular, the result provides a direct explanation of the $1/d$-scaling in the latter case discovered by other means in Section 2 of [10]. Many aspects of the proof are applicable in dimensions 6 and higher as well.

2.2. Preliminaries

Let X be an ideal Calabi–Yau 5-fold. The strata of $\overline{\mathfrak{M}}_1(X,\beta)$ consisting of maps to rational curves can be described by *decorated graphs*

$$\mathcal{T} = \big(\mathrm{Ver}, \mathrm{Edg}, \mathfrak{d}, \underline{\beta}, \kappa, i^*\big),$$

where

(D1) $\Gamma = (\mathrm{Ver}, \mathrm{Edg})$ is a connected graph containing either exactly one loop or a distinguished vertex, but not both,

(D2) $\underline{\beta} = (\beta_i)_{i\in[m]}$ is an m-tuple of elements of $H_+(X)$, with $m \in \{1,2,3\}$,

(D3) $\mathfrak{d} : \mathrm{Ver} \longrightarrow \mathbb{Z}^{\geq 0}$ is a map, $\kappa : \mathfrak{d}^{-1}(\mathbb{Z}^+) \longrightarrow [m]$ is a surjective map,

(D4) $i^* \in \{\star\} \cup [m]$.

The irreducible components and the nodes of the domain Σ of every element $[\Sigma, u]$ of $\mathcal{Z}_\mathcal{T}$ correspond to the sets Ver and Edg respectively. If $v \in \mathrm{Ver}$ is not the distinguished vertex of Γ, the corresponding component Σ_v of Σ is a $\mathbb{P}^1$. Otherwise, Σ_v is nonsingular of genus 1. If $v \in \mathrm{Ver}$, the restriction of u to Σ_v is constant if $\mathfrak{d}(v) = 0$. If $\mathfrak{d}(v) \neq 0$, $u|_{\Sigma_v}$ is a $\mathfrak{d}(v)\!:\!1$ cover of the component $\mathcal{C}_{\kappa(v)}$ of $\mathcal{C}$. If $\mathfrak{d}$ does not vanish identically of the loop in the graph (Ver, Edg) or on the distinguished vertex, i^* is set to $\star$. If $\mathfrak{d}$ vanishes identically on the loop or on the distinguished vertex, the corresponding components of Σ are mapped by u to a point on the i^*-component of $\mathcal{C}$. Since u is continuous, $\mathcal{Z}_\mathcal{T} = \emptyset$ unless κ satisfies certain combinatorial conditions.[5]

Given a generic deformation of ν of the $\bar{\partial}$-operator as in (2.1) and sufficiently small $t \in \mathbb{R}^+$, we will determine the number of solutions $[\Sigma, u]$ of

$$\bar{\partial} u + t\nu(u) = 0, \tag{2.6}$$

with u close to the stratum $\mathcal{Z}_\mathcal{T}$. The assumption that ν is generic implies that all solutions of (2.6) are maps from nonsingular genus 1 curves. The arguments follow [19, 20]. In particular, the gluing construction for $\mathcal{Z}_\mathcal{T}$ will be performed on a family of representatives (Σ, u) for the elements $[\Sigma, u]$ in $\mathcal{Z}_\mathcal{T}$, see Section 2.2 of [20]. Our treatment here is less explicit in order to streamline the discussion.

---

[5]The strata $\mathcal{Z}_\mathcal{T}$ as defined above intersect if $m \geq 2$ and $\mathfrak{d}$ vanishes on the loop or the distinguished vertex of (Ver, Edg). The issue can be easily addressed by allowing i^* to take values in $\{\star\} \cup [m] \cup \{(1,2),(2,3)\}$. However, equation (2.6) will be shown to have no solutions near $\mathcal{Z}_\mathcal{T}$ for a good choice of ν if $m \geq 2$, so further discussion is not needed.

For the rest of Section 2, we fix a decorated graph $\mathcal{T}$ as above. We define

$$|\underline{\beta}| = \sum_{i=1}^{m} \beta_i \in H_+(X).$$

With notation as in Section 1.1, let

$$\mathcal{M}_{\underline{\beta}} = \mathfrak{M}^*_{0,\emptyset}(X, \underline{\beta}) \qquad \text{and} \qquad \overline{\mathcal{M}}_{\underline{\beta}} = \overline{\mathfrak{M}}^*_{0,\emptyset}(X, \underline{\beta}).$$

We denote by $\mathcal{M}_{\underline{\beta},1}$ and $\overline{\mathcal{M}}_{\underline{\beta},1}$ the spaces of pairs $(\mathcal{C}, x)$ such that $\mathcal{C} \in \mathcal{M}_{\underline{\beta}}$ and $x \in \mathcal{C}$ is a nonsingular point of $\mathcal{C}$ in the first case and $\mathcal{C} \in \overline{\mathcal{M}}_{\underline{\beta}}$ and $x \in \mathcal{C}$ is any point of $\mathcal{C}$ in the second case.

Let $\mathcal{S} \longrightarrow \mathcal{M}_{\underline{\beta}}$ be a family of deformations in X of curves in $\mathcal{M}_{\underline{\beta}}$. In other words, the fiber $\mathcal{S}_{\mathcal{C}}$ of $\mathcal{S}$ over $\mathcal{C} \in \mathcal{M}_{\underline{\beta}}$ contains $\mathcal{C}$ and

$$\dim \mathcal{S}_{\mathcal{C}} = \dim \mathcal{M}_{|\underline{\beta}|,1} - \dim \mathcal{M}_{\underline{\beta}} = m.$$

There is a fibration

$$\pi_{\mathcal{C}} : \mathcal{S}_{\mathcal{C}} \longrightarrow \Delta \subset \mathbb{C}^{m-1} \tag{2.7}$$

giving the universal family of deformations of $\mathcal{C}$. If $m = 1$, then $\mathcal{S} = \mathcal{M}_{|\underline{\beta}|,1}$. If $m=3$, $\mathcal{S}$ is a small neighborhood of $\mathcal{M}_{\underline{\beta},1}$ in $\overline{\mathcal{M}}_{|\underline{\beta}|,1}$.

If $\mathrm{ev} : \mathcal{M}_{\underline{\beta},1} \longrightarrow X$ is the evaluation map at the marked point, the bundle

$$Q = \mathrm{ev}^* TX / T\mathcal{S} \longrightarrow \mathcal{M}_{\underline{\beta},1} \tag{2.8}$$

extends naturally over $\overline{\mathcal{M}}_{\underline{\beta},1}$ so that there is an exact sequence

$$0 \longrightarrow f^* T\overline{\mathcal{M}}_{\underline{\beta}} \longrightarrow Q \longrightarrow \mathcal{N}_{|\underline{\beta}|} \longrightarrow 0, \tag{2.9}$$

where $f : \overline{\mathcal{M}}_{\underline{\beta},1} \longrightarrow \overline{\mathcal{M}}_{\underline{\beta}}$ is the forgetful map and $\mathcal{N}_{|\underline{\beta}|}$ is the normal bundle to the family of simple curves of class $|\underline{\beta}|$.

Similarly to Section 3.3 in [12], we choose a family of "exponential" maps

$$\exp^{\mathcal{C}} : TX \longrightarrow X \text{ such that } \exp^{\mathcal{C}}_x(v) \in \mathcal{S}_{\mathcal{C}} \text{ if } x \in \mathcal{C},\ v \in T_x\mathcal{S}_{\mathcal{C}},\ |v| < \delta(\mathcal{C}), \tag{2.10}$$

for some $\delta \in C^\infty(\mathcal{M}_{\bar{\Gamma}}; \mathbb{R}^+)$. Below we will place additional assumptions on $\exp^{\mathcal{C}}$ as needed.

For an ideal Calabi–Yau n-fold with $n \geq 6$, the above stratification would need to be refined further based on the deviation of the normal bundles of curves in $\mathcal{M}_{\underline{\beta},1}$ from balanced splitting. The arguments in Sections 2.3–2.5 below apply to the strata with balanced splitting with minor changes. The main change here is that the map ev is no longer an immersion, and one would need to pass to a blowup of $\overline{\mathcal{M}}_{\underline{\beta},1}$ to obtain analogues of the vector bundle Q and the short exact sequence (2.10). The strata with unbalanced splittings need to be treated separately, with the conclusion that they do not contribute to the genus 1 Gromov–Witten invariants under certain assumptions on X.

2.3. Strata with ghost principal component I

Here we describe the contribution to $N_{1,*}$ from a stratum $\mathcal{Z}_{\mathcal{T}}$ consisting of maps $u:\Sigma\longrightarrow X$ that are constant on the principal, genus-carrying, component(s) Σ_P of Σ. We show $\mathcal{Z}_{\mathcal{T}}$ does not contribute to $N_{1,*}$ unless $\mathcal{Z}_{\mathcal{T}}$ is of type (B2).

For each $m\in\mathbb{Z}^+$, let $\overline{\mathcal{M}}_{1,m}$ be the moduli space of stable curves of genus 1 with m marked points. Let $\mathbb{E}\longrightarrow\overline{\mathcal{M}}_{1,m}$ be the Hodge line bundle of holomorphic differentials. For each $i\in[m]$, denote by $L_i\longrightarrow\overline{\mathcal{M}}_{1,m}$ the universal tangent line bundle at the ith marked point. Let

$$s_i\in\Gamma\big(\overline{\mathcal{M}}_{1,m},\mathrm{Hom}(L_i,\mathbb{E}^*)\big)$$

be the homomorphism induced by the natural pairing of tangent and cotangent vectors at the ith marked point. Denote by

$$\mathcal{M}_{1,m},\mathcal{M}_{1,m}^{\mathrm{eff}}\subset\overline{\mathcal{M}}_{1,m}$$

the subspaces consisting of nonsingular curves and of curves $\mathcal{C}$ with no bubble components ($\mathcal{C}$ is either a nonsingular genus 1 curve or is a circle of rational curves).

Let $L_1\longrightarrow\overline{\mathfrak{M}}_{0,1}(X,\beta)$ be the universal tangent line bundle at the marked point. Denote by

$$\mathcal{D}_1\in\Gamma\big(\overline{\mathfrak{M}}_{0,1}(X,\beta),\mathrm{Hom}(L_1,\mathrm{ev}_1^*TX)\big)$$

the natural homomorphism induced by the derivative of the map at the marked point. For $m\in\mathbb{Z}^+$, let

$$\overline{\mathfrak{M}}_{(0,m)}(X,\beta)=\Big\{(b_i)_{i\in[m]}\in\prod_{i=1}^{m}\overline{\mathfrak{M}}_{0,\{0\}}(X,\beta_i):\beta_i\in H_+(X),\ \sum_{i=1}^{m}\beta_i=\beta,\ \mathrm{ev}_0(b_i)=\mathrm{ev}_0(b_{i'})\ \forall i,i'\in[m]\Big\}.$$

There is a well-defined evaluation map

$$\mathrm{ev}_0:\overline{\mathfrak{M}}_{(0,m)}(X,\beta)\longrightarrow X,\qquad (b_i)_{i\in[m]}\longrightarrow\mathrm{ev}_0(b_i),$$

which is independent of the choice of i. Let

$$\pi_i : \overline{\mathfrak{M}}_{(0,m)}(X,\beta) \longrightarrow \bigsqcup_{\beta_i \in H_+(X)} \overline{\mathfrak{M}}_{0,\{0\}}(X,\beta_i)$$

be the projection onto the ith component. Denote by

$$\mathfrak{M}^{\text{eff}}_{(0,m)}(X,\beta) \subset \overline{\mathfrak{M}}_{(0,m)}(X,\beta)$$

the subset consisting of the tuples $(u_i)_{i\in[m]}$ such that for each $i\in[m]$ the restriction of u_i to the domain component carrying the marked point 0 is not constant.

The stratum $\mathcal{Z}_\mathcal{T}$ admits a decomposition

$$\mathcal{Z}_\mathcal{T} = \big(\mathcal{Z}_{\mathcal{T},P} \times \mathcal{Z}_{\mathcal{T},PB} \times \mathcal{Z}_{\mathcal{T},B}\big)/S_{m_B}, \tag{2.11}$$

where $\mathcal{Z}_{\mathcal{T},P}$ is a stratum of $\mathcal{M}^{\text{eff}}_{1,m_P}$ for some $m_P\in\mathbb{Z}^+$, $\mathcal{Z}_{\mathcal{T},B}$ is a stratum of $\mathfrak{M}^{\text{eff}}_{(0,m_B)}(X,\beta)$ for some $m_B\in\mathbb{Z}^+$, and $\mathcal{Z}_{\mathcal{T},PB}$ is a product of moduli spaces of irreducible stable genus 0 curves. The stratum $\mathcal{Z}_{\mathcal{T},P}$ consists of curves of a fixed topological type, while the elements of $\mathcal{Z}_{\mathcal{T},B}$ are tuples of stable maps from domains of fixed topological types so that the image of the restriction of the map to each component is of a specified homology class and multiplicity. The requirement that

$$\mathcal{Z}_{\mathcal{T},P} \subset \mathcal{M}^{\text{eff}}_{1,m_P} \qquad \text{and} \qquad \mathcal{Z}_{\mathcal{T},B} \subset \mathfrak{M}^{\text{eff}}_{(0,m_B)}(X,\beta)$$

implies that the decomposition (2.11) is well-defined. Let

$$\pi_P, \pi_B : \mathcal{Z}_{\mathcal{T},P} \times \mathcal{Z}_{\mathcal{T},PB} \times \mathcal{Z}_{\mathcal{T},B} \longrightarrow \mathcal{Z}_{\mathcal{T},P}, \mathcal{Z}_{\mathcal{T},B}$$

denote the projection maps. The quotient is by the automorphism groups S_{m_B} of the data.

If X is an ideal CY 5-fold, $\mathcal{Z}_{\mathcal{T},B}$ is smooth. The cokernels of the linearizations D_b of the $\bar\partial$-operator along $\mathcal{Z}_\mathcal{T}$ form the obstruction bundle

$$\mathfrak{O} = \mathfrak{O}_{PB} \oplus \pi_B^*\mathfrak{O}_B = \pi_P^*\mathbb{E}^* \otimes \pi_B^*\text{ev}_0^*TX \oplus \pi_B^*\mathfrak{O}_B, \tag{2.12}$$

where $\mathfrak{O}_B \longrightarrow \mathcal{Z}_{\mathcal{T},B}$ is the obstruction bundle associated with the moduli space $\overline{\mathfrak{M}}_{(0,m)}(X,\beta)$. Let $\bar\nu\in\Gamma(\mathcal{Z}_\mathcal{T},\mathfrak{O})$ be the section induced by ν: $\bar\nu(b)$ is the projection of $\nu(b)$ to the cokernel of D_b. We write

$$\bar\nu_{PB}, \bar\nu_B \in \Gamma\big(\mathcal{Z}_\mathcal{T},\mathfrak{O}_{PB}\big), \Gamma\big(\mathcal{Z}_\mathcal{T},\mathfrak{O}_B\big)$$

for the two components of $\bar\nu$.

There is a natural projection map

$$\bar{\pi}: \mathcal{Z}_{\mathcal{T},B} \longrightarrow \mathcal{M}_{\underline{\beta},1},$$

sending $\pi_B([\Sigma, u])$ to $(u(\Sigma), u(\Sigma_P))$. Denote by

$$\bar{\nu}_{PB}^{\perp} \in \Gamma\big(\mathcal{Z}_{\mathcal{T}}, \pi_P^*\mathbb{E}^* \otimes \pi_B^*\bar{\pi}^*Q\big)$$

the image of $\bar{\nu}_{PB}$ under the natural projection map. Let

$$f\colon \overline{\mathcal{M}}_{1,m_P} \longrightarrow \overline{\mathcal{M}}_{1,1}$$

be the forgetful map, dropping all but the first marked point. The restriction of the bundle

$$\pi_1^*\mathbb{E}^* \otimes \pi_2^*Q \longrightarrow \overline{\mathcal{M}}_{1,1} \times \overline{\mathcal{M}}_{\underline{\beta},1} \tag{2.13}$$

to any boundary stratum $\mathcal{Z}_\Gamma$ contains a subbundle $\mathfrak{O}_\Gamma$ such that

$$\operatorname{rk}\mathfrak{O}_\Gamma - \dim \mathcal{Z}_\Gamma > \operatorname{rk}\big(\pi_1^*\mathbb{E}^* \otimes \pi_2^*Q\big) - \dim\big(\overline{\mathcal{M}}_{1,1} \times \overline{\mathcal{M}}_{\underline{\beta},1}\big) = 0 \tag{2.14}$$

and $\big\{(f\circ\pi_P)\times(\bar{\pi}\circ\pi_B)\big\}^*\mathfrak{O}_\Gamma$ is a quotient of the cokernel bundle over a boundary stratum of $\overline{\mathcal{Z}}_{\mathcal{T}}$. Thus, we can choose a section $\bar{\nu}_{\underline{\beta}}$ of (2.13) with all zeros transverse and contained in $\mathcal{M}_{1,1}\times\mathcal{M}_{\underline{\beta},1}$ and such that there exists ν as above satisfying

$$\bar{\nu}_{PB}^{\perp} = \big\{(f\circ\pi_P)\times(\bar{\pi}\circ\pi_B)\big\}^*\bar{\nu}_{\underline{\beta}}.$$

It is shown in the next section that the contribution of $\mathcal{Z}_{\mathcal{T}}$ to $N_{1,*}$ comes from $\bar{\nu}_{PB}^{\perp\,-1}(0)$. Thus, if $\mathcal{Z}_{\mathcal{T},P} \not\subset \mathcal{M}_{1,m_P}$, then $\bar{\nu}_{PB}^{\perp\,-1}(0)$ is empty for a good choice of ν by (2.14) and the stratum $\mathcal{Z}_{\mathcal{T}}$ does not contribute to $N_{1,*}$. Otherwise, $\bar{\nu}_{PB}^{\perp\,-1}(0)$ is the preimage of a finite subset in $\mathcal{M}_{1,1}\times\mathcal{M}_{\underline{\beta},1}$. It decomposes into connected components

$$\bar{\nu}_{PB}^{\perp\,-1}(0) = \bigsqcup_{(\mathcal{C},x)\in\pi_2(\bar{\nu}_{\underline{\beta}}^{-1}(0))} \mathcal{Z}_{\mathcal{C},x}, \tag{2.15}$$

where $\mathcal{C}$ is a $\underline{\beta}$-curve and x is a nonsingular point of $\mathcal{C}$. Then, $\mathsf{C}_{\mathcal{T}}(\beta)$ is the number of zeros of a map $\varphi_{t\nu}$ from the vector bundle F of gluing parameters to $\mathfrak{O}$ over each of the components $\mathcal{Z}_{\mathcal{C},x}$. The projection of $\varphi_{t\nu}$ in the decomposition (2.12) onto $\pi_P^*\mathbb{E}\otimes T_x\mathcal{S}_{\mathcal{C}}/T_x\mathcal{C}$ is essentially the same as the projection of $t\nu$, which we denote by $t\tilde{\nu}$. Since $\tilde{\nu}$ is a section of a trivial bundle over $\mathcal{Z}_{\mathcal{C},x}$, it can be chosen not to vanish if $m>1$. Thus, $\mathsf{C}_{\mathcal{T}}(\beta)=0$ if $m>1$. On the other hand, the second component of $\varphi_{t\nu}$ with respect to the decomposition (2.12) is essentially $t\bar{\nu}_B$. It does not vanish on $\bar{\nu}_{PB}^{\perp-1}(0)$ for dimensional reasons if $m=1$, but $|\mathrm{Ver}|>2$. Thus, $\mathrm{C}_{\mathcal{T}}(\beta)=0$ if $\mathcal{T}$ is not basic.

Finally, if $\mathcal{T}$ is basic, the principal component of every element of $\mathcal{Z}_{\mathcal{C},x}$ is a fixed nonsingular genus 1 curve Σ_P with one special point z_1 and

$$\mathcal{Z}_{\mathcal{C},x} \approx \mathfrak{M}_{0,1}(\mathbb{P}^1_p, d),$$

where $\mathfrak{M}_{0,1}(\mathbb{P}^1_p, d) \subset \overline{\mathfrak{M}}_{0,1}(\mathbb{P}^1, d)$ is the subspace of elements $[\Sigma, u]$ such that Σ is nonsingular and $\mathrm{ev}_1([\Sigma, u])=p$ for a fixed $p\in\mathbb{P}^1$. Let

$$\mathfrak{D} \in \Gamma\big(\overline{\mathcal{M}}_{1,1}\times\overline{\mathfrak{M}}_{0,1}(\mathbb{P}^1, d), \mathrm{Hom}(\pi_1^*L_1\otimes\pi_2^*L_1, \pi_1^*\mathbb{E}^*\otimes\pi_2^*\mathrm{ev}_1^*T\mathbb{P}^1)\big)$$

be given by

$$\mathfrak{D}(v\otimes w) = s_1(v)\otimes\mathcal{D}_1(w). \tag{2.16}$$

The first component of $\varphi_{t\nu}$ with respect to the decomposition (2.12) is essentially

$$F = \pi_1^*L_1|_{z_1}\otimes\pi_2^*L_1 \longrightarrow \mathfrak{O}_{PB} = \mathbb{E}^*_{\Sigma_P}\otimes T_x\mathcal{C}, \qquad v\longrightarrow \mathfrak{D}(v)+t\bar{\nu}_{PB}. \tag{2.17}$$

Let $\mathfrak{U}$ be the universal curve over $\overline{\mathfrak{M}}_{0,1}(\mathbb{P}^1, d)$, with structure map π and evaluation map ev:

$$\begin{array}{ccc} \mathfrak{U} & \xrightarrow{\ \mathrm{ev}\ } & \mathbb{P}^1 \\ \big\downarrow{\scriptstyle\pi} & & \\ \overline{\mathfrak{M}}_{0,1}(\mathbb{P}^1, d). & & \end{array} \tag{2.18}$$

The restriction of $\bar{\nu}_B$ to $\bar{\nu}_{PB}^{\perp-1}(0)$ is a section of

$$\mathfrak{O}_B = R^1\pi_*\mathrm{ev}_*\big(\mathcal{O}(-1)\oplus\mathcal{O}(-1)\big) \longrightarrow \overline{\mathfrak{M}}_{0,1}(\mathbb{P}^1_p, d).$$

Thus, by the Aspinwall–Morrison and divisor formulas, as in Section 1.1 in [10],

$$^{\pm}\big|\bar{\nu}_B^{-1}(0)\big| = \frac{1}{d^2}. \tag{2.19}$$

On the other hand, $\bar{\nu}_{PB}^{\perp}$ is a section of

$$\pi_1^*\mathbb{E}^*\otimes\pi_2^*\big(f^*T\overline{\mathcal{M}}_\beta\oplus\mathcal{N}_\beta\big) \longrightarrow \overline{\mathcal{M}}_{1,1}\times\overline{\mathcal{M}}_{\beta,1},$$

see (2.8). Therefore,

$$^{\pm}\big|\bar{\nu}_{PB}^{\perp-1}(0)\big| = -\frac{1}{24}\int_{\overline{\mathcal{M}}_{\beta,1}}\big(f^*c_2(\overline{\mathcal{M}}_\beta)\mathrm{ev}_1^*c_1(\mathcal{N}_\beta)+f^*c_1(\overline{\mathcal{M}}_\beta)\,c_2(\mathcal{N}_\beta)\big). \tag{2.20}$$

Since (2.17) has a unique zero in every fiber of F over $\bar{\nu}_{PB}^{\perp\,-1}(0)\cap\bar{\nu}_B^{-1}(0)$ and the restriction of $\mathcal{N}_\beta$ to a fiber of f is of degree -2, equations (2.19) and (2.20) imply

(2.21)
$$\mathsf{C}_{\mathcal{T}_{\mathrm{gh}}(\beta,d)}(d\beta)=\frac{1}{24d^2}\bigg(\int_{\overline{\mathcal{M}}_\beta}2c_2(\overline{\mathcal{M}}_\beta)-\int_{\overline{\mathcal{M}}_{\beta,1}}f^*c_1(\overline{\mathcal{M}}_\beta)c_2(\mathcal{N}_\beta)\bigg).$$

We have proved the second scaling identity in (2.3). The equation

$$\mathsf{C}_{\mathcal{T}_{\mathrm{gh}}(\beta,1)}(\beta)=\frac{1}{24}\int_{\overline{\mathcal{M}}_\beta}2c_2(\overline{\mathcal{M}}_\beta)-c_1^2(\overline{\mathcal{M}}_\beta) \tag{2.22}$$

is obtained from (2.21) from relations (1.14) and (1.15).

2.4. Strata with ghost principal component II

We continue with the setup of Section 2.3. For each $[\Sigma_u,u]\in\mathcal{Z}_{\mathcal{T}}$, denote by $\Sigma_u^0\subset\Sigma_u$ the largest union of irreducible components of Σ_u that contains the principal component(s) of Σ_u and on which u is constant. The topological types of Σ_u and Σ_u^0 are independent of the choice of $[\Sigma_u,u]\in\mathcal{Z}_{\mathcal{T}}$.

The bundle of gluing parameters (or smoothing of the nodes) over $F\longrightarrow\mathcal{Z}_{\mathcal{T}}$ is a direct sum of line bundles (up to a quotient by a finite group). Let $F^\emptyset\subset F$ be the subspace of smoothings with all components nonzero, smoothings that do not leave any nodes. If $v\in F_u^\emptyset$ is sufficiently small, there is a C^∞-map

$$q_v:\Sigma_v\longrightarrow\Sigma_u,$$

where Σ_u is the domain of u and Σ_v is a genus 1 Riemann surface with thin necks replacing the nodes of Σ_u, see Section 2.2 of [18]. This map determines Riemannian metrics and weights on Σ_v which induce the L_1^p- and L^p Sobolev norms, $\|\cdots\|_{v,p,1}$ and $\|\cdots\|_{v,p}$, with $p>2$, appearing below, see Section 3.3 in [18]. These norms are equivalent to the ones used in Section 3 of [11]. Let $\Sigma_v^0=q_v^{-1}(\Sigma_u^0)$.

We take the approximately holomorphic map corresponding to $v\in F_u$ to be

$$u_v=u\circ q_v:\Sigma_v\longrightarrow X.$$

The map satisfies

$$\big\|\bar\partial u_v\big\|_{v,p}\le C(u)|v|^{1/p}. \tag{2.23}$$

Let

$$D_v:\Gamma(v)=\Gamma(\Sigma_v,u_v^*TX)\longrightarrow\Gamma^{0,1}(v)=\Gamma(\Sigma_v,T^*\Sigma_v^{0,1}\otimes u_v^*TX)$$

be the linearization of the $\bar{\partial}$-operator at u_υ defined using the Levi-Civita connection of a Kähler metric $g_{X,u}$ on X. As in Sections 2 and 4.1 in [20], we can construct splittings

(2.24)
$$\Gamma(\upsilon) = \Gamma_-(\upsilon) \oplus \Gamma_+(\upsilon) \text{ and } \Gamma^{0,1}(\upsilon) = \Gamma^{0,1}_{-;PB}(\upsilon) \oplus \Gamma^{0,1}_{-;B}(\upsilon) \oplus \Gamma^{0,1}_{+}(\upsilon),$$

and isomorphisms

$$R_\upsilon : \mathfrak{O}_{PB} \oplus \pi_B^*\mathfrak{O}_B \longrightarrow \Gamma^{0,1}_{-;PB}(\upsilon) \oplus \Gamma^{0,1}_{-;B}(\upsilon) \tag{2.25}$$

with the following properties:

(G1) $D_\upsilon : \Gamma_+(\upsilon) \longrightarrow \Gamma^{0,1}_+(\upsilon)$ is an isomorphism with the norm of the inverse bounded independently of $\upsilon \in F^\emptyset_u$ (but depending on $[\Sigma, u]$),

(G2) the elements of $\Gamma^{0,1}_{-;PB}(\upsilon)$ are supported on a small neighborhood of Σ^0_υ,

(G3) if $\pi^{0,1}_{-;PB} : \Gamma^{0,1}(\upsilon) \longrightarrow \Gamma^{0,1}_{-;PB}(\upsilon)$ is the projection in the second decomposition (2.24),

$$\big\|\pi^{0,1}_{-;PB} D_\upsilon \xi\big\|_{\upsilon,2} \le C(u)|\upsilon|\|\xi\|_{\upsilon,p,1}, \ \forall \xi \in \Gamma(\upsilon), \tag{2.26}$$

(G4) if $|\mathrm{Ver}|=2$ (and thus $\mathcal{Z}_\mathcal{T}$ is basic),

$$\pi^{0,1}_{-;PB} \bar{\partial} u_\upsilon = R_\upsilon \mathfrak{D}\upsilon, \tag{2.27}$$

with $\mathfrak{D}$ as in (2.16),

(G5) every map $\tilde{u} : \Sigma \longrightarrow X$, where Σ is a smooth genus-one Riemann surface, that lies in a small neighborhood of $\mathcal{Z}_\mathcal{T}$ can be written uniquely as $\tilde{u} = \exp_{u_\upsilon} \xi$ for small $\upsilon \in F^\emptyset$ and $\xi \in \Gamma_+(\upsilon)$.

Let

$$\pi^{0,1}_+ : \Gamma^{0,1}(\upsilon) \longrightarrow \Gamma^{0,1}_+(\upsilon) \qquad \text{and} \qquad \pi^{0,1}_{-;B} : \Gamma^{0,1}(\upsilon) \longrightarrow \Gamma^{0,1}_{-;B}(\upsilon)$$

be the component projections in the second decomposition (2.24).

The relation (2.6) for $\tilde{u} = \exp_{u_\upsilon} \xi$ is equivalent to

$$\bar{\partial} u_\upsilon + D_\upsilon \xi + t\nu_\upsilon + N_\upsilon(\xi) + tN_{\nu,\upsilon}(\xi) = 0, \tag{2.28}$$

with N_υ and $N_{\nu,\upsilon}$ satisfying

$$\begin{aligned} \big\|N_\upsilon(\xi) - N_\upsilon(\xi')\big\|_{\upsilon,p} &\le C(u)\big(\|\xi\|_{\upsilon,p,1} + \|\xi'\|_{\upsilon,p,1}\big)\big\|\xi - \xi'\big\|_{\upsilon,p,1}, \\ \big\|N_{\nu,\upsilon}(\xi) - N_{\nu,\upsilon}(\xi')\big\|_{\upsilon,p} &\le C(u)\big\|\xi - \xi'\big\|_{\upsilon,p,1}, \end{aligned} \tag{2.29}$$

if $\upsilon \in F_u^\emptyset$. For a good choice of identifications,

$$\pi^{0,1}_{-;PB} N_\upsilon \xi = 0, \ \forall \xi \in \Gamma(\upsilon). \tag{2.30}$$

By the Contraction Principle and (G1), the equation

$$\pi^{0,1}_{+}\big(\bar{\partial} u_\upsilon + D_\upsilon \xi + t\nu_\upsilon + N_\upsilon(\xi) + tN_{\nu,\upsilon}(\xi)\big) = 0$$

has a unique small solution $\xi_{t\nu}(\upsilon) \in \Gamma_+(\upsilon)$. By (2.23), it satisfies

$$\big\|\xi_{t\nu}(\upsilon)\big\|_{\upsilon,p,1} \le C(u)\big(|\upsilon|^{1/p} + t\big). \tag{2.31}$$

Thus, the number of solutions of (2.6) near $\mathcal{Z}_\mathcal{T}$ is the number of solutions of the equation

(2.32)
$$\bar{\partial} u_\upsilon + D_\upsilon \xi_{t\nu}(\upsilon) + t\nu_\upsilon + N_\upsilon\big(\xi_{t\nu}(\upsilon)\big) + tN_{\nu,\upsilon}\big(\xi_{t\nu}(\upsilon)\big) = 0 \in \Gamma^{0,1}_{-;PB}(\upsilon) \oplus \Gamma^{0,1}_{-;B}(\upsilon).$$

This is an equation on $\upsilon \in F_u^\emptyset$ with $|\upsilon| < \delta(u)$ for some $\delta \in C^\infty(\mathcal{Z}_\mathcal{T}; \mathbb{R}^+)$.

For each $[\Sigma, u] \in \mathcal{Z}_\mathcal{T}$, let $\mathcal{C}_u = u(\Sigma)$ and $\mathcal{S}_u = \mathcal{S}_{\mathcal{C}_u}$, see the end of Section 2.2. The pregluing map u_υ satisfies

$$u_\upsilon(\Sigma_\upsilon) \subset \mathcal{C}_u \subset \mathcal{S}_u.$$

We can choose the splittings (2.24) so that they restrict to splittings for vector fields and (0,1)-forms along u_υ with values in $T\mathcal{S}_u$ and (G1) holds when restricted to $T\mathcal{S}_u$. If $\exp_{u_\upsilon} \xi$ is defined using the "exponential" $\exp^{\mathcal{C}_u}$, the operators D_υ and N_υ in (2.28) preserve $T\mathcal{S}_u$ as well. Therefore,

(2.33)
$$\xi_0(\upsilon) \in \Gamma(\Sigma_\upsilon, u_\upsilon^* T\mathcal{S}_u), \ D_\upsilon \xi_0(\upsilon), N_\upsilon\big(\xi_0(\upsilon)\big) \in \Gamma\big(\Sigma_\upsilon, T^*\Sigma_\upsilon^{0,1} \otimes u_\upsilon^* T\mathcal{S}_u\big).$$

On the other hand, by (G1) and (2.29),

$$\big\|\xi_{t\nu}(\upsilon) - \xi_0(\upsilon)\big\|_{\upsilon,p,1} \le C(u)t, \tag{2.34}$$

if $\upsilon \in F_u$ is sufficiently small. Taking the projection $\pi^{\perp}_{-;PB}$ of (2.32) to $\pi_P^* \mathbb{E}^* \otimes \bar{\pi}^* Q$, we thus find that any solution $\upsilon \in F_u$ of (2.32) satisfies

$$\big\|\bar{\nu}^{\perp}_{PB}(u)\big\| \le \varepsilon(t, \upsilon)$$

for some function $\varepsilon \colon \mathbb{R} \times F^\emptyset \longrightarrow \mathbb{R}^+$ approaching 0 as (t, υ) approaches 0. Therefore, all solutions of (2.6) lie in a small neighborhood of $\bar{\nu}^{\perp\,-1}_{PB}(0) \subset \mathcal{Z}_\mathcal{T}$, as claimed in Section 2.3.

If $m=1$, for a good choice of R_υ on $\pi_B^*\mathfrak{O}_B$

$$(2.35)\quad \langle\!\langle \eta, \eta_- \rangle\!\rangle_{\upsilon,2} = 0, \ \forall\ \eta \in \Gamma\big(\Sigma_\upsilon, T^*\Sigma_\upsilon^{0,1} \otimes u_\upsilon^* T\mathcal{S}_u\big),\ \eta_- \in \Gamma_{-;B}^{0,1}(\upsilon).$$ [6]

Taking the projection of (2.32) onto $\Gamma_{-;B}^{0,1}(\upsilon)$ and using (2.29), (2.33), (2.34), and (2.35), we obtain

$$(2.36)\qquad t\bar{\nu}_B(u) + t\eta(t,\upsilon) = 0,$$

for some $\eta(t,\upsilon)$ approaching 0 as (t,υ) approaches 0. Since u is $d\!:\!1$ cover of the smooth curve $\mathcal{C}_u$, the dimension of the projection of $\bar{\nu}_{PB}^{\perp\,-1}(0)$ onto the third component in the decomposition (2.11) is of dimension at most $2d-2m_B$.[7] Since the rank of $\mathfrak{O}_B$ is $2d-2$, (2.36) has no solutions for a generic choice of ν unless $\mathcal{Z}_\mathcal{T}$ is described by (B2) of Section 2.1. If $\mathcal{Z}_\mathcal{T}$ is of type (B2), the number of solutions of (2.32) is the same as the number of small solutions of

$$(2.37)\qquad \mathfrak{D}\upsilon + t\bar{\nu}_{PB}(u) + \eta(t,\upsilon) = 0 \in \mathbb{E}^* \otimes T_{u(\Sigma_P)}\mathcal{C}_u$$
$$\upsilon \in \pi_1^* L_1 \otimes \pi_2^* L_1|_u, \quad u \in \bar{\nu}_{PB}^{\perp\,-1}(0) \cap \bar{\nu}_B^{-1}(0) \subset \mathcal{M}_{1,1} \times \mathfrak{M}_{0,1}(\mathbb{P}^1; d),$$

with the error term $\eta(t,\upsilon)$ satisfying

$$\big\|\eta(t,\upsilon)\big\|_{\upsilon,2} \leq \varepsilon(t,\upsilon)\big(t + |\upsilon|\big);$$

see (2.26), (2.27), (2.29), and (2.31). If ν is generic, $\mathcal{D}_1$ and thus $\mathfrak{D}$ are nowhere zero on the finite set $\bar{\nu}_{PB}^{\perp\,-1}(0) \cap \bar{\nu}_B^{-1}(0)$. By the same rescaling and cobordism argument as in Section 3.1 of [19], the number of small solutions of (2.37) is the same as the number of solutions of

$$\mathfrak{D}\upsilon + \bar{\nu}_{PB}(u) = 0, \qquad \upsilon \in \pi_1^* L_1 \otimes \pi_2^* L_1\big|_{\bar{\nu}_{PB}^{\perp\,-1}(0) \cap \bar{\nu}_B^{-1}(0)}.$$

There is one solution for each of the elements of $\bar{\nu}_{PB}^{\perp\,-1}(0) \cap \bar{\nu}_B^{-1}(0)$. This concludes the consideration of the $m=1$ case.

We will next show that (2.32) has no solution if $m>1$. Let $\mathcal{Z}_{\mathcal{C},x}$ be as in (2.15). Since x is a nonsingular point of $\mathcal{C}$, on a neighborhood U of x in $\mathcal{C}$ there is an orthogonal decomposition

$$(2.38)\qquad T\mathcal{S}_\mathcal{C}|_U = T\mathcal{C}|_U \oplus \mathcal{N}\mathcal{C}|_U.$$

[6] In this case, $\mathfrak{O}_B$ is isomorphic to the cokernel of a $\bar{\partial}$-operator on $\mathcal{N}_{|\underline{\beta}|}$.

[7] This is the dimension of the space of degree-d covers of $\mathbb{P}^1$ by m_B copies of $\mathbb{P}^1$. The dimension is less than $2d-2m_B$ unless $\mathcal{Z}_{\mathcal{T},B}$ is the main stratum of $\overline{\mathfrak{M}}_{(0,m_B)}(X,\beta)$.

We can assume that the "exponential" map $\exp^{\mathcal{C}}$ satisfies

$$\pi_{\mathcal{C}}\big(\exp_y^{\mathcal{C}} v\big) = d\pi_{\mathcal{C}}|_x v, \ \forall y \in U, \, v \in T_y\mathcal{S}_{\mathcal{C}}, \, |v| < \delta, \tag{2.39}$$

with $\pi_{\mathcal{C}}$ as in (2.7). For any $[\Sigma, u] \in \mathcal{Z}_{\mathcal{T}}$ in a small neighborhood of $\mathcal{Z}_{\mathcal{C},x}$, let $W_u = u^{-1}(U)$ be an open neighborhood of Σ_u^0 in Σ_0. We can assume that every element η of $\Gamma^{0,1}_{-;PB}(v)$ is supported in $W_v = q_v^{-1}(W_u)$, whenever $v \in F_u^{\emptyset}$ is sufficiently small. With D_v^* denoting the formal adjoint of D_v with respect to the inner-product $\langle\!\langle \cdot, \cdot \rangle\!\rangle_{v,2}$, let

(2.40)
$$\Gamma_{+-}(v) = \big\{\xi \in \Gamma_+(v) \colon \langle\!\langle \xi, D_v^* R_v \eta \rangle\!\rangle = 0, \forall \eta \in \mathbb{E}_{\Sigma_P}^* \otimes \mathcal{N}_{u(\Sigma_P)}\mathcal{C}_u\big\} \subset \Gamma(v),$$
$$\Gamma_{++}(v) = \big\{D_v^* R_v \eta \colon \eta \in \mathbb{E}_{\Sigma_P}^* \otimes \mathcal{N}_{u(\Sigma_P)}\mathcal{C}_u\big\} \subset D_v^* \Gamma^{0,1}_{-;PB}(v) \subset \Gamma(v).$$

An explicit expression for $D_v^* R_v \eta$ is given in the proof of Lemma 2.2 in [19]. Section 2.3 of [19] implies that we can take

$$\Gamma_+(v) = \Gamma_{++}(v) \oplus \Gamma_{+-}(v). \tag{2.41}$$

In particular, the proof of Lemma 2.6 shows that the limits of the spaces $\Gamma_{++}(v)$ as $v \longrightarrow 0$ are orthogonal to the limits of the spaces $\Gamma_-(v)$. The decomposition (2.41) is L^2-orthogonal by (2.40) and

$$\|\xi\|_{v,p,1} \le C(u)\|\xi\|_{v,2}, \ \forall \xi \in \Gamma_{++}(v), \tag{2.42}$$

see the proof of Lemma 2.2 in [19].

Let $\xi_{t\nu}^+(v)$ and $\xi_{t\nu}^-(v)$ be the components of $\xi_{t\nu}(v)$ with respect to the decomposition (2.41). Denote by $\tilde{\nu}(u) \in \mathbb{E}_{\Sigma_P}^* \otimes \mathcal{N}_{u(\Sigma_P)}\mathcal{C}_u$ the projection of $\nu(u)$ to $\mathbb{E}_{\Sigma_P}^* \otimes \mathcal{N}_{u(\Sigma_P)}\mathcal{C}_u$. Since $\tilde{\nu}$ is a section of a trivial bundle near $\mathcal{Z}_{\mathcal{C},x}$, we can assume that it has no zeros on $\mathcal{Z}_{\mathcal{C},x}$. In the next paragraph we will show

$$\big\|\xi_{t\nu}^+(v)\big\|_{v,p,1} \le C(u)t. \tag{2.43}$$

Assuming this is the case, we project both sides of (2.32) onto

$$R_v\big(\mathbb{E}_{\Sigma_P}^* \otimes \mathcal{N}_{u(\Sigma_P)}\mathcal{C}_u\big) \subset \Gamma^{0,1}_{-;PB}(v)$$

and take the preimage under R_v. Since the projections of $\bar{\partial} u_v$ and $N_v(\xi_{t\nu(v)})$ vanish, using the first equation in (2.40), (2.43), (2.27), and (2.29), we obtain

$$t\tilde{\nu}(u) + \eta(t, v) = 0$$

with $\eta(t, v)$ satisfying

$$\big\|\eta(t, v)\big\|_{v,2} \le \varepsilon(t, u)t.$$

However, this is impossible if t and υ are sufficiently small ("small" depending continuously on u), since $\tilde{\nu}$ has no zeros over $\mathcal{Z}_{\mathcal{C},x}$.

We now verify (2.43). Let

$$\tilde{\xi}_{t\nu}(\upsilon)=\pi_{\mathcal{C}_u}\circ\exp_{u_\upsilon}\xi_{t\nu}(\upsilon)\colon\Sigma_\upsilon\longrightarrow\mathbb{C}^{m-1},\qquad \tilde{\xi}^{\pm}_{t\nu}(\upsilon)=d\pi_{\mathcal{C}_u}\circ\xi^{\pm}_{t\nu}(\upsilon).$$

Since $\xi^{+}_{t\nu}(\upsilon)$ is supported on W_u, by (2.39)

$$\big\langle\tilde{\xi}^{+}_{t\nu}(\upsilon),\tilde{\xi}^{\pm}_{t\nu}(\upsilon)\big\rangle_z=\big\langle\xi^{+}_{t\nu}(\upsilon),\xi^{\pm}_{t\nu}(\upsilon)\big\rangle_z,\ \forall\, z\in\Sigma_\upsilon. \tag{2.44}$$

By (2.39), we also have

$$\tilde{\xi}_\upsilon|W_\upsilon=\tilde{\xi}^{+}_\upsilon|W_\upsilon+\tilde{\xi}^{-}_\upsilon|W_\upsilon\,. \tag{2.45}$$

Since (2.32) is equivalent to (2.6) for $\tilde{u}=\exp_{u_\upsilon}\xi_{t\nu}(\upsilon)$,

$$\big\|\bar{\partial}\tilde{\xi}_{t\nu}(\upsilon)\big\|_{\upsilon,p}\le C(u)t. \tag{2.46}$$

Since the operator

$$L^p_1(\Sigma_\upsilon,\mathbb{C}^{m-1})\to L^p(\Sigma_\upsilon,T^*\Sigma^{0,1}_\upsilon\mathbb{C}^{m-1})\oplus\mathbb{C}^{m-1},\ \tilde{\xi}\to\bigg(\bar{\partial}\tilde{\xi},\int_{\Sigma_\upsilon}\tilde{\xi}\,dvol_{\Sigma_\upsilon}\bigg),$$

is an isomorphism with the norm of the inverse bounded independently of υ (but depending on u), (2.46) implies that

$$\big\|\tilde{\xi}_{t\nu}(\upsilon)-A_{t\nu}(\upsilon)\big\|_{\upsilon,p,1}\le C(u)t \tag{2.47}$$

for some $A_{t\nu}(\upsilon)\in\mathbb{C}^{m-1}$. Since $\xi^{+}_{t\nu}(\upsilon)$ is supported on W_u, by (2.39) and (2.40),

$$\big\langle\!\big\langle\tilde{\xi}^{+}_{t\nu}(\upsilon),A_{t\nu}(\upsilon)\big\rangle\!\big\rangle_{\upsilon,2}=0.$$

Thus, by (2.44), (2.45), and (2.47),

$$\begin{aligned}\big\|\xi^{+}_{t\nu}(\upsilon)\big\|_{\upsilon,2}=\big\|\tilde{\xi}^{+}_{t\nu}(\upsilon)|W_\upsilon\big\|_{\upsilon,2}&\le\big\|(\tilde{\xi}_{t\nu}(\upsilon)-A_{t\nu}(\upsilon))|W_\upsilon\big\|_{\upsilon,2}\\&\le\big\|\tilde{\xi}_{t\nu}(\upsilon)-A_{t\nu}(\upsilon)\big\|_{\upsilon,p,1}\le C'(u)t.\end{aligned}$$

The estimate (2.43) now follows from (2.42).

Finally, we comment on the choices made in (2.24) and (2.25). Choosing the splittings (2.24) so that (G1) and (G5) hold is essentially equivalent to choosing approximate kernel and cokernel for D_υ that vary smoothly with υ. This is easily accomplished in many possible ways, including via the construction in Section 3 of [11]. In order to ensure that (G2)–(G4) hold, R_υ on $\mathfrak{O}_{PB}$ is constructed by pushing harmonic forms on Σ_P over a small neighborhood of Σ^0_υ, see Section 2.2 of [19]. Finally, in order to obtain (2.30), define $\exp_{u_\upsilon}$ and parallel transport using a Kähler metric which is flat near $u(\Sigma_P)$, as in Section 2.1 of [19].

2.5. Strata with effective principal component

We determine here the contribution to $N_{1,*}$ from a stratum $\mathcal{Z}_\mathcal{T}$ consisting of maps $u\colon \Sigma \longrightarrow X$ that are not constant on the principal, genus-carrying, component(s) Σ_P of Σ. We show that $\mathcal{Z}_\mathcal{T}$ does not contribute to $N_{1,*}$ unless $\mathcal{Z}_\mathcal{T}$ is of type (B1).

Let $\mathfrak{O} \longrightarrow \mathcal{Z}_\mathcal{T}$ and $F \longrightarrow \mathcal{Z}_\mathcal{T}$ be the obstruction bundle and the bundle of gluing parameters as before. The projection map $\mathrm{ev}^*TX \longrightarrow Q$ induces a surjective homomorphism

$$\pi^\perp\colon \mathfrak{O} \longrightarrow \mathfrak{O}^\perp, \tag{2.48}$$

where $\mathfrak{O}^\perp|_{[\Sigma,u]}$ is the cokernel of the $\bar\partial$-operator on Q induced by the $\bar\partial$-operator D_b on TX. By a gluing and obstruction bundle analysis similar to Section 2.4, $\mathsf{C}_\mathcal{T}(*)$ is the number of zeros of a bundle map

$$\varphi_{t\nu}\colon F \longrightarrow \mathfrak{O}$$

over $\mathcal{Z}_\mathcal{T}$ for t sufficiently small. As in the previous case, all zeros of $\varphi_{t\nu}$ arise from the zeros of

$$\bar\nu^\perp = \pi^\perp \circ \bar\nu,$$

where $\bar\nu \in \Gamma(\mathcal{Z}_\mathcal{T};\mathfrak{O})$ is the section induced by ν. The homomorphism (2.48) extends to a surjective homomorphism from the cokernel bundles over $\bar{\mathcal{Z}}_\mathcal{T}$. In the next two paragraphs, we show that $\mathfrak{O}^\perp \longrightarrow \bar{\mathcal{Z}}_\mathcal{T}$ contains a trivial C^∞-subbundle unless $|\mathrm{Ver}|=1$. Therefore, $\mathsf{C}_\mathcal{T}(*)=0$ if $\mathcal{Z}_\mathcal{T}$ is not of type (B1).

Suppose first that $(\mathrm{Ver},\mathrm{Edg})$ contains a loop $L\subset \mathrm{Ver}$ and κ is not constant on L. Then, the image of the principal components Σ_P of any element $[\Sigma,u]$ of $\mathcal{Z}_\mathcal{T}$ contains at least two curves in X. Then, $\mathfrak{O}^\perp$ contains a pull-back of the bundle

$$\mathbb{E}^*\otimes f^*T\overline{\mathcal{M}}_{\underline{\beta}} \oplus s^*\mathcal{N}_{|\underline{\beta}|} \longrightarrow \overline{\mathcal{M}}_{1,1}\times \overline{\mathcal{M}}_{\underline{\beta}},$$

where $s\colon \overline{\mathcal{M}}_{\underline{\beta}} \longrightarrow \overline{\mathcal{M}}_{\underline{\beta},1}$ is the bundle section taking each curve $\mathcal{C}$ to one of the nodes. The bundle contains a trivial C^∞-subbundle for dimensional reasons.

We next consider the remaining cases. Let $P\in[m]$ be the component of the curves in $\overline{\mathcal{M}}_{\underline{\beta}}$ containing the image of the principal component Σ_P of any element $[\Sigma,u]$ of $\mathcal{Z}_\mathcal{T}$. Denote by $\overline{\mathcal{M}}^P_{\underline{\beta},1}\subset\overline{\mathcal{M}}_{\underline{\beta},1}$ the component consisting of the curves $\mathcal{C}_P$, with $\mathcal{C}\in\overline{\mathcal{M}}_{\underline{\beta}}$. If X is an ideal Calabi–Yau 5-fold and $m>1$ (implying $\mathcal{C}\neq\mathcal{C}_P$), the restriction of $\mathcal{N}_{|\underline{\beta}|}$ to $\mathcal{C}_P\approx\mathbb{P}^1$ splits as either $\mathcal{O}\oplus\mathcal{O}$ or $\mathcal{O}\oplus\mathcal{O}(-1)$. If $m=1$, the splitting is $\mathcal{O}(-1)\oplus\mathcal{O}(-1)$.

Case 1: $m>1$ and the restriction of $\mathcal{N}_{|\underline{\beta}|}$ to $\mathcal{C}_P$ splits as $\mathcal{O}\oplus\mathcal{O}$. Here, $Q|_{\overline{\mathcal{M}}^P_{\underline{\beta},1}}=f^*\bar{Q}$ for a bundle $\bar{Q}\longrightarrow\overline{\mathcal{M}}_{\underline{\beta}}$. Since the restriction of $f^*\bar{Q}$ to $u(\Sigma_P)$ is trivial, $\mathfrak{O}^\perp$ contains the subbundle $\mathbb{E}^*\otimes f^*\bar{Q}$, where $\mathbb{E}\longrightarrow\overline{\mathfrak{M}}_1(X,\beta)$ is the Hodge line bundle. The subbundle is a pull-back of the bundle

$$\mathbb{E}^*\otimes\bar{Q}\longrightarrow\overline{\mathcal{M}}_{1,1}\times\overline{\mathcal{M}}_{\underline{\beta}},$$

which contains a trivial C^∞-subbundle for dimensional reasons by (2.8).

Case 2: $m>1$ and the restriction of $\mathcal{N}_{|\underline{\beta}|}$ to $\mathcal{C}_P$ splits as $\mathcal{O}\oplus\mathcal{O}(-1)$. Here, $Q|_{\overline{\mathcal{M}}^P_{\underline{\beta},1}}$ contains a subbundle $f^*\bar{Q}'$ of co-rank 1 for a bundle $\bar{Q}'\longrightarrow\overline{\mathcal{M}}_{\underline{\beta}}$. Since the restriction of $f^*\bar{Q}'$ to $u(\Sigma_P)$ is trivial, $\mathfrak{O}^\perp$ contains the subbundle $\mathbb{E}^*\otimes f^*\bar{Q}'$, which is a pull-back of the bundle

$$\mathbb{E}^*\otimes\bar{Q}'\longrightarrow\overline{\mathcal{M}}_{1,1}\times\overline{\mathcal{M}}_{\underline{\beta}}.$$

Thus, the subbundle admits a section s such that $s^{-1}(0)$ is contained in the union of the spaces $\overline{\mathfrak{M}}_1(\mathcal{C}_i,A_i)$ taken over finitely many $\underline{\beta}$-curves $\mathcal{C}_i$. Since the restriction of $\mathcal{N}_{|\underline{\beta}|}$ to $\mathcal{C}_P$ contains $\mathcal{O}(-1)$, $\mathfrak{O}^\perp$ also contains a line subbundle is isomorphic to $\mathrm{ev}^*_{z_1}L$, where $\mathrm{ev}_{z_1}:\bar{\mathcal{Z}}_\mathcal{T}\longrightarrow X$ is the evaluation map sending $[\Sigma,u]$ to the value of u at a node of Σ taken to a node of $\mathcal{C}_P$. The restriction of this subbundle to $s^{-1}(0)$ is trivial.

Case 3: $m=1$. Here, $\mathfrak{O}^\perp=\mathfrak{O}$ is a bundle of the same rank as the dimension of $\overline{\mathfrak{M}}_1(\overline{\mathcal{M}}_{\beta,1},d)$ for some $d\in\mathbb{Z}^+$. Thus, if $\mathcal{Z}_\mathcal{T}$ is not the main stratum of $\overline{\mathfrak{M}}_1(\overline{\mathcal{M}}_{\beta,1},d)$, the restriction of $\mathfrak{O}^\perp$ to $\bar{\mathcal{Z}}_\mathcal{T}$ contains a trivial C^∞-subbundle.

It remains to consider the case $\mathcal{Z}_\mathcal{T}=\mathcal{Z}_{\mathcal{T}_{\mathrm{eff}}(\beta,d)}$ with $|\mathrm{Ver}|=1$. Then, $\varphi_{t\nu}$ is a generic section of

$$\mathfrak{O}=\mathfrak{O}^\perp=\mathbb{E}^*\otimes f^*T\overline{\mathcal{M}}_\beta\oplus R^1\pi_*\mathrm{ev}^*\mathcal{N}_\beta\longrightarrow\overline{\mathfrak{M}}_1^0(\overline{\mathcal{M}}_{\beta,1},d),$$

where $\overline{\mathfrak{M}}_1^0(\overline{\mathcal{M}}_{\beta,1},d)\subset\overline{\mathfrak{M}}_1(\overline{\mathcal{M}}_{\beta,1},d)$ is the closure of the space of maps with smooth domains, π is the structure map for the universal curve over $\overline{\mathfrak{M}}_1(\overline{\mathcal{M}}_{\beta,1},d)$, and ev is the corresponding evaluation map, see (2.18).

Thus,

(2.49)
$$\begin{aligned}\mathsf{C}_{\mathcal{T}_{\mathrm{eff}}(\beta,d)}(d\beta) &= \big\langle e\big(\mathbb{E}^*\otimes f^*T\overline{\mathcal{M}}_\beta\big)e\big(R^1\pi_*\mathrm{ev}^*\mathcal{N}_\beta\big), \overline{\mathfrak{M}}_1^0(\overline{\mathcal{M}}_{\beta,1},d)\big\rangle \\ &= \big\langle c_2\big(\overline{\mathcal{M}}_\beta\big),\overline{\mathcal{M}}_\beta\big\rangle \int_{\overline{\mathfrak{M}}_1^0(\mathbb{P}^1,d)} e\big(R^1\pi_*\mathrm{ev}^*\big(\mathcal{O}(-1)\oplus\mathcal{O}(-1)\big)\big) \\ &\quad - \big\langle \lambda\, f^*c_1\big(\overline{\mathcal{M}}_\beta\big)e\big(R^1\pi_*\mathrm{ev}^*\mathcal{N}_\beta\big), \overline{\mathfrak{M}}_1^0(\overline{\mathcal{M}}_{\beta,1},d)\big\rangle,\end{aligned}$$

where $\lambda = c_1(\mathbb{E})$. Using the Atiyah–Bott Localization Theorem of [2] as in Section 27.5 of [8], we find

(2.50)
$$\int_{\overline{\mathfrak{M}}_1^0(\mathbb{P}^1,d)} e\big(R^1\pi_*\mathrm{ev}^*\big(\mathcal{O}(-1)\oplus\mathcal{O}(-1)\big)\big) = \frac{d-1}{12d^2}\,.$$ [8]

In the next paragraph, we will obtain

(2.51)
$$\begin{aligned}&\big\langle \lambda\, f^*c_1\big(\overline{\mathcal{M}}_\beta\big)e\big(R^1\pi_*\mathrm{ev}^*\mathcal{N}\big), \overline{\mathfrak{M}}_1^0(\overline{\mathcal{M}}_{\beta,1},d)\big\rangle \\ &\qquad = \frac{d-1}{24d^2}\int_{\overline{\mathcal{M}}_{\beta,1}} c_2(\mathcal{N}_\beta)\, f^*c_1(\overline{\mathcal{M}}_\beta).\end{aligned}$$

Along with (2.49), (2.50), and (2.21), we conclude the first identity in (2.3).

With $\mathfrak{M}_{0,2}(\overline{\mathcal{M}}_{\beta,1},d) \subset \overline{\mathfrak{M}}_{0,2}(\overline{\mathcal{M}}_{\beta,1},d)$ denoting the locus of maps with nonsingular domains, let

$$\mathfrak{M}_{0,1=2}(\overline{\mathcal{M}}_{\beta,1},d) = \big\{b\in\mathfrak{M}_{0,2}(\overline{\mathcal{M}}_{\beta,1},d)\colon \mathrm{ev}_1(b)=\mathrm{ev}_2(b)\big\}.$$

Denote by $\overline{\mathfrak{M}}_{0,1=2}^0(\overline{\mathcal{M}}_{\beta,1},d)$ the closure of $\mathfrak{M}_{0,1=2}^0(\overline{\mathcal{M}}_{\beta,1},d)$ in $\overline{\mathfrak{M}}_{0,2}(\overline{\mathcal{M}}_{\beta,1},d)$. Let

$$\overline{\Delta}_1^0(\overline{\mathcal{M}}_{\beta,1},d) \subset \overline{\mathfrak{M}}_1^0(\overline{\mathcal{M}}_{\beta,1},d)$$

[8] The space $\overline{\mathfrak{M}}_1^0(\mathbb{P}^1,d)$ is singular, and thus [2] is not generally applicable. However, with the setup of Section 27.5 in [8], the restrictions of $e\big(R^1\pi_*\mathrm{ev}^*\big(\mathcal{O}(-1)\oplus\mathcal{O}(-1)\big)\big)$ to all fixed loci, with the exception of the simplest 1-edged ones, vanish. Furthermore, $\overline{\mathfrak{M}}_1^0(\mathbb{P}^1,d)$ is smooth along the 1-edged fixed loci. Therefore, the usual Atiyah–Bott localization formula applies. The normal bundles to the only contributing loci are the same as the normal bundles in the desingularization $\widetilde{\mathfrak{M}}_1^0(\mathbb{P}^1,d)$ of $\overline{\mathfrak{M}}_1^0(\mathbb{P}^1,d)$ constructed in [17] and described in Section 1.4 in [17].

be the subspace consisting of the stable maps $[\Sigma, u]$ such that the principal component Σ_P of Σ is singular. There is a natural node-identifying immersion

$$\iota\colon \overline{\mathfrak{M}}^0_{0,1=2}(\overline{\mathcal{M}}_{\beta,1}, d)/S_2 \longrightarrow \overline{\Delta}^0_1(\overline{\mathcal{M}}_{\beta,1}, d),$$

which is an embedding outside of a divisor. Note

$$\iota^* e\big(R^1\pi_*\mathrm{ev}^*\mathcal{N}_\beta\big) = c_2(\mathcal{N}_\beta)\, e\big(R^1\pi_*\mathrm{ev}^*\mathcal{N}_\beta\big).$$

If $\Delta_1 \subset \overline{\mathcal{M}}_{1,1}$ is the locus of the nodal elliptic curve,

$$\lambda = \frac{1}{12}\Delta_1 \in H^2\big(\overline{\mathcal{M}}_{1,1}\big).$$

Therefore,

$$\begin{aligned}
&\big\langle \lambda f^* c_1\big(\overline{\mathcal{M}}_\beta\big) e\big(R^1\pi_*\mathrm{ev}^*\mathcal{N}_\beta\big), \overline{\mathfrak{M}}^0_1(\overline{\mathcal{M}}_{\beta,1}, d)\big\rangle \\
&\quad = \frac{1}{24}\big\langle f^* c_1\big(\overline{\mathcal{M}}_\beta\big)\, c_2(\mathcal{N}_\beta) e\big(R^1\pi_*\mathrm{ev}^*\mathcal{N}_\beta\big), \overline{\mathfrak{M}}^0_{0,1=2}(\overline{\mathcal{M}}_{\beta,1}, d)\big\rangle \\
&\quad = \frac{1}{24}\big\langle c_2(\mathcal{N}_\beta) f^* c_1\big(\overline{\mathcal{M}}_\beta\big), \overline{\mathcal{M}}_{\beta,1}\big\rangle \\
&\qquad\qquad \times \int_{\overline{\mathfrak{M}}^0_{0,1=2}(\mathbb{P}^1_p, d)} e\big(R^1\pi_*\mathrm{ev}^*\big(\mathcal{O}(-1)\oplus\mathcal{O}(-1)\big)\big).
\end{aligned} \tag{2.52}$$

Using the localization as in Section 27.5 of [8] once again, we find

$$\int_{\overline{\mathfrak{M}}^0_{0,1=2}(\mathbb{P}^1_p, d)} e\big(R^1\pi_*\mathrm{ev}^*\big(\mathcal{O}(-1)\oplus\mathcal{O}(-1)\big)\big) = \frac{d-1}{d^2}.$$ [9]

Along with (2.52), this identity implies (2.51).

§3. Local $\mathbb{P}^2$

3.1. Gromov–Witten invariants

We consider here the local Calabi–Yau 5-fold given by the total space

$$X = \mathcal{O}(-1) \oplus \mathcal{O}(-1) \oplus \mathcal{O}(-1) \longrightarrow \mathbb{P}^2. \tag{3.1}$$

There are only two primary Gromov–Witten invariants in each degree d:

$$N_{0,d} = N_{0,d}(H^2, H^2) \quad \text{and} \quad N_{1,d},$$

[9] Similarly to the situation discussed for (2.50), $\overline{\mathfrak{M}}^0_{0,1=2}(\mathbb{P}^1_p, d)$ has singularities but is nonsingular along the only fixed locus to which $e\big(R^1\pi_*\mathrm{ev}^*\big(\mathcal{O}(-1)\oplus\mathcal{O}(-1)\big)\big)$ restricts non-trivially.

where H is the hyperplane class in $H^2(X,\mathbb{Z}) = H^2(\mathbb{P}^2,\mathbb{Z})$. We compute both Gromov–Witten invariants by localization[10] and then state a conjectural formula found by Martin for the integer counts $n_{1,d}$.

Lemma 3.1. *For* $d\in\mathbb{Z}^+$,

$$N_{0,d} = \frac{(-1)^{d-1}}{d} \qquad \textit{and} \qquad N_{1,d} = \frac{(-1)^d}{8d}.$$

Proof. Let (a,b,c) be the weights of the torus action on the vector space $\mathbb{C}^3$. The weights of the torus action on $T\mathbb{P}^2$ at the fixed points are then

$$\begin{aligned} P_1=[1,0,0]: &\qquad b-a,\, c-a,\\ P_2=[0,1,0]: &\qquad a-b,\, c-b,\\ P_3=[0,0,1]: &\qquad a-c,\, b-c. \end{aligned}$$

We choose linearizations on the 3 bundles $\mathcal{O}(-1)$ with the following weights at the fixed points:

	$\mathcal{O}(-1)$	$\mathcal{O}(-1)$	$\mathcal{O}(-1)$
P_1 :	0	$a-b$	$a-c$
P_2 :	$b-a$	0	$b-c$
P_3 :	$c-a$	$c-b$	$0.$

In order to compute the numbers $N_{0,d}$, we choose the points P_1 and P_2 for the insertions and integrate over

$$\overline{\mathfrak{M}} = \big\{b\in\overline{\mathfrak{M}}_{0,2}(\mathbb{P}^2,d)\colon \mathrm{ev}_1(b)=P_1,\ \mathrm{ev}_2(b)=P_2\big\}.$$

By the choice of the weights and the points, there is a unique fixed locus with non-zero contribution, see Section 27.5 in [8] for a similar situation. The locus consists of the d-fold cover u of the line

$$\mathbb{P}^1_{12}=\overline{P_1P_2}$$

branched over only P_1 and P_2 and with the marked points 1 and 2 mapped to P_1 and P_2, respectively. The weights of the fibers of the

[10] In the genus 0 case, the moduli space $\overline{\mathfrak{M}}_{0,2}(\mathbb{P}^2,d)$ is a nonsingular stack and the usual Atiyah–Bott localization formula applies. In the genus 1 case, the virtual localization formula of [7] is used.

relevant bundles at the fixed locus are given by

$$\begin{aligned}
&H^1(u^*\mathcal{O}(-1)): && \tfrac{(-1)^{d-1}(d-1)!}{d^{d-1}}(a-b)^{d-1}\\
&H^1(u^*\mathcal{O}(-1)): && \tfrac{(-1)^{d-1}(d-1)!}{d^{d-1}}(b-a)^{d-1}\\
&H^1(u^*\mathcal{O}(-1)): && (-1)^{d-1}\textstyle\prod_{r=1}^{d-1}\left(c-\tfrac{(d-r)a+rb}{d}\right)\\
&T\overline{\mathfrak{M}}: && \tfrac{(-1)^{d-1}(d-1)!^2}{d^{2(d-1)}}(a-b)^{2(d-1)}\textstyle\prod_{r=1}^{d-1}\left(c-\tfrac{(d-r)a+rb}{d}\right),
\end{aligned}$$

see Section 27.2 in [8]. The number $N_{0,d}$ is the ratio of the product of the first three expressions and the last expression, divided by d for the stack automorphism factor.

We next compute the number $N_{1,d}$. There are now 6 fixed loci with nonzero contribution: the three d-fold Galois covers of the three lines together with a choice of vertex for the contracted elliptic component. By symmetry, the contribution of the d-fold cover of $\mathbb{P}^1_{12}$ with the contracted component at P_1 determines the other cases. The weights of the fibers of the relevant bundles at the d-fold cover of $\mathbb{P}^1_{12}$ are given by

$$\begin{aligned}
&H^1(u^*\mathcal{O}(-1)): && \tfrac{(-1)^{d-1}(d-1)!}{d^{d-1}}(a-b)^{d-1}(-\lambda)\\
&H^1(u^*\mathcal{O}(-1)): && \tfrac{(-1)^{d-1}(d-1)!}{d^{d-1}}(b-a)^{d-1}(a-b-\lambda)\\
&H^1(u^*\mathcal{O}(-1)): && (-1)^{d-1}\textstyle\prod_{r=1}^{d-1}\left(c-\tfrac{(d-r)a+rb}{d}\right)(a-c-\lambda)\\
&Obs(\mathbb{P}^2): && (b-a-\lambda)(c-a-\lambda)\\
&T\overline{\mathfrak{M}}_1(\mathbb{P}^2,d): && \tfrac{(-1)^{d}d!^2}{d^{2d-1}}(a-b)^{2d-1}\textstyle\prod_{r=0}^{r=d}\left(c-\tfrac{(d-r)a+rb}{d}\right)\left(\tfrac{b-a}{d}-\psi\right),
\end{aligned}$$

where λ is the first chern class of the Hodge line bundle $\mathbb{E}\longrightarrow\overline{\mathcal{M}}_{1,1}$. The contribution of the locus to $N_{1,d}$ is the ratio of the product of the first four expressions and the last expression, divided by the stack factor d, and integrated over $\overline{\mathcal{M}}_{1,1}$,

$$\mathrm{Cont}(a,b)=\frac{(-1)^d}{24d}\,\frac{c-a}{c-b}\,.$$

Symmetrizing over a, b, and c, we obtain $N_{1,d}$. Q.E.D.

By Lemma 3.1 and the $n=2$ case of (0.2), the genus 0 counts for X are given by

$$n_{0,d}=n_{0,d}(H^2,H^2)=\begin{cases}1, & \text{if } d=1;\\ -1, & \text{if } d=2;\\ 0, & \text{if } d\geq 3.\end{cases}$$

Using the algorithm of Section 1.2, we have computed the genus 1 count $n_{1,d}$ for X for $d \leq 200$. All are integers.

3.2. Martin's conjecture

Recall the definition of the Möbius μ-function,

$$\mu\colon \mathbb{Z}^+ \to \{0, \pm 1\},$$

$$\mu(d) = \begin{cases} (-1)^r, & \text{if } d \text{ is the product of } r \text{ distinct primes,} \\ 0, & \text{otherwise.} \end{cases}$$

Define a sign function $S(d)$ and an absolute value function $V(d)$ as follows:

$$S(d) = \begin{cases} \mu(d), & \text{if } d \not\cong 4 \ (\mathrm{mod}\, 8), \\ \mu(d/4), & \text{if } d \cong 4 \ (\mathrm{mod}\, 8), \end{cases}$$

$$V(d) = \frac{k^2-1}{8} \times \begin{cases} \frac{k^2-1}{8}, & \text{if } d=k,\ 2 \nmid k, \\ \frac{17k^2+7}{8}, & \text{if } d=2k,\ 2 \nmid k, \\ 2k^2+1, & \text{if } d=4k,\ 2 \nmid k. \end{cases}$$

Conjecture 2 (G. Martin). *For every $d \in \mathbb{Z}^+$, the genus 1 degree d count for the local Calabi–Yau 5-fold $\mathbb{P}^2$ is given by*

$$n_{1,d} = S(d)V(d). \tag{3.2}$$

If $8|d$, then $S(d)$ vanishes and a definition of $V(d)$ is not required for (3.2). As our method for computing the numbers $n_{1,d}$ from $n_{0,d}$ and $N_{1,d}$ is completely explicit and the starting data is fairly simple, a verification of Conjecture 2 by elementary identities may be possible. Unfortunately, the algorithm involves a significant number of simultaneous recursions.[11]

Geometric consequences are easily obtained from the conjecture. For example, since $n_{1,d}$ is predicted to vanish whenever $8|d$, we expect Calabi–Yau 5-folds obtained from suitably generic deformations of the local $\mathbb{P}^2$ geometry to contain *no embedded elliptic curves of degrees divisible by 8.* Is there a simple symplectic reason for this?

References

[1] P. Aspinwall and D. Morrison, Topological field theory and rational curves, Comm. Math. Phys., **151** (1993), 245–262.

---

[11] Explicit forms of these recursions can be found in the appendix to this paper available from the authors' websites.

[2] M. Atiyah and R. Bott, The moment map and equivariant cohomology, Topology, **23** (1984), 1–28.

[3] G. Ellingsrud and S. Strömme, Bott's formula and enumerative geometry, J. Amer. Math. Soc., **9** (1996), 175–193.

[4] K. Fukaya and K. Ono, Arnold conjecture and Gromov–Witten invariant, Topology, **38** (1999), 933–1048.

[5] R. Gopakumar and C. Vafa, M-theory and topological Strings I, arXiv:hep-th/9809187.

[6] R. Gopakumar and C. Vafa, M-theory and topological strings II, arXiv:hep-th/9812127.

[7] T. Graber and R. Pandharipande, Localization of virtual classes, Invent. Math., **135** (1999), 487–518.

[8] K. Hori, S. Katz, A. Klemm, R. Pandharipande, R. Thomas, C. Vafa, R. Vakil and E. Zaslow, Mirror Symmetry, Clay Math. Inst., Amer. Math. Soc., 2003.

[9] S. Katz, On the finiteness of rational curves on quintic threefolds, Compositio Math., **60** (1986), 151–162.

[10] A. Klemm and R. Pandharipande, Enumerative geometry of Calabi–Yau 4-folds, Comm. Math. Phys., **281** (2008), 621–653.

[11] J. Li and G. Tian, Virtual moduli cycles and Gromov–Witten invariants of general symplectic manifolds, In: Topics in Symplectic 4-Manifolds, First Int. Press Lect. Ser., I, Internat. Press, 1998, pp. 47–83.

[12] J. Li and A. Zinger, On Gromov–Witten invariants of a quintic threefold and a rigidity conjecture, Pacific J. Math., **233** (2007), 417–480.

[13] R. Pandharipande, Intersection of $\mathbb{Q}$-divisors on Kontsevich's moduli space $\bar{M}_{0,n}(\mathbb{P}^r, d)$ and enumerative geometry, Trans. Amer. Math. Soc., **351** (1999), 1481–1505.

[14] R. Pandharipande, Hodge integrals and degenerate contributions, Comm. Math. Phys., **208** (1999), 489–506.

[15] R. Pandharipande and R. Thomas, Curve counting via stable pairs in the derived category, Invent. Math., **178** (2009), 407–447.

[16] R. P. Thomas, A holomorphic Casson invariant for Calabi–Yau 3-folds and bundles on $K3$ fibrations, J. Differential Geom., **54** (2000), 367–438.

[17] R. Vakil and A. Zinger, A desingularization of the main component of the moduli space of genus-1 stable maps into $\mathbb{P}^n$, Geom. Topol., **12** (2008), 1–95.

[18] A. Zinger, Enumerative vs. symplectic invariants and obstruction bundles, J. Symplectic Geom., **2** (2004), 445–543.

[19] A. Zinger, Enumeration of genus-2 curves with a fixed complex structure in $\mathbb{P}^2$ and $\mathbb{P}^3$, J. Differential Geom., **65** (2003), 341–467.

[20] A. Zinger, A sharp compactness theorem for genus-1 pseudo-holomorphic maps, Geom. Topol., **13** (2009), 2427–2522.

[21] A. Zinger, Reduced genus-1 Gromov–Witten invariants, J. Differential Geom., **83** (2009), 407–460.

[22] A. Zinger, Genus-0 2-point hyperplane integrals in Gromov–Witten Theory, arXiv:math/0705.2725.

[23] A. Zinger, Standard vs. reduced genus-1 Gromov–Witten invariants, Geom. Topol., **12** (2008), 1203–1241.

Rahul Pandharipande
Department of Mathematics
Princeton University
Princeton, NJ 08544
USA
E-mail address: rahulp@math.princeton.edu

Aleksey Zinger
Department of Mathematics
SUNY Stony Brook
Stony Brook, NY 11794
USA
E-mail address: azinger@math.sunysb.edu

Advanced Studies in Pure Mathematics 59, 2010
New Developments in Algebraic Geometry,
Integrable Systems and Mirror Symmetry (Kyoto, 2008)
pp. 289–347

Fourier–Laplace transform of a variation of polarized complex Hodge structure, II

Claude Sabbah

Abstract.

We show that the limit, by rescaling, of the 'new supersymmetric index' attached to the Fourier–Laplace transform of a polarized variation of Hodge structure on a punctured affine line is equal to the spectral polynomial attached to the same object. We also extend the definition by Deligne of a Hodge filtration on the de Rham cohomology of a exponentially twisted polarized variation of complex Hodge structure and prove a E_1-degeneration property for it.

§ Introduction

The purpose of this article, mainly concerned with exhibiting properties of the Fourier–Laplace transform of a variation of Hodge structure, is twofold.

(1) Let X be a compact Riemann surface, let S be a finite set of points on X. We will denote by $j : U = X \smallsetminus S \hookrightarrow X$ the inclusion. Let $f : X \to \mathbb{P}^1$ be a meromorphic function on X which is holomorphic on U and let $(V, {}^V\nabla)$ be a holomorphic bundle on U equipped with a holomorphic connection. We denote by $\mathcal{M}$ the locally free $\mathscr{O}_X(*S)$-module of finite rank with a connection having regular singularities at each point of S and such that $\mathcal{M}_{|U} = (V, {}^V\nabla)$ (Deligne's meromorphic extension of $(V, {}^V\nabla)$).

Received September 9, 2008.
2000 *Mathematics Subject Classification.* Primary 14D07; Secondary 32G20, 32S40.
Key words and phrases. Flat bundle, variation of Hodge structure, polarization, twistor $\mathscr{D}$-module, Fourier–Laplace transform, supersymmetric index, spectral polynomial.
Part of this work has been achieved during and shortly after the conference "New developments in Algebraic Geometry, Integrable Systems and Mirror symmetry", while the author was visiting RIMS, Kyoto. The author thanks this institution for hospitality and Professor M.-H. Saito for the invitation and financial support by JSPS Grant-in Aid for Scientific Research (S-19104002).

In [4], P. Deligne defines a Hodge filtration on the de Rham cohomology of the exponentially twisted connection $H^*_{\mathrm{DR}}(X, \mathcal{M}\otimes\mathcal{E}^f)$, i.e., that of the meromorphic bundle $\mathcal{M}$ with the twisted connection ${}^V\nabla + df\wedge$, at least when the monodromy of $(V, {}^V\nabla)$ is unitary (and thus corresponds to a variation of polarized Hodge structure of type $(0,0)$).

This Hodge filtration is indexed by real numbers, and Deligne proves a E_1-degeneration property for the de Rham complex. It has a good behaviour with respect to duality.

One is naturally led to the following questions:

- In what sense do we get a Hodge filtration, i.e., what are the underlying Hodge properties?
- Why are the jumps of this Hodge filtration related to the eigenvalues of the monodromy of f around $f = \infty$ (more precisely, the spectrum of f at infinity relative to $(V, {}^V\nabla)$)?
- Is there a possible extension of this construction without the unitarity assumption, when $(V, {}^V\nabla)$ is only assumed to underlie a polarized variation of Hodge structure?

In §6, we extend the construction by Deligne of a filtration on the twisted de Rham cohomology $H^*_{\mathrm{DR}}(X, \mathcal{M}\otimes\mathcal{E}^f)$ when $(V, {}^V\nabla)$ underlies a variation of polarized complex Hodge structure and give an answer to the previous questions. However, for simplicity, we restrict to the case where $X = \mathbb{P}^1 = \mathbb{A}^1 \cup \{\infty\}$ and f is the coordinate function on $\mathbb{A}^1$.

(2) Let H be a complex vector space equipped with a positive definite Hermitian form h (that we call a Hermitian metric) and two endomorphisms $\mathcal{U}$ and $\mathcal{Q}$, where $\mathcal{Q}$ is selfadjoint with respect to h. The other purpose of this article is to give a relation between polynomials of degree $\dim H$ attached to this situation:

- On the one hand, the characteristic polynomial of $\mathcal{Q}$, denoted by $\mathrm{Susy}_{(H,h,\mathcal{U},\mathcal{Q})}(T)$.
- On the other hand, the spectral polynomials. The spectral polynomial at infinity, as defined in §1.a below, is attached to the holomorphic bundle with a meromorphic connection having a pole of order two associated to $\mathcal{U}$ and $\mathcal{Q}$ (cf. §1.d), and denoted by $\mathrm{SP}^\infty_{(H,h,\mathcal{U},\mathcal{Q})}(T)$. With a supplementary assumption called "no ramification", one can also define the spectral polynomial at the origin $\mathrm{SP}^0_{(H,h,\mathcal{U},\mathcal{Q})}(T)$ (cf. §1.b).

There is a rescaling operator μ^*_τ, parametrized by $\tau\in\mathbb{C}^*$, acting on the data $(H, h, \mathcal{U}, \mathcal{Q})$ (more precisely and more accurately, on the associated integrable twistor structure, cf. Appendix B).

The other main result of this article (Theorem 7.1) is to prove, under some conditions made explicit below (namely, $(H, h, \mathscr{U}, \mathscr{Q})$ is the de Rham cohomology of the exponential twist of a variation of polarized Hodge structure on a punctured line, in particular, the "no ramification" condition holds), a relation conjectured by C. Hertling:

$$(0.1)\qquad \begin{aligned} \lim_{\tau\to 0} \mathrm{Susy}_{\mu_\tau^*(H,h,\mathscr{U},\mathscr{Q})}(T) &= \mathrm{SP}^{\infty}_{(H,h,\mathscr{U},\mathscr{Q})}(T), \\ \lim_{\tau\to \infty} \mathrm{Susy}_{\mu_\tau^*(H,h,\mathscr{U},\mathscr{Q})}(T) &= \mathrm{SP}^{0}_{(H,h,\mathscr{U},\mathscr{Q})}(T). \end{aligned}$$

A similar relation was first proved by C. Hertling (cf. [7, Th. 7.20]) when the connection ∇ has a regular singularity at $z = 0$.

The relation between the two approaches (1) and (2) above is made explicit in Remark 7.2 below. Both questions rely on a detailed analysis of the Fourier–Laplace transform of a variation of polarized complex Hodge structure on the punctured affine line.

Remark. In a recent preprint [13], T. Mochizuki extends the limit theorems 3.1 and 3.5 in the higher dimensional case and gives applications to a characterization of nilpotent orbits.

Acknowledgements. I thank Claus Hertling for useful discussions on this subject and for his comments on a preliminary version of this article. I thank the referee for his careful reading of the manuscript and useful comments.

§1. Connections with a pole of order two

Let Ω be an open disc centered at the origin of $\mathbb{C}$ with coordinate z and let $\mathscr{H}$ be a $\mathscr{O}_\Omega$-locally free sheaf with a meromorphic connection ∇ having a pole of order two at the origin and no other pole (one can consider a more general situation, but we will restrict to this setting). We will moreover assume that the eigenvalues of the monodromy operator and of the formal monodromy operator have absolute value equal to one (so that the V-filtrations below are indexed by real numbers; here also, a more general situation could be considered, but we will restrict to this setting).

1.a. Spectrum at infinity

There exists a unique locally free $\mathscr{O}_{\mathbb{P}^1}(*\infty)$-module $\widetilde{\mathscr{H}}$ equipped with a meromorphic connection ∇ with poles at 0 and ∞ only, such that ∞ is a regular singularity, and which coincides with $\mathscr{H}$ when restricted to Ω (it is called the *meromorphic Deligne extension* of $(\mathscr{H}, \nabla)$ at infinity, cf. [2]). Let us denote by $z' = 1/z$ the coordinate at infinity on $\mathbb{P}^1$. For any $\gamma \in \mathbb{R}$, the γ-Deligne extension of $\widetilde{\mathscr{H}}$ at infinity is the locally free $\mathscr{O}_{\mathbb{P}^1}$-module $V^\gamma\widetilde{\mathscr{H}}$ on which the connection has a logarithmic pole at infinity with residue having eigenvalues in $[\gamma, \gamma+1)$. According to the Birkhoff–Grothendieck theorem, $V^\gamma\widetilde{\mathscr{H}}$ decomposes as the direct sum of rank-one locally free $\mathscr{O}_{\mathbb{P}^1}$-modules $V^\gamma\widetilde{\mathscr{H}} = \mathscr{O}_{\mathbb{P}^1}(a_1) \oplus \cdots \oplus \mathscr{O}_{\mathbb{P}^1}(a_{\operatorname{rk}\mathscr{H}})$ with $a_1 \geqslant a_2 \geqslant \cdots$. We denote by v_γ the number of such line bundles which are $\geqslant 0$ and by ν_γ the difference $v_\gamma - v_{>\gamma}$.

We can express these numbers a little differently. We have a natural morphism

$$\mathscr{O}_{\mathbb{P}^1} \otimes_{\mathbb{C}} \Gamma(\mathbb{P}^1, V^\gamma\widetilde{\mathscr{H}}) \longrightarrow V^\gamma\widetilde{\mathscr{H}} \tag{1.1}$$

whose image is denoted by $\widetilde{\mathscr{V}}^\gamma$. This is a subbundle of $V^\gamma\widetilde{\mathscr{H}}$ in the sense that $V^\gamma\widetilde{\mathscr{H}}/\widetilde{\mathscr{V}}^\gamma$ is also a locally free sheaf of $\mathscr{O}_{\mathbb{P}^1}$-modules; more precisely, fixing a Birkhoff–Grothendieck decomposition as above, we have $\widetilde{\mathscr{V}}^\gamma = \bigoplus_{i|a_i \geqslant 0} \mathscr{O}_{\mathbb{P}^1}(a_i)$ (indeed, for any line bundle $\mathscr{O}_{\mathbb{P}^1}(k)$, $\mathscr{O}_{\mathbb{P}^1} \otimes_{\mathbb{C}} \Gamma(\mathbb{P}^1, \mathscr{O}_{\mathbb{P}^1}(k)) \to \mathscr{O}_{\mathbb{P}^1}(k)$ is onto if $k \geqslant 0$ and 0 if $k < 0$) so $\widetilde{\mathscr{V}}^\gamma$ is a direct summand of $V^\gamma\widetilde{\mathscr{H}}$ of rank v_γ. Restricting to Ω, we get a decreasing filtration $\mathscr{V}^\bullet$ of $\mathscr{H}$ indexed by $\mathbb{R}$. The graded pieces $\operatorname{gr}^\gamma_{\mathscr{V}}\mathscr{H} := \mathscr{V}^\gamma/\mathscr{V}^{>\gamma}$ are locally free $\mathscr{O}_\Omega$-modules (being isomorphic to the kernel of $\mathscr{H}/\mathscr{V}^{>\gamma} \to \mathscr{H}/\mathscr{V}^\gamma$), and $\nu_\gamma = \operatorname{rk}\operatorname{gr}^\gamma_{\mathscr{V}}\mathscr{H}$.

Let us recall the definition of the spectral polynomial $\mathrm{SP}^\infty_{\mathscr{H}}$ (cf. [17] or [14, §III.2.b]).

Definition 1.2 (Spectrum at infinity). The spectral polynomial of $\mathscr{H}$ at infinity is the polynomial $\mathrm{SP}^\infty_{\mathscr{H}}(T) = \prod_\gamma (T-\gamma)^{\nu_\gamma}$ with (for any $z_o \in \Omega$)

$$\nu_\gamma = \operatorname{rk}\operatorname{gr}^\gamma_{\mathscr{V}}\mathscr{H} = \dim i^*_{z_o}\operatorname{gr}^\gamma_{\mathscr{V}}\mathscr{H} = \dim \mathscr{V}^\gamma/(\mathscr{V}^{>\gamma} + (z-z_o)\mathscr{V}^\gamma).$$

In the following, we will often use an algebraic version of the previous construction, which is obtained as follows: set $G_0 = \Gamma(\mathbb{P}^1, \widetilde{\mathscr{H}})$, which is a free $\mathbb{C}[z]$-module of finite rank; the (decreasing) Deligne V-filtration of $G := \mathbb{C}[z, z^{-1}] \otimes_{\mathbb{C}[z]} G_0$ at $z = \infty$ is a filtration $V^\bullet G$ of G by $\mathbb{C}[z']$-free submodules; in particular, $z'V^\gamma G = V^{\gamma+1}G$ and $z\partial_z + \gamma = -z'\partial_{z'} + \gamma$ is nilpotent on $\operatorname{gr}^\gamma_V G := V^\gamma G/V^{>\gamma}G$. Then, $\Gamma(\mathbb{P}^1, V^\gamma\widetilde{\mathscr{H}}) = V^\gamma G \cap G_0 =: V^\gamma G_0$.

When tensored with $\mathscr{O}_{\mathbb{P}^1}(*\infty)$ and after taking global sections, (1.1) is the inclusion morphism

$$\mathbb{C}[z] \cdot V^\gamma G_0 \hookrightarrow G_0.$$

As $\mathbb{C}[z] \cdot V^\gamma G_0$ is a direct summand in G_0, this inclusion induces an inclusion of fibres at z_o for any $z_o \in \mathbb{C}$, and so

$$\begin{aligned} V^\gamma\big(G_0/(z-z_o)G_0\big) &:= V^\gamma G_0/[(z-z_o)G_0 \cap V^\gamma G] \\ &= \big[\mathbb{C}[z]\cdot V^\gamma G_0 + (z-z_o)G_0\big]/(z-z_o)G_0 \end{aligned}$$

has dimension v_γ. Then we also have

(1.3)
$$\begin{aligned} \nu_\gamma(G_0) &= \dim \operatorname{gr}_V^\gamma(G_0/(z-z_o)G_0) \\ &= \dim(G_0 \cap V^\gamma G)/\big[(G_0 \cap V^{>\gamma}G) + ((z-z_o)G_0 \cap V^\gamma G)\big]. \end{aligned}$$

Example 1.4. In [17] (where an increasing version of the V-filtration is used, hence the change of sign below), this polynomial is denoted by $\mathrm{SP}_\psi(G, G_0)$. Correspondingly, the set of pairs $(-\gamma, \nu_\gamma)$ above is called the spectrum (at infinity) of (G, G_0). When G_0 is the Brieskorn lattice attached to a cohomologically tame function on an affine smooth variety of dimension $n+1$ (cf. loc. cit.), the spectrum at infinity is symmetric with respect to $(n+1)/2$ and the numbers $-\gamma$ belong to $[0, n+1] \cap \mathbb{Q}$.

1.b. Spectrum at the origin

We now make a supplementary assumption on $(\mathscr{H}, \nabla)$. Let us denote by $\mathscr{H}[1/z]$ the locally free $\mathscr{O}_\Omega[1/z]$-module $\mathscr{O}_\Omega[1/z] \otimes_{\mathscr{O}_\Omega} \mathscr{H}$, with its natural meromorphic connection. By the Levelt–Turrittin theorem, the associated formal module $\mathbb{C}[\![z]\!][1/z] \otimes_{\mathscr{O}_\Omega} \mathscr{H}$ can be decomposed, after a suitable ramification of z, as the direct sum of meromorphic connections which are tensor product of a rank-one irregular connection with a regular one. Here, we make the assumption that *no ramification is needed to get the Levelt–Turrittin decomposition* (cf. Appendix B for the need of such a condition). One can formulate this condition in terms of Laplace transforms, in the coordinate $z' := 1/z$:

Lemma 1.5. *The "no ramification" condition is equivalent to saying that the Laplace transform of the $\mathbb{C}[z']\langle\partial_{z'}\rangle$-module G associated with $\mathscr{H}$ has only regular singularities (included at infinity).*

Sketch of proof. This follows from the slope correspondence in the Fourier–Laplace transform (cf. [10, Chap. V]). We will not distinguish between the Laplace transform and the inverse Laplace transform. Assume that G is the Laplace transform of M. The formal part of M at the origin produces the formal part of slope < 1 of G at $z' = \infty$. By assumption, only the slope 0 can appear, so M is regular at the origin.

A similar reasoning can be done at each singular point of M by twisting by a suitable exponential term, showing that M has only regular singularities at finite distance. The part of slope < 1 of M at infinity produces the formal part of G at $z' = 0$, and as G is regular at $z' = 0$, only the slope 0 occurs as a slope < 1 for M at infinity. Slopes equal to 1 for M at infinity would produce singular points of G at finite distance, and not equal to $z' = 0$. There are none. Lastly, slopes > 1 for M at infinity would produce slopes > 1 for G at $z' = \infty$. There are none. Hence M has to be regular (slope 0) at infinity. Q.E.D.

Let us set $\mathscr{H}^\wedge = \mathbb{C}[\![z]\!] \otimes_{\mathscr{O}_\Omega} \mathscr{H}$. When the "no ramification" condition is fulfilled, the Levelt–Turrittin decomposition for $\mathscr{H}^\wedge[1/z]$ already exists for $\mathscr{H}^\wedge$. There exists then a finite number of pairwise distinct complex numbers c_i ($i \in I$) and a finite number of free $\mathbb{C}\{z\}$-modules $\mathscr{H}_i$ with a regular meromorphic connection having a pole of order at most two, such that

$$\mathscr{H}^\wedge \simeq \bigoplus_{i\in I} (\mathscr{H}_i \otimes \mathscr{E}^{c_i/z})^\wedge, \tag{1.6}$$

where $\mathscr{E}^{c_i/z}$ is $\mathbb{C}\{z\}$ equipped with the connection $d - c_i dz/z^2$. Each $\mathscr{H}_i$ is equipped with a regular meromorphic connection ∇. The free $\mathbb{C}\{z\}[1/z]$-module $\mathscr{H}_i[1/z]$ has a canonical decreasing Deligne filtration $V^\bullet \mathscr{H}_i[1/z]$ indexed by real numbers (by our assumption) so that $z\partial_z - \gamma$ is nilpotent on the vector space $\mathrm{gr}_V^\gamma \mathscr{H}_i[1/z]$.

Definition 1.7 (Spectrum at the origin). For any $i \in I$, the spectral polynomial of the regular meromorphic connection $\mathscr{H}_i$ at the origin is the polynomial $\mathrm{SP}^0_{\mathscr{H}_i}(T) = \prod_\gamma (T+\gamma)^{\mu_{i,\gamma}}$, with

$$\mu_{i,\gamma} = \dim \frac{\mathscr{H}_i \cap V^\gamma \mathscr{H}_i[1/z]}{\mathscr{H}_i \cap V^{>\gamma} \mathscr{H}_i[1/z] + z\mathscr{H}_i \cap V^\gamma \mathscr{H}_i[1/z]},$$

and we set $\mathrm{SP}^0_{\mathscr{H}}(T) = \prod_i \mathrm{SP}^0_{\mathscr{H}_i}(T)$.

(The choice $T + \gamma$ is done in order to have similar formulas for SP^0 and SP^∞.)

Example 1.8. When $\mathscr{H}$ is the analytization of the Brieskorn lattice G_0 of a cohomologically tame function on a smooth affine variety (cf. Example 1.4), then $\mathscr{H}_i \neq 0$ only if $-c_i$ is a critical value of this function, and $\mathscr{H}_i^\wedge$ is the formal (with respect to $\partial_t^{-1} = z$) local Brieskorn lattice at this critical value (this follows from [14, Prop. V.3.6] for instance). The set of pairs $(\gamma, \mu_{i,\gamma})$ is the spectrum at this critical value (with a shift by one with respect to the definition of [23]): it is symmetric with respect to $(n+1)/2$ and the numbers γ belong to $(0, n+1)$. (See also [23], [22] and [6, Chap. 10–11] and the references therein for detailed results in the case of a singularity germ.)

Remark 1.9. Assume that ∇ has a pole of order one on $\mathscr{H}$. Then $\mathrm{SP}^0_{\mathscr{H}}$ is the characteristic polynomial of $-\operatorname{Res}\nabla$ (residue of the connection at $z=0$). We also have $\mathrm{SP}^\infty_{\mathscr{H}} = \mathrm{SP}^0_{\mathscr{H}}$ (cf. e.g. [14, Ex. III.2.6]). We conclude that in an exact sequence of logarithmic connections, $\mathrm{SP}^\infty_{\mathscr{H}}$ and $\mathrm{SP}^0_{\mathscr{H}}$ behave multiplicatively.

1.c. Connection with a pole of order two by Laplace transform

Let us recall the notion of Laplace transform of a filtered $\mathbb{C}[t]\langle\partial_t\rangle$-module (cf. [14, §V.2.c] or [19, §1.d]). Let $\mathbb{A}^1$ be the affine line with coordinate t and let M be a holonomic $\mathbb{C}[t]\langle\partial_t\rangle$-module. We set $G := M[\partial_t^{-1}] = \mathbb{C}[t]\langle\partial_t, \partial_t^{-1}\rangle \otimes_{\mathbb{C}[t]\langle\partial_t\rangle} M$ (it is known that G is also holonomic as a $\mathbb{C}[t]\langle\partial_t\rangle$-module) and we denote by $\widehat{\mathrm{loc}} : M \to G$ the natural morphism (the kernel and cokernel of which are isomorphic to powers of $\mathbb{C}[t]$ with its natural structure of left $\mathbb{C}[t]\langle\partial_t\rangle$-module). For any lattice L of M, i.e., a $\mathbb{C}[t]$-submodule of finite type such that $M = \mathbb{C}[\partial_t]\cdot L$, we set

$$G_0^{(L)} = \sum_{j\geqslant 0} \partial_t^{-j}\widehat{\mathrm{loc}}(L). \tag{1.10}$$

This is a $\mathbb{C}[\partial_t^{-1}]$-submodule of G. Moreover, because of the relation $[t, \partial_t^{-1}] = (\partial_t^{-1})^2$, it is naturally equipped with an action of $\mathbb{C}[t]$. If M has a regular singularity at infinity, then $G_0^{(L)}$ has finite type over $\mathbb{C}[\partial_t^{-1}]$ (cf. [14, Th. V.2.7]). We have $G = \mathbb{C}[\partial_t]\cdot G_0^{(L)}$.

Let us now assume that M is equipped with a good filtration $F_\bullet M$. In the following, in order to keep the correspondence with Hodge theory, we will work with decreasing filtrations $F^\bullet M$, the correspondence being given by $F^pM := F_{-p}M$. Let $p_0 \in \mathbb{Z}$. We say that *$F^\bullet M$ is generated by $F^{p_0}M$* if, for any $\ell \geqslant 0$, we have $F^{p_0-\ell}M = F^{p_0}M + \cdots + \partial_t^\ell F^{p_0}M$. The $\mathbb{C}[\partial_t^{-1}]$-module $\partial_t^{p_0}G_0^{(F^{p_0})}$ does not depend on the choice of the index p_0, provided that the generating assumption is satisfied (cf. [19, §1.d]). We thus define the *Brieskorn lattice of the filtration $F^\bullet M$* as

$$G_0^{(F)} = \partial_t^{p_0}G_0^{(F^{p_0})} \quad \text{for some (or any) index } p_0 \text{ of generation.} \tag{1.11}$$

If we also set $z = \partial_t^{-1}$, then $G_0^{(F)}$ is a free $\mathbb{C}[z]$-module which satisfies $G = \mathbb{C}[z, z^{-1}] \otimes_{\mathbb{C}[z]} G_0^{(F)}$ and which is stable by the action of $z^2\partial_z := t$.

For any p, we have

$$\widehat{\mathrm{loc}}(F^pM) \subset z^p G_0^{(F)}. \tag{1.12}$$

Indeed, $z^p G_0^{(F)} = \sum_{j\geqslant 0} \partial_t^{-j-p+p_0} \widehat{\mathrm{loc}}(F^{p_0}M)$; if $p \geqslant p_0$, we have $\partial_t^{p-p_0} F^pM \subset F^{p_0}M$, hence the desired inclusion after applying $\widehat{\mathrm{loc}}$; if $p \leqslant p_0$, we have $F^pM = F^{p_0} + \cdots + \partial_t^{p_0-p} F^{p_0}M$, and the result is clear.

1.d. Integrable twistor structures

Let H be a finite dimensional complex vector space equipped with a Hermitian metric h and of two endomorphisms $\mathscr{U}$ and $\mathscr{Q}$, with $\mathscr{Q}$ being selfadjoint with respect to h. Let $\mathscr{U}^\dagger$ be the h-adjoint of $\mathscr{U}$. Let Ω_0 be an open neighbourhood of the closed disc $|z| \leqslant 1$ in $\mathbb{C}$ and let us set $\mathscr{H}' = \mathscr{O}_{\Omega_0} \otimes_{\mathbb{C}} H$, equipped with the meromorphic connection $\nabla = d + (z^{-2}\mathscr{U} - z^{-1}\mathscr{Q} - \mathscr{U}^\dagger)dz$.

We will denote by $\mathscr{T} = (\mathscr{H}', \mathscr{H}', \mathscr{C}_{\mathrm{S}})$ the associated twistor structure (as defined in §2.b below, by taking X to be a point).

We will denote by $\mathrm{SP}^\infty_{\mathscr{T}}(T)$ or by $\mathrm{SP}^\infty_{(H,h,\mathscr{U},\mathscr{Q})}(T)$ the spectral polynomial at infinity $\mathrm{SP}^\infty_{\mathscr{H}'}(T)$. On the other hand, if ∇ has no ramification at the origin, we will denote by $\mathrm{SP}^0_{\mathscr{T}}(T)$ or by $\mathrm{SP}^0_{(H,h,\mathscr{U},\mathscr{Q})}(T)$ the spectral polynomial at the origin $\mathrm{SP}^0_{\mathscr{H}'}(T)$.

§2. A review on integrable twistor $\mathscr{D}$-modules

In this section and in Section 4, we gather the notation and results needed for the proofs of the main theorems of this article. We refer to [16, 19, 18] for details.

2.a. Integrable harmonic Higgs bundles

Let X be a complex manifold and let E be a holomorphic bundle on X, equipped with a Hermitian metric h. For any operator P acting linearly on E, we will denote by $P^\dagger$ its adjoint with respect to h.

Let θ be a holomorphic Higgs field on E, that is, an $\mathscr{O}_X$-linear morphism $E \to \Omega^1_X \otimes_{\mathscr{O}_X} E$ satisfying the "integrability relation" $\theta \wedge \theta = 0$. We then say that (E, θ) is a Higgs bundle (cf. [26]).

Let E be a holomorphic bundle with a Hermitian metric h and a holomorphic Higgs field θ. Let H be the associated C^∞ bundle, so that $E = \operatorname{Ker} d''$, let $D = D' + D''$, with $D'' = d''$, be the Chern connection of h. We say, after [26], that (E, h, θ) is a *harmonic Higgs bundle* (or that h is *Hermite–Einstein* with respect to (E, θ)) if ${}^V\!D := D + \theta + \theta^\dagger$ is an integrable connection on H. The holomorphic bundle $V = \operatorname{Ker}(d'' + \theta^\dagger)$

is then equipped with a flat holomorphic connection ${}^V\nabla$, which is the restriction of ${}^VD' := D' + \theta$ to V.

We say (cf. [7], see also [16, Chap. 7]) that it is *integrable* if there exist two endomorphisms $\mathscr{U}$ and $\mathscr{Q}$ of H such that

$$\mathscr{U} \text{ is holomorphic, i.e., } d''(\mathscr{U}) = 0, \tag{2.1}$$

$$\mathscr{Q}^\dagger = \mathscr{Q}, \tag{2.2}$$

$$[\theta, \mathscr{U}] = 0, \tag{2.3}$$

$$D'(\mathscr{U}) - [\theta, \mathscr{Q}] + \theta = 0, \tag{2.4}$$

$$D'(\mathscr{Q}) + [\theta, \mathscr{U}^\dagger] = 0. \tag{2.5}$$

Remark 2.6. Let us note that $(\mathscr{U} + c\,\mathrm{Id}, \mathscr{Q} + \lambda\,\mathrm{Id})$ satisfy the same equations for any $c \in \mathbb{C}$ and $\lambda \in \mathbb{R}$. One way to fix λ is to impose a compatibility condition with a given supplementary real structure. This would impose that $\mathscr{Q}$ is purely imaginary (cf. [7]). We then denote by $\mathscr{Q}^{\mathrm{Hert}}$ this choice, which is the only one among the $\mathscr{Q} + \lambda\,\mathrm{Id}$, $\lambda \in \mathbb{R}$, to be purely imaginary. With respect to the symmetric nondegenerate bilinear form deduced from the Hermitian metric and the real structure, $\mathscr{Q}^{\mathrm{Hert}}$ is skewsymmetric, hence its characteristic polynomial satisfies $\mathrm{Susy}_{(H,h,\mathscr{U},\mathscr{Q}^{\mathrm{Hert}})}(-T) = (-1)^{\dim H}\,\mathrm{Susy}_{(H,h,\mathscr{U},\mathscr{Q}^{\mathrm{Hert}})}(T)$.

For any $x \in X$, the Hermitian vector space (H_x, h_x) decomposes with respect to the eigenvalues of $\mathscr{Q}_x$. However, these eigenvalues, which are real, may vary with x.

Example 2.7 (Polarized variation of complex Hodge structure)**.** If $\mathscr{U} = 0$ (or $\mathscr{U} = c\,\mathrm{Id}$), then, according to (2.5) and $(2.5)^\dagger$, $D(\mathscr{Q}) = 0$ and, working in a local h-orthonormal frame where $\mathscr{Q}$ is diagonal, this implies that the eigenvalues of $\mathscr{Q}$ are constant. Let H^p denote the eigen subbundle corresponding to the eigenvalue $p \in \mathbb{R}$. Then $D'H^p \subset \Omega^1_X \otimes H^p$, $D''H^p \subset \Omega^1_{\overline{X}} \otimes H^p$ and (2.4) implies $\theta H^p \subset \Omega^1_X \otimes H^{p-1}$. The decreasing filtration (indexed by $\mathbb{R}$) defined by $F^pH = \bigoplus_{p' \geqslant p} H^{p'}$ is stable by ${}^VD''$, hence induces a filtration $F^\bullet V$ of the holomorphic bundle $V := \mathrm{Ker}\,{}^VD''$ by holomorphic subbundles, which satisfies ${}^V\nabla F^pV \subset F^{p-1}V$. Moreover, if we choose a sign $\varepsilon_p \in \{\pm 1\}$ in such a way that, for any $p \in \mathbb{R}$, $\varepsilon_{p+1} = -\varepsilon_p$, the nondegenerate sesquilinear form k defined by the properties that the decomposition $\bigoplus_{p\in\mathbb{R}} H^p$ is k-orthogonal and $k_{|H^p} = \varepsilon_p h_{|H^p}$, is ${}^VD$-flat. We thus recover the standard notion of a polarized variation of complex Hodge structure of weight 0, if we accept filtrations indexed by real numbers, and if we set $H^p = H^{p,-p}$.

2.b. Variations of twistor structures

The notion of an integrable variation of twistor structures (and, more generally, that of an integrable twistor $\mathscr{D}$-module) is a convenient way to handle integrable harmonic Higgs bundles. It was introduced in [27]. The presentation given here follows [16], and the reader can also refer to [11, Chap. 3].

Notation. If X is a complex manifold, $\overline{X}$ will denote the conjugate manifold (with structure sheaf $\overline{\mathscr{O}_X}$), and $X_{\mathbb{R}}$ will denote the underlying real-analytic or C^∞-manifold. We will denote by $\mathbb{P}^1$ the Riemann sphere, covered by the two affine charts $\simeq \mathbb{A}^1$ with coordinate z and $1/z$, and by $p : X \times \mathbb{P}^1 \to X$ the projection.

The coordinate z being fixed, we denote by $\mathbf{S}$ the circle $|z| = 1$, by Ω_0 an open neighbourhood of the closed disc $\Delta_0 := \{|z| \leqslant 1\}$ and by Ω_∞ an open neighbourhood of the closed disc $\Delta_\infty := \{|z| \geqslant 1\}$.

We will denote by $\sigma : \mathbb{P}^1 \to \overline{\mathbb{P}^1}$ the anti-holomorphic involution $z \mapsto -1/\overline{z}$. We assume that $\Omega_\infty = \sigma(\Omega_0)$. We denote by $\iota : \mathbb{P}^1 \to \mathbb{P}^1$ the holomorphic involution $z \mapsto -z$.

It will be convenient to use the notation $\mathscr{X}$ for $X \times \Omega_0$ and $\overline{\mathscr{X}}$ for $\overline{X} \times \Omega_\infty$. Let us introduce the notion of twistor conjugation. Let $\mathscr{H}''$ be a holomorphic vector bundle on $\mathscr{X}$. Then $\overline{\mathscr{H}''}$ is a holomorphic bundle on the conjugate manifold $\overline{\mathscr{X}} := \overline{X} \times \overline{\Omega_0}$ and $\sigma^*\overline{\mathscr{H}''}$ is a holomorphic bundle on $\overline{\mathscr{X}} = \overline{X} \times \Omega_\infty$ (i.e., is an anti-holomorphic family of holomorphic bundles on Ω_∞). We will set $\overline{\mathscr{H}''} := \sigma^*\overline{\mathscr{H}''}$.

By a C^∞ family of holomorphic vector bundles on $\mathbb{P}^1$ parametrized by $X_{\mathbb{R}}$ we will mean the data of a triple $(\mathscr{H}', \mathscr{H}'', \mathscr{C}_{\mathbf{S}})$ consisting of holomorphic vector bundle $\mathscr{H}', \mathscr{H}''$ on $X \times \Omega_0$ and a nondegenerate $\mathscr{O}_{X\times\mathbf{S}} \otimes_{\mathscr{O}_{\mathbf{S}}} \mathscr{O}_{\overline{X}\times\mathbf{S}}$-linear morphism

$$\mathscr{C}_{\mathbf{S}} : \mathscr{H}'_{|\mathbf{S}} \otimes_{\mathscr{O}_{\mathbf{S}}} \overline{\mathscr{H}''_{|\mathbf{S}}} \longrightarrow \mathscr{C}^{\infty,\mathrm{an}}_{X_{\mathbb{R}}\times\mathbf{S}},$$

where $\mathscr{C}^{\infty,\mathrm{an}}_{X_{\mathbb{R}}\times\mathbf{S}}$ is the sheaf of C^∞ functions on $X_{\mathbb{R}} \times \mathbf{S}$ which are real analytic with respect to $z \in \mathbf{S}$. The nondegeneracy condition means that $\mathscr{C}_{\mathbf{S}}$ defines a $C^{\infty,\mathrm{an}}$-gluing between the dual $\mathscr{H}'^\vee$ of $\mathscr{H}'$ and $\overline{\mathscr{H}''}$, giving rise to a $\mathscr{C}^{\infty,\mathrm{an}}_{X_{\mathbb{R}}\times\mathbb{P}^1}$-locally free sheaf of finite rank that we denote by $\widetilde{\mathscr{H}}$.

Variations of twistor structures. By a C^∞ variation of twistor structure on X we mean the data of a triple $(\mathscr{H}', \mathscr{H}'', \mathscr{C}_{\mathbf{S}})$ defining a C^∞ family of holomorphic bundles on $\mathbb{P}^1$ as above, such that each of the holomorphic bundles $\mathscr{H}', \mathscr{H}''$ is equipped with a *relative* holomorphic connection

$$\nabla : \mathscr{H}'('') \longrightarrow \frac{1}{z}\Omega^1_{\mathscr{X}/\Omega_0} \otimes_{\mathscr{O}_{X\times\Omega_0}} \mathscr{H}'('') \tag{2.8}$$

which has a pole along $z = 0$ and is integrable. Moreover, the pairing $\mathscr{C}_{\mathbf{S}}$ has to be compatible (in the usual sense) with the connections, i.e.,

$$d'\mathscr{C}_{\mathbf{S}}(m', \overline{m''}) = \mathscr{C}_{\mathbf{S}}(\nabla m', \overline{m''}) \quad \text{and} \quad d''\mathscr{C}_{\mathbf{S}}(m', \overline{m''}) = \mathscr{C}_{\mathbf{S}}(m', \overline{\nabla m''}).$$

Let us note that we can define $\overline{\nabla}$ as

$$\overline{\nabla} : \overline{\mathscr{H}''} \longrightarrow z\Omega^1_{\overline{\mathscr{X}}/\Omega_\infty} \otimes_{\mathscr{O}_{\overline{\mathscr{X}}}} \overline{\mathscr{H}''}.$$

If we regard $\mathscr{C}_{\mathbf{S}}$ as a $C^{\infty,\mathrm{an}}$-linear isomorphism

$$\mathscr{C}_{\mathbf{S}} : \mathscr{C}^{\infty,\mathrm{an}}_{X_{\mathbb{R}}\times\mathbf{S}} \otimes_{\mathscr{O}_{\overline{X}\times\mathbf{S}}} \overline{\mathscr{H}''_{|\mathbf{S}}} \xrightarrow{\sim} \mathscr{C}^{\infty,\mathrm{an}}_{X_{\mathbb{R}}\times\mathbf{S}} \otimes_{\mathscr{O}_{X\times\mathbf{S}}} \mathscr{H}'^{\vee}_{|\mathbf{S}}, \tag{2.9}$$

the compatibility with ∇ means that $\mathscr{C}_{\mathbf{S}}$ is compatible with the connection $d' + \overline{\nabla}$ on the left-hand side and $\nabla^\vee + d''$ on the right-hand side, where d', d'' are the standard differentials with respect to X only.

- The adjoint $(\mathscr{H}', \mathscr{H}'', \mathscr{C}_{\mathbf{S}})^\dagger$ is defined as $(\mathscr{H}'', \mathscr{H}', \mathscr{C}^\dagger_{\mathbf{S}})$, with

$$\mathscr{C}^\dagger_{\mathbf{S}}(m'', \overline{m'}) := \overline{\mathscr{C}_{\mathbf{S}}(m', \overline{m''})}.$$

  With respect to (2.9), we can write $\mathscr{C}^\dagger_{\mathbf{S}} = \overline{\mathscr{C}_{\mathbf{S}}}^\vee$.
- If $k \in \frac{1}{2}\mathbb{Z}$, the Tate twist (k) is defined by $\mathscr{T}(k) := (\mathscr{H}', \mathscr{H}'', (iz^{-2k})\mathscr{C}_{\mathbf{S}})$.

We say that the variation is

- *Hermitian* if $\mathscr{H}'' = \mathscr{H}'$ and $\mathscr{C}_{\mathbf{S}}$ is "Hermitian", i.e., $\mathscr{C}^\dagger_{\mathbf{S}} = \mathscr{C}_{\mathbf{S}}$,
- *pure of weight* 0 if the restriction to each $x \in X$ defines a *trivial* holomorphic bundle on $\mathbb{P}^1$,
- *polarized and pure of weight* 0 if it is pure of weight 0, Hermitian, and the Hermitian form on the bundle $\overline{H} := p_*\widetilde{\mathscr{H}}$ is positive definite, i.e., is a Hermitian metric.

Lemma 2.10 (C. Simpson [27]). *We have an equivalence between variations of polarized pure twistor structures of weight* 0 *and harmonic Higgs bundles, by taking* $\mathbb{P}^1$*-global sections.*

Remark 2.11. There is a natural notion of a (polarized) variation of twistor structure of weight $w \in \mathbb{Z}$ (cf. [27]). We say that $\mathscr{T} = (\mathscr{H}', \mathscr{H}'', \mathscr{C}_{\mathbf{S}})$ is pure of weight w if the restriction to each $x \in X$ defines a bundle isomorphic to $\mathscr{O}_{\mathbb{P}^1}(w)^d$. A Hermitian duality is an isomorphism $\mathscr{S} = (S', S'') : \mathscr{T} \to \mathscr{T}^\dagger(-w)$. We say that $\mathscr{S}$ is a polarization if the Tate twisted object $(\mathscr{T}, \mathscr{S})(w/2)$ is polarized (cf. [16] for details, see also [11]).

2.c. Integrable variations of twistor structures

A variation of twistor structure $(\mathscr{H}',\mathscr{H}'',\mathscr{C}_{\mathbf{S}})$ is *integrable* if the relative connection ∇ on $\mathscr{H}',\mathscr{H}''$ comes from an absolute connection also denoted by ∇, which has Poincaré rank one (cf. [16, Chap. 7]). In other words, $z\nabla$ should be an integrable meromorphic z-connection on $\mathscr{H}',\mathscr{H}''$ with a logarithmic pole along $z=0$. We also ask for a supplementary compatibility property of the absolute connection with the pairing in the following way:

$$z\frac{\partial}{\partial z}\mathscr{C}_{\mathbf{S}}(m',\overline{m''}) = \mathscr{C}_{\mathbf{S}}(z\nabla_{\partial_z}m',\overline{m''}) - \mathscr{C}_{\mathbf{S}}(m',\overline{z\nabla_{\partial_z}m''}).$$

[Here, we regard $\mathscr{C}^{\infty,\mathrm{an}}_{X_{\mathbb{R}}\times\mathbf{S}}$ as the sheaf of germs along $X\times\mathbf{S}$ of C^∞ functions which are holomorphic with respect to z; when considering it as the sheaf of C^∞ functions which are real analytic with respect to $z\in\mathbf{S}$, one should replace the operator $z\frac{\partial}{\partial z}$ with $z\frac{\partial}{\partial z}-\overline{z}\frac{\partial}{\partial\overline{z}}$.]

Lemma 2.10 can be extended to integrable variations:

Lemma 2.12 (C. Hertling [7], cf. also [16, Cor. 7.2.6]). *The equivalence of Lemma* 2.10 *specializes to an equivalence between integrable variations of pure polarized twistor structures of weight* 0 *and integrable harmonic Higgs bundles, i.e., harmonic Higgs bundles equipped with endomorphisms* $\mathscr{U},\mathscr{Q}$ *satisfying* (2.1)–(2.5).

Let us indicate one direction of the correspondence. Starting from $(H,h,\theta,\mathscr{U},\mathscr{Q})$, we construct an integrable variation of Hermitian twistor structures $(\mathscr{H}',\mathscr{H}'',\mathscr{C}_{\mathbf{S}})$ by setting $\mathscr{H}=p^*H$ on $\mathscr{X}$, with the d''-operator $D''_{\mathscr{H}}:=D''+z\theta^\dagger$ and we set $\mathscr{H}'=\operatorname{Ker}D''_{\mathscr{H}}=\mathscr{H}''$. The relative connection (2.8) is defined as the restriction of $D'_{\mathscr{H}}:=D'+z^{-1}\theta$ to $\mathscr{H}'$. The absolute connection is obtained by adding to the relative connection ∇ the connection in the z-variable $d'_z+(z^{-2}\mathscr{U}-z^{-1}\mathscr{Q}-\mathscr{U}^\dagger)dz$ (cf. [7] or [16, §7.2.c] for more details).

Remark 2.13. Given an integrable variation of twistor structure $\mathscr{T}=(\mathscr{H}',\mathscr{H}'',\mathscr{C}_{\mathbf{S}})$, we will say that the structure obtained by changing the action of $z^2\nabla_{\partial_z}$ on $\mathscr{H}'$ to $z^2\nabla_{\partial_z}-\lambda z$ and that on $\mathscr{H}''$ to $z^2\nabla_{\partial_z}-\overline{\lambda}z$ ($\lambda\in\mathbb{C}$) is *equivalent* to the previous one. If the variation of twistor structure is Hermitian, the equivalent structure is still compatible with $\mathscr{S}$ iff $\lambda\in\mathbb{R}$.

Remark 2.14 (Tate twist and integrability). The effect of the Tate twist (k) (with $k\in\frac{1}{2}\mathbb{Z}$) on the action of $z^2\partial_z$ is a shift between $\mathscr{H}'$ and $\mathscr{H}''$ by $2kz$. In this article, it will be convenient to choose a nonsymmetric shift, namely, the action on $\mathscr{H}''$ is unchanged, and that on $\mathscr{H}'$ is changed into $z^2\partial_z-2kz$.

Remark 2.15. In the previous correspondence, we identify the bundle E on X with the restriction $\mathscr{H}''/z\mathscr{H}''$. Then $\mathscr{U}$ is induced by the action of $z^2\nabla_{\partial_z}$. If $\mathscr{U} = 0$ (the case of a variation of Hodge structure), then $\mathscr{H}''$ is stable by $z\nabla_{\partial_z}$, and the characteristic polynomial of $\mathscr{Q}$ is equal to that of the restriction of $-z\nabla_{\partial_z}$ to E.

2.d. The 'new supersymmetric index'

We assume in this paragraph that X is a point, so we work with twistor structures.

Definition 2.16 (cf. [1] and [7]). Let $\mathscr{T} = (\mathscr{H}', \mathscr{H}', \mathscr{C}_{\mathbf{S}})$ be an integrable polarized pure twistor structure of weight 0 with polarization $\mathscr{S} = (\mathrm{Id}, \mathrm{Id})$. Let $\mathscr{U}, \mathscr{Q}$ be the associated endomorphisms of the corresponding finite dimensional complex vector space with positive definite Hermitian form. The endomorphism $\mathscr{Q}$ is called the 'new supersymmetric index' attached to $\mathscr{T}$. We denote by $\mathrm{Susy}_{\mathscr{T}}(T)$ its characteristic polynomial.

It will be convenient to extend to any weight the previous definition.

Definition 2.17. Let $\mathscr{T} = (\mathscr{H}', \mathscr{H}'', \mathscr{C}_{\mathbf{S}})$ be an integrable pure twistor structure of weight w with polarization $\mathscr{S}$. We set

$$\mathrm{Susy}_{\mathscr{T}}(T) := \mathrm{Susy}_{\mathscr{T}(w/2)}(T).$$

Similarly, we define the spectral polynomials:

Definition 2.18. Let $\mathscr{T} = (\mathscr{H}', \mathscr{H}'', \mathscr{C}_{\mathbf{S}})$ be an integrable twistor structure (not necessarily pure or polarized). We define (if the "no ramification" condition is fulfilled, for SP^0):

$$\mathrm{SP}^{\infty}_{\mathscr{T}}(T) = \mathrm{SP}^{\infty}_{\mathscr{H}''}(T) \quad \text{and} \quad \mathrm{SP}^{0}_{\mathscr{T}}(T) = \mathrm{SP}^{0}_{\mathscr{H}''}(T).$$

According to Definition 2.17 and to Remark 2.14, we have, for any $k \in \frac{1}{2}\mathbb{Z}$,

$$\mathrm{Susy}_{\mathscr{T}(k)} = \mathrm{Susy}_{\mathscr{T}}, \quad \mathrm{SP}^{\infty}_{\mathscr{T}(k)} = \mathrm{SP}^{\infty}_{\mathscr{T}}, \quad \mathrm{SP}^{0}_{\mathscr{T}(k)} = \mathrm{SP}^{0}_{\mathscr{T}}. \tag{2.19}$$

2.e. Twistor $\mathscr{D}$-modules

In order to allow singularities in variations of twistor structure, we have introduced in [16] the notion of polarizable twistor $\mathscr{D}$-module (see also [11] for an extension of this notion with parabolic weights). We will briefly recall this notion.

We first introduce the sheaf $\mathscr{R}_{\mathscr{X}}$ of differential operators, locally isomorphic to $\mathscr{O}_{\mathscr{X}}\langle \eth_{x_1}, \dots, \eth_{x_n}\rangle$, by setting $\eth_{x_i} = z\partial_{x_i}$. A left $\mathscr{R}_{\mathscr{X}}$-module is nothing else but a $\mathscr{O}_{\mathscr{X}}$-module with a flat z-connection.

The category $\mathscr{R}$- Triples(X) consists of triples $(\mathscr{M}', \mathscr{M}'', \mathscr{C}_{\mathbf{S}})$, where $\mathscr{M}', \mathscr{M}''$ are left $\mathscr{R}_{\mathscr{X}}$-modules and $\mathscr{C}_{\mathbf{S}} : \mathscr{M}'_{|\mathbf{S}} \otimes_{\mathscr{O}_{|\mathbf{S}}} \overline{\mathscr{M}''} \to \mathfrak{Db}_{X_{\mathbb{R}}\times\mathbf{S}/\mathbf{S}}$ is a $\mathscr{R}_{\mathscr{X}|\mathbf{S}} \otimes_{\mathscr{O}_{|\mathbf{S}}} \overline{\mathscr{R}_{\mathscr{X}|\mathbf{S}}}$ linear morphism with values in the sheaf $\mathfrak{Db}_{X_{\mathbb{R}}\times\mathbf{S}/\mathbf{S}}$ of distributions on $X \times \mathbf{S}$ which are continuous with respect to $z \in \mathbf{S}$. There is a natural notion of morphism. This category has Tate twists by $\frac{1}{2}\mathbb{Z}$, and a notion of adjunction (cf. [16, §1.6]). Restricting $\mathscr{M}'$ or $\mathscr{M}''$ to $z = 1$ gives left $\mathscr{D}_X$-modules, while restricting them to $z = 0$ gives $\mathscr{O}_X$-modules with a Higgs field.

There is a notion of direct image, hence of de Rham cohomology when taking the direct image by the constant map.

Supplementary properties are introduced in order to define the notion of polarized twistor $\mathscr{D}$-module of some weight. We will not recall them here and refer to [16] for further details.

2.f. Specialization and integrability (the tame case)

In the remaining part of Section 2, we assume that X is a disc with coordinate x and we denote by $j : X^* \hookrightarrow X$ the inclusion of the punctured disc $X \smallsetminus \{0\}$ in X.

Let $\mathscr{T} = (\mathscr{M}', \mathscr{M}'', \mathscr{C}_{\mathbf{S}})$ be a regular twistor $\mathscr{D}$-module of weight w on X polarized by $\mathscr{S} = (S' = (-1)^w S'', S'')$ (cf. [16]) with only singularity at $x = 0$, so that $\mathscr{T}_{|X^*}$ is a polarized variation of twistor structures of weight w, which has a tame behaviour near the singularity, in the sense of [25] (cf. [16, 11]).

For any $\beta \in \mathbb{C}$ with $\operatorname{Re}\beta \in (-1, 0]$, the nearby cycle functor[1] Ψ_x^β sends such a triple $\mathscr{T}$ to a triple $\Psi_x^\beta \mathscr{T} \in \mathscr{R}$- Triples$(\{0\})$, equipped with a morphism $\mathscr{N} : \Psi_x^\beta \mathscr{T} \in \mathscr{R}$- Triples$(\{0\}) \to \Psi_x^\beta \mathscr{T}(-1) \in \mathscr{R}$- Triples$(\{0\})$. If $\mathrm{M}_\bullet$ denotes the monodromy filtration, then the graded object $\mathrm{gr}^{\mathrm{M}}_\bullet \Psi_x^\beta \mathscr{T}$, equipped with the morphism $\mathrm{gr}^{\mathrm{M}}_{-2} \mathscr{N}$, is a graded Lefschetz twistor $\mathscr{D}$-module of weight w and type $\varepsilon = -1$. Moreover, $\Psi_x^\beta \mathscr{S}$ induces, by grading, a polarization of this object.

We have a similar result for vanishing cycles: $(\mathrm{gr}^{\mathrm{M}}_\bullet \phi_x^{-1} \mathscr{T}, \mathrm{gr}^{\mathrm{M}}_{-2} \mathscr{N})$ is an object of $\mathrm{MLT}^{(\mathrm{r})}(X, w; -1)$ and $\phi_x^{-1} \mathscr{S}$ induces, by grading, a polarization (this follows from [16, Cor. 4.1.17]).

Let us moreover assume that $(\mathscr{T}, \mathscr{S})$ is integrable (cf. [16, Chap. 7], where one should modify (7.1.2) by using the operator $z\partial/\partial z - \overline{z}\partial/\partial\overline{z}$ (standard conjugation) on the left-hand side, as (7.1.2) was mistakenly written there for distributions which are holomorphic with respect to z), so that, in particular, the corresponding variation is integrable on X^*.

[1] In [16] it is defined with the increasing convention for the V-filtration. Here we use the decreasing one. The correspondence is $\Psi_x^\beta = \Psi_{x,\alpha}$ with $\beta = -\alpha - 1$.

Then from loc. cit. we know that $\Psi_x^\beta \mathscr{M} \neq 0$ only if β is real, so that $\beta \star z = \beta z$ above, there is no difference between the notations $\Psi_x^\beta \mathscr{T}$ and $\psi_x^\beta \mathscr{T}$ of loc. cit., and moreover, using the induced action of $z^2\partial_z$, $\Psi_x^\beta \mathscr{T}$ remains integrable. Going then to $\mathrm{gr}_\ell^{\mathrm{M}} \Psi_x^\beta \mathscr{T}$, we get an integrable polarized twistor structure of weight $w+\ell$ (cf. [16, Lemma 7.3.8]). The action of $z^2\partial_z$ on $\mathrm{gr}_\ell^{\mathrm{M}} \Psi_x^\beta \mathscr{M}''$ is the action naturally induced by $z^2\partial_z$ on $\mathscr{M}''$. A similar result holds for $\mathrm{gr}_\ell^{\mathrm{M}} \phi_x^{-1} \mathscr{T}$.

2.g. Specialization and integrability (the wild case)

We keep the notation of §2.f, but we will consider the more general case of a polarized wild twistor $\mathscr{D}$-module $\mathscr{T} = (\mathscr{M}', \mathscr{M}'', \mathscr{C}_{\mathrm{S}})$ of weight w with polarization $\mathscr{S}$, for which we refer to [20, 12]. We will also assume that the "no ramification" condition is fulfilled, that is, we assume that [20, Prop. 4.5.4] holds with ramification index q equal to one. Therefore, setting $\mathscr{M} = \mathscr{M}'$ or $\mathscr{M}''$, we have a formal decomposition

$$\text{(DEC}^\wedge\text{)} \qquad \widetilde{\mathscr{M}}^\wedge \xrightarrow{\sim} \bigoplus_i (\widetilde{\mathscr{R}}_i^\wedge \otimes \mathscr{E}^{\varphi_i/z}), \qquad \varphi_j \in x^{-1}\mathbb{C}[x^{-1}].$$

Let us also notice that, when we restrict to $z=0$, the decomposition holds at the level of $\widetilde{\mathscr{M}/z\mathscr{M}}$ (cf. [20, Rem. 4.5.5]).

For any $\varphi \in x^{-1}\mathbb{C}[x^{-1}]$ and any $\beta \in \mathbb{C}$ with $\operatorname{Re}\beta \in (-1,0]$, we set $\Psi_x^{\varphi,\beta} \widetilde{\mathscr{M}} := \Psi_x^\beta(\widetilde{\mathscr{M}} \otimes \mathscr{E}^{-\varphi/z})$. We can then define the objects $\Psi_x^{\varphi,\beta}\widetilde{\mathscr{T}}$, equipped with $\mathscr{N} : \Psi_x^{\varphi,\beta}\widetilde{\mathscr{T}} \to \Psi_x^{\varphi,\beta}\widetilde{\mathscr{T}}(-1)$. The condition of being a polarized wild twistor $\mathscr{D}$-module of weight w at $x=0$ means that, for all φ, β as above, $(\mathrm{gr}_\bullet^{\mathrm{M}} \Psi_x^{\varphi,\beta}\widetilde{\mathscr{T}}, \mathrm{gr}_{-2}^{\mathrm{M}} \mathscr{N})$, equipped with the naturally induced sesquilinear duality, is a graded Lefschetz twistor structure of weight w, in the sense of [16, §2.1.e]. As a consequence, if $\varphi = 0$, the vanishing cycles $(\mathrm{gr}_\bullet^{\mathrm{M}} \phi_x^{0,-1}\widetilde{\mathscr{T}}, \mathrm{gr}_{-2}^{\mathrm{M}} \mathscr{N})$ are of the same kind.

For any $\varphi \in x^{-1}\mathbb{C}[x^{-1}]$, the exponentially twisted $\mathscr{R}_{\mathscr{X}}[x^{-1}]$-module $\widetilde{\mathscr{M}} \otimes \mathscr{E}^{-\varphi/z}$ remains integrable, if $\widetilde{\mathscr{M}}$ is so, hence, according to [16, Prop. 7.3.1], so are the modules $\Psi_x^{\varphi,\beta}\widetilde{\mathscr{M}}$ ($\beta \in (-1,0]$) and $\phi_x^{0,-1}\widetilde{\mathscr{M}}$. Moreover, the formal module $\widetilde{\mathscr{M}}^\wedge$ is clearly integrable.

Lemma 2.20. *Each* $\widetilde{\mathscr{R}}_i^\wedge$ *entering in the decomposition* (DEC$^\wedge$) *is integrable.*

Proof. Firstly, the irregular part $\widetilde{\mathscr{M}}_{\mathrm{irr}}^\wedge$ of $\widetilde{\mathscr{M}}^\wedge$ (i.e., corresponding in (DEC$^\wedge$) to the sum over the nonzero φ_i) remains integrable, as, near any $z_o \in \Omega_0$, it can be realized as the intersection $\bigcap_k V_{(z_o)}^k \widetilde{\mathscr{M}}^\wedge$, where $V^\bullet \widetilde{\mathscr{M}}^\wedge$ is the V-filtration of $\widetilde{\mathscr{M}}^\wedge$ (defined near z_o), and we know that each step of the V-filtration is integrable ([16, Prop. 7.4.1]).

Let $i_0 \in I$ be such that $\varphi_{i_0} = 0$. We claim that $\widetilde{\mathscr{R}}_{i_0}^{\wedge}$ is integrable: because of the previous remark applied to $\widetilde{\mathscr{M}}^{\wedge} \otimes \mathscr{E}^{-\varphi_i/z}$ for $i \neq i_0$, $(z^2\partial_z + \varphi_i)\widetilde{\mathscr{R}}_{i_0}^{\wedge}$—hence $z^2\partial_z\widetilde{\mathscr{R}}_{i_0}^{\wedge}$—has no component on the regular part of $\widetilde{\mathscr{M}}^{\wedge} \otimes \mathscr{E}^{-\varphi_i/z}$, that is, on $\widetilde{\mathscr{R}}_i^{\wedge}$; thus $\widetilde{\mathscr{R}}_{i_0}^{\wedge}$ is stable by $z^2\partial_z$. The same result applies to any $\widetilde{\mathscr{R}}_i^{\wedge}$, by globally twisting $\widetilde{\mathscr{M}}^{\wedge}$ by $\mathscr{E}^{-\varphi_i/z}$, hence the lemma. Q.E.D.

By assumption on $\widetilde{\mathscr{M}}$, each $\widetilde{\mathscr{R}}_i^{\wedge}$ is strictly specializable. Applying [16, Lemma 7.3.7], we find that $\Psi_x^{\varphi,\beta}\widetilde{\mathscr{M}} \neq 0 \Rightarrow \beta \in \mathbb{R}$.

§3. Specialization of the new supersymmetric index

In this section, X denotes a disc with coordinate x and j denotes the inclusion of the punctured disc $X^* := X \smallsetminus \{0\}$ into X.

3.a. The tame case

In this subsection, we keep the setting of §2.f and we assume that $(\mathscr{T}, \mathscr{S})$ is integrable.

Theorem 3.1. *We have the following correspondence between* Susy *polynomials:*

$$(3.1)(*) \qquad \lim_{x\to 0} \operatorname{Susy}_{\mathscr{T}_x}(T) = \prod_{\beta\in(-1,0]} \prod_{\ell\geqslant 0} \operatorname{Susy}_{\operatorname{gr}_\ell^{\mathrm{M}} \Psi_x^\beta \mathscr{T}}(T).$$

Remark 3.2. If $\mathscr{T}_{|X^*}$ consists of a polarized variation of Hodge structures of weight w, then $\operatorname{Susy}_{\mathscr{T}_x}(T)$ is constant (cf. Lemma 5.4 below). In general, however, the eigenvalues of $\mathscr{Q}_x$ do vary (see the example in [7, (7.115)] for instance).

Proof of Theorem 3.1. For simplicity, we will assume $w = 0$ (this can be obtained by a Tate twist $(w/2)$) and that $\mathscr{M}' = \mathscr{M}''$ and $\mathscr{S} = (\mathrm{Id}, \mathrm{Id})$. Let us fix $\beta \in (-1, 0]$. By assumption, $(\operatorname{gr}_\bullet^{\mathrm{M}} \Psi_x^\beta \mathscr{T}, \mathscr{N})$ is a graded Lefschetz twistor structure which is polarized and of weight 0 (cf. [16, §2.1.e]). It thus corresponds to a Hermitian vector space H with a $\mathrm{SL}_2(\mathbb{R})$-action (cf. [16, Rem. 2.1.15]), hence a graded vector space $H = \bigoplus_\ell H_\ell$ with a nilpotent endomorphism of degree -2. We denote the standard action of the generators of $\mathfrak{sl}_2(\mathbb{R})$ by $\mathrm{X}, \mathrm{Y}, \mathrm{H}$, so that H_ℓ is the eigenspace of H for the eigenvalue ℓ. Then, for $\ell \geqslant 0$, a basis $\boldsymbol{e}^o_{\beta,\ell,\ell}$ of the primitive subspace $\mathrm{P}H_\ell$ defines a global frame of $\mathrm{P}_\ell^\beta \mathscr{M} := \mathrm{P}\operatorname{gr}_\bullet^{\mathrm{M}} \Psi_x^\beta \mathscr{M}$, which is orthonormal for $\mathrm{P}_\ell^\beta \mathscr{C}_{\mathrm{S}}$ (the sesquilinear form of $\mathrm{P}\operatorname{gr}_\bullet^{\mathrm{M}} \Psi_x^\beta \mathscr{T}$) and in which the matrix of $z^2\partial_z$ takes the form ${}^{\mathrm{p}}\mathscr{U}_{\beta,\ell} - {}^{\mathrm{p}}\mathscr{Q}_{\beta,\ell} z - {}^{\mathrm{p}}\mathscr{U}_{\beta,\ell}^\dagger z^2$.

We can assume that ${}^p\mathscr{Q}_{\beta,\ell}$ is diagonal, being selfadjoint with respect to the positive definite Hermitian form on $\mathrm{P}H_\ell$.

The construction done in [16, §5.4.c] extends this family of frames first to a frame $\boldsymbol{e}^o_{\beta,\ell,k}$ ($\ell \in \mathbb{N}$, $k = 0, \dots, \ell$) of $\mathrm{gr}^{\mathrm{M}}_{\ell-2k} \Psi^\beta_x \mathscr{M}$ and then to a local frame $\boldsymbol{e}_\beta$ of $V^\beta \mathscr{M}$ (the local construction near each z_o done in loc. cit. is not needed here as the V-filtration is globally defined with respect to z, cf. [16, Rem. 3.3.6(2)]).

The action of $z^2\partial_z$ leaves the V-filtration invariant (cf. [16, Prop. 7.3.1]), as well as the lift of the M-filtration on each V^β (cf. [16, Lemma 7.3.8]). Therefore, the matrix B of $z^2\partial_z$ in the frame $\boldsymbol{e}$, which is holomorphic, is "triangular" up to powers of x with respect to $\mathrm{M}_\bullet V^\bullet$, i.e., can be written as

$$B = \bigoplus_\beta \Big[B_{\beta,\beta,0} \oplus B_{\beta,\beta,\leqslant -1} \oplus \bigoplus_{\beta' \neq \beta} B_{\beta',\beta} \Big] \tag{3.3}$$

with $B_{\beta',\beta}/x$ holomorphic if $\beta' < \beta$, and where the index j in $B_{\beta,\beta,j}$ denotes the weight with respect to H, so that $[\mathrm{H}, B_{\beta,\beta,j}] = jB_{\beta,\beta,j}$. Moreover, the matrix $B_{\beta,\beta,0}$ can be written as $\mathscr{U}_{\beta,\beta,0} - \mathscr{Q}_{\beta,\beta,0} z - \mathscr{U}^\dagger_{\beta,\beta,0} z^2$, and is block-diagonal with respect to the previous decomposition (ℓ, k) of the frame $\boldsymbol{e}_\beta$, and the diagonal (ℓ,k)-block of $\mathscr{Q}_{\beta,\beta,0}$ is ${}^p\mathscr{Q}_{\beta,\ell} + (-k+\ell/2)\,\mathrm{Id}$. In particular, the characteristic polynomial of $\bigoplus_{\beta\in(-1,0]} \mathscr{Q}_{\beta,\beta,0}$ is the right-hand side in (3.1)(∗).

Let us denote by $A(x,z)$ the matrix $\bigoplus_{\beta\in(-1,0]} |x|^\beta \mathrm{L}(x)^{\mathrm{H}/2}$, with $\mathrm{L}(x) := |\log|x|^2|$. By [16, Lemma 5.4.7],[2] there exists on $X^* \times \Omega_0$ (up to shrinking X and Ω_0, as defined in §1.d) a matrix $S(x,z)$ with $\lim_{x\to 0} S(x,z) = 0$ uniformly with respect to z, such that the frame

$$\boldsymbol{\varepsilon} := \boldsymbol{e} \cdot A(x,z)^{-1}(\mathrm{Id} + S(x,z))$$

is an orthonormal frame for $\mathscr{C}_{\mathbf{S}}$. The matrix of $z^2\partial_z$ in this frame will enable us to compute the left-hand side in the theorem.

This matrix is equal to

$$(\mathrm{Id} + S)^{-1} ABA^{-1} (\mathrm{Id} + S) + (\mathrm{Id} + S)^{-1} A z^2 \partial_z [A^{-1}(\mathrm{Id} + S)].$$

The second term is a multiple of z^2 and will not contribute to $\mathscr{Q}_x$.

Let us note that the block $(ABA^{-1})_{\beta',\beta}$ ($\beta' \neq \beta$) is equal to

$$|x|^{\beta'-\beta} \mathrm{L}(x)^{\mathrm{H}/2} B_{\beta',\beta} \mathrm{L}(x)^{-\mathrm{H}/2}, \tag{3.4}$$

[2]In loc. cit., the matrix A is multiplied by e^{-zX}; this is in fact not needed in the argument.

and tends to 0 when $x \to 0$ (since, when $\beta' < \beta$ and $\beta, \beta' \in (-1, 0]$, $B_{\beta',\beta}/x$ is locally bounded and $1 + \beta' - \beta > 0$). Then so does its conjugate by $\mathrm{Id} + S$.

A similar reasoning can be done for $A_\beta B_{\beta,\beta,\leqslant -1} A_\beta^{-1}$, which gives a decay at least like $\mathrm{L}(x)^{-1/2}$ when $x \to 0$.

Now, $A_\beta B_{\beta,\beta,0} A_\beta^{-1} = B_{\beta,\beta,0}$, and the coefficient of $-z$ is $\mathscr{Q}_{\beta,\beta,0}$, which is thus equal to $\lim_{x\to 0} \mathscr{Q}_x$. This gives the conclusion. Q.E.D.

3.b. The wild case

We now consider the setting of §2.g. We then have the following generalization of Theorem 3.1:

Theorem 3.5. *Let $\mathscr{T} = (\mathscr{M}', \mathscr{M}'', \mathscr{C}_{\mathrm{S}})$ be a wild twistor $\mathscr{D}$-module of weight w polarized by $\mathscr{S}$, satisfying the "no ramification" condition. We have the following correspondence between* Susy *polynomials:*

$$(3.5)(*) \quad \lim_{x\to 0} \mathrm{Susy}_{\mathscr{T}_x}(T) = \prod_{\varphi \in x^{-1}\mathbb{C}[x^{-1}]} \prod_{\beta \in (-1,0]} \prod_{\ell \geqslant 0} \mathrm{Susy}_{\mathrm{gr}_\ell^{\mathrm{M}} \Psi_x^{\varphi,\beta} \mathscr{T}}(T).$$

Proof. As in the proof of Theorem 3.1, we will assume $w = 0$, $\mathscr{M}' = \mathscr{M}''$ and $\mathscr{S} = (\mathrm{Id}, \mathrm{Id})$. We will make an extensive use of [20, §§5.2 & 5.4]. As in the tame case, we start with a frame $\boldsymbol{e}^o_{\varphi,\beta,\ell}$ of $\mathrm{P}_\ell^{\varphi,\beta} \mathscr{M} := \mathrm{P}\,\mathrm{gr}_\ell^{\mathrm{M}} \Psi_x^{\varphi,\beta} \mathscr{M}$ which is orthonormal for $\mathrm{P}_\ell^{\varphi,\beta} \mathscr{C}_{\mathrm{S}}$, for any φ, β, ℓ. The matrix of $z^2 \partial_z$ in this frame takes the form ${}^{\mathrm{p}}\mathscr{U}_{\varphi,\beta,\ell} - {}^{\mathrm{p}}\mathscr{Q}_{\varphi,\beta,\ell} z - {}^{\mathrm{p}}\mathscr{U}^\dagger_{\varphi,\beta,\ell} z^2$. The constructions of loc. cit. produce a frame $\widetilde{\boldsymbol{\varepsilon}}$ of $\widetilde{\mathscr{M}}_{|X^*}$ which is orthonormal with respect to $\mathscr{C}_{\mathrm{S}}$ (cf. [20, Cor. 5.4.3]), and we wish to compute the matrix of $z^2 \partial_z$ in this frame. Let us recall the steps going from $\boldsymbol{e}$ to $\widetilde{\boldsymbol{\varepsilon}}$.

(1) We first lift, exactly as in the tame case, each of the frames $\boldsymbol{e}^o_{\varphi_i,\beta}$ to a frame $(\widehat{\mathbf{e}}_{\varphi_i,\beta})_\beta$ of $\widetilde{\mathscr{R}}_i^\wedge$. Arguing as in the tame case, the matrix $\widehat{B}$ of $z^2 \partial_z$ in the frame $\widehat{\mathbf{e}}$ takes the form $\bigoplus_i \widehat{B}_{ii}$, where $\widehat{B}_{ii}$ decomposes as in (3.3). As remarked in §2.g (after (DEC$^\wedge$)), we can assume that, when restricted to $z = 0$, the frame $\widehat{\mathbf{e}}_{|z=0}$ is a frame of $\widetilde{\mathscr{M}}/z\widetilde{\mathscr{M}}$, and is compatible with the corresponding φ-decomposition, so $\widehat{B}_{ii}(x, 0)$ is convergent.

(2) We then work locally with respect to z_o and in small sectors in the variable x. Let $\pi : Y \to X$ be the real oriented blow up of X at the origin, with $S^1 = \pi^{-1}(0)$, and let us set $\mathscr{Y} = Y \times \Omega_0$. Let us denote by $\mathscr{A}_{\mathscr{Y}}$ the sheaf of C^∞ functions on $\mathscr{Y}$ which are holomorphic with respect to z and holomorphic on $X^* \times \Omega_0$. We then lift the frame $\widehat{\mathbf{e}}$ to a $\mathscr{A}_{\mathscr{Y},\xi_o,z_o}$-frame, for any $\xi_o \in S^1$ and $z_o \in \Omega_0$, and we get frames

${}^{\mathscr{A}}\boldsymbol{e}^{(\xi_o,z_o)} = ({}^{\mathscr{A}}\boldsymbol{e}_{\varphi_i}^{(\xi_o,z_o)})_i$. We can assume that, when restricted to $z=0$, the frame ${}^{\mathscr{A}}\boldsymbol{e}_{|z=0}^{(\xi_o,0)}$ comes from a frame of $\widetilde{\mathscr{M}}/z\widetilde{\mathscr{M}}$ compatible with the φ-decomposition. The matrix ${}^{\mathscr{A}}B^{(\xi_o,z_o)}$ of $z^2\partial_z$ satisfies the following properties (according to [20, Lemma 5.2.6]):

(a) if $i,j \in I$ are distinct, the term ${}^{\mathscr{A}}B_{ij}^{(\xi_o,z_o)}$ is infinitely flat along $S^1\times\Omega_0$ in a neighbourhood of (ξ_o,z_o) and, for $z_o=0$, ${}^{\mathscr{A}}B_{ij}^{(\xi_o,0)}(x,0)\equiv 0$,

(b) for any $i\in I$, the term ${}^{\mathscr{A}}B_{ii}^{(\xi_o,z_o)}$ has an asymptotic expansion equal to $\widehat{B}_{ii}$ when $x\to 0$ near the direction ξ_o, uniformly with respect to $z\in\mathrm{nb}(z_o)$ and, for $z_o=0$, ${}^{\mathscr{A}}B_{ii}^{(\xi_o,0)}(x,0)$ does not depend on ξ_o and is holomorphic with respect to x (it takes the form (3.3) at $z=0$).

It is then clear (after the tame case) that the limit, when $x\to 0$ in the neighbourhood of the direction ξ_o and $z\in\mathrm{nb}(z_o)$, of the characteristic polynomial of the coefficient of $-z$ in ${}^{\mathscr{A}}B^{(\xi_o,z_o)}$ is equal to the RHS in (3.5)(∗).

(3) We now define the local untwisted C^∞ frame $\boldsymbol{\varepsilon}^{(\xi_o,z_o)} = {}^{\mathscr{A}}\boldsymbol{e}^{(\xi_o,z_o)}\cdot A^{-1}(x,z)$, where $A=\bigoplus_i A_{ii}$ and each A_{ii} is as in the tame case. Let ${}^{\infty}B^{(\xi_o,z_o)}$ be the matrix of $z^2\partial_z$ in this frame. The non-diagonal blocks ${}^{\infty}B_{ij}^{(\xi_o,z_o)}$ for $i\neq j$ remain infinitely flat when $x\to 0$ in the direction ξ_o, as A and A^{-1} have moderate growth. Moreover, the z-constant term ${}^{\infty}B^{(\xi_o,0)}(x,0)$ of ${}^{\infty}B^{(\xi_o,0)}(x,z)$ still satisfies ${}^{\infty}B^{(\xi_o,0)}(x,0)_{ij}=0$ if $i\neq j$, as A is diagonal with respect to the φ-decomposition. Moreover, as in the tame case, ${}^{\infty}B^{(\xi_o,0)}(x,0)_{ii}(x,0)$ has a limit when $x\to 0$. Then the same argument as in the tame case shows that the limit of the characteristic polynomial of the coefficient of $-z$ in ${}^{\infty}B^{(\xi_o,z_o)}$ is the same as for ${}^{\mathscr{A}}B^{(\xi_o,z_o)}$, hence is equal to the RHS in (3.5)(∗).

(4) We globalize the construction, by using a partition of unity with respect to ξ_o and by using the argument of [16, Lemma 5.4.6] (cf. [20, Lemma 5.2.11]), to get a frame $\boldsymbol{\varepsilon}$. The base change from any $\boldsymbol{\varepsilon}^{(\xi_o,z_o)}$ to $\boldsymbol{\varepsilon}$ takes the form $\mathrm{Id}+R^{(\xi_o,z_o)}(x,z)$, with $R^{(\xi_o,z_o)}$ satisfying $\lim_{x\to 0}\mathrm{L}(x)^\delta R^{(\xi_o,z_o)}=0$ uniformly with respect to $z\in\mathrm{nb}(z_o)$, for some $\delta>0$. We also note that we can achieve $R^{(\xi_o,0)}(x,0)\equiv 0$ in the base change, as the frame ${}^{\mathscr{A}}\boldsymbol{e}_{|z=0}^{(\xi_o,0)}$ is already globally defined with respect to ξ, and so does the frame $\boldsymbol{\varepsilon}_{|z=0}^{(\xi_o,0)}$; moreover, the argument of [16, Lemma 5.4.6] gives a contribution equal to Id at $z_o=0$ for the base change. Therefore, the conclusion of (3) holds for the matrix ${}^{\infty}B$ of $z^2\partial_z$ in the frame $\boldsymbol{\varepsilon}$.

(5) Now, the base change from ε to $\widetilde{\varepsilon}$ given by [20, Prop. 5.4.1 and (5.3.2)] takes the form

$$\widetilde{\varepsilon} = \varepsilon \cdot (\mathrm{Id} + S'(x,z))^{-1}(\mathrm{Id} + U_0(x))^{-1} \operatorname{diag}(e^{z\overline{\varphi_i}} \mathrm{Id}),$$

where $S'(x,z)$ is continuous and holomorphic with respect to z on $X^* \times \mathrm{nb}(\{z \leqslant 1\})$, and satisfies $S'(x,0) \equiv 0$, and $U_0(x)$ is continuous with respect to $x \in X$, $U_0(0) = 0$, and U_0 is diagonal with respect to the φ-decomposition. As we are only interested in the coefficient of $-z$ in the matrix $\widetilde{{}^{\infty}B}$ of $z^2\partial_z$ in the frame $\widetilde{\varepsilon}$, and as the matrix of the base change is holomorphic with respect to z, it is enough to consider the corresponding coefficients in the conjugate matrix

$$\begin{aligned}\operatorname{diag}(e^{-z\overline{\varphi_i}} \mathrm{Id})(\mathrm{Id} + U_0(x))(\mathrm{Id} + S'(x,z)) \cdot {}^{\infty}B(x,z)\\ \cdot (\mathrm{Id} + S'(x,z))^{-1}(\mathrm{Id} + U_0(x))^{-1} \operatorname{diag}(e^{z\overline{\varphi_i}} \mathrm{Id}).\end{aligned}$$

Let us set ${}^{\infty}B(x,z) = {}^{\infty}B^{(0)}(x) - z\,{}^{\infty}B^{(1)}(x) + \cdots$ and $S'(x,z) = zS'^{(1)}(x) + \cdots$. On the one hand, we know that ${}^{\infty}B^{(0)}(x)$ is diagonal with respect to the φ-decomposition and has a limit when $x \to 0$, and the limit when $x \to 0$ of the characteristic polynomial of ${}^{\infty}B^{(1)}$ is the RHS in (3.5)(*).

On the other hand, S' defines a continuous map $X \to \mathrm{Mat}_d(L^2(\mathbf{S}))$, and, as such, $S'(0) = 0$ (cf. [20, Prop. 5.4.1]). Therefore, the (Fourier) coefficient $S'^{(1)}(x)$ is a continuous function of x and has limit 0 when $x \to 0$. We thus have

$$\begin{aligned}(\mathrm{Id} + U_0(x))&(\mathrm{Id} + S'(x,z)) \cdot {}^{\infty}B \cdot (\mathrm{Id} + S'(x,z))^{-1}(\mathrm{Id} + U_0(x))^{-1}\\ &= (\mathrm{Id} + U_0(x))\,{}^{\infty}B^{(0)}(\mathrm{Id} + U_0(x))^{-1}\\ &\quad - z \cdot (\mathrm{Id} + U_0(x))\big({}^{\infty}B^{(1)} + [{}^{\infty}B^{(0)}, S'^{(1)}]\big)(\mathrm{Id} + U_0(x))^{-1} + \cdots\end{aligned}$$

As the z-constant term above is diagonal with respect to the φ-decomposition, it commutes with $\operatorname{diag}(e^{z\overline{\varphi_i}} \mathrm{Id})$ and therefore is not altered by the conjugation by this matrix. It follows that the coefficient of $-z$ in $\widetilde{{}^{\infty}B}$ is

$$(\mathrm{Id} + U_0(x))\big({}^{\infty}B^{(1)} + [{}^{\infty}B^{(0)}, S'^{(1)}]\big)(\mathrm{Id} + U_0(x))^{-1}.$$

As $\lim_{x\to 0}[{}^{\infty}B^{(0)}, S'^{(1)}] = 0$, the limit, when $x \to 0$, of its characteristic polynomial (that is, the LHS in (3.5)(*)), is thus equal to the limit, when $x \to 0$, of the characteristic polynomial of ${}^{\infty}B^{(1)}(x)$, which we know to be the RHS in (3.5)(*).

Q.E.D.

§4. A review on exponential twist and Fourier–Laplace transform

In this section, we review some results of [19]. The base manifold X will be $\mathbb{P}^1$ with its two affine charts having coordinates t and t'. We will denote by $\mathscr{P}^1$ the corresponding manifold $\mathscr{X}$ as in the notation of §2.b.

4.a. De Rham cohomology with exponential twist for twistor $\mathscr{D}$-modules

Although we do not gain much by simply attaching a polarized variation of twistor structure to a polarized variation of Hodge structure, the advantage is clearer when we apply an exponential twist and integrate.

De Rham cohomology with exponential twist. Let $\mathcal{M}$ be a $\mathscr{D}_{\mathbb{P}^1}$-module and $\widetilde{\mathcal{M}} = \mathcal{M}(*\infty)$. The exponentially twisted de Rham cohomology is the hypercohomology on $\mathbb{P}^1$ of the complex

$$\mathrm{DR}(\mathcal{M} \otimes \mathcal{E}^{-t}) := \{0 \longrightarrow \widetilde{\mathcal{M}} \xrightarrow{\nabla - dt} \widetilde{\mathcal{M}} \longrightarrow 0\},$$

that we denote $H^*_{\mathrm{DR}}(\mathbb{P}^1, \mathcal{M} \otimes \mathcal{E}^{-t})$. If we assume $\mathcal{M}$ to be $\mathscr{D}_{\mathbb{P}^1}$-holonomic, then $M := \Gamma(\mathbb{P}^1, \widetilde{\mathcal{M}})$ is a holonomic $\mathbb{C}[t]\langle\partial_t\rangle$-module and the previous hypercohomology is the cohomology of the complex

$$0 \longrightarrow M \xrightarrow{\partial_t - 1} M \longrightarrow 0 \tag{4.1}$$

and has cohomology in degree one only, this cohomology being a finite dimensional $\mathbb{C}$-vector space. Its dimension is computed in [10, Prop. 1.5, p. 79]. If $\mathcal{M}$ has a regular singularity at infinity, this dimension is equal to the sum (over the singular points at finite distance) of the dimension of vanishing cycles of $\mathrm{DR}\,\mathcal{M}$.

Exponential twist of a twistor $\mathscr{D}$-module. We will use the notation of §2.b. Let $\mathscr{M}$ be a left $\mathscr{R}_{\mathscr{P}^1}$-module ($\mathscr{P}^1 = \mathbb{P}^1 \times \Omega_0$). We denote by $\widetilde{\mathscr{M}}$ the localized module $\mathscr{R}_{\mathscr{P}^1}(*\infty) \otimes_{\mathscr{R}_{\mathscr{P}^1}} \mathscr{M}$. We set $\mathscr{E}^{-t/z} = \mathscr{O}_{\mathscr{P}^1}(*\infty)$ with z-connection $zd - dt$. The exponentially twisted $\mathscr{R}_{\mathscr{P}^1}$-module ${}^F\!\mathscr{M}$ is $\mathscr{E}^{-t/z} \otimes_{\mathscr{O}_{\mathscr{P}^1}(*\infty)} \widetilde{\mathscr{M}}$ equipped with its natural z-connection.

It is useful to introduce the category $\widetilde{\mathscr{R}}$- Triples($\mathbb{P}^1$), whose objects $(\widetilde{\mathscr{M}}', \widetilde{\mathscr{M}}'', \widetilde{\mathscr{C}}_{\mathbf{S}})$ consist of $\mathscr{R}_{\mathscr{P}^1}(*\infty)$-modules with a pairing taking values in the sheaf of distributions on $(\mathbb{P}^1 \smallsetminus \{\infty\}) \times \mathbf{S}$ which have moderate growth at $\{\infty\} \times \mathbf{S}$ (i.e., which can be extended as distributions on $\mathbb{P}^1 \times \mathbf{S}$) and depend continuously on $z \in \mathbf{S}$.

If we remark that, for $z \in \mathbf{S}$, the C^∞ function $e^{-t/z} \cdot \overline{e^{-t/z}} = e^{z\bar{t} - t/z}$ has moderate growth as well as all its derivatives with respect to $t, \bar{t}$

when $t \to \infty$, we can associate to an object $\mathscr{T} = (\mathscr{M}', \mathscr{M}'', \mathscr{C}_{\mathrm{S}})$ of $\mathscr{R}$-Triples$(\mathbb{P}^1)$ the object ${}^F\mathscr{T} = ({}^F\mathscr{M}', {}^F\mathscr{M}'', e^{z\bar{t}-t/z}\mathscr{C}_{\mathrm{S}})$ of $\widetilde{\mathscr{R}}$-Triples$(\mathbb{P}^1)$.

Let now $\mathscr{T}$ be a polarized twistor $\mathscr{D}$-module on $\mathbb{P}^1$. Then the previous construction can be refined to give an object ${}^F\mathscr{T} = ({}^F\mathscr{M}', {}^F\mathscr{M}'', {}^F\mathscr{C}_{\mathrm{S}})$ of $\mathscr{R}$-Triples$(\mathbb{P}^1)$. The regularization ${}^F\mathscr{C}_{\mathrm{S}}$ of $e^{z\bar{t}-t/z}\mathscr{C}_{\mathrm{S}}$ is obtained by specializing ${}^{\mathscr{F}}\mathscr{C}_{\mathrm{S}}$, whose construction is recalled in §4.b, at $\tau = 1$. Let a be the constant map on $\mathbb{P}^1$. The following is proved in [15] (and its erratum):

Theorem 4.2 (Exponentially twisted Hodge theorem). *If $(\mathscr{T}, \mathscr{S})$ is a polarized regular twistor $\mathscr{D}$-module of weight w on $\mathbb{P}^1$, then $\mathscr{H}^0 a_+ {}^F\mathscr{T}$ is a polarized twistor structure of weight w.*

4.b. Fourier–Laplace transform (cf. [16, Appendix])

We continue to work with the projective line $\mathbb{P}^1$ equipped with its two charts having coordinates t and t', and we consider another copy of it, denoted by $\widehat{\mathbb{P}}^1$, having coordinates τ, τ'. We will set $\infty = \{t' = 0\}$ and $\widehat{\infty} = \{\tau' = 0\}$. We consider the diagram

(4.3)

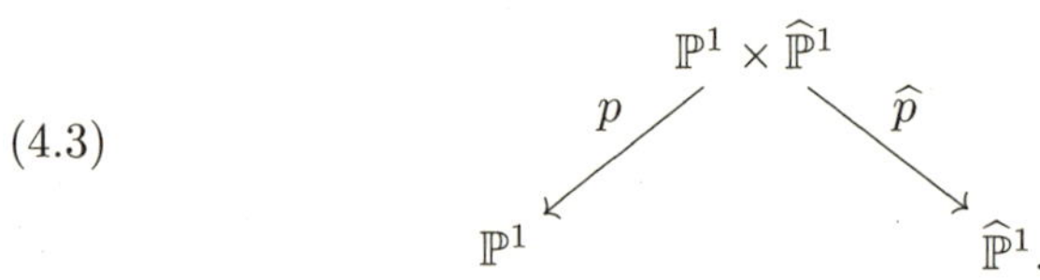

Let $\mathscr{M}$ be a good $\mathscr{R}_{\mathscr{P}^1}$-module (in the sense of [16, §1.1.c]). We set ${}^{\mathscr{F}}\mathscr{M} := p^+\widetilde{\mathscr{M}}(*\widehat{\infty}) \otimes \mathcal{E}^{-t\tau/z}$ (cf. [16, §A.2]). We know (cf. [16, Prop. A.2.7]) that ${}^{\mathscr{F}}\mathscr{M}$ is a good $\mathscr{R}_{\mathscr{Z}}(*\widehat{\infty})$-module, where $Z = \mathbb{P}^1 \times \widehat{\mathbb{P}}^1$ and $\mathscr{Z} = Z \times \Omega_0$. Taking direct images, $\widehat{p}_+ {}^{\mathscr{F}}\mathscr{M}$ is a coherent $\mathscr{R}_{\widehat{\mathscr{P}}^1}(*\widehat{\infty})$-module.

Let $\mathscr{T} = (\mathscr{M}', \mathscr{M}'', \mathscr{C}_{\mathrm{S}})$ be an object of $\mathscr{R}$-Triples$(\mathbb{P}^1)$, such that $\mathscr{M}', \mathscr{M}''$ are $\mathscr{R}_{\mathscr{P}^1}$-good. Then ${}^{\mathscr{F}}\mathscr{T}$ is defined as $({}^{\mathscr{F}}\mathscr{M}', {}^{\mathscr{F}}\mathscr{M}'', {}^{\mathscr{F}}\mathscr{C}_{\mathrm{S}})$, where ${}^{\mathscr{F}}\mathscr{M}', {}^{\mathscr{F}}\mathscr{M}''$ are as above and ${}^{\mathscr{F}}\mathscr{C}_{\mathrm{S}}$ is defined in [16, p. 196] (note that the twist for $\mathscr{C}_{\mathrm{S}}$ needs some care). The fibre at $\tau = 1$ (suitably defined as nearby cycles) of ${}^{\mathscr{F}}\mathscr{T}$ is identified with ${}^F\mathscr{T}$. The Fourier–Laplace transform $\widehat{\mathscr{T}}$ of $\mathscr{T}$ is defined as the direct image of ${}^{\mathscr{F}}\mathscr{T}$ by $\widehat{p}$.

Let us assume that $\mathscr{T}$ is integrable. Then (cf. [16, Rems. A.2.9 & A.2.15]), ${}^{\mathscr{F}}\mathscr{T}$ is also integrable.

Lemma 4.4. *The action of $z^2\eth_z$ on ${}^{\mathscr{F}}\mathscr{M} = p^+\widetilde{\mathscr{M}}(*\widehat{\infty}) \otimes \mathscr{E}^{-t\tau/z}$ satisfies, for any local section of $\widetilde{\mathscr{M}}$,*

$$(z^2\eth_z + \tau\eth_\tau)(m \otimes \mathscr{E}^{-t\tau/z}) = (z^2\eth_z m) \otimes \mathscr{E}^{-t\tau/z}.$$

Proof. This directly follows from the definition of the actions (cf. [16, A.2.2 & A.2.3]). Q.E.D.

Let us also notice that one gets a similar relation with the coordinate τ' by using the relation $\tau'\eth_{\tau'} = -\tau\eth_\tau$.

4.c. Fourier–Laplace transformation of twistor $\mathscr{D}$-modules

Let $(\mathscr{T},\mathscr{S})$ be a polarized regular twistor $\mathscr{D}$-module of weight w (in the sense of [16] or [11]) on $\mathbb{P}^1$. The Fourier–Laplace transform $(\widehat{\mathscr{T}},\widehat{\mathscr{S}})$ is an object of the same kind on the *analytic* affine line $\widehat{\mathbb{A}}^1$ with coordinate τ, after [15] and [18]. Moreover, it is smooth on the punctured line $\widehat{\mathbb{A}}^1 \smallsetminus \{\tau = 0\}$, and its restriction at $\tau = 1$ is naturally identified with $\mathscr{H}^0 a_+{}^F\mathscr{T}$.

In [18, Cor. 5.20 & Prop. 5.23], we also show an "inverse stationary phase formula" computing the nearby and vanishing cycles of $\widehat{\mathscr{T}}$ at $\tau = 0$ in terms of the nearby cycles at $t' = 0$ of $\mathscr{T}$.

Let us moreover assume that $(\mathscr{T},\mathscr{S})$ is integrable. Recall that a denotes the constant map $\mathbb{P}^1 \to \mathrm{pt}$. The basic comparison result [18, Cor. 5.20 and Prop. 5.23], together with Lemma 4.4, gives (cf. §2.f for the notation):

Proposition 4.5. *We have natural isomorphisms of integrable polarized pure twistor structures of weight $w+\ell$ ($\ell \in \mathbb{Z}$, $\beta \in (-1,0)$ for the first line):*

$$(4.5)(*)\qquad \begin{aligned} (\mathrm{gr}_\ell^{\mathrm{M}}\,\Psi_\tau^\beta\widehat{\mathscr{T}}, z^2\partial_z + \beta z) &\simeq (\mathrm{gr}_\ell^{\mathrm{M}}\,\Psi_{t'}^\beta\mathscr{T}, z^2\partial_z),\\ (\mathrm{gr}_\ell^{\mathrm{M}}\,\phi_\tau^{-1}\widehat{\mathscr{T}}, z^2\partial_z - z) &\simeq (\mathrm{gr}_\ell^{\mathrm{M}}\,\Psi_{t'}^0\mathscr{T}, z^2\partial_z)\end{aligned}$$

and we also have

$$(4.5)(**)\qquad (\mathrm{P}\,\mathrm{gr}_0^{\mathrm{M}}\,\Psi_\tau^0\widehat{\mathscr{T}}, z^2\partial_z) \simeq (\mathscr{H}^0 a_+\mathscr{T}, z^2\partial_z).$$

In Appendix A (Theorem A.1), we show that the Fourier–Laplace transform $\widehat{\mathscr{T}}$ on $\widehat{\mathbb{A}}^1$ naturally extends as a wild twistor $\mathscr{D}$-module (in the sense of [20], cf. also [12]) near $\widehat{\infty} \in \widehat{\mathbb{P}}^1$ and we relate the corresponding nearby cycles with the vanishing cycles of $\mathscr{T}$ at its critical points ("stationary phase formula"):

Corollary 4.6 (of Theorem A.1, (A.11) and (A.12))**.** *For any $c \in \mathbb{C}$, we have natural isomorphisms of integrable polarized pure twistor structures of weight $w+\ell$ ($\ell \in \mathbb{Z}$, $\beta \in (-1,0)$ for the first line):*

$$(4.6)(*)\qquad \begin{aligned} (\mathrm{gr}_\ell^{\mathrm{M}}\,\Psi_{\tau'}^{c/\tau',\beta}\widehat{\mathscr{T}}, z^2\partial_z - (\beta+1)z) &\simeq (\mathrm{gr}_\ell^{\mathrm{M}}\,\Psi_{t+c}^\beta\mathscr{T}, z^2\partial_z),\\ (\mathrm{gr}_\ell^{\mathrm{M}}\,\Psi_{\tau'}^{c/\tau',0}\widehat{\mathscr{T}}, z^2\partial_z - z) &\simeq (\mathrm{gr}_\ell^{\mathrm{M}}\,\phi_{t+c}^{-1}\mathscr{T}, z^2\partial_z - z).\end{aligned}$$

§5. Twistor structures and Hodge structures

In this section, we make explicit the functor Tw which associates to any polarized complex Hodge structure (resp. variation of Hodge structure, resp. polarized complex mixed Hodge structure) an integrable polarized twistor structure (resp. ...). In this section, Y will denote a complex manifold, X will denote a disc with coordinate x and X^* will denote the punctured disc $X \smallsetminus \{0\}$.

5.a. The integrable variation attached to a polarizable variation of Hodge structures

Let $(V, {}^V\nabla)$ be a holomorphic vector bundle with an integrable holomorphic connection on a complex manifold Y. Let us assume that $(V, {}^V\nabla)$ underlies a polarized variation of Hodge structures of weight w. The C^∞-bundle H associated to V comes equipped with a flat C^∞ connection $D = {}^V\nabla + d''$ and a decomposition $H = \oplus_p H^{p,w-p}$ indexed by integers. There is a D-flat sesquilinear pairing k on H such that the decomposition is k-orthogonal, and the sesquilinear pairing h such that the decomposition is h-orthogonal and $h = i^{-w}(-1)^p k$ on $H^{p,w-p}$ is Hermitian positive definite. As usual, we set $F^pV = \bigoplus_{r \geqslant p} H^{r,w-r}$.

For any $j \in \frac{1}{2}\mathbb{Z}$, the Tate twist is defined as

$$(V, {}^V\nabla, F^\bullet V, k, w)(j) := (V, {}^V\nabla, F^\bullet V, i^{-2j}k, w - 2j).$$

We denote by $\mathscr{H}' = R_{F[w]}V$, $\mathscr{H}'' = R_F V$, the Rees modules associated to $F[w]^\bullet V := F^{w+\bullet}V$ and $F^\bullet V$, that is:

$$\tag{5.1} \begin{aligned} \mathscr{H}' &:= \bigoplus_p F[w]^p z^{-p} = \bigoplus_r z^{w-r} H^{r,w-r}[z], \\ \mathscr{H}'' &:= \bigoplus_p F^p z^{-p} = \bigoplus_r z^{-r} H^{r,w-r}[z]. \end{aligned}$$

We denote by $R_F k$ the map naturally induced by k on $R_{F[w]}V \otimes_{\mathbb{C}[z,z^{-1}]} \overline{R_F V}$ with values in $\mathscr{C}_Y^\infty[z, z^{-1}]$. We associate to this variation the triple $\mathscr{T} = (R_{F[w]}V, R_F V, R_F k)$. We set $\mathscr{S} = (S', S'')$ with $S', S'' : \mathscr{H}'' \to \mathscr{H}'$, S'' is the multiplication by z^w and S' by $(-z)^w$.

The integrable connection ∇ is defined as ${}^V\nabla + d'_z$. We note that $R_{F'}V$ and $R_{F''}V$ are stable by $z\nabla_{\partial_z}$ (reflecting the fact that $\mathscr{U} = 0$). In particular, the action of $z^2\partial_z$ enables one to recover the grading of $R_F V$, hence the filtration $F^\bullet V$. The following is easy:

Lemma 5.2. *The object*

$$\mathrm{Tw}(V, F^\bullet V, k, w) := \big(\mathscr{T} = (R_{F[w]}V, R_F V, R_F k), \mathscr{S}, z^2\partial_z\big)$$

is an integrable polarized variation of twistor structures of weight w.

Remark 5.3. According to the convention made in Remark 2.14, the functor Tw is compatible with Tate twist (that is, $[\mathrm{Tw}(V, F^\bullet V, k, w)](j)$ is canonically isomorphic to $\mathrm{Tw}[(V, F^\bullet V, k, w)(j)]$.

Lemma 5.4. *Let $(\mathscr{T}, \mathscr{S})$ be the integrable polarized twistor structure of weight w attached to a polarized Hodge structure of weight w (in particular, the "no ramification" condition is fulfilled). Then*

$$\mathrm{SP}^\infty_{\mathscr{T}}(T) = \mathrm{Susy}_{\mathscr{T}}(T) = \mathrm{SP}^0_{\mathscr{T}}(T).$$

Proof. According to (2.19) one can assume $w = 0$. On the one hand, $\mathscr{U} = 0$, so $\mathscr{Q}$ is conjugate to the opposite of the residue at $z = 0$ of ∂_z acting on $R_F V$. As $z\partial_z$ acts as $-p\,\mathrm{Id}$ on $F^p z^{-p}$, we find that $\mathrm{Susy}_{\mathscr{T}}(T) = \prod_p (T - p)^{\dim \mathrm{gr}^p_F}$.

We now have $G = \mathbb{C}[z, z^{-1}] \otimes_{\mathbb{C}} H$ and $V^p G = z^{-p}\mathbb{C}[z^{-1}] \otimes_{\mathbb{C}} H$ (for the V-filtration at $z = \infty$). Then (5.1) shows that (using the notation in Definition 1.2) $\nu_p = \dim H^{p,-p} = \dim \mathrm{gr}^p_F$, hence the first equality. On the other hand, the V-filtration at $z = 0$ is given by $V^p G = z^p\mathbb{C}[z] \otimes_{\mathbb{C}} H$ and G has a regular singularity at $z = 0$, so there is no nontrivial exponential term in the decomposition (1.6). We then have (using the notation in Definition 1.7) $\mu_{0,p} = \dim H^{-p,p}$, hence the second equality. Q.E.D.

5.b. Integrable twistor structure attached to a polarized complex mixed Hodge structure

Let V_o be a complex vector space equipped with a filtration $F^\bullet V_o$, a nilpotent endomorphism N_o and a sesquilinear pairing k_o. We denote by $\mathrm{M}_\bullet$ the monodromy filtration of V_o associated to N_o. Let $w \in \mathbb{Z}$. We say (cf. [24, 9]) that $(V_o, F^\bullet V_o, k_o, \mathrm{N}_o)$ is a polarized complex mixed Hodge structure of weight w if the following conditions are fulfilled:

(1) k_o is $(-1)^w$-Hermitian and N_o is skew-adjoint with respect to k_o,
(2) $\mathrm{N}_o F^\bullet V_o \subset F^{\bullet-1} V_o$,
(3) if we set $\overline{F}^p V_o = \overline{(F^{w-p+1} V_o)^\perp}$ (which also satisfies $N\overline{F}^p V_o \subset \overline{F}^{p-1} V_o$), then $(F^\bullet V_o, \overline{F}^\bullet V_o, \mathrm{M}_\bullet)$ is a mixed Hodge structure of weight w,
(4) the object $(\mathrm{P}\,\mathrm{gr}^{\mathrm{M}}_\ell V_o, F^\bullet \mathrm{P}\,\mathrm{gr}^{\mathrm{M}}_\ell V_o, k_o(\bullet, \overline{\mathrm{N}^\ell_o \bullet}))$ is a polarized complex Hodge structure of weight $w + \ell$.

Remark 5.5 (cf. [22, Lemma 2.8]). If $(V_o, F^\bullet V_o, k_o, \mathrm{N}_o)$ is a polarized complex mixed Hodge structure of weight w, then there exists an increasing filtration $\widetilde{F}_\bullet V_o$ which is opposite to $F^\bullet V_o$ (i.e., V_o decomposes as $\bigoplus_p F^p \cap \widetilde{F}_p$) and which satisfies $\mathrm{N}_o \widetilde{F}_\bullet V_o \subset \widetilde{F}_{\bullet-1} V_o$ (in particular, $\widetilde{F}_\bullet V_o$ is stable by N_o). Indeed, V_o is bigraded by Deligne's $I^{p,q}$ (cf. [3]) with

$$I^{p,q} = (F^p \cap W_{p+q}) \cap \Big(\overline{F}^q \cap W_{p+q} + \textstyle\sum_{j\geqslant 1} \overline{F}^{q-j} \cap W_{p+q-j-1}\Big),$$

where $W_\ell := \mathrm{M}_{w+\ell}$, and $F^p V_o = \bigoplus_{p'\geqslant p} \bigoplus_q I^{p',q}$. We can set $\widetilde{F}_p V_o = \bigoplus_{p'\leqslant p} \bigoplus_q I^{p',q}$.

Definition 5.6. For a polarized complex mixed Hodge structure $(V_o, F^\bullet, k_o, \mathrm{N}_o)$ of weight w, we set $\mathrm{Tw}(V_o, F^\bullet, k_o, \mathrm{N}_o) := (\mathscr{T}, \mathscr{S}, \mathscr{N}, z^2\partial_z)$ with

(1) $\mathscr{T} = (R_{F[w]} V_o, R_F V_o, R_F k_o)$ (an object of $\mathscr{R}$-Triples(pt)),
(2) $\mathscr{S} = ((-z)^w, z^w)$ (a sesquilinear duality of $\mathscr{T}$ of weight w),
(3) $\mathscr{N} : \mathscr{T} \to \mathscr{T}(-1)$ defined as $\mathscr{N} = (z\mathrm{N}_o, -z\mathrm{N}_o)$,
(4) $z^2\partial_z$ is the natural derivation on $R_{F[w]} V_o, R_F V_o$.

Lemma 5.7. *If $(V_o, F^\bullet, k_o, \mathrm{N}_o)$ is a polarized complex mixed Hodge structure of weight w, the monodromy filtration of $z\mathrm{N}_o$ on $R_F V_o$ is such that $\mathrm{gr}_\ell^{\mathrm{M}(z\mathrm{N}_o)} R_F V_o = R_F \,\mathrm{gr}_\ell^{\mathrm{M}(\mathrm{N}_o)} V_o$. Moreover, the object $(\mathrm{gr}_\bullet^{\mathrm{M}} \mathscr{T}, \mathrm{gr}_\bullet^{\mathrm{M}} \mathscr{S})$ is a graded Lefschetz twistor structure of weight w (cf. [16, §2.1.e]). Last, we have a canonical isomorphism of objects of weight $w+\ell$ ($\ell \geqslant 0$):*

$$\mathrm{P}\,\mathrm{gr}_\ell^{\mathrm{M}} \mathrm{Tw}(V_o, F^\bullet, k_o, \mathrm{N}_o) \xrightarrow{\sim} \mathrm{Tw}[\mathrm{P}\,\mathrm{gr}_\ell^{\mathrm{M}}(V_o, F^\bullet, k_o, \mathrm{N}_o)].$$

Proof. Let us indicate the proof for the last part. We can reduce to weight 0 by twisting by $w/2$, and also to $\mathscr{S} = (\mathrm{Id}, \mathrm{Id})$. The left-hand side in the formula is by definition (cf. [16, Example 2.1.14]) given by

$$\mathscr{T}_\ell = ((z\mathrm{N}_o)^\ell R_F \mathrm{P}\,\mathrm{gr}_\ell^{\mathrm{M}} V_o, R_F \mathrm{P}\,\mathrm{gr}_\ell^{\mathrm{M}} V_o, R_F k_o), \quad \mathscr{S}_\ell = ((z\mathrm{N}_o)^\ell, (-z\mathrm{N}_o)^\ell),$$

and the action of $z^2\partial_z$ is the natural one. According to Lemma 5.2, the right-hand side is given by

$$\widetilde{\mathscr{T}_\ell} = (R_{F[\ell]} \mathrm{P}\,\mathrm{gr}_\ell^{\mathrm{M}} V_o, R_F \mathrm{P}\,\mathrm{gr}_\ell^{\mathrm{M}} V_o, R_F k_o(\bullet, \overline{\mathrm{N}_o^\ell \bullet})), \quad \widetilde{\mathscr{S}_\ell} = ((-z)^\ell, z^\ell),$$

and the action of $z^2\partial_z$ is the natural one. If one notices that $R_{F[\ell]} \mathrm{P}\,\mathrm{gr}_\ell^{\mathrm{M}} V_o = z^\ell R_F \mathrm{P}\,\mathrm{gr}_\ell^{\mathrm{M}} V_o$, then one checks that $\varphi := ((-\mathrm{N}_o)^\ell, \mathrm{Id}) : (\mathscr{T}_\ell, \mathscr{S}_\ell) \to (\widetilde{\mathscr{T}_\ell}, \widetilde{\mathscr{S}_\ell})$ is an isomorphism. Q.E.D.

Let us note that, by definition, for $(\mathscr{T},\mathscr{S},\mathscr{N},z^2\partial_z)$ as in Definition 5.6, the object $(\operatorname{gr}^{\mathrm{M}}_{\bullet}\mathscr{T},\operatorname{gr}^{\mathrm{M}}_{\bullet}\mathscr{S},\operatorname{gr}^{\mathrm{M}}_{-2}\mathscr{N},z^2\partial_z)$ is a polarized graded Lefschetz twistor structure of weight w and type -1 (cf. [16, §2.1.e]). For such an object, there is a reduction to weight 0 and type 0 (cf. loc. cit.) giving rise to a polarized (graded Lefschetz) twistor structure of weight 0 (and type 0). This structure remains integrable and therefore comes equipped with a Susy polynomial. We denote it by $\mathrm{Susy}_{\mathrm{Tw}(V_o,F^\bullet,k_o,\mathrm{N}_o)}(T)$.

Lemma 5.8. *Let $(V_o,F^\bullet,k_o,\mathrm{N}_o)$ be a polarized complex mixed Hodge structure of weight w. Then*

$$\mathrm{SP}^{\infty}_{\mathrm{Tw}(V_o,F^\bullet,k_o,\mathrm{N}_o)}(T)=\mathrm{Susy}_{\mathrm{Tw}(V_o,F^\bullet,k_o,\mathrm{N}_o)}(T)=\mathrm{SP}^{0}_{\mathrm{Tw}(V_o,F^\bullet,k_o,\mathrm{N}_o)}(T).$$

Proof. Each $\operatorname{gr}^{\mathrm{M}}_{\ell}\mathscr{T}$ comes equipped with a polarization $\mathscr{S}_\ell$ defined from that on the various $\mathrm{P}\operatorname{gr}^{\mathrm{M}}_{\ell'}\mathscr{T}$ by using the Lefschetz decomposition, making it a polarized twistor structure of weight $w+\ell$ (cf. [16, Rem. 2.1.15]), and one has

$$\mathrm{Susy}_{\mathrm{Tw}(V_o,F^\bullet,k_o,\mathrm{N}_o)}(T)=\prod_{\ell\in\mathbb{Z}}\mathrm{Susy}_{\operatorname{gr}^{\mathrm{M}}_{\ell}\mathrm{Tw}(V_o,F^\bullet,k_o,\mathrm{N}_o)}(T).$$

On the other hand, because each $\operatorname{gr}^{\mathrm{M}}_{\ell}R_FV_o$ is a free $\mathbb{C}[z]$-module (being equal to $R_F\operatorname{gr}^{\mathrm{M}}_{\ell}V_o$) we have such a product formula for SP^0 and SP^∞, according to Remark 1.9. Then Lemma 5.4 applies. Q.E.D.

Definition 5.9 (Vanishing cycles, cf. [9, Prop. 2.1.3]). Consider a polarized complex mixed Hodge structure $(V_o,F^\bullet,k_o,\mathrm{N}_o)$ of weight w. The vanishing cycle polarized complex mixed Hodge structure $(\widetilde{V}_o,\widetilde{F}^\bullet,\widetilde{k}_o,\widetilde{\mathrm{N}}_o)$ of weight $w+1$ attached to it is defined as follow:

$$\widetilde{V}_o=\mathrm{N}_oV_o,\quad \widetilde{F}^\bullet=\mathrm{N}_oF^\bullet,\quad \widetilde{k}_o(\mathrm{N}_ox,\overline{\mathrm{N}_oy})=k_o(x,\overline{\mathrm{N}_oy}),\quad \widetilde{\mathrm{N}}_o=N_{o|\widetilde{V}_o}.$$

5.c. Extension of Tw through a singularity

Let $(V,{}^V\nabla)$ be a holomorphic bundle with connection on the punctured disc X^* underlying a polarized variation of Hodge structure of weight w. We are in the situation considered in §5.a. According to Lemma 5.2, $\big(\mathscr{T}=(R_{F[w]}V,R_FV,R_Fk),\mathscr{S}\big)$ is a polarized variation of twistor structures of weight w on X^*. We will indicate how to extend it as a polarized twistor $\mathscr{D}$-module on X.

According to Schmid [24], the $\mathscr{O}_X[x^{-1}]$-submodule $\widetilde{\mathcal{M}}$ of j_*V consisting of sections whose h-norm has moderate growth at the origin is locally free and the connection ∇ extends to it with regular singularities. We denote by $\mathcal{M}$ the $\mathscr{D}_X$-submodule of $\widetilde{\mathcal{M}}$ generated by local sections v

whose h-norm is bounded by $C|x|^{-1+\varepsilon}$ for some $C > 0$ and $\varepsilon > 0$. It is known that $\mathcal{M}$ is a regular holonomic $\mathscr{D}_X$-module, which coincides with the *minimal extension* of $\widetilde{\mathcal{M}}$, so $\mathrm{DR}\,\mathcal{M}$ is the *intermediate extension* (or intersection complex) of the local system $\operatorname{Ker}{}^V\nabla$ on X^*. Moreover, the filtration $F^\bullet V$ extends to a filtration of $\widetilde{\mathcal{M}}$ by holomorphic locally free $\mathscr{O}_X$-modules, and then to a filtration of $\mathcal{M}$, which is a good filtration when we consider it as an increasing filtration. Lastly, the flat sesquilinear form k defined from the metric h extends as a $\mathscr{D}_X \otimes_{\mathbb{C}} \overline{\mathscr{D}_X}$-linear pairing $k : \mathcal{M} \otimes_{\mathbb{C}} \overline{\mathcal{M}} \to \mathfrak{Db}_X$.

We can apply the Rees construction R_F to these data.

Proposition 5.10 (cf. [19, §3.g]). *The object*

$$(\mathscr{T} = (R_{F[w]}\mathcal{M}, R_F\mathcal{M}, R_F k), \mathscr{S})$$

is an integrable polarized twistor $\mathscr{D}$-module of weight w on X.

Definition 5.11. We will call such an object a *polarized complex Hodge $\mathscr{D}$-module of weight w.*

5.d. Nearby and vanishing cycles

We will set $F^pV^\beta\mathcal{M} := F^p\mathcal{M} \cap V^\beta\mathcal{M}$. In particular, for any $k \geqslant 0$, $x^kF^pV^\beta\mathcal{M} \subset F^pV^{\beta+k}\mathcal{M}$.

Assume that $(\mathcal{M}, F^\bullet\mathcal{M})$ underlies a polarized complex Hodge $\mathscr{D}$-module (cf. Definition 5.11), then (cf. [21, §3.2] and [19, §3.d])

$$\begin{aligned} \forall\,\beta > -1,\ \forall\,p \in \mathbb{Z},\quad F^pV^\beta\mathcal{M} &= j_*j^*F^p \cap V^\beta\mathcal{M} \\ F^p\mathcal{M} &= \sum_{j \geqslant 0} \eth_x^j F^{p+j}V^{>-1}\mathcal{M}. \end{aligned} \tag{5.12}$$

In particular, as a consequence of the first line of (5.12), we have

$$\forall\,\beta > -1,\ \forall\,p,\ \forall\,k \geqslant 0,\quad x^kF^pV^\beta\mathcal{M} = F^pV^{\beta+k}\mathcal{M}. \tag{5.13}$$

Moreover, $x\eth_x : \mathrm{gr}_V^\beta\,\mathcal{M} \to \mathrm{gr}_V^\beta\,\mathcal{M}$ strictly shifts the filtration $F^\bullet$ by -1. It is an isomorphism if $\beta \neq 0$.

As a consequence of the results recalled in §2.f, we find:

Corollary 5.14. *If $(\mathscr{T}, \mathscr{S})$ is a polarized complex Hodge $\mathscr{D}$-module of weight w then, for any $\beta \in (-1, 0]$,*

1. $\Psi_x^\beta\mathscr{T} = (R_{F[w]}\psi_x^\beta\mathcal{M}, R_F\psi_x^\beta\mathcal{M}, R_F\psi_x^\beta k)$.
2. *This equality is compatible with the natural actions of $z^2\eth_z$ on both terms, and therefore $z\eth_z$ acts on $\Psi_x^\beta\mathscr{T}$.*
3. *$(\Psi_x^\beta\mathscr{T}, \Psi_x^\beta\mathscr{S}, \mathscr{N}, z^2\eth_z)$ is a polarized complex mixed Hodge structure of weight w.*

Let us notice that 5.14(3) can be regarded as a reformulation of Theorem (6.16) in [24], and is similar to Corollary 1 of [21] on a punctured disc. Notice also that, compared to [21], the choice of the behaviour of weights by taking nearby/vanishing cycles is not the same here and in [16], as we are working with complex Hodge structures, and we can use Tate twists by half-integers (see also the vanishing cycles below, which also has to be compared with Corollary 1 of [21]).

For vanishing cycles (cf. [16, §3.6.b]) we find:

Corollary 5.15. *For $(\mathscr{T},\mathscr{S})$ as above, $(\phi_x^{-1}\mathscr{T},\phi_x^{-1}\mathscr{S},\mathscr{N},z^2\partial_z-z)$ is a polarized complex mixed Hodge structure of weight w.*

Sketch of proof. According to [16, Cor. 4.1.17] (and to an easy consequence of §4.2 of loc. cit. for the polarization), the object $(\phi_x^{-1}\mathscr{T}(-1/2),\phi_x^{-1}\mathscr{S}(-1/2),\mathscr{N})$ is isomorphic to the image of $\mathscr{N}:\mathscr{T}\to\mathscr{T}(-1)$ and gives rise, after grading with respect to the monodromy filtration, to a graded Lefschetz twistor structure of weight $w+1$. In order that the morphism $\mathscr{C}an$ of [16, Lemma 3.6.21] is compatible with the action of $z\partial_z$ (giving the grading), we should equip $\phi_x^{-1}\mathscr{T}(-1/2)$ with the shifted naturally induced action $z\partial_z-1$. Then $(\phi_x^{-1}\mathscr{T}(-1/2),\phi_x^{-1}\mathscr{S}(-1/2),\mathscr{N})$ is isomorphic to Tw of the vanishing cycles (as defined in 5.9) of the polarized complex mixed Hodge structure $\mathrm{Tw}^{-1}(\Psi_x^\beta\mathscr{T},\Psi_x^\beta\mathscr{S},\mathscr{N})$. Applying a Tate twist (1/2) we find, according to our convention on Tate twist (Remark 2.14), that $(\phi_x^{-1}\mathscr{T},\phi_x^{-1}\mathscr{S},\mathscr{N},z^2\partial_z-z)$ is (isomorphic to Tw of) a polarized complex mixed Hodge structure of weight w. Q.E.D.

5.e. Exponential twist of an integrable twistor $\mathscr{D}$-module

Starting from a variation of polarized Hodge structures $(V,F^\bullet V)$ on $U\subset\mathbb{A}^1$, Proposition 5.10 produces a complex Hodge $\mathscr{D}$-module $\mathscr{T}=\big((R_{F[w]}\mathcal{M},R_F\mathcal{M},R_Fk),\mathscr{S}\big)$. Localizing away from ∞ and taking global sections produces a filtered $\mathbb{C}[t]\langle\partial_t\rangle$-module $(M,F^\bullet M)$ with a pairing taking values in tempered distributions on $\mathbb{A}^1$ depending continuously on $z\in\mathbf{S}$.

As integrability is preserved by direct images (cf. [16, Prop. 7.1.4]), we can apply Theorem 4.2 together with Proposition 5.10:

Corollary 5.16. *If $(\mathscr{T},\mathscr{S})$ is a polarized complex Hodge $\mathscr{D}$-module of weight w on $\mathbb{P}^1$, then $\mathscr{H}^0a_+{}^F\mathscr{T}$ is an integrable polarized twistor structure of weight w.*

Although we did not give the precise definition of the direct image functor a_+ (cf. [16, §1.6.d] for more details), one can notice that, according to the strictness property of polarized twistor $\mathscr{D}$-modules, the restriction to $z = 1$ commutes with taking a_+. In other words, the vector space corresponding to the polarized pure twistor structure $\mathscr{H}^0 a_+{}^F\mathscr{T}$ is the cokernel of $\partial_t - 1 : M \to M$.

Another expression of the exponentially twisted de Rham cohomology. One can give a more explicit expression for $\mathscr{H}^0 a_+{}^F\mathscr{T}$ considered in Corollary 5.16. We will recall it below. For simplicity, we will assume that $w = 0$.

We can extend the correspondence of §1.c to objects with a sesquilinear pairing as follows. Let $(M, F^\bullet M)$ be as in §1.c and let us moreover assume that M comes equipped with a $\mathbb{C}[t]\langle\partial_t\rangle \otimes_{\mathbb{C}} \overline{\mathbb{C}[t]\langle\partial_t\rangle}$-linear pairing $k : M \otimes_{\mathbb{C}} \overline{M} \to \mathscr{S}'(\mathbb{A}^1)$ with values in the Schwartz space of temperate distributions on $\mathbb{A}^1$. To $(M, F^\bullet M, k)$ we associate a Hermitian twistor structure $(\mathscr{H}', \mathscr{H}', \mathscr{C}_{\mathbf{S}})$:

- we set $\mathscr{H}' = G_0^{(F),\mathrm{an}}$ (the analytization of the object defined by (1.11));
- composing k with the Fourier transform of temperate distributions with kernel $e^{\overline{t\tau} - t\tau} \frac{i}{2\pi} dt \wedge d\overline{t}$ induces, by restriction to $\mathbf{S} := \{|\tau| = 1\} = \{|z| = 1\}$, a sesquilinear pairing $\mathscr{C}_{\mathbf{S}} : \mathscr{H}'_{|\mathbf{S}} \otimes_{\mathscr{O}_{|\mathbf{S}}} \overline{\mathscr{H}'_{|\mathbf{S}}} \to \mathscr{O}_{|\mathbf{S}}$.

Moreover, this twistor structure is integrable (by using the action of $t = z^2\partial_z$ on $G_0^{(F)}$).

Lemma 5.17 (cf. [19, Lemma 2.1 and §2.c]). *The twistor structure $(\mathscr{H}', \mathscr{H}', \mathscr{C}_{\mathbf{S}})$ is the exponentially twisted de Rham cohomology of the object $(R_F M, R_F M, R_F k)$ of $\widetilde{\mathscr{R}}$-Triples$(\mathbb{P}^1)$.*

In the case where $(M, F^\bullet M, k)$ comes from a polarized variation of Hodge structures of weight 0 on U as explained at the beginning of this paragraph, we get from Corollary 5.16:

Corollary 5.18 (cf. [19, Cor. 3.15]). *Under the previous assumption, the integrable twistor structure $(\mathscr{H}', \mathscr{H}', \mathscr{C}_{\mathbf{S}})$ associated to $(M, F^\bullet M, k)$ is pure of weight 0 and polarized.*

Remark 5.19. The previous description makes it clear how to compute the conjugacy class of the endomorphism $\mathscr{U}$ of §2.a: indeed, this is the conjugacy class of the restriction of $z^2\partial_z$ to $\mathscr{H}'/z\mathscr{H}'$. It is therefore equal to the conjugacy class of t acting on $G_0^{(F)}/zG_0^{(F)}$. Its eigenvalues are the singular points of M (at finite distance). Therefore, in general, it is not a multiple of Id, and the integrable twistor structure $(\mathscr{H}', \mathscr{H}', \mathscr{C}_{\mathbf{S}})$ does not correspond in the usual way to a polarized Hodge structure.

5.f. Fourier–Laplace transformation of variations of polarized Hodge structures

We will make explicit the behaviour of the functor Tw under Laplace transform. We will work with the associated $\mathbb{C}[t]\langle\partial_t\rangle$-modules.

Let $(M, F^\bullet M)$ be a regular holonomic $\mathbb{C}[t]\langle\partial_t\rangle$-module with good filtration. Let us consider the Rees module $R_F M$, which is a $\mathbb{C}[t,z]\langle\eth_t\rangle$-module, and its Laplace transform $\widehat{R_F M}$, which is a $\mathbb{C}[\tau,z]\langle\eth_\tau\rangle$-module. Recall that $\widehat{R_F M} = R_F M$ as a $\mathbb{C}[z]$-module and that τ acts as $\eth_t$ and $\eth_\tau$ as $-t$. Notice that $\widehat{R_F M}$ can be obtained as the cokernel of

$$\mathbb{C}[\tau]\otimes_{\mathbb{C}} R_F M \xrightarrow{\eth_t - \tau} \mathbb{C}[\tau]\otimes_{\mathbb{C}} R_F M$$

by the map $\sum_{k\geqslant 0}\tau^k\otimes m_k \mapsto \sum_{k\geqslant 0}\eth_t^k m_k$, and the natural action of $\mathbb{C}[\tau,z]\langle\eth_\tau\rangle$, as well as the action of $z^2\partial_z$, are obtained by conjugating the usual actions by $\mathscr{E}^{-t\tau/z}$. In particular, the action of $z\partial_z$ on $\widehat{R_F M}$ coming from the identification with $R_F M$ and which gives the grading of this Rees module corresponds to the action denoted $z\partial_z\otimes 1$ in Lemma 4.4, and the natural action of $z^2\partial_z$ is that given by this lemma.

Let $G_0^{(F)}$ be the Brieskorn lattice of the filtration $F^\bullet M$ (cf. (1.11)), that we will denote by G_0 for short.

We will be mainly concerned with $\widehat{R_F M}_{\mathrm{loc}} := \mathbb{C}[\tau,\tau^{-1},z]\otimes_{\mathbb{C}[\tau,z]}\widehat{R_F M}$. Recall (cf. [19, Lemma 2.1]) that $\widehat{R_F M}_{\mathrm{loc}} \simeq \mathbb{C}[\tau,\tau^{-1}]\otimes_{\mathbb{C}} G_0$, where, on the right-hand side, the $\mathbb{C}[\tau,\tau^{-1},z]\langle\eth_\tau\rangle$-action is given as follows:

- the $\mathbb{C}[\tau,\tau^{-1}]$-structure is the natural one,
- the action of z is by $\tau\otimes\partial_t^{-1}$,
- the action of $\eth_\tau$ is by $z\cdot(\partial_\tau\otimes 1) - 1\otimes t$.

Recall that $G := \widehat{M}[\partial_t^{-1}]$ is a $\mathbb{C}[\partial_t,\partial_t^{-1}]$-module with connection. In the following we will use the notation $\theta = \partial_t$, $\theta' = \partial_t^{-1}$ (with the identification above, $\theta' = z\tau'$ with $\tau' = \tau^{-1}$). We will denote by $V_\theta^\bullet G$ the V-filtration of G at $\partial_t = 0$ and we will set $\psi_\theta^\gamma G := \mathrm{gr}_{V_\theta}^\gamma G$. Similarly, we denote by $V_{\theta'}^\bullet G$ the V-filtration at $\partial_t^{-1} = 0$ and we set $\psi_{\theta'}^{0,\gamma} G := \mathrm{gr}_{V_{\theta'}}^\gamma G$. For any $c\in\mathbb{C}$, we also set $\psi_{\theta'}^{c/\theta',\gamma} G := \mathrm{gr}_{V_{\theta'}}^\gamma (G\otimes\mathcal{E}^{-c/\theta'})$. As G has a regular singularity at $\theta = 0$, each $V_\theta^\gamma G$ is $\mathbb{C}[\theta]$-free of finite type, while, as the singularity at $\theta' = 0$ is usually irregular, each $V_{\theta'}^\gamma G$ has finite type over $\mathbb{C}[\theta']\langle\theta'\partial_{\theta'}\rangle$. If the $G_i^\wedge$ are the regular formal modules entering in the decomposition analogous to (1.6) for $G^\wedge$, and if i_0 is the index i such that $c_i = 0$, then $\mathrm{gr}_{V_{\theta'}}^\gamma G = \mathrm{gr}_{V_{\theta'}}^\gamma G_{i_0}^\wedge$.

Lemma 5.20. *The $\mathbb{C}[\tau,\tau^{-1},z]\langle\eth_\tau\rangle$-module $\widehat{R_FM}_{\mathrm{loc}}$ is strictly specializable at $\tau=0$ and $\tau=\infty$. The V-filtration is given by*

$$(*)_\infty \qquad V_\tau^\gamma\widehat{R_FM}_{\mathrm{loc}} = \bigoplus_{k\in\mathbb{Z}} \tau^k\otimes(G_0\cap V_\theta^{\gamma-k}G) \simeq R_{G(F)}V_\theta^\gamma G,$$

$$(*)_0 \qquad V_{\tau'}^\gamma\widehat{R_FM}_{\mathrm{loc}} = \bigoplus_{k\in\mathbb{Z}} \tau'^k\otimes(G_0\cap V_{\theta'}^{\gamma-k}G) \simeq R_{G(F)}V_{\theta'}^\gamma G.$$

Proof. Let us denote by $U_\tau^\gamma\widehat{R_FM}_{\mathrm{loc}}$ the right-hand side in $(*)_\infty$. Let us set $G_k=\theta'^{-k}G_0\subset G$. This is an increasing filtration of G by $\mathbb{C}[\theta']$-submodules. Let us fix $\gamma\in\mathbb{R}$. Then, for $k\ll 0$, we have $G_k\cap V_\theta^\gamma G=\{0\}$ and, for $k\gg 0$, $G_k\cap V_\theta^\gamma G=G_{k-1}\cap V_\theta^\gamma G+G_k\cap V_\theta^{\gamma+1}G$ (the last equality expresses that $\nu_\alpha(G_0)=0$ for $\alpha\ll 0$, cf. [17]). If we consider $G_\bullet\cap V_\theta^\gamma G$ as a filtration of the $\mathbb{C}[\theta]$-module $V_\theta^\gamma G$ compatible with the filtration $\deg_\bullet\mathbb{C}[\theta]$ by the degree in θ, these two properties are equivalent to saying that $G_\bullet\cap V_\theta^\gamma G$ is a *good* filtration, or equivalently that the Rees module $R_{G(F)}V_\theta^\gamma G:=\bigoplus_{k\in\mathbb{Z}}(G_k\cap V_\theta^\gamma G)z^k$ is a $R_{\deg}\mathbb{C}[\theta]$-module of finite type. According to the definition of the action of z above, we identify $R_{\deg}\mathbb{C}[\theta]$ with $\mathbb{C}[\tau,z]$ and $R_{G(F)}V_\theta^\gamma G$ with $U_\tau^\gamma\widehat{R_FM}_{\mathrm{loc}}$, hence the finiteness of $U_\tau^\gamma\widehat{R_FM}_{\mathrm{loc}}$ over $\mathbb{C}[\tau,z]$.

Moreover, we get in the same way an identification of $\mathrm{gr}_{U_\tau}^\gamma\widehat{R_FM}_{\mathrm{loc}}$ with the Rees module $R_{G(F)}\,\mathrm{gr}_{V_\theta}^\gamma G$. In particular it is $\mathbb{C}[z]$-free of finite rank, hence the strictness property.

As $tV_\theta^\gamma G\subset\partial_t^{-1}V_\theta^\gamma G=V_\theta^{\gamma-1}G$, we have

$$\tau\eth_\tau(1\otimes[G_0\cap V_\theta^\gamma G])=-\tau\otimes t[G_0\cap V_\theta^\gamma G]\subset\tau\otimes[G_0\cap V_\theta^{\gamma-1}G],$$

showing that $U_\tau^\gamma\widehat{R_FM}_{\mathrm{loc}}$ is stable by $\tau\eth_\tau$. Similarly, one shows that, for $N\gg 0$, $(\tau\eth_\tau-\gamma z)^N U_\tau^\gamma\widehat{R_FM}_{\mathrm{loc}}\subset U_\tau^{>\gamma}\widehat{R_FM}_{\mathrm{loc}}$. This gives $(*)_\infty$ (cf. [16, Lemma 3.3.4]).

For $(*)_0$, the argument is similar. It is easy to check that $R_{G(F)}V_{\theta'}^\gamma G$ is a $\mathbb{C}[\tau',z]\langle\tau'\eth_{\tau'}\rangle$-module, and that $\tau'\eth_{\tau'}-\gamma z$ is nilpotent on $R_{G(F)}\,\mathrm{gr}_{V_{\theta'}}^\gamma G$, which has no $\mathbb{C}[z]$-torsion by definition. The only new point is to check that $R_{G(F)}V_{\theta'}^\gamma G$ has finite type over $\mathbb{C}[\tau',z]\langle\tau'\eth_{\tau'}\rangle$.

Let k_0 be such that $G_{k_0}\subset V_{\theta'}^\gamma G$ and, for any $k\geqslant k_0$, let $\boldsymbol{e}_k$ be a finite system of $\mathbb{C}[\theta']$-generators of $G_k\cap V_{\theta'}^\gamma G$. Recalling that θ' acts as $z\tau'$ on $R_{G(F)}V_{\theta'}^\gamma G$, we find that, for any $k_1\geqslant k_0$, $\bigoplus_{k\leqslant k_1}z^k(G_k\cap V_{\theta'}^\gamma G)$ is contained in the $\mathbb{C}[\tau',z]$-submodule of $R_{G(F)}V_{\theta'}^\gamma G$ generated by the $z^j\boldsymbol{e}_j$, $j=k_0,\dots,k_1$.

On the other hand, the formula for the action of $\eth_\tau$ given above implies that $\tau'\eth_{\tau'}$ acts on $R_{G(F)}V_{\theta'}^\gamma G$ by $z\cdot(\theta'\partial_{\theta'}+k)$ on $z^k(G_k\cap V_{\theta'}^\gamma G)$.

We will show that, for k_1 large enough and any $k \geqslant k_1$, $z^k(G_k \cap V_{\theta'}^{\gamma} G)$ is contained in the $\mathbb{C}[\tau', z]\langle \tau' \eth_{\tau'}\rangle$-module generated by $z^{k_1} e_{k_1}$.

We claim that

(a) There exists k_1 such that, for any $k \geqslant k_1$,

$$G_1 \cap V_{\theta'}^{\gamma+k} G = \theta' \partial_{\theta'}(G_0 \cap V_{\theta'}^{\gamma+k} G) + G_0 \cap V_{\theta'}^{\gamma+k} G.$$

Note that this is equivalent to $G_{k+1} \cap V_{\theta'}^{\gamma} G = \theta' \partial_{\theta'}(G_k \cap V_{\theta'}^{\gamma} G) + G_k \cap V_{\theta'}^{\gamma} G$. If $k = k_1$, such an equality implies $z^{k_1+1}(G_{k_1+1} \cap V_{\theta'}^{\gamma} G) \subset (\mathbb{C}[\tau', z] + \mathbb{C}[\tau', z]\tau' \eth_{\tau'}) z^{k_1} e_{k_1}$. Iterating the argument gives the desired inclusion.

We will prove Claim (a) by working at the formal level. As it is clearly true away from $\theta' = 0$, it is enough to prove (a)$^\wedge$, that is, (a) after tensoring with $\mathbb{C}[\![\theta']\!]$.

Firstly, by uniqueness of the $V_{\theta'}^{\bullet}$-filtration, we have $(V_{\theta'}^{\gamma} G)^\wedge = V_{\theta'}^{\gamma}(G^\wedge)$. Moreover, $(V_{\theta'}^{\gamma} G \cap G_0)^\wedge = (V_{\theta'}^{\gamma} G)^\wedge \cap G_0^\wedge$ in $G^\wedge$ [indeed, use that this is clearly true for $+$ instead of $\cap$ and that $\mathbb{C}[\![\theta']\!]$ is flat over $\mathbb{C}[\theta']$, and apply this to $(V_{\theta'}^{\gamma} G + G_0)^\wedge / G_0^\wedge = (V_{\theta'}^{\gamma} G)^\wedge / (V_{\theta'}^{\gamma} G \cap G_0)^\wedge$]. Therefore, it is enough to prove (a)$^\wedge$.

Notice now that Claim (a) is equivalent to

(b) There exists k_1 such that, for any $k \geqslant k_1$,

$$\theta'^2 \partial_{\theta'} : V_{\theta'}^{\gamma+k}(G_0/\theta' G_0) \longrightarrow V_{\theta'}^{\gamma+k+1}(G_0/\theta' G_0)$$

is onto.

Similarly, (a)$^\wedge$ is equivalent to (b)$^\wedge$. Recall that $G^\wedge$ decomposes as $G^\wedge_{\mathrm{reg}} \oplus G^\wedge_{\mathrm{irr}}$ and that $V_{\theta'}^{\gamma} G^\wedge$ and G_0 decompose correspondingly. It is thus enough to prove (b)$^\wedge$ on each term. On the regular part, there exists k_1 such that $V_{\theta'}^{\gamma+k_1}(G^\wedge_{\mathrm{reg},0}/\theta' G^\wedge_{\mathrm{reg},0}) = 0$ (because $V_{\theta'}^{\gamma+k} G^\wedge_{\mathrm{reg}}$ has finite type over $\mathbb{C}[\![\theta']\!]$), hence both terms are 0 in (b)$^\wedge$. On the purely irregular part, $V_{\theta'}^{\gamma} G^\wedge_{\mathrm{irr}} = G^\wedge_{\mathrm{irr}}$ for any γ, and $\theta'^2 \partial'_\theta$ does not have the eigenvalue 0 on $G^\wedge_{\mathrm{irr},0}/\theta' G^\wedge_{\mathrm{irr},0}$ (cf. Remark 5.19), hence it is onto. Q.E.D.

For the remaining of this section, we assume that $(M, F^\bullet M)$ is also equipped with a sesquilinear pairing k such that $(M, F^\bullet M, k)$ comes from a polarized complex Hodge $\mathscr{D}$-module on $\mathbb{P}^1$.

Corollary 5.21. $\prod_{\beta \in (-1,0]} \prod_{\ell \in \mathbb{Z}} \mathrm{SP}^\infty_{\mathrm{gr}_\ell^{\mathrm{M}} \Psi_\tau^\beta \widehat{R_F M}}(T) = \mathrm{SP}^\infty_{G_0^{(F)}}(T)$.

Proof. For $\beta \in (-1, 0]$, let $G^{(F),\bullet} \psi_\theta^\beta G$ be the filtration naturally induced by $G^{(F),\bullet}$. As a consequence of the identification $V_\tau^\gamma \widehat{R_F M}_{\mathrm{loc}}$ with $R_{G^{(F)}} V_\theta^\gamma G$, we get, for any $\beta \in (-1, 0]$,

$$\Psi_\tau^\beta \widehat{R_F M} = R_{G^{(F)}} \psi_\theta^\beta G. \tag{5.22}$$

Note however that the natural action of $z^2\partial_z$ on the left-hand side differs by $\tau\eth_\tau$ from the action on the right-hand side defined from the z-grading (cf. Lemma 4.4). Nevertheless, by our assumption on $(M, F^\bullet M)$, the graded pieces of $\Psi_\tau^\beta\widehat{R_F M}$ with respect to the monodromy filtration are strict (i.e., $\mathbb{C}[z]$-free), and at the level of $\mathrm{gr}_\bullet^{\mathrm{M}}$, $z^2\partial_z$ on the left-hand side differs by βz from the action on the right-hand side. Because of freeness and uniqueness of the monodromy filtration, we have

$$\mathrm{gr}_\ell^{\mathrm{M}}\Psi_\tau^\beta\widehat{R_F M} = R_{G(F)}\,\mathrm{gr}_\ell^{\mathrm{M}}\psi_\theta^\beta G.$$

On the other hand, as the action of ∂_z has a simple pole on $R_{G(F)}\psi_\theta^\beta G$, we can apply Remark 1.9 to get

$$\prod_{\ell\in\mathbb{Z}}\mathrm{SP}^\infty_{\mathrm{gr}_\ell^{\mathrm{M}}\Psi_\tau^\beta\widehat{R_F M}}(T+\beta) = \prod_{\ell\in\mathbb{Z}}\mathrm{SP}^\infty_{R_{G(F)}\,\mathrm{gr}_\ell^{\mathrm{M}}\psi_\theta^\beta G}(T) = \mathrm{SP}^\infty_{R_{G(F)}\psi_\theta^\beta G}(T).$$

Recall that, if we set

$$\nu_{\beta,p} = \dim\frac{V_\theta^\beta G\cap G^{(F),p}}{V_\theta^{>\beta}G\cap G^{(F),p} + V_\theta^\beta G\cap G^{(F),p+1}}$$

we have, as in the proof of Lemma 5.8, and by Definition 1.2 and (1.3),

$$\mathrm{SP}^\infty_{R_{G(F)}\psi_\theta^\beta G}(T) = \prod_{p\in\mathbb{Z}}(T-p)^{\nu_{\beta,p}}$$

and

$$\mathrm{SP}^\infty_{G_0^{(F)}}(T) = \prod_{\beta\in(-1,0]}\prod_{p\in\mathbb{Z}}(T-\beta-p)^{\nu_{\beta,p}}.$$

This gives the desired equality. Q.E.D.

We now consider the specialization at $\partial_t^{-1} = 0$. Let $G_i^\wedge$ be the formal microlocalized module attached to M at $-c_i$. Let p_i be such that $F^{p_i}\mathcal{M}$ generates $\mathcal{M}$ as a $\mathscr{D}$-module near $-c_i$, let $G_{i,0}^{(F^{p_i})}$ be the saturation by $\theta' := \partial_t^{-1}$ of the image of $F^{p_i}\mathcal{M}$ in $G_i^\wedge$ (by tensoring with formal microlocal differential operators of order zero), and let us set $G_{i,0}^{(F)} = \theta'^{-p_i}G_{i,0}^{(F_{p_i})}$, which is independent of the generating index p_i (cf. §1.c). Then it is known (cf. e.g. [14, Prop. V.3.6]) that the Levelt–Turrittin decomposition (1.6) for $G_0^{(F)}$ has components $\mathscr{H}_i' = G_{i,0}^{(F)}$. We have $G_i^{(F),\bullet}\psi_{\theta'}^\gamma G_i^\wedge = G^{(F),\bullet}\psi_{\theta'}^{c_i/\theta',\gamma}G$.

Corollary 5.23. *For any i,*

$$\prod_{\beta\in(-1,0]}\prod_{\ell\in\mathbb{Z}}\mathrm{SP}^0_{\mathrm{gr}_\ell^{\mathrm{M}}\Psi_{\tau'}^{c_i/\tau',\beta}\widehat{R_F M}}(T) = \mathrm{SP}^0_{G_{i,0}^{(F)}}(T).$$

Proof. We will show the result for $c_i = 0$. The same argument applies for any c_i after twisting by $\mathcal{E}^{-c_i/\theta'}$ or $\mathscr{E}^{-c_i/z\tau'}$. We denote by i_0 the index i such that $c_{i_0} = 0$.

From $(*)_0$ we get, for any $\beta \in (-1, 0]$,

$$\Psi^{0,\beta}_{\tau'}\widehat{R_F M} = R_{G^{(F)}}\psi^{0,\beta}_{\theta'}G = R_{G^{(F)}_{i_0}}\psi^{\beta}_{\theta'}G^{\wedge}_{i_0},$$

and this equality is compatible with the action of $z^2\partial_z - \tau'\eth_{\tau'}$ on the left-hand side and that of $z^2\partial_z$ on the right-hand side (cf. Lemma 4.4). By our assumption on $(M, F^\bullet M)$, we know from Appendix A that the graded pieces of $\Psi^{c_{i_0}/\tau',\beta}_{\tau'}\widehat{R_F M}$ with respect to the monodromy filtration are strict (i.e., $\mathbb{C}[z]$-free). The same property holds for the right-hand side above, and going to the graded pieces, we find that the equality holds with $z^2\partial_z$ action on the left-hand side shifted by $-\beta z$. Arguing as for Corollary 5.21, we find

$$\prod_{\ell\in\mathbb{Z}} \mathrm{SP}^0_{\mathrm{gr}^{\mathrm{M}}_\ell \Psi^{0,\beta}_{\tau'}\widehat{R_F M}}(T-\beta) = \mathrm{SP}^0_{R_{G^{(F)}_{i_0}}\psi^{\beta}_{\theta'}G^{\wedge}_{i_0}}(T).$$

Setting now

$$\mu_{i_0,\beta,p} = \dim \frac{V^{\beta}_{\theta'}G^{\wedge}_{i_0} \cap G^{(F),p}_{i_0}}{V^{>\beta}_{\theta'}G^{\wedge}_{i_0} \cap G^{(F),p}_{i_0} + V^{\beta}_{\theta'}G^{\wedge}_{i_0} \cap G^{(F),p+1}_{i_0}},$$

we have

$$\mathrm{SP}^0_{R_{G^{(F)}_{i_0}}\psi^{\beta}_{\theta'}G^{\wedge}_{i_0}}(T) = \prod_{p\in\mathbb{Z}}(T-p)^{\mu_{i_0,\beta,p}}$$

and

$$\mathrm{SP}^0_{G^{(F)}_{i_0,0}}(T) = \prod_{\beta\in(-1,0]}\prod_{p\in\mathbb{Z}}(T+\beta-p)^{\mu_{i_0,\beta,p}},$$

hence the result. Q.E.D.

§6. Deligne's filtration

In this section, we will be concerned with the first point considered in the introduction. Let us consider the setting of §1.c, that is, a holonomic $\mathbb{C}[t]\langle\partial_t\rangle$-module equipped with a good filtration $F^\bullet M$. Recall that $\widetilde{\mathcal{M}}$ denotes the associated $\mathscr{D}_{\mathbb{P}^1}(*\infty)$-module with connection and $\mathcal{M}$ denotes its minimal extension across ∞. We will now assume that M has only regular singularities at finite distance and at infinity.

We will keep the notation of §1.c, but we will simply denote by G_0 the $\mathbb{C}[\partial_t^{-1}]$-module $G_0^{(F)}$ defined by (1.11). The spectral polynomial

$\mathrm{SP}^{\infty}_{G_0}(T)$ is determined as soon as we determine the number $\nu_\gamma(G_0)$ for any $\gamma \in \mathbb{R}$.

In §6.c, we will define Deligne's filtration $F^{\bullet}_{\mathrm{Del}}$ (indexed by $\mathbb{R}$) on $\mathcal{M} \otimes \mathcal{E}^{-t}$, then on the corresponding de Rham complex, and then on its hypercohomology. The main result of this section will be:

Theorem 6.1. *Assume that $(M, F^{\bullet}M)$ underlies a polarized complex Hodge $\mathscr{D}$-module (cf. Definition 5.11). Then,*

(1) *the spectral sequence associated to the hypercohomology of the filtered de Rham complex $F^{\bullet}_{\mathrm{Del}} \mathrm{DR}(\widetilde{\mathcal{M}} \otimes \mathcal{E}^{-t})$ on $\mathbb{P}^1$ degenerates at E_1;*

(2) *for any $\gamma \in \mathbb{R}$, $\nu_\gamma(G_0) = \dim \mathbb{H}^1\big(\mathbb{P}^1, \mathrm{gr}^{\gamma+1}_{F_{\mathrm{Del}}} \mathrm{DR}(\widetilde{\mathcal{M}} \otimes \mathcal{E}^{-t})\big)$.*

Let us remark that the $\mathscr{O}_{\mathbb{P}^1}$-coherent sheaf $\mathrm{gr}^{\gamma}_{F_{\mathrm{Del}}}(\widetilde{\mathcal{M}} \otimes \mathcal{E}^{-t})$ is supported at infinity if $\gamma \notin \mathbb{Z}$.

6.a. Laplace transform

We denote by $\widehat{\mathbb{P}}^1$ the projective line with coordinates θ, θ' (that we do not denote by τ, τ' as above at the moment) and by $\widehat{\mathbb{A}}^1$ its chart with coordinate θ. Recall that $G = \mathbb{C}[\theta', \theta'^{-1}] \otimes_{\mathbb{C}[\theta']} G_0$ ($\theta' = \partial_t^{-1}$ as above) is equipped with a connection having a regular singularity at $\theta = 0$, defined as the multiplication by $-t$. We denote by $V^{\bullet}_{\theta} G$ the corresponding V-filtration, that we assume to be indexed by $\mathbb{R}$ (this assumption is implied by the assumption in Theorem 6.1 that $(M, F^{\bullet}M)$ underlies a polarized complex Hodge $\mathscr{D}$-module).

The exponentially twisted de Rham complex (4.1) is quasi-isomorphic to

$$0 \longrightarrow G \xrightarrow{\partial_t - 1} G \longrightarrow 0,$$

which is quasi-isomorphic to

$$0 \longrightarrow G \xrightarrow{\partial_t^{-1} - 1} G \longrightarrow 0,$$

which in turn is quasi-isomorphic to

$$0 \longrightarrow G_0 \xrightarrow{\partial_t^{-1} - 1} G_0 \longrightarrow 0.$$

In other words, the hypercohomology $H^1_{\mathrm{DR}}(\mathbb{P}^1, \mathcal{M} \otimes \mathcal{E}^{-t})$ is identified with the fibre at $\theta' = 1$ of the free $\mathbb{C}[\theta']$-module G_0.

The V-filtration $V^{\bullet}_{\theta} G$ enables one to define, in a natural way, a filtration $V^{\bullet} H^1_{\mathrm{DR}}(\mathbb{P}^1, \mathcal{M} \otimes \mathcal{E}^{-t})$ by setting, for any $\gamma \in \mathbb{R}$,

$$V^\gamma H^1_{\mathrm{DR}}(\mathbb{P}^1, \mathcal{M}\otimes\mathcal{E}^{-t}) = V^\gamma(G_0/(\theta'-1)G_0)$$
$$:= \mathrm{image}\big[G_0\cap V^\gamma_\theta G \to G_0/(\theta'-1)G_0\big].$$

According to (1.3), we have

$$\nu_\gamma(G_0) = \dim \mathrm{gr}^\gamma_V H^1_{\mathrm{DR}}(\mathbb{P}^1, \mathcal{M}\otimes\mathcal{E}^{-t}). \tag{6.2}$$

Example 6.3. Let us consider the case where $M = \mathbb{C}[t]\langle\partial_t\rangle/(t\partial_t - \alpha)$ for some $\alpha\in(0,1)$. We regard M as corresponding to a variation of Hodge structure V of type $(0,0)$ on $\mathbb{A}^1\smallsetminus\{0\}$ with filtration $F^\bullet V$ given by $F^0V = V$ and $F^1V = 0$. Let $V^\bullet_t M$ denote the V-filtration of M at $t=0$. Then we set (cf. (5.12)) $F^0M = V_t^{>-1}M$, $F^1M = 0$ and, for $\ell\geqslant 0$, $F^{-\ell}M := V_t^{>-1}M + \cdots + \partial_t^\ell V_t^{>-1}M = V_t^{>-\ell-1}M$.

Denoting by $[\cdot]$ the class in M, we have $[1]\in V^\alpha_t M$, and $F^0M = \mathbb{C}[t]\cdot[\partial_t]$. We also have $\widehat{M} = G = \mathbb{C}[\theta]\langle\partial_\theta\rangle/(\partial_\theta\theta+\alpha)$ which is free of rank one over $\mathbb{C}[\theta,\theta^{-1}]$, and $G^{(F)}_0 = \mathbb{C}[\theta^{-1}]\cdot[\theta]$. As $[\theta]$ is in $V^{-\alpha}_\theta G$, we finally get

$$\nu_\gamma(G^{(F)}_0) = \begin{cases} 1 & \text{if } \gamma = -\alpha,\\ 0 & \text{otherwise.}\end{cases}$$

Let us consider the V-filtration $V^\bullet\widehat{R_FM}_{\mathrm{loc}}$ (cf. Lemma 5.20). Notice that, for any $\gamma\in\mathbb{R}$, the multiplication by $\tau - z$ is injective on $V^\gamma\widehat{R_FM}_{\mathrm{loc}}$. Indeed, let us use as in Lemma 5.20 the identification $V^\gamma\widehat{R_FM}_{\mathrm{loc}} = R_{G^{(F)}}V^\gamma_\theta G$. Then, the localization with respect to z gives $\mathbb{C}[z,z^{-1}]\otimes_{\mathbb{C}} V^\gamma_\theta G$, where the action of τ is induced by $z\otimes\theta$. In particular, it is $\mathbb{C}[\tau,z,z^{-1}]$-free and the multiplication by $\tau - z$ is injective on this module. Therefore, so is the multiplication by $\tau - z$ on the $\mathbb{C}[\tau,z]$-submodule $R_{G^{(F)}}V^\gamma_\theta G$. We will compute its cokernel $V^\gamma\widehat{R_FM}_{\mathrm{loc}}/(\tau-z)V^\gamma\widehat{R_FM}_{\mathrm{loc}}$.

Recall (cf. [5, Def. B.1]) that a V-solution to the Birkhoff problem for G_0 is a free $\mathbb{C}[\theta]$-submodule G'^0 of G, which is stable by $\theta\partial_\theta$, which generates G over $\mathbb{C}[\theta,\theta^{-1}]$, and such that, for any $\gamma\in\mathbb{R}$,

$$G_0\cap V^\gamma_\theta G = \bigoplus_{j\geqslant 0}\theta'^j(G_0\cap G'^0\cap V^{\gamma+j}_\theta G). \tag{6.4}$$

(Each term in the sum, as well as $G_0\cap G'^0$ and $G_0\cap V^\gamma_\theta G$, is a finite dimensional $\mathbb{C}$-vector space, and the sum is finite, as $G_0\cap V^{\gamma+j}_\theta G = 0$ for $j\gg 0$; moreover, $G'^0\subset V^\gamma_\theta G$ for $\gamma\ll 0$.) By definition of a solution to Birkhoff's problem, a $\mathbb{C}$-basis $G_0\cap G'^0$ is a $\mathbb{C}[\theta']$-basis of G_0; therefore, the natural morphism $G_0\cap G'^0\to G_0/(\theta'-1)G_0$ is an isomorphism.

Lemma 6.5. *If the Birkhoff problem for G_0 has a V-solution G'^0, then, for any $\gamma \in \mathbb{R}$, $V^\gamma \widehat{R_F M}_{\mathrm{loc}}/(\tau - z)V^\gamma \widehat{R_F M}_{\mathrm{loc}}$ is identified with the Rees module of the filtration $V^{\gamma+\bullet} H^1_{\mathrm{DR}}(\mathbb{P}^1, \mathcal{M} \otimes \mathcal{E}^{-t})$.*

Proof. We note that $(*)_\infty$ in Lemma 5.20 gives (as $z = \tau \otimes \theta'$):

$$V^\gamma \widehat{R_F M}_{\mathrm{loc}}/(\tau - z)V^\gamma \widehat{R_F M}_{\mathrm{loc}} = \bigoplus_k \tau^k \otimes \Big[(G_0 \cap V_\theta^{\gamma-k} G)/(\theta' - 1)(G_0 \cap V_\theta^{\gamma-k+1} G)\Big].$$

As G'^0 is a V-solution, the natural inclusion

$$(\theta' - 1)(G_0 \cap V_\theta^{\gamma+1} G) \subset [(\theta' - 1)G_0] \cap (G_0 \cap V_\theta^\gamma G)$$

is an equality for any $\gamma \in \mathbb{R}$: indeed, since $G_0 = \bigcup_{\ell \in \mathbb{Z}} (G_0 \cap V_\theta^{\gamma-\ell} G)$, an element in the RHS can be written both as a polynomial $(\theta' - 1)\sum_{j \geqslant 0} a_j \theta'^j$ with $a_j \in G_0 \cap G'^0 \cap V_\theta^{\gamma-\ell+j} G$ for some fixed $\ell \in \mathbb{Z}$, and as a polynomial $\sum_{j \geqslant 0} b_j \theta'^j$ with $b_j \in G_0 \cap G'^0 \cap V_\theta^{\gamma+j} G$, according to (6.4); the assertion follows by considering the term of highest degree with respect to θ' and by a straightforward induction. As a consequence,

$$(G_0 \cap V_\theta^\gamma G)/(\theta' - 1)(G_0 \cap V_\theta^{\gamma+1} G)$$

is equal to the image of $G_0 \cap V_\theta^\gamma G$ in $G_0/(\theta' - 1)G_0$ for any γ. The result follows. Q.E.D.

The previous proof also shows that the morphism $G_0 \cap V_\theta^\gamma G \to G_0/(\theta' - 1)G_0$ induces an isomorphism

$$G_0 \cap G'^0 \cap V_\theta^\gamma G \xrightarrow{\sim} V^\gamma(G_0/(\theta' - 1)G_0) \simeq V^\gamma H^1_{\mathrm{DR}}(\mathbb{P}^1, \mathcal{M} \otimes \mathcal{E}^{-t}).$$

As a consequence of the lemma, for any $\beta \in (-1, 0]$,

$$V^\beta \widehat{R_F M}_{\mathrm{loc}}/(\tau - z)V^\beta \widehat{R_F M}_{\mathrm{loc}} = R_{V^{\beta+\bullet}} H^1_{\mathrm{DR}}(\mathbb{P}^1, \mathcal{M} \otimes \mathcal{E}^{-t}). \tag{6.6}$$

Lemma 6.7. *If $(M, F_\bullet M)$ underlies a polarizable complex Hodge $\mathcal{D}$-module, then the Birkhoff problem for the Brieskorn lattice of $(M, F_\bullet M)$ has a V-solution.*

Proof. We follow the argument of [22, Lemma 2.8]. According to [5, Prop. B.3(1)], giving a V-solution to the Brieskorn problem for $G_0^{(F)}$ is equivalent to giving, for any $\beta \in (-1, 0]$, a filtration of $\psi_\theta^\beta G$ which is opposite to $G^{(F),\bullet}\psi_\theta^\beta G$ and which is stable by the nilpotent operator N_θ induced by $-(\theta\partial_\theta - \beta)$. According to (5.22) and Corollary

5.14(3), $G^{(F),\bullet}\psi_\theta^\beta G$ is the Hodge filtration of a polarized complex mixed Hodge structure for which the nilpotent endomorphism is a nonzero multiple of N_θ. Remark 5.5 gives then a convenient opposite filtration.

Q.E.D.

6.b. Deligne's filtration on the exponentially twisted $\mathscr{D}_X$-module

In this subsection, we denote by X an open disc in $\mathbb{C}$ with a coordinate x centered at its origin. Let $\mathcal{M}$ be a regular holonomic $\mathscr{D}_X$-module equipped with a good filtration $F_\bullet\mathcal{M}$. We will make the following assumptions:

(1) $\mathcal{M}$ has a singularity at $x = 0$ at most and is the minimal extension of its localized module $\widetilde{\mathcal{M}} := \mathscr{O}_X[1/x] \otimes_{\mathscr{O}_X} \mathcal{M}$.

(2) The eigenvalues of the monodromy of the local system $\mathrm{Ker}\,[\partial_x : \mathcal{M}_{|X^*} \to \mathcal{M}_{|X^*}]$ have an absolute value equal to 1. (Hence, the decreasing Kashiwara–Malgrange filtration $V^\bullet\mathcal{M}$ of $\mathcal{M}$ at the origin is indexed by a finite set of real numbers translated by $\mathbb{Z}$.)

We denote by $\mathcal{M}\otimes\mathcal{E}^{-1/x}$ the $\mathscr{O}_X[1/x]$-module $\widetilde{\mathcal{M}}$ equipped with the twisted connection $\nabla - d(1/x)$ (i.e., if e is the generator of the rank one $\mathscr{O}_X[1/x]$-module $\mathcal{E}^{-1/x}$, we have $\partial_x(e\otimes m) = e\otimes((\partial_x + x^{-2})m))$. For any $\gamma\in\mathbb{R}$, we denote by $\lceil\gamma\rceil$ the smallest integer $\geqslant\gamma$, so that $\gamma - \lceil\gamma\rceil \in (-1,0]$. Deligne's filtration is defined for $\gamma\in\mathbb{R}$ by:

$$F^\gamma_{\mathrm{Del}}(\widetilde{\mathcal{M}}\otimes\mathcal{E}^{-1/x}) := \sum_{k\geqslant 0}\partial_x^k x^{-1}(F^{\lceil\gamma\rceil+k}V^{\gamma-\lceil\gamma\rceil}\mathcal{M}\otimes\mathcal{E}^{-1/x}).$$

(The usefulness of the shift by x^{-1} will appear later.) From now on, we will skip the term $\otimes\mathcal{E}^{-1/x}$, so we will write

$$F^\gamma_{\mathrm{Del}}\widetilde{\mathcal{M}} := \sum_{k\geqslant 0}(\partial_x + x^{-2})^k x^{-1}F^{\lceil\gamma\rceil+k}V^{\gamma-\lceil\gamma\rceil}\mathcal{M}.$$

The sum above consists of a finite number of terms since, for $\beta\in(-1,0]$ fixed, $F^rV^\beta\mathcal{M} = 0$ for $r\gg 0$. Therefore, each $F^\gamma_{\mathrm{Del}}\widetilde{\mathcal{M}}$ is a locally free $\mathscr{O}_X$-module of finite rank. Moreover, we clearly have the transversality property $(\partial_x + x^{-2})F^\gamma_{\mathrm{Del}}\widetilde{\mathcal{M}} \subset F^{\gamma-1}_{\mathrm{Del}}\widetilde{\mathcal{M}}$.

Let us show that the filtration is decreasing. Assume that $\gamma' := \beta' + p' \geqslant \gamma := \beta + p$, with $\beta,\beta'\in(-1,0]$ and $p,p'\in\mathbb{Z}$. The inclusion $F^{\beta'+p'}_{\mathrm{Del}} \subset F^{\beta+p}_{\mathrm{Del}}$ is clear if $\beta'\geqslant\beta$ (so $p'\geqslant p$). It remains to consider the case where $\beta'<\beta$ and $p'\geqslant p+1$ and it is enough to assume $p' = p+1$.

Writing $1 = (\partial_x + x^{-2})x^2 - \partial_x x^2$, we get, for $k \geqslant 0$,

$$(\partial_x + x^{-2})^k x^{-1} F^{p+1+k} V^{\beta'} \mathcal{M} \subset (\partial_x + x^{-2})^{k+1} x^{-1} F^{p+1+k} V^{\beta'+2} \mathcal{M} + (\partial_x + x^{-2})^k \partial_x x F^{p+1+k} V^{\beta'} \mathcal{M}.$$

In the right-hand side, the first term is contained in $F_{\mathrm{Del}}^{\beta+p}$ as $\beta' + 2 \geqslant \beta$. For the second one, we note that $\partial_x x F^{p+1+k} V^{\beta'} \subset F^{p+k} V^{\beta'} \subset x^{-1} F^{p+k} V^{\beta'+1}$, so the second term is also contained in $F_{\mathrm{Del}}^{\beta+p}$, as $\beta' + 1 \geqslant \beta$.

Example 6.8. Let us assume, as in [4], that (5.12) holds and that $j^* F^\bullet \mathcal{M}$ has only one jump at $p = 0$, so $j^* F^1 \mathcal{M} = 0$ and $j^* F^0 \mathcal{M} = j^* \mathcal{M}$. Then, for $\beta \in (-1, 0]$ and $p \in \mathbb{Z}$,

$$F_{\mathrm{Del}}^{\beta+p} \widetilde{\mathcal{M}} = \begin{cases} 0 & \text{if } p \geqslant 1, \\ x^{2p-1} V^\beta & \text{if } p \leqslant 0. \end{cases}$$

Indeed, if $p = -1$ for instance,

$$F_{\mathrm{Del}}^{\beta-1} \widetilde{\mathcal{M}} = (\partial_x + x^{-2}) x^{-1} V^\beta + x^{-1} V^\beta = x^{-3} V^\beta,$$

as $x^2 \partial_x + 1$ is invertible on $V^\beta \mathcal{M}$ (as $\mathcal{M}$ has a regular singularity at $x = 0$, it is enough to check this on modules like $\mathscr{O}_X\langle\partial_x\rangle/(x\partial_x - \alpha)^\mu$).

6.c. Deligne's filtration on the de Rham complex

We now consider the case where $(\mathcal{M}, F^\bullet)$ is a filtered $\mathscr{D}$-module on the projective line $\mathbb{P}^1$. We denote by t a fixed affine coordinate on the affine line $\mathbb{A}^1 = \mathbb{P}^1 \smallsetminus \{\infty\}$. We define the Deligne filtration on $\mathcal{M} \otimes \mathcal{E}^{-t}$, with $\widetilde{\mathcal{M}} := \mathscr{O}_{\mathbb{P}^1}(*\infty) \otimes_{\mathscr{O}_{\mathbb{P}^1}} \mathcal{M}$, by the following formulas (for simplicity, we identify $\mathcal{M} \otimes \mathcal{E}^{-t}$ with the $\mathscr{O}_{\mathbb{P}^1}(*\infty)$-module $\widetilde{\mathcal{M}}$ with twisted connection $\nabla - dt$):

- Away from ∞, we set $F_{\mathrm{Del}}^{\beta+p} \widetilde{\mathcal{M}} = F^p \mathcal{M}$ for $\beta \in (-1, 0]$, and $p \in \mathbb{Z}$,
- near ∞, we set $x = 1/t$ and use the definition of §6.b.

The de Rham complex $\mathrm{DR}(\widetilde{\mathcal{M}} \otimes \mathcal{E}^{-t})$ is filtered by setting, for each $\gamma \in \mathbb{R}$,

$$F_{\mathrm{Del}}^\gamma \mathrm{DR}(\widetilde{\mathcal{M}} \otimes \mathcal{E}^{-t}) = \{0 \to F_{\mathrm{Del}}^\gamma(\widetilde{\mathcal{M}} \otimes \mathcal{E}^{-t}) \xrightarrow{\nabla} \Omega_{\mathbb{P}^1}^1 \otimes F_{\mathrm{Del}}^{\gamma-1}(\widetilde{\mathcal{M}} \otimes \mathcal{E}^{-t}) \to 0\},$$

a complex which is also written as

$$\{0 \to F_{\mathrm{Del}}^\gamma \widetilde{\mathcal{M}} \xrightarrow{\nabla - dt} \Omega_{\mathbb{P}^1}^1 \otimes F_{\mathrm{Del}}^{\gamma-1} \widetilde{\mathcal{M}} \to 0\}.$$

Example 6.9. Let us consider the case of Example 6.3. Using the computation in Example 6.8, we find, on $\mathbb{P}^1 \setminus \{t=0\}$ with the coordinate t',

$$F_{\mathrm{Del}}^{\gamma}(\widetilde{\mathcal{M}} \otimes \mathcal{E}^{-t}) = \begin{cases} 0 & \text{if } \gamma > 0, \\ \mathbb{C}[t']t'^{-2p}[1] & \text{if } \gamma = -p,\ p \in \mathbb{N}, \\ \mathbb{C}[t']t'^{-(2p+1)}[1] & \text{if } \gamma = -\alpha - p,\ p \in \mathbb{N}. \end{cases}$$

In particular, the complex $\mathrm{gr}_{F_{\mathrm{Del}}}^{1-\alpha} \mathrm{DR}(\widetilde{\mathcal{M}} \otimes \mathcal{E}^{-t})$ reduces to the complex having only $\mathrm{gr}_{F_{\mathrm{Del}}}^{-\alpha}(\widetilde{\mathcal{M}} \otimes \mathcal{E}^{-t})$ in degree one, and we get 6.1(2) in that case.

Theorem 6.1 is a consequence of the following proposition, together with Lemma 6.7 and (6.2).

Proposition 6.10. *If* $(\mathcal{M}, F^{\bullet})$ *underlies a polarized complex Hodge* $\mathscr{D}$*-module on* $\mathbb{P}^1$*, then the filtered de Rham complex* $\boldsymbol{R}\Gamma\,\mathrm{DR}(\widetilde{\mathcal{M}} \otimes \mathcal{E}^{-t}, F_{\mathrm{Del}}^{\bullet})$ *is strict, that is, for any* $\gamma \in \mathbb{R}$*, the natural morphism*

$$\mathbb{H}^*\big(\mathbb{P}^1, F_{\mathrm{Del}}^{\gamma}\,\mathrm{DR}(\widetilde{\mathcal{M}} \otimes \mathcal{E}^{-t})\big) \longrightarrow \mathbb{H}^*\big(\mathbb{P}^1, \mathrm{DR}(\widetilde{\mathcal{M}} \otimes \mathcal{E}^{-t})\big)$$

is injective, with image $V^{\gamma-1}\mathbb{H}^1\big(\mathbb{P}^1, \mathrm{DR}(\widetilde{\mathcal{M}} \otimes \mathcal{E}^{-t})\big)$ *(when* $* = 1$*).*

Proof. We assume for simplicity that $R_F\mathcal{M}$ underlies a polarized Hodge $\mathscr{D}$-module of weight 0, with polarization $(\mathrm{Id}, \mathrm{Id})$, that we denote $\mathscr{T} = (R_F\mathcal{M}, R_F\mathcal{M}, R_F k) = (\mathscr{M}, \mathscr{M}, C)$ (cf. [19]). Let us consider a new copy of the affine line, that we denote by $\widehat{\mathbb{A}}^1$ with coordinate τ, and let us denote by $p : \mathbb{P}^1 \times \widehat{\mathbb{A}}^1 \to \mathbb{P}^1$ the projection. Let us set ${}^{\mathscr{F}}\mathscr{T} = ({}^{\mathscr{F}}\mathscr{M}, {}^{\mathscr{F}}\mathscr{M}, {}^{\mathscr{F}}C)$ with

$${}^{\mathscr{F}}\mathscr{M} := \mathscr{E}^{-t\tau/z} \otimes p^+ \widetilde{\mathscr{M}}$$

and ${}^{\mathscr{F}}C$ defined as in [18, §3]. If $V^{\bullet}\widetilde{\mathscr{M}}$ denotes the V-filtration along $t' = 0$, we have $V_{t'}^{\beta}\widetilde{\mathscr{M}} = R_F V_{t'}^{\beta}\mathcal{M}$ for $\beta > -1$. The V-filtration of ${}^{\mathscr{F}}\mathscr{M}$ along $\tau = 0$ is computed in [18]. In a chart near $(t = \infty, \tau = 0)$, setting $t' = 1/t$, it is given by the formula (if $\beta \in (-1, 0]$)

$$V_{\tau}^{\beta}\,{}^{\mathscr{F}}\mathscr{M} = \sum_{k \geqslant 0} \eth_{t'}^{k}\big(p^* R_F V_{t'}^{\beta-1}\widetilde{\mathcal{M}} \otimes \mathscr{E}^{-t\tau/z}\big).$$

(There is also a formula for $V_{\tau}^{\gamma}\,{}^{\mathscr{F}}\mathscr{M}$ for any $\gamma \in \mathbb{R}$, but it will be needed here.) This can be rewritten as

$$V_{\tau}^{\beta}\,{}^{\mathscr{F}}\mathscr{M} = \sum_{k \geqslant 0} (1 \otimes z\partial_{t'} + \tau \otimes t'^{-2})^k \big(\mathbb{C}[\tau] \otimes_{\mathbb{C}} t'^{-1} R_F V_{t'}^{\beta}\widetilde{\mathcal{M}}\big).$$

Let us now consider the multiplication by $\tau - z$. It is clearly injective on ${}^{\mathcal{F}}\!\mathcal{M}$, hence on each $V_\tau^\beta {}^{\mathcal{F}}\!\mathcal{M}$. The cokernel is given by

$$V_\tau^\beta {}^{\mathcal{F}}\!\mathcal{M}/(\tau - z)V_\tau^\beta {}^{\mathcal{F}}\!\mathcal{M} = \sum_{k\geqslant 0}(\partial_{t'} + t'^{-2})^k t'^{-1} z^k \mathbb{C}[z] R_F V_{t'}^\beta \widetilde{\mathcal{M}}$$

that is, for any $\beta \in (-1, 0]$,

$$V_\tau^\beta {}^{\mathcal{F}}\!\mathcal{M}/(\tau - z)V_\tau^\beta {}^{\mathcal{F}}\!\mathcal{M} = R_{F_{\mathrm{Del}}^{\beta+\bullet}}(\widetilde{\mathcal{M}} \otimes \mathcal{E}^{-1/t'}). \tag{6.11}$$

In a chart away from $t = \infty$ (and near $\tau = 0$), we have, for $\beta \in (-1, 0]$,

$$V_\tau^\beta {}^{\mathcal{F}}\!\mathcal{M} = {}^{\mathcal{F}}\!\mathcal{M},$$

and therefore (6.11) remains valid in this chart.

Let $\widehat{p} : \mathbb{P}^1 \times \widehat{\mathbb{A}}^1 \to \widehat{\mathbb{A}}^1$ denote the projection. Then, according to [16, Th. 3.3.15], [18, Prop. 4.1(ii) and Cor. 5.9] and [16, Th. 6.1.1], the filtered complex $\widehat{p}_+ V_\tau^\bullet {}^{\mathcal{F}}\!\mathcal{M}$ is strict, that is, $\mathcal{H}^j \widehat{p}_+ V_\tau^\beta {}^{\mathcal{F}}\!\mathcal{M} \to \mathcal{H}^j \widehat{p}_+ {}^{\mathcal{F}}\!\mathcal{M}$ is injective for any j and β, therefore $\mathcal{H}^j \widehat{p}_+ V_\tau^\beta {}^{\mathcal{F}}\!\mathcal{M} = 0$ for any $j \neq 0$ and any β, and $\mathcal{H}^0 \widehat{p}_+ V_\tau^\beta {}^{\mathcal{F}}\!\mathcal{M}$ is identified with $V_\tau^\beta \mathcal{H}^0 \widehat{p}_+ {}^{\mathcal{F}}\!\mathcal{M} = V_\tau^\beta \widehat{R_F M} = V_\tau^\beta \widehat{R_F M}_{\mathrm{loc}}$ (because $\beta > -1$).

We thus have an exact sequence

$$0 \to \mathcal{H}^0 \widehat{p}_+ V_\tau^\beta {}^{\mathcal{F}}\!\mathcal{M} \xrightarrow{\ \tau - z\ } \mathcal{H}^0 \widehat{p}_+ V_\tau^\beta {}^{\mathcal{F}}\!\mathcal{M} \to \mathcal{H}^0 \widehat{p}_+ R_{F_{\mathrm{Del}}^{\beta+\bullet}}(\widetilde{\mathcal{M}} \otimes \mathcal{E}^{-t}) \to 0$$

which is identified with the exact sequence (cf. (6.6))

$$0 \to V_\tau^\beta \widehat{R_F M}_{\mathrm{loc}} \xrightarrow{\ \tau - z\ } V_\tau^\beta \widehat{R_F M}_{\mathrm{loc}} \to R_{V^{\beta+\bullet}} H^1_{\mathrm{DR}}(\mathbb{P}^1, \widetilde{\mathcal{M}} \otimes \mathcal{E}^{-t}) \to 0.$$

The identification of the action of $z^2 \partial_z$, i.e., the grading with respect to the filtrations involved, gives

$$\mathcal{H}^0 \widehat{p}_+ R_{F_{\mathrm{Del}}^{\beta+\bullet}}(\widetilde{\mathcal{M}} \otimes \mathcal{E}^{-t}) = \mathbb{H}^1\big(\mathbb{P}^1, R_{F_{\mathrm{Del}}^{\beta+\bullet+1}} \mathrm{DR}(\widetilde{\mathcal{M}} \otimes \mathcal{E}^{-t})\big).$$

Fixing the power of z shows therefore that, for any $p \in \mathbb{Z}$, the morphism

$$\mathbb{H}^1\big(\mathbb{P}^1, F_{\mathrm{Del}}^{\beta+p} \mathrm{DR}(\widetilde{\mathcal{M}} \otimes \mathcal{E}^{-t})\big) \longrightarrow \mathbb{H}^1\big(\mathbb{P}^1, \mathrm{DR}(\widetilde{\mathcal{M}} \otimes \mathcal{E}^{-t})\big)$$

induces an isomorphism onto $V^{\beta+p-1} H^1_{\mathrm{DR}}(\mathbb{P}^1, \widetilde{\mathcal{M}} \otimes \mathcal{E}^{-t})$, as was to be proved. Q.E.D.

Remark 6.12. It is possible to define the Deligne filtration as a filtration on G and not only on its fibre at $\theta = 1$. For any $\gamma \in \mathbb{R}$, one sets

$$F_{\mathrm{Del}}^\gamma G = \mathbb{C}[\theta, \theta^{-1}] \cdot (G_0 \cap V_\theta^\gamma G) \subset G.$$

One easily checks that Griffiths transversality holds for this filtration (i.e., $t \cdot F^{\gamma}_{\mathrm{Del}} G \subset F^{\gamma-1}_{\mathrm{Del}} G$). Moreover, the limit filtration when $\theta \to 0$ is the Hodge filtration on $\bigoplus_{\beta\in(-1,0]} \mathrm{gr}^{\beta}_{V_\theta} G$, that is, for any $\beta \in (-1, 0]$,

$$F^{\gamma}_{\mathrm{Del}} \mathrm{gr}^{\beta}_{V_\theta} G := (F^{\gamma}_{\mathrm{Del}} G \cap V^{\beta}_{\theta} G)/(F^{\gamma}_{\mathrm{Del}} G \cap V^{>\beta}_{\theta} G)$$

jumps at most at $\gamma = \beta + p$, $p \in \mathbb{Z}$, and

$$F^{\beta+p}_{\mathrm{Del}} \mathrm{gr}^{\beta}_{V_\theta} G = G^p \, \mathrm{gr}^{\beta}_{V_\theta} G := (G^p \cap V^{\beta}_{\theta} G)/(G^p \cap V^{>\beta}_{\theta} G),$$

where we recall that $G^p = \theta'^p G_0$. These properties are checked by using a V-solution of the Birkhoff problem for G_0, as in Lemma 6.5.

§7. The new supersymmetric index and the spectrum

Let $(\mathscr{T}, \mathscr{S})$ be a polarized complex Hodge $\mathscr{D}$-module of weight w on $\mathbb{P}^1$ (cf. Definition 5.11). Its exponentially twisted de Rham cohomology $\mathscr{H}^0 a_+{}^F\mathscr{T}$ is an integrable polarized twistor structure of weight w, according to Corollary 5.16, which is identified to the fibre $(\widehat{\mathscr{T}}, \widehat{\mathscr{S}})_1$ of the Fourier–Laplace transform $(\widehat{\mathscr{T}}, \widehat{\mathscr{S}})$ at $\tau = 1$.

Recall (cf. Appendix B) that we have a rescaling action with respect to $\tau \in \mathbb{C}^*$, that we denote by μ^*_τ, on integrable twistor structures. Applying it to $\widehat{\mathscr{T}_1}$, we get a family $\mathrm{Susy}_{\mu^*_\tau \widehat{\mathscr{T}_1}}(T)$ of polynomials in T with coefficients depending on $\tau \in \mathbb{C}^*$.

Theorem 7.1. *If $(\mathscr{T}, \mathscr{S})$ is a polarized complex Hodge $\mathscr{D}$-module of weight w, then*

$$\lim_{\tau\to 0} \mathrm{Susy}_{\mu^*_\tau \widehat{\mathscr{T}_1}}(T) = \mathrm{SP}^{\infty}_{\widehat{\mathscr{T}_1}}(T),$$
$$\lim_{\tau\to \infty} \mathrm{Susy}_{\mu^*_\tau \widehat{\mathscr{T}_1}}(T) = \mathrm{SP}^{0}_{\widehat{\mathscr{T}_1}}(T).$$

Remark 7.2. It follows that the eigenvalues of the new supersymmetric index of $\mu^*_\tau \widehat{\mathscr{T}_1}$ interpolate, when τ varies between 0 and ∞, between the spectrum at ∞, which gives, by exponentiating, the eigenvalues of the monodromy of the original variation of Hodge structure around $t = \infty$, and the spectrum at 0, which gives, by exponentiating, the eigenvalues of the monodromy of the original variation near the singular points at finite distance. In particular, in general, $\mathrm{Susy}_{\mu^*_\tau \widehat{\mathscr{T}_1}}(T)$ is far from being constant with respect to τ.

As we will indicate below, $\mu^*_\tau \widehat{\mathscr{T}_1}$ is nothing but the fibre $\widehat{\mathscr{T}_\tau}$. If $w = 0$, $\widehat{\mathscr{T}}_{|\tau\neq 0}$ is a variation of polarized pure twistor structure of weight 0, hence corresponds to a flat bundle with harmonic metric $\widehat{h}$. The flat bundle is nothing else but G^{an} with its connection. Now, $\mathrm{Susy}_{\widehat{\mathscr{T}}_\tau}(T)$ is the characteristic polynomial of the selfadjoint operator $\mathscr{Q}$ acting on the C^∞ bundle associated to G. Therefore, at each $\tau^o\in\mathbb{C}^*$, the fibre G_{τ^o} has a $\widehat{h}$-orthogonal decomposition indexed by the eigenvalues of $\mathscr{Q}_{\tau^o}$. This decomposition does not give rise to a filtration of G^{an} indexed by a discrete set of $\mathbb{R}$, however. On the other hand, Deligne's filtration introduced in Remark 6.12 is a Hodge-type filtration, but does not correspond, in general, to the previous 'Hodge' decomposition. Nevertheless, this difference disappears asymptotically when $\tau\to 0$, according to the theorem.

Example 7.3. Let f be a cohomologically tame function on an smooth complex affine variety U as in Examples 1.4 and 1.8 and let G_0 be the corresponding Brieskorn lattice. In [19, Th. 4.10] we have defined (following a conjecture of C. Hertling) a sesquilinear pairing $\widehat{C}$ on G_0 and proved that $\mathscr{G}:=(G_0,G_0,\widehat{C})$ is a pure twistor structure of weight 0 polarized by $(\mathrm{Id},\mathrm{Id})$. It is moreover integrable, and can be obtained as the direct image by the constant map of the exponential twist of the minimal extension of a suitable variation of Hodge structure (namely, the intermediate direct image by f of $\mathscr{O}_U$, with a suitable Tate twist).

In such a case, there is a natural real structure coming from the real structure on the cohomology $H^{\dim U}(U,f^{-1}(t))$ for a regular value $t\in\mathbb{C}$ of f, and we can define $\mathscr{Q}^{\mathrm{Hert}}$ as in Remark 2.6, whose eigenvalues are symmetric with respect to 0. According to the symmetry, mentioned in Examples 1.4 and 1.8, of the spectrum at the origin or at infinity with respect to $\frac12\dim U$, Theorem 7.1 reads:

$$\lim_{\tau\to 0}\mathrm{Susy}^{\mathrm{Hert}}_{\mu_\tau^*\mathscr{G}}(T)=\mathrm{SP}^\infty_{G_0}(T-\tfrac12\dim U),$$
$$\lim_{\tau\to \infty}\mathrm{Susy}^{\mathrm{Hert}}_{\mu_\tau^*\mathscr{G}}(T)=\mathrm{SP}^0_{G_0}(T-\tfrac12\dim U).$$

Proof of Theorem 7.1. It will have three steps.

Step 1. According to Proposition B.2 in the appendix, $\mu_\tau^*\widehat{\mathscr{T}_1}$ is regarded as the fibre at τ of the Fourier–Laplace transform $\widehat{\mathscr{T}}$ of $\mathscr{T}$, whose definition is recalled in §4.b.

Step 2. According to [19] (cf. §4.c), $\widehat{\mathscr{T}}$ is an integrable polarized regular twistor $\mathscr{D}$-module of weight w on the analytic affine line with coordinate τ, and according to Theorem A.1 of the appendix A, it is an integrable wild twistor $\mathscr{D}$-module of weight w at $\tau=\infty$ (this could also be deduced from [12]). We can therefore apply Theorem 3.1 at $\tau=0$ and Theorem 3.5 at $\tau=\infty$ to compute the left-hand sides in Theorem 7.1.

Step 3 for $\tau \to 0$. From Corollary 5.14 applied to the right-hand sides in (4.5)(∗), we conclude from Lemma 5.8 that the Susy polynomials are equal to the corresponding SP^∞ polynomials. It follows that such a property holds for the left-hand sides, as the shift by β is the same for Susy and SP^∞. By Corollary 5.16 and Lemma 5.8, $\mathscr{H}^0 a_+ \mathscr{T}$ satisfies Susy $= \mathrm{SP}^\infty$, hence so does the left-hand side in (4.5)(∗∗). It follows that $(\Psi^0_\tau \widetilde{\mathscr{T}}, \Psi^0_\tau \mathscr{S}, \mathscr{N}, z^2\partial_z)$ is Tw of a polarized complex mixed Hodge structure of weight w, hence also satisfies Susy $= \mathrm{SP}^\infty$. Therefore,

$$\begin{aligned}\prod_{\beta\in(-1,0]}\prod_{\ell\in\mathbb{Z}} \mathrm{Susy}_{\mathrm{gr}^{\mathrm{M}}_\ell \Psi^\beta_\tau \widehat{\mathscr{T}}}(T) &= \prod_{\beta\in(-1,0]}\prod_{\ell\in\mathbb{Z}} \mathrm{SP}^\infty_{\mathrm{gr}^{\mathrm{M}}_\ell \Psi^\beta_\tau \widehat{\mathscr{T}}}(T)\\ &= \mathrm{SP}^\infty_{G_0^{(F)}}(T) \quad \text{after Corollary 5.21}\\ &= \mathrm{SP}^\infty_{\widehat{\mathscr{T}}_1}(T) \quad \text{by definition.}\end{aligned}$$

On the other hand, Theorem 3.1 gives

$$\lim_{\tau\to 0} \mathrm{Susy}_{\widehat{\mathscr{T}}_\tau}(T) = \prod_{\beta\in(-1,0]}\prod_{\ell\in\mathbb{Z}} \mathrm{Susy}_{\mathrm{gr}^{\mathrm{M}}_\ell \Psi^\beta_\tau \widehat{\mathscr{T}}}(T).$$

Step 3 for $\tau \to \infty$. We argue similarly at $\tau = \infty$. We apply Corollaries 5.14 and 5.15 to the right-hand sides in (4.6)(∗) and get, according to Lemma 5.8, the equality between the Susy polynomial and the SP^0 polynomial. This equality then also holds for the left-hand side, as the same shift of $-(\beta+1)$ applies to both. We conclude:

$$\begin{aligned}\prod_i \prod_{\beta\in(-1,0]}\prod_{\ell\in\mathbb{Z}} \mathrm{Susy}_{\mathrm{gr}^{\mathrm{M}}_\ell \Psi^{c_i/\tau',\beta}_{\tau'} \widehat{\mathscr{T}}}(T) &= \prod_i \prod_{\beta\in(-1,0]}\prod_{\ell\in\mathbb{Z}} \mathrm{SP}^0_{\mathrm{gr}^{\mathrm{M}}_\ell \Psi^{c_i/\tau',\beta}_{\tau'} \widehat{\mathscr{T}}}(T)\\ &= \prod_i \mathrm{SP}^0_{G^{(F)}_{i,0}}(T) \quad \text{after Corollary 5.23}\\ &= \mathrm{SP}^0_{\widehat{\mathscr{T}}_1}(T) \quad \text{by definition.}\end{aligned}$$

On the other hand, Theorem 3.5 gives

$$\lim_{\tau'\to 0} \mathrm{Susy}_{\widehat{\mathscr{T}}_{\tau'}}(T) = \prod_i \prod_{\beta\in(-1,0]}\prod_{\ell\in\mathbb{Z}} \mathrm{Susy}_{\mathrm{gr}^{\mathrm{M}}_\ell \Psi^{c_i/\tau',\beta}_{\tau'} \widehat{\mathscr{T}}}(T).$$

Q.E.D.

§Appendix A. Stationary phase formula for polarized twistor $\mathscr{D}$-modules

Let $(\mathscr{T}, \mathscr{S})$ be a polarized regular twistor $\mathscr{D}$-module of weight w (in the sense of [16] or [11]) on $\mathbb{P}^1$. The Fourier–Laplace transform

$(\widehat{\mathscr{T}},\widehat{\mathscr{S}})$ is an object of the same kind on the *analytic* affine line $\widehat{\mathbb{A}}^1$ with coordinate τ, after [15] and [18]. The purpose of this appendix A is to show that this Fourier–Laplace transform naturally extends as a wild twistor $\mathscr{D}$-module (in the sense of [20], cf. also [12]) near $\widehat{\infty}\in\widehat{\mathbb{P}}^1$ and to relate the corresponding nearby cycles with the vanishing cycles of $(\mathscr{T},\mathscr{S})$ at its critical points (stationary phase formula). While the first goal could directly be obtained from recent work of T. Mochizuki [12], we follow here the method of [18] in order to get in the same way the stationary phase formula. We will use the notation introduced in §2.b and in §4.b.

Theorem A.1. *Let $(\mathscr{T},\mathscr{S})$ be a polarized regular twistor $\mathscr{D}$-module of weight w on $\mathbb{P}^1$. Then its Fourier–Laplace transform $(\widehat{\mathscr{T}},\widehat{\mathscr{S}})$ is a polarized wild twistor $\mathscr{D}$-module of weight w on $\widehat{\mathbb{P}}^1$ and, for any $c\in\mathbb{C}$, we have functorial isomorphisms in $\mathscr{R}$-*Triples(pt) *compatible with the polarizations induced by $\widehat{\mathscr{S}}$ and $\mathscr{S}$ respectively:*

$$(*)\qquad \begin{aligned}(\Psi_{\tau'}^{c/\tau',\beta}\widehat{\mathscr{T}},\mathscr{N}_{\tau'})&\simeq(\Psi_{t+c}^{\beta}\mathscr{T},\mathscr{N}_{t+c})\quad \textit{if } \operatorname{Re}\beta\in(-1,0),\\ \operatorname{gr}^{\mathrm{M}}_{\bullet}\Psi_{\tau'}^{c/\tau',\beta}\widehat{\mathscr{T}}&\simeq\operatorname{gr}^{\mathrm{M}}_{\bullet}\Psi_{t+c}^{\beta}\mathscr{T}\quad \textit{if } \beta\in i\mathbb{R}^*,\end{aligned}$$

$$(**)\qquad (\Psi_{\tau'}^{c/\tau',0}\widehat{\mathscr{T}},\mathscr{N}_{\tau'})\simeq(\phi_{t+c}^{-1}\mathscr{T},\mathscr{N}_{t+c}).$$

Remark A.2. In [18] and [16, Appendix], the distinction between the two lines in the analogue of A.1($*$) was mistakenly forgotten in the corresponding statements (see Footnote 3 below).

Proof. According to the results of [15] and [18], it is enough to prove the conditions on the wild specialization at $\widehat{\infty}$, that is, ($*$) and ($**$) with possible ramification at $\tau'=0$. We will denote by i_0 the inclusion $\{0\}\hookrightarrow\mathbb{P}^1$. We can reduce to the case $w=0$ by Tate twist by $(w/2)$, and assume that $\mathscr{S}=(\mathrm{Id},\mathrm{Id})$, so $\widehat{\mathscr{S}}=(\mathrm{Id},\mathrm{Id})$. The compatibility with polarizations will then be clear from the proof.

As in [18, Prop. 4.1], we will denote by D_β the divisor $1\cdot i$ if $\beta\in i\mathbb{R}^*_-$ and $1\cdot(-i)$ if $\beta\in i\mathbb{R}^*_+$, and $D_\beta=0$ otherwise. For a $\mathscr{R}$-module $\mathscr{N}$, the $\mathscr{R}$-module $\mathscr{N}(D_\beta)$ is defined as usual as $\mathscr{O}_{\Omega_0}(D_\beta)\otimes_{\mathscr{O}_{\Omega_0}}\mathscr{N}$. We denote the monodromy filtration of a nilpotent endomorphism by $\mathrm{M}_\bullet$. Q.E.D.

Lemma A.3. *Let $\mathscr{M}$ be a coherent $\mathscr{R}_{\mathscr{P}^1}$-module which is strictly specializable at $t=0$. Then the $\mathscr{R}_{\mathscr{X}}(*\widehat{\infty})$-module ${}^{\mathscr{F}}\!\mathscr{M}$ is strictly specializable along $\tau'=0$ and we have natural functorial isomorphisms of*

$\mathscr{R}_{\mathscr{P}^1}$*-modules with nilpotent endomorphism*

$$(*)\quad \begin{aligned}(\Psi^{0,\beta}_{\tau'}{}^{\mathscr{F}}\mathscr{M}_{|\Delta_0}, \mathrm{N}_{\tau'}) &\simeq i_{0,+}(\Psi^{\beta}_t \mathscr{M}_{|\Delta_0}, \mathrm{N}_t) \quad \textit{if } \operatorname{Re}\beta \in (-1,0),\\ \operatorname{gr}^{\mathrm{M}}_{\bullet} \Psi^{0,\beta}_{\tau'}{}^{\mathscr{F}}\mathscr{M}_{|\Delta_0} &\simeq i_{0,+}(\operatorname{gr}^{\mathrm{M}}_{\bullet} \Psi^{\beta}_t \mathscr{M}(D_\beta)_{|\Delta_0}) \quad \textit{if } \beta \in i\mathbb{R}^*,\end{aligned}$$

$$(**)\quad (\Psi^{0,0}_{\tau'}{}^{\mathscr{F}}\mathscr{M}, \mathrm{N}_{\tau'}) \simeq i_{0,+}(\phi_t^{-1}\mathscr{M}, \mathrm{N}_t).$$

Recall (cf. [16, §3.6.b]) that $\phi_t^{-1}\mathscr{M}$ is also denoted $\psi_t^{-1}\mathscr{M}$ (or $\psi_{t,0}\mathscr{M}$ if one uses the increasing notation), but later we will extend the correspondence with sesquilinear pairings, where the distinction between ϕ and ψ is important. Recall also (cf. [20]) that the notation $\Psi^{0,\beta}_{\tau'}{}^{\mathscr{F}}\mathscr{M}$ has the same meaning as $\Psi^{\beta}_{\tau'}{}^{\mathscr{F}}\mathscr{M}$ (as used on the right-hand side), but we mean here that the possible exponential factor is zero, for later use.

Proof. We will consider the two charts (t,τ') and (t',τ'). Let us first start with the second one. Let $(m_i)_{i\in I}$ be a finite set of $\mathscr{R}_{\mathscr{P}^1,(t'_o,z_o)}$-generators of $\mathscr{M}_{t'_o,z_o}$. Then, from the formulas (cf. [16, (A.2.5)])

$$\begin{aligned}m/t'\tau' \otimes \mathcal{E}^{-t\tau/z} &= \tau'\eth_{\tau'}(m\otimes \mathcal{E}^{-t\tau/z})\\ \eth_{t'}(m\otimes \mathcal{E}^{-t\tau/z}) &= (\eth_{t'}m)\otimes \mathcal{E}^{-t\tau/z} + \tau'(\tau'\eth_{\tau'})^2(m\otimes \mathcal{E}^{-t\tau/z}),\end{aligned}$$

we conclude that, in this chart, ${}^{\mathscr{F}}\mathscr{M}$ is $V_0\mathscr{R}_{\mathscr{X}}$-coherent (where the V-filtration on $\mathscr{X}$ is relative to $\tau'=0$) and that it is strictly specializable along $\tau'=0$ with a constant V-filtration.

Let us now consider the chart (t,τ') and the corresponding formulas [16, (A.2.4)]

$$\begin{aligned}\eth_t(m\otimes \mathcal{E}^{-t\tau/z}) &= [(\eth_t - 1/\tau')m]\otimes \mathcal{E}^{-t\tau/z},\\ \eth_{\tau'}(m\otimes \mathcal{E}^{-t\tau/z}) &= tm/\tau'^2 \otimes \mathcal{E}^{-t\tau/z}.\end{aligned}$$

The proof is very similar to that of [18, Prop. 4.1]. It will be simpler to work with the algebraic version of ${}^{\mathscr{F}}\mathscr{M}$, that is, to consider the projection p in the algebraic sense, so $p^+\widetilde{\mathscr{M}}(*\widehat{\infty}) = \mathbb{C}[\tau',\tau'^{-1}]\otimes_{\mathbb{C}}\widetilde{\mathscr{M}}$. Moreover, as we work in the (analytic) chart with coordinate t, there is no difference between $\widetilde{\mathscr{M}}$ and $\mathscr{M}$. We will exhibit the V-filtration of ${}^{\mathscr{F}}\mathscr{M}$ along $\tau'=0$. As such a filtration is only locally defined with respect to z, we fix z_o and work with in some neighbourhood of z_o. We will forget z_o in the notation of the V-filtration. For any $b\in\mathbb{R}$, let us set

$$U^b\,{}^{\mathscr{F}}\mathscr{M} := \sum_{\ell\geqslant 0}\eth_t^\ell \sum_{k\in\mathbb{Z}} \tau'^{-k}(V^{b+k}\mathscr{M}\otimes \mathcal{E}^{-t/z\tau'}) \subset \mathscr{M}[\tau',\tau'^{-1}]\otimes \mathcal{E}^{-t/z\tau'},$$

where $V^\bullet\mathscr{M}$ is the V-filtration (near z_o) of $\mathscr{M}$ along $t=0$. The following properties are easily checked:

• $U^\bullet\mathscr{F}\!\mathscr{M}$ is a decreasing filtration of $\mathscr{F}\!\mathscr{M}$ by $\mathscr{R}_{\mathscr{P}^1}[\tau']\langle\tau'\eth_{\tau'}\rangle$-modules.

• $\tau' U^b\mathscr{F}\!\mathscr{M} = U^{b+1}\mathscr{F}\!\mathscr{M}$ (so τ' induces $\mathrm{gr}_U^b\mathscr{F}\!\mathscr{M} \xrightarrow{\sim} \mathrm{gr}_U^{b+1}\mathscr{F}\!\mathscr{M}$).

• From the strict specializability of $\mathscr{M}$ (cf. [16, Def. 3.3.8 & Rem. 3.3.9(2)]) we get, for $b \in (-1,0]$ and $k \in \mathbb{N}$, $V^{b+k}\mathscr{M} = t^k V^b\mathscr{M}$ and $V^{b-1-k}\mathscr{M} = \sum_{j=0}^{k-1} \eth_t^j V^{-1}\mathscr{M} + \eth_t^k V^{b-1}\mathscr{M}$. If we write, according to [16, (A.2.4)],

$$U^b\mathscr{F}\!\mathscr{M} = \sum_{\ell\geqslant 0}\eth_t^\ell\Big(\sum_{k\geqslant 0}(\tau'\eth_{\tau'})^k(V^b\mathscr{M}\otimes\mathcal{E}^{-t/z\tau'}) + \tau'\sum_{k\geqslant 0}\tau'^k(V^{b-1-k}\mathscr{M}\otimes\mathcal{E}^{-t/z\tau'})\Big)$$

and, for $k \geqslant 0$,

$$\tau'^k(V^{b-1-k}\mathscr{M}\otimes\mathcal{E}^{-t/z\tau'}) = (\tau'\eth_t+1)^k(V^{b-1}\mathscr{M}\otimes\mathcal{E}^{-t/z\tau'}) + \sum_{j=0}^{k-1}\tau'^j(\tau'\eth_t+1)^{k-j}(V^{-1}\mathscr{M}\otimes\mathcal{E}^{-t/z\tau'}),$$

we find that $U^b\mathscr{F}\!\mathscr{M}$ is $\mathscr{R}_{\mathscr{P}^1}[\tau']\langle\tau'\eth_{\tau'}\rangle$-coherent, and locally generated as such by $m_i\otimes\mathcal{E}^{-t/z\tau'}$ and $\tau'(n_j\otimes\mathcal{E}^{-t/z\tau'})$, if m_i (resp. n_j) are local $\mathscr{O}_{\mathscr{P}^1}\langle t\eth_t\rangle$-generators of $V^b\mathscr{M}$ (resp. $V^{b-1}\mathscr{M}$).

• If $B_b(s)$ is the minimal polynomial of $t\eth_t$ on $\mathrm{gr}_V^b\mathscr{M}$, then, for any local section m of $V^b\mathscr{M}$, as $(t\eth_t+\tau'\eth_{\tau'})(m\otimes\mathcal{E}^{-t/z\tau'}) = (t\eth_t m)\otimes\mathcal{E}^{-t/z\tau'}$ and as $\eth_t t(m\otimes\mathcal{E}^{-t/z\tau'}) \in U^{b+1}\mathscr{F}\!\mathscr{M}$, we find $B_b(\tau'\eth_{\tau'}-1)(m\otimes\mathcal{E}^{-t/z\tau'}) \in U^{>b}\mathscr{F}\!\mathscr{M}$.

It follows from these properties that $V^b\mathscr{F}\!\mathscr{M} := U^{b-1}\mathscr{F}\!\mathscr{M}$ is a good candidate for being the V-filtration of $\mathscr{F}\!\mathscr{M}$. It remains to check the strict specializability, by computing the graded modules.

For any $b\in\mathbb{R}$, the map (where η is a new variable)

$$\begin{aligned} V^b\mathscr{M}[\eta] &\longrightarrow U^b\mathscr{F}\!\mathscr{M} = V^{b+1}\mathscr{F}\!\mathscr{M}\\ \sum_p m_p\eta^p &\longmapsto \sum_p \eth_t^p(m_p\otimes\mathcal{E}^{-t/z\tau'})\end{aligned}$$

induces a mapping

$$(\mathrm{gr}_V^b\mathscr{M}[\eta], t\eth_t) \longrightarrow (\mathrm{gr}_U^b\mathscr{F}\!\mathscr{M}, \tau'\eth_{\tau'}-1) \xrightarrow[\sim]{\tau'^{-1}} (\mathrm{gr}_V^b\mathscr{F}\!\mathscr{M}, \tau'\eth_{\tau'}), \tag{A.4}$$

and we have a commutative diagram

(A.5)
$$\begin{array}{ccc} \mathrm{gr}_V^b \mathscr{M}[\eta] & \longrightarrow & \mathrm{gr}_V^b {}^{\mathscr{F}}\!\mathscr{M} \\ {\scriptstyle \eth_t}\downarrow & & \downarrow{\scriptstyle \tau'^{-1}} \\ \mathrm{gr}_V^{b-1} \mathscr{M}[\eta] & \longrightarrow & \mathrm{gr}_V^{b-1} {}^{\mathscr{F}}\!\mathscr{M}. \end{array}$$

Moreover, if we identify in a natural way $\mathrm{gr}_V^b \mathscr{M}[\eta]$ with $i_{0,+}\,\mathrm{gr}_V^b \mathscr{M}$, this map is a morphism of $\mathscr{R}_{\mathscr{P}^1}$-modules. Q.E.D.

Lemma A.6. *If* $b < 0$, (A.4) *is an isomorphism of* $\mathscr{R}_{\mathscr{P}^1}$*-modules.*

Proof. If $b \in < 0$, we assert that

$$U^b\,{}^{\mathscr{F}}\!\mathscr{M} = U^{>b}\,{}^{\mathscr{F}}\!\mathscr{M} + \sum_{\ell\geqslant 0} \eth_t^\ell(V^b\mathscr{M} \otimes \mathcal{E}^{-t/z\tau'}),$$

which implies that the morphism (A.4) is onto. Indeed, on the one hand, using the formulas [16, (A.2.4)] recalled above and iterating the inclusion

$$\begin{aligned}\tau'^{-1}(V^{b+1}\mathscr{M} \otimes \mathcal{E}^{-t/z\tau'}) &\subset \partial_t(V^{b+1}\mathscr{M} \otimes \mathcal{E}^{-t/z\tau'}) - (\partial_t V^{b+1}\mathscr{M} \otimes \mathcal{E}^{-t/z\tau'}) \\ &\subset U^{b+1}\,{}^{\mathscr{F}}\!\mathscr{M} + (V^b\mathscr{M} \otimes \mathcal{E}^{-t/z\tau'}),\end{aligned}$$

we get

$$\sum_{k\geqslant 0} \tau'^{-k}(V^{b+k}\mathscr{M} \otimes \mathcal{E}^{-t/z\tau'}) \subset (V^b\mathscr{M} \otimes \mathcal{E}^{-t/z\tau'}) + U^{b+1}\,{}^{\mathscr{F}}\!\mathscr{M}.$$

On the other hand, if $b < 0$, for any $k \geqslant 1$ we have $V^{b-k}\mathscr{M} = \eth_t^k V^b\mathscr{M} + V^{>b-k}\mathscr{M}$ and

$$\tau'^k(V^{b-k}\mathscr{M} \otimes \mathcal{E}^{-t/z\tau'}) \subset (\tau'\eth_t + 1)^k(V^b\mathscr{M} \otimes \mathcal{E}^{-t/z\tau'}) + U^{>b}\,{}^{\mathscr{F}}\!\mathscr{M}.$$

Notice then that, for $j \geqslant 1$, $\eth_t^j(V^b\mathscr{M} \otimes \mathcal{E}^{-t/z\tau'}) \subset U^b\,{}^{\mathscr{F}}\!\mathscr{M}$ and $\tau'^j\eth_t^j(V^b\mathscr{M} \otimes \mathcal{E}^{-t/z\tau'}) \subset U^{b+j}\,{}^{\mathscr{F}}\!\mathscr{M} \subset U^{>b}\,{}^{\mathscr{F}}\!\mathscr{M}$.

The morphism (A.4) is also injective: one remarks that, given local sections n_j of $\mathscr{M}$, if $\sum_j \tau'^j n_j \otimes \mathcal{E}^{-t/z\tau'}$ is a local section of $U^b\,{}^{\mathscr{F}}\!\mathscr{M}$, then the dominant coefficient with respect to τ'^{-1} belongs to $V^b\mathscr{M}$ (by considering the dominant coefficient with respect to τ'^{-1} in an expression like $\sum_p \eth_t^p(m_p \otimes \mathcal{E}^{-t/z\tau'})$); arguing as in [18, Proof of Prop. 4.1, §(ii)(6)], one gets the injectivity. Q.E.D.

At this point, we have proved the strict specializability of $\mathscr{F}\mathscr{M}$ along $\tau' = 0$. We now prove the existence of the isomorphisms $(*)$ and $(**)$ of Lemma A.3.

Let β be such that $\operatorname{Re}\beta \in (-1,0]$. The morphism (A.4) induces, near z_o, a morphism $(i_{0,+}\psi_t^\beta \mathscr{M}, \mathrm{N}_t) \to (\psi_{\tau'}^\beta \mathscr{F}\mathscr{M}, \mathrm{N}_{\tau'})$. One can show that these locally defined morphisms glue together. Setting $\ell_{z_o}(\beta) = \operatorname{Re}\beta - (\operatorname{Im} z_o)(\operatorname{Im}\beta)$ (cf. [16, p. 17]), if $\ell_{z_o}(\beta) < 0$, it is an isomorphism near z_o, according to the Lemma A.6 with $b = \ell_{z_o}(\beta)$. Let us first show that such remains the case if $\ell_{z_o}(\beta) = 0$. In this case, by definition of strict specializability (cf. [16, Def. 3.3.8(c)], we know that $\eth_t : \psi_t^\beta \mathscr{M} \to \psi_t^{\beta-1}\mathscr{M}$ is an isomorphism near z_o, so we conclude using (A.5).

Let us now assume that $\ell_{z_o}(\beta) > 0$ and let us choose $k \in \mathbb{N}$ such that $\ell_{z_o}(\beta - k) \in (-1,0]$. Near z_o, we get from (A.5) a commutative diagram

$$\begin{array}{ccc} i_{0,+}\psi_t^\beta \mathscr{M} & \longrightarrow & \psi_{\tau'}^\beta \mathscr{F}\mathscr{M} \\ {\scriptstyle i_{0,+}\eth_t^k}\big\downarrow & & \big\downarrow{\scriptstyle \tau'^{-k}} \\ i_{0,+}\psi_t^{\beta-k}\mathscr{M} & \longrightarrow & \psi_{\tau'}^{\beta-k}\mathscr{F}\mathscr{M} \end{array}$$

where the right vertical map is an isomorphism, by definition, and the choice of k implies that the lower horizontal map is an isomorphism. On the other hand, by the strict specializability of $\mathscr{M}$ at $t = 0$ (cf. [16, Def. 3.3.8(1b) and Rem. 3.3.9(2)]) and the choice of k, the map $t^k : \psi_t^{\beta-k}\mathscr{M} \to \psi_t^\beta \mathscr{M}$ is an isomorphism, and $\eth_t^k t^k : \psi_t^{\beta-k}\mathscr{M} \to \psi_t^{\beta-k}\mathscr{M}$ is equal to $\prod_{j=0}^{k-1}\big[(\beta - j)\star z + \mathrm{N}_t\big]$. We note that $(\beta - j)\star z + \mathrm{N}_t$ is invertible near z_o unless $(\beta - j)\star z_o = 0$. With the conditions $z_o \in \Delta_0$, $\beta \neq 0$, $\operatorname{Re}(\beta - j) \leqslant 0$, $j = 0, \dots, k-1$, $\ell_{z_o}(\beta - j) > 0$, this vanishing only occurs if $j = 0$, $\operatorname{Re}\beta = 0$ and $z_o = i$ if $\operatorname{Im}\beta < 0$, $z_o = -i$ if $\operatorname{Im}\beta > 0$. Therefore, on the one hand, the natural morphism (A.4) induces

$$\text{(A.7)} \qquad i_{0,+}\psi_t^\beta \mathscr{M}_{|\Delta_0} \xrightarrow{\sim} \psi_{\tau'}^\beta \mathscr{F}\mathscr{M}_{|\Delta_0} \quad \text{if } \operatorname{Re}\beta \in (-1,0).$$

On the other hand, if $\beta \in i\mathbb{R}^*$, one checks similarly that, grading first[3] by the monodromy filtration in order to kill N_t, (A.4) induces an isomorphism

$$\text{(A.8)} \qquad i_{0,+}\operatorname{gr}_\bullet^{\mathrm{M}} \psi_t^\beta \mathscr{M}_{|\Delta_0} \xrightarrow{\sim} \operatorname{gr}_\bullet^{\mathrm{M}} \psi_{\tau'}^\beta \mathscr{F}\mathscr{M}_{|\Delta_0}(-D_\beta).$$

Lastly, if $\beta = 0$, we consider the isomorphism

$$\text{(A.9)} \qquad i_{0,+}\psi_t^{-1}\mathscr{M}_{|\Delta_0} \xrightarrow{\sim} \psi_{\tau'}^{-1}\mathscr{F}\mathscr{M}_{|\Delta_0} \xrightarrow[\sim]{\tau'} \psi_{\tau'}^0 \mathscr{F}\mathscr{M}_{|\Delta_0}.$$

[3]This grading was forgotten in [18].

Notice that, by definition, $\Psi^{\beta}_{\tau'}{}^{\mathscr{F}}\!\mathscr{M} = \psi^{\beta}_{\tau'}{}^{\mathscr{F}}\!\mathscr{M}$. So, at this point, we have obtained A.3(∗∗). In order to get A.3(∗), it remains to compare $\psi_t^{\beta}\mathscr{M}_{|\Delta_0}$ with $\Psi_t^{\beta}\mathscr{M}_{|\Delta_0}$. Arguing in a way similar to that of [18, Lemma 4.18], we conclude that, for $\operatorname{Re}\beta \in (-1,0]$, the natural inclusion $\psi_t^{\beta}\mathscr{M}_{|\Delta_0} \hookrightarrow \Psi_t^{\beta}\mathscr{M}_{|\Delta_0}$ is an isomorphism. This gives A.3(∗).

Finally, the functoriality of the isomorphisms A.3(∗) and (∗∗) is clear from the construction.

Let $\mathscr{T} = (\mathscr{M}', \mathscr{M}'', \mathscr{C}_{\mathbf{S}})$ be an object of $\mathscr{R}$-Triples$(\mathbb{P}^1)$, such that $\mathscr{M}', \mathscr{M}''$ are $\mathscr{R}_{\mathscr{P}^1}$-coherent and strictly specializable along $t = 0$. Then ${}^{\mathscr{F}}\!\mathscr{T}$ is defined as $({}^{\mathscr{F}}\!\mathscr{M}', {}^{\mathscr{F}}\!\mathscr{M}'', {}^{\mathscr{F}}\mathscr{C}_{\mathbf{S}})$, where ${}^{\mathscr{F}}\!\mathscr{M}', {}^{\mathscr{F}}\!\mathscr{M}''$ are as above and ${}^{\mathscr{F}}\mathscr{C}_{\mathbf{S}}$ is defined in [16, p. 196].

Lemma A.10. *Let* $\mathscr{T}$ *be as above. Then the isomorphisms of Lemma* A.3 *extend as isomorphisms*

$$
\begin{aligned}
(*)\qquad & (\Psi^{0,\beta}_{\tau'}{}^{\mathscr{F}}\!\mathscr{T}, \mathscr{N}_{\tau'}) \simeq i_{0,+}(\Psi_t^{\beta}\mathscr{T}, \mathscr{N}_t) \quad \textit{if } \operatorname{Re}\beta \in (-1,0),\\
& \operatorname{gr}^{\mathrm{M}}_{\bullet}\Psi^{0,\beta}_{\tau'}{}^{\mathscr{F}}\!\mathscr{T} \simeq i_{0,+}(\operatorname{gr}^{\mathrm{M}}_{\bullet}\Psi_t^{\beta}\mathscr{T}) \quad \textit{if } \beta \in i\mathbb{R}^*,\\
(**)\qquad & (\Psi^{0,0}_{\tau'}{}^{\mathscr{F}}\!\mathscr{T}, \mathscr{N}_{\tau'}) \simeq i_{0,+}(\phi_t^{-1}\mathscr{T}, \mathscr{N}_t).
\end{aligned}
$$

Proof. The point is to prove the compatibility of the corresponding sesquilinear pairings under A.3(∗) and (∗∗), up to Γ-factors that we will analyse, as in [18, Proof of Prop. 5.8].

Let us fix $z_o \in \mathbf{S}$ and let us work in the neighbourhood of z_o. For $\beta \neq 0$ with $\operatorname{Re}\beta \in (-1,0]$, let us set $\alpha = -\beta - 1$ and $b = \ell_{z_o}(\beta)$. Let m', m'' be local section near $(t = 0, z_o)$ of $V^b\mathscr{M}', V^b\mathscr{M}''$ inducing sections $[m'], [m'']$ of $\psi_t^{\beta}\mathscr{M}', \psi_t^{\beta}\mathscr{M}''$ on $\operatorname{gr}_V^b\mathscr{M}', \operatorname{gr}_V^b\mathscr{M}''$ (recall that $\psi_t^{\beta}\mathscr{M}_{|\mathbf{S}} = \Psi_t^{\beta}\mathscr{M}_{|\mathbf{S}}$, cf. [16, Lemma 3.4.2(2)]). We regard $[m'], [m'']$ as sections of $i_{0,+}\psi_t^{\beta}\mathscr{M}', i_{0,+}\psi_t^{\beta}\mathscr{M}''$ (degree 0 with respect to η). Following (A.4), they correspond to sections $[\tau'^{-1}m' \otimes \mathcal{E}^{-t/z\tau'}], [\tau'^{-1}m'' \otimes \mathcal{E}^{-t/z\tau'}]$ of $\Psi^{0,\beta}_{\tau'}{}^{\mathscr{F}}\!\mathscr{M}', \Psi^{0,\beta}_{\tau'}{}^{\mathscr{F}}\!\mathscr{M}''$. By definition, we have, for any C^∞ form $\varphi(t)$ of type $(1,1)$ on $\mathbb{P}^1$ with compact support in the chart t,

$$
\begin{aligned}
&\big\langle \psi^{\beta}_{\tau'}{}^{\mathscr{F}}\mathscr{C}_{\mathbf{S}}([\tau'^{-1}m' \otimes \mathcal{E}^{-t/z\tau'}], \overline{[\tau'^{-1}m'' \otimes \mathcal{E}^{-t/z\tau'}]}), \varphi\big\rangle\\
&\qquad = \operatorname{Res}_{s=\alpha\star z/z}\Big\langle {}^{\mathscr{F}}\mathscr{C}_{\mathbf{S}}(m' \otimes \mathcal{E}^{-t/z\tau'}, \overline{m'' \otimes \mathcal{E}^{-t/z\tau'}}),\\
&\qquad\qquad\qquad \varphi \wedge |\tau'|^{2(s-1)}\widehat{\chi}(\tau')\tfrac{i}{2\pi}d\tau' \wedge d\overline{\tau'}\Big\rangle\\
&\qquad = \operatorname{Res}_{s=\alpha\star z/z}\big\langle \mathscr{C}_{\mathbf{S}}(m', \overline{m''}), I_{\widehat{\chi}}(t,s,z)\varphi\big\rangle,
\end{aligned}
$$

where $\widehat{\chi}$ is C^∞ with compact support in $\widehat{\mathbb{P}}^1$, $\equiv 1$ near $\tau'=0$ and $\equiv 0$ far from $\tau'=0$, and $I_{\widehat{\chi}}$ is defined for $\operatorname{Re} s>0$ as in [16, §3.6.b] by

$$I_{\widehat{\chi}}(t,s,z)=\int e^{z\overline{t}/\overline{\tau'}-t/z\tau'}|\tau'|^{2(s-1)}\widehat{\chi}(\tau')\tfrac{i}{2\pi}d\tau'\wedge d\overline{\tau'}.$$

The last equality above means that $\langle \mathscr{C}_{\mathbf{S}}(m',\overline{m''}),I_{\widehat{\chi}}(t,s,z)\varphi\rangle$ is holomorphic with respect to s for $\operatorname{Re} s>0$ and extends as a meromorphic function of s, of which we take the residue at $s=\alpha\star z/z$.

One first proves, as in [16, Lemma 3.6.6] that $\langle \mathscr{C}_{\mathbf{S}}(m',\overline{m''}),I_{\widehat{\chi}}(t,s,z)\varphi\rangle$ has poles on sets $s=\gamma\star z/z$ ($z\in\mathbf{S}$) with $\operatorname{Re}\gamma<\operatorname{Re}\alpha$ or $\gamma=\alpha$. Moreover, only the first case occurs if φ vanishes along $t=0$, and we can thus assume that $\varphi\equiv\frac{i}{2\pi}dt\wedge d\overline{t}$ near $t=0$. As the residue at $s=\alpha\star z/z$ does not depend on the such a φ, we can assume $\varphi=\chi^2\frac{i}{2\pi}dt\wedge d\overline{t}$ with $\chi\equiv 1$ near $t=0$. We will compare $\langle \mathscr{C}_{\mathbf{S}}(m',\overline{m''}),I_{\widehat{\chi}}(t,s,z)\chi\rangle$ and $\langle \mathscr{C}_{\mathbf{S}}(m',\overline{m''}),|t|^{2s}\chi^2\rangle$ before taking their residue.

If we denote by T the distribution $\chi\mathscr{C}_{\mathbf{S}}(m',\overline{m''})$, and by $\mathscr{F}$ the Fourier transform with kernel $e^{\overline{t\tau}z-t\tau/z}\frac{i}{2\pi}d\tau\wedge d\overline{\tau}$ (cf. [16, Rem. 3.6.17]), these functions are respectively written as

$$\int \mathscr{F}T(\tau,z)|\tau|^{-2(s+1)}\widehat{\chi}(\tau)\tfrac{i}{2\pi}d\tau\wedge d\overline{\tau}\ \text{and}\ \int \mathscr{F}T(\tau,z)\widehat{I}_\chi(\tau,s,z)\tfrac{i}{2\pi}d\tau\wedge d\overline{\tau}$$

with $\widehat{I}_\chi(\tau,s,z):=\mathscr{F}^{-1}(|t|^{2s}\chi)$ (this is analogous to (3.6.25) and (3.6.26) in loc. cit.).

We note that $(1-\widehat{\chi}(\tau))\mathscr{F}T(\tau,z)$ is C^∞ with compact support, so its inverse Fourier transform η is in the Schwartz class, and

$$\int \mathscr{F}T(\tau,z)\widehat{I}_\chi(\tau,s,z)(1-\widehat{\chi}(\tau))\tfrac{i}{2\pi}d\tau\wedge d\overline{\tau}$$

reads $\int\eta\chi|t|^{2s}\frac{i}{2\pi}dt\wedge d\overline{t}$, so takes the form $\Gamma(s+1)h(s,z)$, where h is entire with respect to s. Using the computation in [18, Lemma 5.14] for $\widehat{I}_\chi$ (replacing t there with τ here) and in particular [18, (5.17)], we finally find, if $\beta\neq 0$,

$$\frac{\Gamma(1+\alpha\star z/z)}{\Gamma(-\alpha\star z/z)}\operatorname{Res}_{s=\alpha\star z/z}\langle \mathscr{C}_{\mathbf{S}}(m',\overline{m''}),I_{\widehat{\chi}}(t,s,z)\chi\rangle$$
$$=\operatorname{Res}_{s=\alpha\star z/z}\langle \mathscr{C}_{\mathbf{S}}(m',\overline{m''}),|t|^{2s}\chi\rangle.$$

Let us write $\Gamma(1+\alpha\star z/z)/\Gamma(-\alpha\star z/z)=\mu\overline{\mu}$ as in [18, Lemma 5.5], and $D_\mu=-D_\beta$. Using (A.7) or (A.8) we find (grading only if $\beta\in i\mathbb{R}^*$, that is, if $D_\mu \neq \varnothing$)

$$(\mathrm{gr}^{\mathrm{M}}_{\bullet})(\Psi^{\beta}_{\tau'}{}^{\mathscr{F}}\!\mathscr{M}'(D_\mu), \Psi^{\beta}_{\tau'}{}^{\mathscr{F}}\!\mathscr{M}''(D_\mu), \mu\overline{\mu}{}^{\mathscr{F}}\mathscr{C}_{\mathbf{S}}) \simeq i_{0,+}(\mathrm{gr}^{\mathrm{M}}_{\bullet})(\Psi^{\beta}_t\mathscr{M}', \Psi^{\beta}_t\mathscr{M}'', \mathscr{C}_{\mathbf{S}}).$$

A.10(∗) follows then from [18, Lemma 5.6].

Arguing similarly for $\beta = 0$, we note that A.10(∗∗) is by definition (cf. [16, (3.6.18) & Rem. 3.6.20]).

Arguing as in [18] by applying [16, §6.3], we get A.1(∗) and (∗∗) for $c = 0$. It is then not difficult to check that replacing t with $t + c$ corresponds to twisting $\widehat{\mathscr{T}}$ by $\mathcal{E}^{c/z\tau'}$, and to get (∗) and (∗∗) for any $c \in \mathbb{C}$ in the same way. One checks that A.1(∗) and (∗∗) hold after ramification by using [20, Rem. 2.3.3]. Q.E.D.

The integrable case. Let us now assume that $(\mathscr{T}, \mathscr{S})$ is integrable. We will describe the compatibility between the actions of $z^2\partial_z$ in Theorem A.1. Recall that, with this assumption, the numbers β such that $\Psi^{\beta}_t\mathscr{T} \neq 0$ are real, and thus so are the numbers β such that $\Psi^{\beta}_t\widehat{\mathscr{T}} \neq 0$. In the computation above, we can set $b = \beta$ and do not worry about the local dependence with respect to z_o (cf. [16, Chap. 7]).

The morphism (A.4) is compatible with the natural action of $z^2\partial_z$ on $i_{0,+}\,\mathrm{gr}^b_V\mathscr{M}$ on the one hand, and the action of $z^2\partial_z - \tau'\partial'_\tau$ on $\mathrm{gr}^V_{b+1}{}^{\mathscr{F}}\!\mathscr{M}$ on the other hand (where the action of $z^2\partial_z$ comes from the natural action of $z^2\partial_z$ on ${}^{\mathscr{F}}\!\mathscr{M}$, cf. Lemma 4.4). As we set $\beta = b$, we thus have, for $\beta < 0$,

$$(i_{0,+}\psi^{\beta}_t\mathscr{M}, z^2\partial_z) \xrightarrow{\sim} (\psi^{\beta+1}_{\tau'}{}^{\mathscr{F}}\!\mathscr{M}, z^2\partial_z - (\beta+1)z - \mathrm{N}_{\tau'}).$$

As multiplication by τ' commutes with $z^2\partial_z$ and $\mathrm{N}_{\tau'}$, we obtain

$$(i_{0,+}\psi^{\beta}_t\mathscr{M}, z^2\partial_z) \xrightarrow{\sim} \begin{cases} (\psi^{\beta}_{\tau'}{}^{\mathscr{F}}\!\mathscr{M}, z^2\partial_z - (\beta+1)z - \mathrm{N}_{\tau'}) & \text{if } \beta \in (-1,0), \\ (\psi^{0}_{\tau'}{}^{\mathscr{F}}\!\mathscr{M}, z^2\partial_z - \mathrm{N}_{\tau'}) & \text{if } \beta = -1. \end{cases}$$

In a way analogous to that of Lemma A.10, we conclude

$$(i_{0,+}\psi^{\beta}_t\mathscr{T}, z^2\partial_z) \xrightarrow{\sim} (\psi^{\beta}_{\tau'}{}^{\mathscr{F}}\mathscr{T}, z^2\partial_z - (\beta+1)z - \mathrm{N}_{\tau'}) \quad \text{if } \beta \in (-1,0),$$
$$(i_{0,+}\phi^{-1}_t\mathscr{T}, z^2\partial_z) \xrightarrow{\sim} (\psi^{0}_{\tau'}{}^{\mathscr{F}}\!\mathscr{M}, z^2\partial_z - \mathrm{N}_{\tau'}).$$

As $\psi^{\beta}_{\tau'}$ commutes with the direct image by $\widehat{p}$ in our context, we get

$$(\Psi^{\beta}_t\mathscr{T}, z^2\partial_z) \xrightarrow{\sim} (\Psi^{0,\beta}_{\tau'}\widehat{\mathscr{T}}, z^2\partial_z - (\beta+1)z - \mathrm{N}_{\tau'}) \quad \text{if } \beta \in (-1,0),$$
$$(\phi^{-1}_t\mathscr{T}, z^2\partial_z) \xrightarrow{\sim} (\Psi^{0,0}_{\tau'}\widehat{\mathscr{T}}, z^2\partial_z - \mathrm{N}_{\tau'}),$$

and, grading with respect to $\mathrm{M}_\bullet$ kills $\mathrm{N}_{\tau'}$ and gives, for any $\ell \in \mathbb{Z}$,

(A.11)
$$(\mathrm{gr}_\ell^{\mathrm{M}} \Psi_t^\beta \mathscr{T}, z^2\partial_z) \xrightarrow{\sim} (\mathrm{gr}_\ell^{\mathrm{M}} \Psi_{\tau'}^{0,\beta} \widehat{\mathscr{T}}, z^2\partial_z - (\beta+1)z) \quad \text{if } \beta \in (-1,0),$$

(A.12)
$$(\mathrm{gr}_\ell^{\mathrm{M}} \phi_t^{-1} \mathscr{T}, z^2\partial_z) \xrightarrow{\sim} (\mathrm{gr}_\ell^{\mathrm{M}} \Psi_{\tau'}^{0,0} \widehat{\mathscr{T}}, z^2\partial_z).$$

Lastly, we get a similar result after translating t by $c \in \mathbb{C}$.

§Appendix B. Rescaling

In this appendix B, we recall the notion of rescaling of an integrable twistor structure considered in [8, Def. 4.1]. We also explain how the "no ramification" condition is related to a good behaviour of the rescaling.

Let $(\mathscr{H}', \mathscr{H}'', \mathscr{C}_{\mathbf{S}}, \nabla)$ be an integrable twistor structure (cf. §2.c with X reduced to a point). In order to clarify notation, we will denote by η the coordinate denoted by z before. By integrability, $\nabla_{\eta^2\partial_\eta}$ acts on $\mathscr{H}', \mathscr{H}''$ in a way compatible with $\mathscr{C}_{\mathbf{S}}$. For the sake of simplicity, we will denote by $\eta^2\partial_\eta$ this action. The bundles $\mathscr{H}', \mathscr{H}''$ are a priori defined on some open neighbourhood of $\{|\eta \leqslant 1|\}$ but, using the gluing defined by $\mathscr{C}_{\mathbf{S}}$ and its compatibility with ∇, we can assume that they are defined, together with the action of $\eta^2\partial_\eta$, on the whole complex line $\mathbb{C}_\eta$ with coordinate η and that $\mathscr{C}_{\mathbf{S}}$ is the restriction to $\mathbf{S}$ of a sesquilinear pairing compatible with ∇

$$\mathscr{C} : \mathscr{H}'_{|\mathbb{C}^*_\eta} \otimes_{\mathscr{O}_{\mathbb{C}^*_\eta}} \overline{\mathscr{H}''_{|\mathbb{C}^*_\eta}} \longrightarrow \mathscr{O}_{\mathbb{C}^*_\eta}.$$

Let us consider the map $\mu : \mathbb{C}_{\tau'} \times \Omega_0 \to \mathbb{C}_\eta$ defined by $\mu(\tau', z) = \eta = \tau' z$ (for $\tau' \neq 0$, we will set $\tau = \tau'^{-1}$; this corresponds to the coordinate τ in §4.b). The pull-backs $\mu^*\mathscr{H}', \mu^*\mathscr{H}''$ are holomorphic bundles on $\mathbb{C}_{\tau'} \times \Omega_0$. When restricted to the open set $\tau' \neq 0$, they are equipped with a flat meromorphic connection having a pole of Poincaré rank one along $z = 0$. We have

$$z^2\partial_z(1 \otimes m) = \tau'^{-1}(1 \otimes \eta^2\partial_\eta m) \quad \text{and} \quad \tau'^2\eth_{\tau'}(1 \otimes m) = 1 \otimes \eta^2\partial_\eta m.$$

For a fixed $\tau_o \in \mathbb{C}^*$, denote by $\mu^*_{\tau_o}$ the composition of μ^* with the restriction to $\tau' = \tau_o^{-1}$. This defines $\mu^*_{\tau_o}\mathscr{H}', \mu^*_{\tau_o}\mathscr{H}''$.

In order to define the rescaling of the sesquilinear pairing, we need to be careful. Indeed, the rescaling is not compatible with twistor conjugation, as we have $\mu^*_{\tau_o}\overline{\mathscr{H}''_{|\mathbb{C}^*}} = \overline{\mu^*_{\overline{\tau}_o^{-1}}\mathscr{H}''_{|\mathbb{C}^*}}$. We therefore need an identification $\mu^*_{\tau_o}\mathscr{H}''_{|\mathbb{C}^*} \simeq \mu^*_{\overline{\tau}_o^{-1}}\mathscr{H}''_{|\mathbb{C}^*}$ in order to get a sesquilinear pairing $\mu^*_{\tau_o}\mathscr{C}$

such that $(\mu_{\tau_o}^*\mathscr{H}', \mu_{\tau_o}^*\mathscr{H}'', \mu_{\tau_o}^*\mathscr{C})$ is a twistor structure. This identification is obtained through the parallel transport with respect to the holomorphic connection on $\mu^*\mathscr{H}''_{|\mathbb{C}^*\times\mathbb{C}^*}$ from $\tau_o \times \mathbb{C}^*$ to $\overline{\tau}_o^{-1} \times \mathbb{C}^*$ along the segment between τ_o and $\overline{\tau}_o^{-1}$.

One can rewrite this definition in a way independent of $\tau_o \in \mathbb{C}^*$. For that purpose, let us denote by $\mathscr{L}', \mathscr{L}''$ the local systems on $\mathbb{C}^*_\eta$ determined by $(\mathscr{H}', \nabla)$ and $(\mathscr{H}'', \nabla)$. We get a pairing $\mathscr{L}'_{\mathbf{S}} \otimes \sigma^{-1}\overline{\mathscr{L}''_{\mathbf{S}}} \to \mathbb{C}_{\mathbf{S}}$ by restricting $\mathscr{C}_{\mathbf{S}}$ to these local systems. We remark that, on $\mathbf{S}$, σ coincides with the involution $\iota : \eta \mapsto -\eta$, and we can extend (by parallel transport) in a unique way $\mathscr{C}_{\mathbf{S}}$ as a pairing

$$\widetilde{\mathscr{C}} : \mathscr{L}' \otimes \iota^{-1}\overline{\mathscr{L}''} \longrightarrow \mathbb{C}_{\mathbb{C}^*_\eta}.$$

Taking the pull-back by μ commutes with ι (with respect to η and to z), and restricting to $\mathbb{C}^*_{\tau'} \times \mathbf{S}$ gives a pairing

$$\mu^{-1}\widetilde{\mathscr{C}} : \mu^{-1}\mathscr{L}'_{\mathbb{C}^*_{\tau'}\times\mathbf{S}} \otimes \iota^{-1}\overline{\mu^{-1}\mathscr{L}''_{\mathbb{C}^*_{\tau'}\times\mathbf{S}}} \longrightarrow \mathbb{C}_{\mathbb{C}^*_{\tau'}\times\mathbf{S}}.$$

Identifying now $\iota_{|\mathbf{S}}$ and $\sigma_{|\mathbf{S}}$ in the z-variable gives the desired sesquilinear pairing $\mu^*\mathscr{C}_{\mathbf{S}}$ at the level of local systems, and thus at the level of holomorphic bundles. Clearly, it is nondegenerate.

Definition B.1 (Rescaling). Let $\mathscr{T} = (\mathscr{H}', \mathscr{H}'', \mathscr{C}_{\mathbf{S}})$ be an integrable twistor structure. The rescaling $\mu^*\mathscr{T}$ is the triple $(\mu^*\mathscr{H}', \mu^*\mathscr{H}'', \mu^*\mathscr{C}_{\mathbf{S}})$ defined as above.

We have the following properties:

• By construction (and because $\mu^*\mathscr{C}_{\mathbf{S}}$ is nondegenerate), $\mu^*\mathscr{T}$ is an integrable variation of twistor structure on $\mathbb{C}^*_{\tau'}$.

• Functoriality: This mainly reduces to showing that, given $\mathscr{T} = (\mathscr{H}', \mathscr{H}'', \mathscr{C}_{\mathbf{S}})$ with $\mathscr{H}', \mathscr{H}''$ defined on some neighbourhood of $\{\eta \leqslant 1\}$, the extension of $\mathscr{H}', \mathscr{H}''$ to $\mathbb{C}_\eta$ by using the gluing is functorial. This is done as follows. The pairing $\mathscr{C}_{\mathbf{S}}$ induces an isomorphism $\mathscr{H}' \simeq \overline{\mathscr{H}''^\vee}$ on some neighbourhood of $\{\eta = 1\}$, which allows one to extend $\mathscr{H}'$ as a bundle on $\mathbb{C}_\eta$. If $\varphi : \mathscr{T}_1 \to \mathscr{T}_2$ is a morphism, the previous isomorphism is compatible with $\varphi' : \mathscr{H}'_2 \to \mathscr{H}'_1$ and $\overline{\varphi}''^\vee : \overline{\mathscr{H}''^\vee_2} \to \overline{\mathscr{H}''^\vee_1}$ by definition. Therefore, it extends to $\mathbb{C}_\eta$.

• Compatibility with adjunction: Restricted to local systems on $\{\eta = 1\}$, the adjoint $\mathscr{C}^*_{\mathbf{S}}$ of $\mathscr{C}_{\mathbf{S}}$ is $\sigma^{-1}\mathscr{C}^\dagger_{\mathbf{S}}$, where $\mathscr{C}^\dagger_{\mathbf{S}}$ is the adjoint with respect to the standard conjugation. So, working on local systems,

$$\mathscr{C}^*_{\mathbf{S}} = \sigma^{-1}\mathscr{C}^\dagger_{\mathbf{S}} = \iota^{-1}\widetilde{\mathscr{C}}^\dagger_{|\eta=1}$$

and

$$\mu^*\mathscr{C}_{\mathbf{S}}^* = (\mu^{-1}\iota^{-1}\widetilde{\mathscr{C}}^\dagger)_{|z=1} = (\iota^{-1}\mu^{-1}\widetilde{\mathscr{C}}^\dagger)_{|z=1}$$
$$= \sigma^{-1}(\mu^{-1}\widetilde{\mathscr{C}})^\dagger_{|z=1} = (\mu^*\mathscr{C}_{\mathbf{S}})^*.$$

- Compatibility with Tate twist: For $k \in \frac{1}{2}\mathbb{Z}$,

$$\mathscr{T}(k) := (\mathscr{H}', \mathscr{H}'', (iz)^{-2k}\mathscr{C}_{\mathbf{S}}),$$

so

$$\mu^*(\mathscr{T}(k)) = (\mu^*\mathscr{H}', \mu^*\mathscr{H}'', \tau'^{-2k}(iz)^{-2k}\mu^*\mathscr{C}_{\mathbf{S}}).$$

We therefore have an isomorphism

$$\varphi : (\mu^*\mathscr{T})(k) \xrightarrow{\sim} \mu^*(\mathscr{T}(k)),$$

with $\varphi = (\varphi', \varphi'')$ and $\varphi' = \tau'^{-2k}\,\mathrm{Id}_{\mu^*\mathscr{H}'}$, $\varphi'' = \mathrm{Id}_{\mu^*\mathscr{H}''}$.

Proposition B.2. *If $(\mathscr{T}, \mathscr{S})$ is the polarized twistor $\mathscr{D}$-module of weight w associated to a variation of Hodge structure of weight w as in Proposition* 5.10, *then, when restricted to $\tau \neq 0, \infty$, the Fourier–Laplace transform $\widehat{\mathscr{T}}$ is identified with the rescaling of its fibre at $\tau = 1$ as defined above.*

Sketch of proof. One first reduces to weight 0, by using the compatibility of rescaling and Fourier–Laplace transform with Tate twist by $w/2$. One can also assume that $\mathscr{S} = (\mathrm{Id}, \mathrm{Id})$. Then the result follows from [19, §2b & 2c] (in particular, Lemma 2.4 in loc. cit.). Q.E.D.

In the remaining part of this section, we explain why a good behaviour at $\tau' = 0$ of the rescaled twistor structure imposes the "no ramification" condition of §1.b. Instead of working on $\{\tau' \neq 0\}$, we now work on the whole line $\mathbb{C}_{\tau'}$. Let us set $\mathscr{H} = \mathscr{H}'$ or $\mathscr{H}''$ and $\widetilde{\mathscr{M}} = \mu^*\mathscr{H}[\tau'^{-1}]$. If we set $X = \mathbb{C}_{\tau'}$, then $\widetilde{\mathscr{M}}$ is an integrable $\mathscr{R}_{\mathscr{X}}[\tau'^{-1}]$-module.

Proposition B.3. *If $\widetilde{\mathscr{M}}$ is strictly specializable with ramification and exponential twist at $\tau' = 0$ (in the sense of* [16] *and* [20]*), then $(\mathscr{H}, \nabla_{\partial_\eta})$ has no ramification.*

Proof. It will be easier to work in an algebraic framework. One can find a free $\mathbb{C}[\eta]$-module H with an algebraic connection having a double pole at $\eta = 0$ and no other pole, such that $(\mathscr{H}, \nabla_{\partial_\eta}) = (\mathscr{O}_{\mathbb{C}_\eta} \otimes_{\mathbb{C}[\eta]} H, \nabla_{\partial_\eta})$. Notice then that $\widetilde{\mathscr{M}}$ is the analytization of $\widetilde{M} \simeq \mathbb{C}[\tau', \tau'^{-1}] \otimes_{\mathbb{C}} H$, where the action of z is defined as $\tau'^{-1} \otimes \eta$.

Let us try to find a Bernstein relation (in the sense of [16] or, more generally with parabolic structure, of [11]) for elements of $\widetilde{M}$ at $\tau' = 0$. Let $m \in H$. The differential equation of minimal degree satisfied by m can be written as $b(\eta\partial_\eta)m = \eta P(\eta, \eta\partial_\eta)m$, where $b \in \mathbb{C}[s] \smallsetminus \{0\}$ and P is an operator in $\eta\partial_\eta$ with coefficients in $\mathbb{C}[\eta]$. Let us set $b(s) = \prod_{\beta\in\mathbb{C}}(s-\beta)^{\nu_\beta}$. One deduces

$$\prod_{\beta\in\mathbb{C}}(\tau'\eth_{\tau'} - \beta z)^{\nu_\beta}(1\otimes m) = z^{\deg b}\cdot(\tau' z)\cdot P(\tau' z, z^{-1}\tau'\eth_{\tau'})(1\otimes m).$$

For such a relation to be a Bernstein relation in the sense of [16, 11] two conditions must be fulfilled.

(1) The right-hand side should have no pole in z; this is possible if and only if the smallest positive slope of the Newton polygon of the equation $b(\eta\partial_\eta) - \eta P(\eta, \eta\partial_\eta)$ is $\geqslant 1$; but we assumed that the order of the pole of the connection is at most two, hence the biggest slope of the Newton polygon (Katz invariant) is $\leqslant 1$. Both conditions imply that the Newton polygon has only the slopes 0 and 1.

(2) For any β with $\nu_\beta \neq 0$, the function $z \mapsto \beta z$ should be written as $z \mapsto \gamma z^2 + bz + \overline{\gamma}$ for some $\gamma \in \mathbb{C}$ and $b \in \mathbb{R}$ (cf. [11]). This implies $\gamma = 0$ and $\beta \in \mathbb{R}$.

We should apply these conditions to any exponentially twisted module $\widetilde{\mathscr{M}} \otimes \mathscr{E}^{-\varphi/z}$. Condition (1) applied to any $\widetilde{\mathscr{M}} \otimes \mathscr{E}^{-c/z\tau'}$ implies

(3) $(\mathscr{H}, \nabla_{\partial_\eta})$ satisfies the "no ramification" condition at $\eta = 0$.

Q.E.D.

References

[1] S. Cecotti, P. Fendley, K. Intriligator and C. Vafa, A new supersymmetric index, Nuclear Phys. B, **386** (1992), 405–452.

[2] P. Deligne, Équations Différentielles à Points Singuliers Réguliers, Lecture Notes in Math., **163**, Springer-Verlag, 1970.

[3] ———, Théorie de Hodge II, Publ. Math. Inst. Hautes Études Sci., **40** (1971), 5–57.

[4] ———, Théorie de Hodge irrégulière (mars 1984 & août 2006), In: Singularités Irrégulières, Correspondance et Documents, Doc. Math., **5**, Soc. Math. France, Paris, 2007, pp. 109–114 & 115–128.

[5] A. Douai and C. Sabbah, Gauss–Manin systems, Brieskorn lattices and Frobenius structures (I), Ann. Inst. Fourier (Grenoble), **53** (2003), 1055–1116.

[6] C. Hertling, Frobenius manifolds and moduli spaces for singularities, Cambridge Tracts in Math., **151**, Cambridge Univ. Press, 2002.

[7] ———, tt^* geometry, Frobenius manifolds, their connections, and the construction for singularities, J. Reine Angew. Math., **555** (2003), 77–161.

[8] C. Hertling and Ch. Sevenheck, Nilpotent orbits of a generalization of Hodge structures, J. Reine Angew. Math., **609** (2007), 23–80; arXiv:math/0603564[math.AG].

[9] M. Kashiwara and T. Kawai, The Poincaré lemma for variations of polarized Hodge structure, Publ. Res. Inst. Math. Sci., **23** (1987), 345–407.

[10] B. Malgrange, Équations différentielles à coefficients polynomiaux, Progr. Math., **96**, Birkhäuser, Basel, Boston, 1991.

[11] T. Mochizuki, Asymptotic behaviour of tame harmonic bundles and an application to pure twistor D-modules, Mem. Amer. Math. Soc., **185**, no. 869-870, Amer. Math. Soc., Providence, RI, 2007; arXiv:math/0312230[math.DG] & arXiv:math/0402122[math.DG].

[12] ———, Wild harmonic bundles and wild pure twistor D-modules, arXiv:0803.1344.

[13] ———, Asymptotic behaviour of variation of pure polarized TERP structures, arXiv:0811.1384.

[14] C. Sabbah, Déformations isomonodromiques et variétés de Frobenius, Savoirs Actuels, CNRS Éditions & EDP Sciences, Paris, 2002; English Transl.: Universitext, Springer & EDP Sciences, 2007.

[15] ———, Fourier–Laplace transform of irreducible regular differential systems on the Riemann sphere, Russian Math. Surveys, **59** (2004), 1165–1180; II, Moscow Math. J., **9** (2009), 885–898.

[16] ———, Polarizable twistor $\mathscr{D}$-modules, Astérisque, **300**, Soc. Math. France, Paris, 2005.

[17] ———, Hypergeometric periods for a tame polynomial, Portugal. Math., **63** (2006), 173–226; arXiv:math/9805077[math.AG].

[18] ———, Monodromy at infinity and Fourier transform II, Publ. Res. Inst. Math. Sci., **42** (2006), 803–835.

[19] ———, Fourier–Laplace transform of a variation of polarized complex Hodge structure, J. Reine Angew. Math., **621** (2008), 123–158; arXiv:math/0508551[math.AG].

[20] ———, Wild twistor $\mathscr{D}$-modules, In: Algebraic Analysis and Around, Adv. Stud. Pure Math., **54**, Math. Soc. Japan, 2009, pp. 293–353; arXiv:0803.0287.

[21] M. Saito, Modules de Hodge polarisables, Publ. Res. Inst. Math. Sci., **24** (1988), 849–995.

[22] ———, On the structure of Brieskorn lattices, Ann. Inst. Fourier (Grenoble), **39** (1989), 27–72.

[23] J. Scherk and J. H. M. Steenbrink, On the mixed Hodge structure on the cohomology of the Milnor fiber, Math. Ann., **271** (1985), 641–655.

[24] W. Schmid, Variation of Hodge structure: the singularities of the period mapping, Invent. Math., **22** (1973), 211–319.

[25] C. Simpson, Harmonic bundles on noncompact curves, J. Amer. Math. Soc., **3** (1990), 713–770.
[26] ———, Higgs bundles and local systems, Publ. Math. Inst. Hautes Études Sci., **75** (1992), 5–95.
[27] ———, Mixed twistor structures, Prépublication Université de Toulouse; arXiv:math/9705006[math.AG].

UMR 7640 du CNRS
Centre de Mathématiques Laurent Schwartz
École polytechnique
F–91128 Palaiseau cedex
France
E-mail address: sabbah@math.polytechnique.fr
URL: http://www.math.polytechnique.fr/~sabbah

Advanced Studies in Pure Mathematics 59, 2010
New Developments in Algebraic Geometry,
Integrable Systems and Mirror Symmetry (Kyoto, 2008)
pp. 349–369

Computation of principal $\mathcal{A}$-determinants through dimer dynamics

Jan Stienstra

Abstract.

$\mathcal{A}$ is a set of N vectors in $\mathbb{Z}^{N-2}$ situated in a hyperplane not through 0 and spanning $\mathbb{Z}^{N-2}$ over $\mathbb{Z}$. Gulotta's algorithm [4] constructs from $\mathcal{A}$ a dimer model. A theorem in [6] states that the principal $\mathcal{A}$-determinant equals the determinant of (a suitable form of) the Kasteleyn matrix of that dimer model. In the present note we translate Gulotta's pictorial description of the algorithm into matrix operations. As a result one obtains an algorithm for computing the principal $\mathcal{A}$-determinant, which is much faster than the algorithm in [5].

§1. Introduction

$\mathcal{A} = \{\mathbf{a}_1, \ldots, \mathbf{a}_N\}$ is a set of vectors in $\mathbb{Z}^{N-2}$ situated in a hyperplane not through 0 and spanning $\mathbb{Z}^{N-2}$ over $\mathbb{Z}$. The principal $\mathcal{A}$-determinant, defined in [2], describes the singularities of Gelfand–Kapranov–Zelevinsky's $\mathcal{A}$-hypergeometric system of partial differential equations [3]. It also describes for which N-tuples of coefficients $u_1, \ldots, u_N \in \mathbb{C}$ the Laurent polynomial $\sum_{j=1}^{N} u_j \mathbf{x}^{\mathbf{a}_j}$ in $N-2$ variables is singular (see [2] for details). It is a polynomial with integer coefficients in the variables $u_1, \ldots, u_N$. The restriction $\operatorname{rank} \mathcal{A} = \sharp\mathcal{A} - 2$ means that the corresponding hypergeometric functions are essentially functions in two variables and that in the Laurent polynomial the number of terms exceeds the number of variables by 2. Even in this case the definition of the principal $\mathcal{A}$-determinant is fairly complicated and only a few of its coefficients could explicitly be calculated in [2]. In [1] Dickenstein and Sturmfels re-examined the definition of the principal $\mathcal{A}$-determinant and related

Received November 14, 2008.
2000 *Mathematics Subject Classification.* Primary 82B20; Secondary 33C70.
Partly supported by JSPS Grant-in-Aid for Scientific Research (S-19104002).

it to Chow forms which is another important concept from [2]. In [6] it was shown how these Chow forms and the principal $\mathcal{A}$-determinant can be easily computed as the determinant of a suitable version of the Kasteleyn matrix of a dimer model associated with $\mathcal{A}$. When writing [6] I had only the algorithm in [5] to construct that dimer model. Later Gulotta gave another algorithm [4] for constructing the appropriate dimer model. He describes the algorithm as a process that transforms certain doubly periodic configurations of curves in the plane. In Sections 2, 3, 4 we give a faithfull reproduction of those configurations of curves by matrices and of Gulotta's algorithm by row and column operations on these matrices. In that form Gulotta's algorithm, which is a fast converging iterative process, is much more efficient than the algorithm in [5], which is a search with many trial-and-errors. Moreover, unlike for the algorithm in [5] for Gulotta's algorithm it can be guaranteed that it finds a desired dimer model. On the other hand there are cases in which [5] yields two different models and [4] gives only one.

In this note we do not need formal definitions of 'dimer model' and 'principal $\mathcal{A}$-determinant'. Dimer models are implicitly present through their patterns of zigzags. This is briefly explained in Remark 2.4. Principal $\mathcal{A}$-determinants appear only in Section 6 in a quotation from [6].

In Section 5 we recall from [5] how a pattern of zigzags (alias dimer model) is faithfully represented by a matrix $\mathbb{K}_{\mathcal{Z}}(\mathbf{z}, \mathbf{u})$, which is in fact a suitable *generalization of the Kasteleyn matrix of the dimer model.* In Section 6 we recall from [6] the theorem that expresses the principal $\mathcal{A}$-determinant as the determinant of $\mathbb{K}_{\mathcal{Z}}(\mathbf{z}, \mathbf{u})$. Sections 5 and 6 can be read immediately after Definitions 2.2 and 2.3.

§2. Patterns of zigzags on the torus

2.1. This note is about patterns of zigzags on the torus $\mathbb{T} = \mathbb{R}^2/\mathbb{Z}^2$. Here *zigzag* means the image (modulo $\mathbb{Z}^2$) of an oriented connected curve C in the plane (i.e. the image of a continuous map from $\mathbb{R}$ to $\mathbb{R}^2$) such that the two coordinate functions restrict to monotone functions on C and such that $\mathbf{t} + C = C$ for some non-zero vector $\mathbf{t} \in \mathbb{Z}^2$.

We denote by ε_1 (resp. ε_2) the zigzags in $\mathbb{T}$ coming from the first (resp. second) coordinate axis of $\mathbb{R}^2$. The homology classes of ε_1 and ε_2 form the standard basis for the homology group $H_1(\mathbb{T}, \mathbb{Z})$. We denote the intersection number of two elements α, $\beta \in H_1(\mathbb{T}, \mathbb{Z})$ by $\alpha \wedge \beta$. For two zigzags Z, Z' on $\mathbb{T}$ we write $Z \wedge Z'$ for the intersection number of their homology classes. For two zigzags Z, Z' which intersect in only a finite number of points and for which the intersections are transverse, each intersection point contributes +1 or −1 to $Z \wedge Z'$ according to the orientation; e.g. $\varepsilon_1 \wedge \varepsilon_1 = \varepsilon_2 \wedge \varepsilon_2 = 0$, $\varepsilon_1 \wedge \varepsilon_2 = -\varepsilon_2 \wedge \varepsilon_1 = 1$.

Writing $[Z]$ for the homology class of a zigzag Z we have

$$[Z] = (Z \wedge \varepsilon_2)[\varepsilon_1] - (Z \wedge \varepsilon_1)[\varepsilon_2]\,. \tag{1}$$

2.2. Definition In this note *pattern of zigzags* means a finite sequence $\mathcal{Z} = (Z_1, \dots, Z_p)$ of zigzags on $\mathbb{T}$ which satisfies the following conditions:

1. The homology classes $[Z_1], \dots, [Z_p]$ span $H_1(\mathbb{T}, \mathbb{Q})$. Their sum is 0.
2. For every j the greatest common divisor of $(Z_j \wedge \varepsilon_1, Z_j \wedge \varepsilon_2)$ is 1.
3. Every point of $\mathbb{T}$ lies on at most two zigzags in $\mathcal{Z}$.
4. Every pair of zigzags Z_i, Z_j intersects in only a finite number of points and the intersections are transverse.
5. The 2-cells (i.e. connected components) of $\mathbb{T} \setminus \bigcup_{i=1}^{p} Z_i$ are divided into three types: those (called +-cells) of which the boundary is positively oriented, those (called −-cells) of which the boundary is negatively oriented and those of which the boundary is not oriented. It is required that for every intersection point x of zigzags in $\mathcal{Z}$ one of the four 2-cells having x in their boundary is a +-cell, one is a −-cell and two have an unoriented boundary; see Figures 1–10.
6. The number of +-cells equals the number of −-cells.

2.3. Definition We say that a pattern of zigzags $\mathcal{Z} = (Z_1, \dots, Z_p)$ is *good* if in addition to the above conditions, it also satisfies:

7. Write ζ_i for the column vector $[Z_i \wedge \varepsilon_1,\ Z_i \wedge \varepsilon_2]^t$. Then the determinants $\det(\zeta_i, \zeta_{i+1})$ for $i = 1, \dots, p-1$ and $\det(\zeta_p, \zeta_1)$ are not negative. Moreover, $\zeta_i \neq -\zeta_{i+1}$ for $i = 1, \dots, p-1$ and $\zeta_p \neq \pm\zeta_1$.
8. Every ordered pair (Z_i, Z_j) of zigzags has the same orientation at all its intersection points. This is equivalent with:

$$\forall i, j : \qquad |Z_i \wedge Z_j| = \sharp(Z_i \cap Z_j)\,.$$

Condition 7 means that the homology classes are ordered counterclockwise with increasing indices and that for every homology class the indices of the zigzags in that class form a connected interval in the index set $\{1, \dots, p\}$. In 2.12 we formulate a condition which also restrains the ordering of zigzags within their homology class.

2.4. Remark The dimer model for a pattern of zigzags is the graph with a node • for every +-cell and a node ∘ for every −-cell. Two nodes are connected by an edge if the cells have a vertex in common. Figure 1 shows an example.

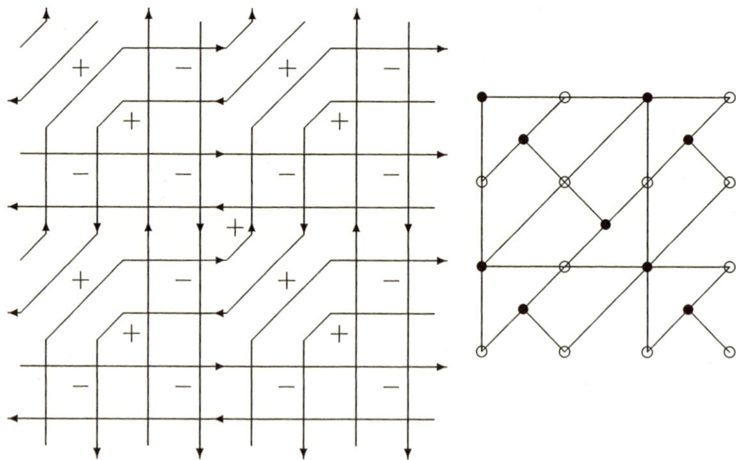

Fig. 1. *Zigzag pattern and corresponding dimer model*

2.5. To a pattern of zigzags $\mathcal{Z} = (Z_1, \dots, Z_p)$ we assign the $2\times p$-matrix

$$B_{\mathcal{Z}} = \begin{bmatrix} Z_1 \wedge \varepsilon_1 & ,\dots, & Z_p \wedge \varepsilon_1 \\ Z_1 \wedge \varepsilon_2 & ,\dots, & Z_p \wedge \varepsilon_2 \end{bmatrix},$$

displaying the intersection numbers of the zigzags in the pattern with the two curves ε_1 and ε_2. We assume that all intersections of zigzags with ε_1 and ε_2 are transverse. The intersection numbers are visible in the pictures as follows. Represent $\mathbb{T}$ by the unit square with opposite sides identified. Then $Z_j \wedge \varepsilon_2$ is the number of times the zigzag Z_j crosses the right-hand vertical edge from left to right minus the number of times it crosses from right to left. Similarly, $Z_j \wedge \varepsilon_1$ is the number of times Z_j crosses the top horizontal edge downwards minus the number of times it crosses upwards.
Condition 2.2.1. is equivalent with

the rank of $B_{\mathcal{Z}}$ is 2 and the sum of its columns is 0.

2.6. From (1) we see:

$$Z_i \wedge Z_j = (Z_i \wedge \varepsilon_1)(Z_j \wedge \varepsilon_2) - (Z_i \wedge \varepsilon_2)(Z_j \wedge \varepsilon_1) = \det \begin{bmatrix} (Z_i \wedge \varepsilon_1) & (Z_j \wedge \varepsilon_1) \\ (Z_i \wedge \varepsilon_2) & (Z_j \wedge \varepsilon_2) \end{bmatrix}.$$

This together with Condition 2.2.2. implies

$$Z_i \wedge Z_j = 0 \qquad \Longleftrightarrow \qquad [Z_i] = \pm[Z_j]\,.$$

And thus, $\operatorname{rank} B_{\mathcal{Z}} = 2$ if and only if $[Z_i] \neq \pm[Z_j]$ for some i, j.

2.7. Let $\mathcal{Z} = \{Z_1, \ldots, Z_p\}$ be a pattern of zigzags on $\mathbb{T}$. Pick a point $\star$ in one of the $-$-cells. To every 2-cell c we associate a row vector in $\mathbb{Z}^p$ as follows. Take any path γ on $\mathbb{T}$ starting at $\star$ and ending in (the interior of) c, such that γ intersects zigzags transversely. Each point in $Z_j \cap \gamma$ contributes, depending on the orientation, $+1$ or -1 to the intersection number $Z_j \wedge \gamma$. Then to c we associate the *vector of intersection numbers*, or briefly *intersection vector*, $[Z_1 \wedge \gamma, \ldots, Z_p \wedge \gamma]$. Choosing another path γ' from $\star$ to c changes this vector by a $\mathbb{Z}$-linear combination of the rows of the matrix $B_{\mathcal{Z}}$.

It follows from Condition 2.1.5 that at an intersection point of zigzags Z_i and Z_j the vectors for the two cells with unoriented boundary and the cell with negatively oriented boundary can be obtained from the vector for the $+$-cell by subtracting 1 from the i-th coordinate, respectively 1 from the j-th coordinate, respectively 1 from both i-th and j-th coordinate. We can thus capture all relevant information of the pattern of zigzags $\mathcal{Z}$ in the matrix $B_{\mathcal{Z}}$ and two additional matrices $I_{\mathcal{Z}}$ and $P_{\mathcal{Z}}$, defined as follows.

2.8. Definition The columns of $I_{\mathcal{Z}}$ and $P_{\mathcal{Z}}$ correspond with the zigzags $Z_1, \ldots, Z_p$. The rows of $I_{\mathcal{Z}}$ and $P_{\mathcal{Z}}$ correspond with the intersection points of pairs of zigzags in $\mathcal{Z}$. The row of matrix $I_{\mathcal{Z}}$ for a point $x \in Z_i \cap Z_j$ has 1 in positions i and j and 0 elsewhere. Matrix $P_{\mathcal{Z}}$ has in the row for intersection point x an intersection vector of the $+$-cell which has x in its boundary; see Figures 2 and 10 for examples.

It is also convenient to have the short notation $Q_{\mathcal{Z}} = P_{\mathcal{Z}} - I_{\mathcal{Z}}$. Then matrix $Q_{\mathcal{Z}}$ has in the row for an intersection point x an intersection vector of the $-$-cell which has x in its boundary.

2.9. Remark Due to the $\mathbb{Z}^2 B_{\mathcal{Z}}$-ambiguity in the choice of the intersection vectors there is also a $\mathbb{Z}^2 B_{\mathcal{Z}}$-ambiguity in the rows of the matrices $P_{\mathcal{Z}}$ and $Q_{\mathcal{Z}}$ in Definition 2.8. In the algorithm we start with well-defined matrices $P_{\mathcal{Z}}$ and $Q_{\mathcal{Z}}$. In the course of the algorithm we only delete rows and perform the same operations on the columns of the matrices $B_{\mathcal{Z}}$, $P_{\mathcal{Z}}$, $Q_{\mathcal{Z}}$ and $I_{\mathcal{Z}}$ simultaneously. So the algorithm is also unambiguous. It does however happen that rows of $P_{\mathcal{Z}}$ (resp. $Q_{\mathcal{Z}}$) which correspond to the same $+$-cell (resp. $-$-cell) are not the same, but differ by a vector in $\mathbb{Z}^2 B_{\mathcal{Z}}$.

In Section 5 we pass to $\mathbb{Z}^p / \mathbb{Z}^2 B_{\mathcal{Z}}$.

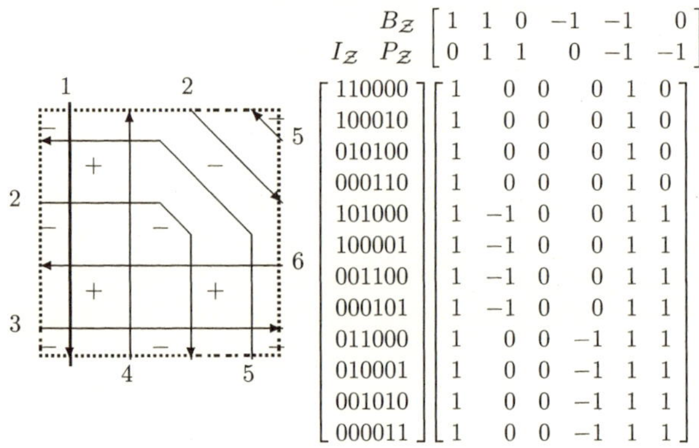

Fig. 2. *Pattern of zigzags and the corresponding matrices*

2.10. Notation For a matrix M we denote j-th column as $M(:,j)$, the i-th row as $M(i,:)$ and the (i,j)-entry as $M(i,j)$.

2.11. Definition Let Z_j and Z_k be two zigzags such that $[Z_j] = -[Z_k]$. We say that $(Z_j,\, Z_k)$ is a *+-opposite pair* (resp. *−-opposite pair*) if

$$Q_{\mathcal{Z}}(:,j) = -Q_{\mathcal{Z}}(:,k) \quad (\text{resp. } P_{\mathcal{Z}}(:,j) = -P_{\mathcal{Z}}(:,k)).$$

Suppose $Z_j \cap Z_k = \emptyset$. Then $(Z_j,\, Z_k)$ is a +-opposite pair (resp. −-opposite pair) if and only if there are between Z_j and Z_k no −-cells (resp. no +-cells). Most pictures in this note contain examples of opposite pairs. The term 'opposite pair' without ± was introduced in [4] §5.3.

2.12. Definition A good pattern of zigzags $\mathcal{Z} = (Z_1, \ldots, Z_p)$ is said to be *very good* if it satisfies:

9. For every homology class $[Z]$ the sequence of all zigzags $(Z_i, \ldots, Z_{i+r})$ in homology class $[Z]$ and the sequence of all zigzags $(Z_j, \ldots, Z_{j+s})$ in homology class $-[Z]$ satisfy:

$$\begin{array}{ll} & \textit{for } 0 \le t \le \min(r,s): \qquad (Z_{i+t}, Z_{j+t}) \textit{ is a +-opposite pair} \\ (2) & \text{and either:} \\ & \quad \textit{for } 0 \le t < \min(r,s): (Z_{i+t+1}, Z_{j+t}) \textit{ is a −-opposite pair} \\ & \text{or: } \textit{for } 0 \le t < \min(r,s): (Z_{i+t}, Z_{j+t+1}) \textit{ is a −-opposite pair} \end{array}$$

§3. The moves in the algorithm

Gulotta's algorithm transforms in an iterative way a very good pattern of zigzags $\mathcal{Z}$ into another very good one $\mathcal{Z}'$. In [4] the algorithm is mainly described by transforming a drawing of $\mathcal{Z}$ into a drawing of $\mathcal{Z}'$. We will present the same algorithm by row and column operations on the matrices $B_{\mathcal{Z}}$, $I_{\mathcal{Z}}$, $P_{\mathcal{Z}}$.

3.1. Merging move *The basic move in the algorithm merges two zigzags Z_i and Z_j which intersect in exactly one point, as shown in Figure 3.* It is evident that the merging moves preserve Conditions 1–6 in 2.2.

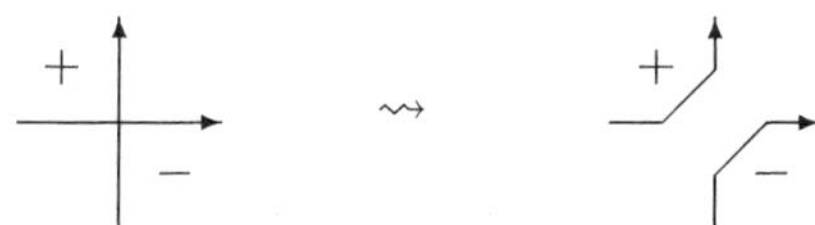

Fig. 3. *Merging move*

It was pointed out in [4], that when two zigzags Z_i and Z_j of a pattern $\mathcal{Z}$ are merged and become one zigzag Z in $\mathcal{Z}'$ then

$$\left[\begin{array}{c} Z \wedge \varepsilon_1 \\ Z \wedge \varepsilon_2 \end{array}\right] = \left[\begin{array}{c} Z_i \wedge \varepsilon_1 \\ Z_i \wedge \varepsilon_2 \end{array}\right] + \left[\begin{array}{c} Z_j \wedge \varepsilon_1 \\ Z_j \wedge \varepsilon_2 \end{array}\right]. \tag{3}$$

Actually this means $[Z] = [Z_i] + [Z_j]$ and, hence, also the column of the matrix $P_{\mathcal{Z}'}$ which corresponds with the zigzag Z is the sum of the i-th and the j-th columns of $P_{\mathcal{Z}}$. Moreover, as the picture indicates, the point of intersection $Z_i \cap Z_j$ disappears. The following statement also specifies where we put the new zigzag Z in the list of zigzags for $\mathcal{Z}'$.
Conclusion: *The merging of Z_i and Z_j for $Z_i \wedge Z_j = 1$ is given by the same column operation on $B_{\mathcal{Z}}$, $I_{\mathcal{Z}}$, $P_{\mathcal{Z}}$, namely: add the j-th column to the i-th and subsequently delete the j-th column. It also deletes from $I_{\mathcal{Z}}$ and $P_{\mathcal{Z}}$ the row for $Z_i \cap Z_j$.*

Merging moves performed on a *very good* pattern of zigzags need not preserve Conditions 2.3.7–8 and 2.12.9. Some repairing may be needed in order to turn the pattern of zigzags produced by the merging moves into a very good one again.

Fig. 4. *Repairing move 1*

3.2. Repairing move 1 *The first type of repairing move is shown in Figure 4.* This is used when $[Z_i] = -[Z_j]$ and $\sharp(Z_i \cap Z_j) = 2$, while the area between the zigzags is just one −-cell (as suggested in the picture) or one +-cell (interchange + and − in the picture). From this picture one immediately comes to the conclusion:

Conclusion: *Repairing move 1 just deletes the rows for the two points of $Z_i \cap Z_j$ from $I_{\mathcal{Z}}$ and $P_{\mathcal{Z}}$.*

3.3. Repairing move 2 *The second type of repairing move is shown in Figure 5.* This is used when $[Z_i] = [Z_j]$ and $\sharp(Z_i \cap Z_j) = 2$. In this case

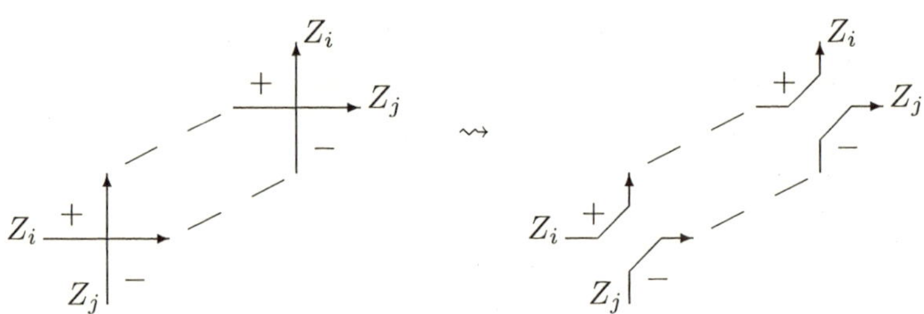

Fig. 5. *Repairing move 2*

one may distinguish three kinds of rows in the matrix $P_{\mathcal{Z}}$, according to whether the i-th entry minus the j-th entry is equal to $m-2$, $m-1$ or m, for some integer m (depending on i and j). The rows of the latter kind correspond in the picture with +-cells in the area between the two zigzags. When the intervals of Z_i and Z_j between the points of $Z_i \cap Z_j$ are swopped (as suggested by the right-hand picture) one must add 1 to the j-th coordinate and subtract 1 from the i-th coordinate in all rows of $P_{\mathcal{Z}}$ corresponding with a +-cell in the area between Z_i and Z_j. One must also interchange the i-th and j-th entries in the rows of $I_{\mathcal{Z}}$ which correspond with intersections with the intervals of Z_i and Z_j between the two points of $Z_i \cap Z_j$. And one must delete from $I_{\mathcal{Z}}$ and $P_{\mathcal{Z}}$ the two rows corresponding to the two points of $Z_i \cap Z_j$.

Conclusion: *Repairing move 2 operates on the columns of $P_{\mathcal{Z}}$ as follows: Write $H(r) = P_{\mathcal{Z}}(r,i) - P_{\mathcal{Z}}(r,j)$ and $m = \max_r(H(r))$. Then*

$$
\begin{array}{llll}
P_{\mathcal{Z}}(r,i) \rightsquigarrow P_{\mathcal{Z}}(r,i) - 1\,, & P_{\mathcal{Z}}(r,j) \rightsquigarrow P_{\mathcal{Z}}(r,j) + 1 & \text{if} & H(r) = m\,, \\
P_{\mathcal{Z}}(r,i) \rightsquigarrow P_{\mathcal{Z}}(r,i)\,, & P_{\mathcal{Z}}(r,j) \rightsquigarrow P_{\mathcal{Z}}(r,j) & \text{if} & H(r) \neq m\,.
\end{array}
$$

It operates on the columns of $I_{\mathcal{Z}}$ by:

$$
\begin{array}{l}
I_{\mathcal{Z}}(r,i) \rightsquigarrow I_{\mathcal{Z}}(r,j)\,, \quad I_{\mathcal{Z}}(r,j) \rightsquigarrow I_{\mathcal{Z}}(r,i) \\
\quad \text{if} \quad H(r) = m \text{ and } I_{\mathcal{Z}}(r,i) = 1, \text{ or } \ H(r) = m-1 \text{ and } I_{\mathcal{Z}}(r,j) = 1, \\
I_{\mathcal{Z}}(r,i) \rightsquigarrow I_{\mathcal{Z}}(r,i)\,, \quad I_{\mathcal{Z}}(r,j) \rightsquigarrow I_{\mathcal{Z}}(r,j) \quad \text{otherwise}\,.
\end{array}
$$

Finally, it deletes from $I_{\mathcal{Z}}$ and $P_{\mathcal{Z}}$ the rows for the points of $Z_i \cap Z_j$.

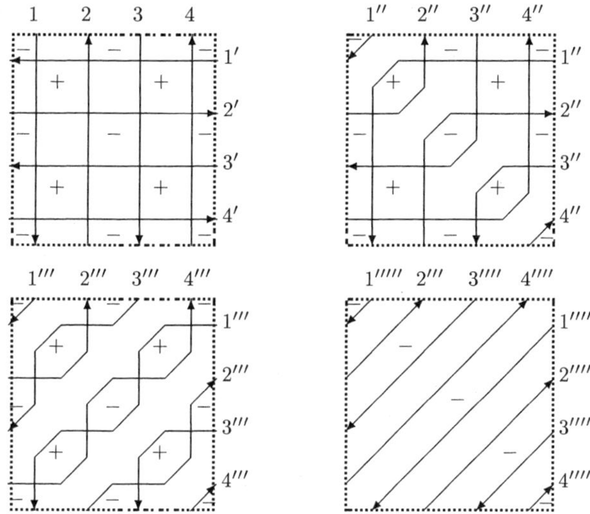

Fig. 6. *Transforming an alternating sequence of opposite pairs*

3.4. Example Let $(Z_1, \dots, Z_{2s})$ and $(Z'_1, \dots, Z'_{2s})$ be two sequences of zigzags in a pattern $\mathcal{Z}^{(1)}$ such that $\sharp(Z_i \cap Z'_j) = 1$ for all i, j and $Z_i \cap Z_j = Z'_i \cap Z'_j = \emptyset$ for all $i \neq j$ and such that (Z_{j-1}, Z_j) and (Z'_{j-1}, Z'_j) are $(-1)^j$-opposite pairs for $j = 2, \dots, 2s$. There are two well controllable cases in which merging of these two sequences followed by repairing moves 2 and 1 yields a sequence of zigzags $(Z''''_1, \dots, Z''''_{2s})$

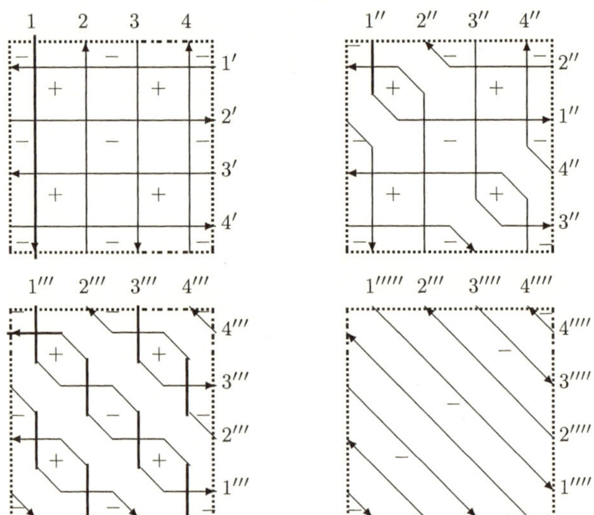

Fig. 7. *Transforming an alternating sequence of opposite pairs*

such that $Z_i'''' \cap Z_j'''' = \emptyset$ for all $i \neq j$ and such that (Z_{j-1}'''', Z_j'''') is a $(-1)^j$-opposite pair for $j = 2, \ldots, 2s$.

Case 1: *Z_j and Z_j' merge for $j = 1, \ldots, 2s$.*

Case 2: *Z_j and $Z_{j-(-1)^j}'$ merge for $j = 1, \ldots, 2s$.*

Figures 6 and 7 show this for $s = 2$ and clearly generalize to arbitrary s.

In either case let $(Z_1'', \ldots, Z_{2s}'')$ be the sequence of zigzags in the pattern $\mathcal{Z}^{(2)}$ which results from the merging. Now transform $\mathcal{Z}^{(2)}$ by applying repairing moves of type 2 at the points of $Z_i'' \cap Z_j''$ for all $i \neq j$ for which $[Z_i''] = [Z_j'']$. The result is the sequence of zigzags $(Z_1''', \ldots, Z_{2s}''')$ in the pattern $\mathcal{Z}^{(3)}$. Then $Z_i''' \cap Z_j''' \neq \emptyset$ only if $[Z_i'''] = -[Z_j''']$. Next transform $\mathcal{Z}^{(3)}$ by applying repairing moves of type 1 at the points of $Z_i''' \cap Z_j'''$ for all i, j. Call the resulting pattern of zigzags $\mathcal{Z}^{(4)}$ and the relevant sequence of zigzags $(Z_1'''', \ldots, Z_{2s}'''')$. In this last sequence $Z_i'''' \cap Z_j'''' = \emptyset$ for all $i \neq j$ and (Z_{j-1}'''', Z_j'''') is a $(-1)^j$-opposite pair for $j = 2, \ldots, 2s$.

It is instructive to perform the moves in Figures 6 and 7 also for the matrices $B_{\mathcal{Z}}$, $I_{\mathcal{Z}}$, $P_{\mathcal{Z}}$. For the top-left picture $B_{\mathcal{Z}}$, $I_{\mathcal{Z}}$, $P_{\mathcal{Z}}$ are given in Figure 10. For the other pictures one may follow the description of the merging and repairing moves.

3.5. Remark In 3.4 we have chosen the labels for the zigzags while drawing the pictures. If one uses the matrix operations instead, the algorithm determines the labels and it may be necessary to reorder the columns of the matrices $P_{\mathcal{Z}}$, $Q_{\mathcal{Z}}$, $I_{\mathcal{Z}}$ to meet the requirements of Equation (2). Example 3.4 shows that such a reordering is always possible.

3.6. Repairing move 3 *The third type of repairing move is shown in Figure 8.* It is used for a zigzag Z_0 and a sequence of zigzags $(Z_1, \ldots, Z_{2s})$ which satisfy the following conditions. Firstly, $Z_i \cap Z_j = \emptyset$ for all $i > j \geq 1$ and (Z_{j-1}, Z_j) is a $(-1)^j$-opposite pair for $j = 2, \ldots, 2s$. Secondly $[Z_0] = \pm[Z_1]$ and $\sharp(Z_0 \cap Z_j) = 2$ for $j = 1, \ldots, 2s$. Reversing if necessary the labeling in the sequence $(Z_1, \ldots, Z_{2s})$ we may without loss of generality assume $[Z_0] = (-1)^j[Z_j]$ for $j = 1, \ldots, 2s$.

Gulotta's instructions (cf. [4] §5.3) in this situation are to remove $(Z_1, \ldots, Z_{2s})$ and to insert a sequence of zigzags $(Z'_1, \ldots, Z'_{2s})$ such that $Z'_i \cap Z'_j = \emptyset$ for all $j > i \geq 0$ and such that (Z'_{j-1}, Z'_j) is a $(-1)^j$-opposite pair for $j = 1, \ldots, 2s$. For notational convenience we write here and below $Z'_0 = Z_0$.

In terms of the matrices $I_{\mathcal{Z}}$ and $P_{\mathcal{Z}}$ this means that we first delete from $I_{\mathcal{Z}}$ and $P_{\mathcal{Z}}$ all rows which correspond with an intersection point on one of the zigzags $Z_1, \ldots, Z_{2s}$ and subsequently replace, for $j = 1, \ldots, 2s$, the column of $P_{\mathcal{Z}}$ which corresponds with the zigzag Z_j by $(-1)^j$ times the column of $P_{\mathcal{Z}}$ which corresponds with the zigzag Z_0.

Next we expand every row of $I_{\mathcal{Z}}$ and $P_{\mathcal{Z}}$ which corresponds with an intersection point of Z_0 and a zigzag $Z_\infty \neq Z_0, Z_1, \ldots, Z_{2s}$ to $1+2s$ rows which correspond with the intersection points of Z_∞ with $Z'_0, Z'_1, \ldots, Z'_{2s}$ (see Figure 9). The columns of $I_{\mathcal{Z}}$ and $P_{\mathcal{Z}}$ are labeled in such a way that the column which originally corresponded to Z_j now corresponds to Z'_j for $j = 0, \ldots, 2s$.

Thus a row of $I_{\mathcal{Z}}$ for an intersection point x of Z_∞ and Z_0 is replaced by $1 + 2s$ rows of which the j-th one (for $j = 0, \ldots, 2s$) has entry 1 in the columns corresponding with the zigzags Z_∞ and Z'_j. Of the $1 + 2s$ new rows of $P_{\mathcal{Z}}$ the 0-th one is equal to the row of $P_{\mathcal{Z}}$ which corresponds to x. Figure 9 shows that for odd j the j-th row is obtained from the $(j-1)$-st one by adding -1 in the column for Z'_{j-1} and $+1$ in the column for Z'_j; for even $j \geq 2$ the j-th row is equal to the $(j-1)$-st one.

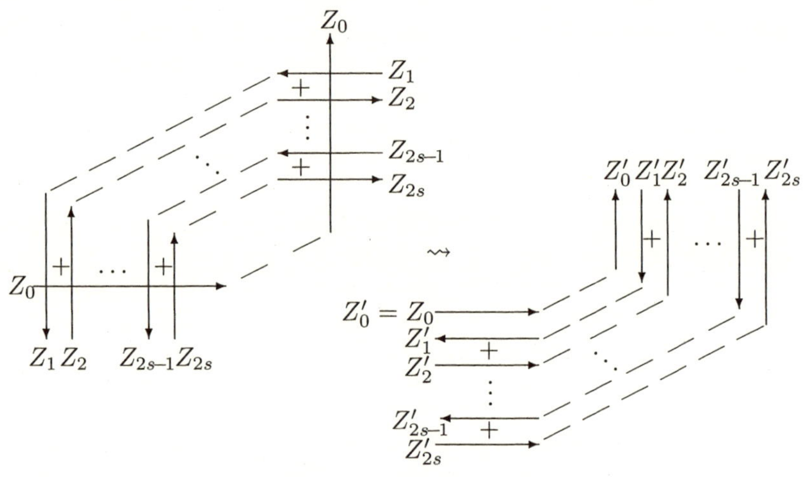

Fig. 8. *Repairing move 3*

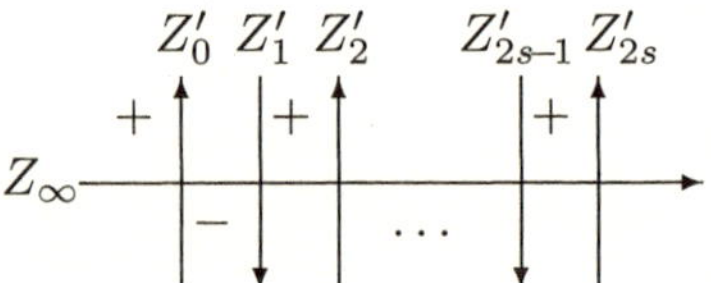

Fig. 9. *Intersections with alternating sequence of opposite pairs*

§4. Running the algorithm

In this Section we translate the algorithm described by Gulotta in terms of pictures, into an iterative process operating on matrices.

4.1. In order to prepare the input for the algorithm from the set $\mathcal{A} = \{\mathbf{a}_1, \dots, \mathbf{a}_N\}$ (see §1) we take a $2 \times N$-matrix $B_{\mathcal{A}}$ such that its rows are a $\mathbb{Z}$-basis for the lattice $\{(\ell_1, \dots, \ell_N) \in \mathbb{Z}^N \mid \ell_1\mathbf{a}_1 + \dots + \ell_N\mathbf{a}_N = 0\}$.

The aim of the algorithm is to create a very good pattern of zigzags $\mathcal{Z}$ such that $B_{\mathcal{Z}}$ results by permuting and splitting up the columns of $B_{\mathcal{A}}$ as follows

$$\text{(4)} \qquad \text{column} \begin{bmatrix} r \\ s \end{bmatrix} \text{ of } B_{\mathcal{A}} \rightsquigarrow d = \text{g.c.d.}(r, s) \text{ columns } \frac{1}{d}\begin{bmatrix} r \\ s \end{bmatrix} \text{ in } B_{\mathcal{Z}}.$$

4.2. The algorithm starts with a very good pattern of zigzags $\mathcal{Z}$ for which $B_{\mathcal{Z}}$ is the $2 \times (2n_1 + 2n_2)$-matrix

$$\left[\begin{array}{cccc} \overbrace{1\ldots1}^{n_1} & \overbrace{0\ldots0}^{n_2} & \overbrace{-1\cdots-1}^{n_1} & \overbrace{0\ldots\ \ 0}^{n_2} \\ 0\ldots0 & 1\ldots1 & 0\ldots\ \ 0 & -1\cdots-1 \end{array}\right],$$

where n_1 (resp. n_2) is the sum of the positive entries in the first (resp. second) row of $B_{\mathcal{A}}$. This matrix is realized by a pattern with straight lines, n_1 vertically down, n_1 vertically up, n_2 horizontally left-to-right and n_2 horizontally right-to-left. To get a very good pattern one takes the vertical lines alternatingly down and up, and the horizontal lines alternatingly left-to-right and right-to-left. The matrices $B_{\mathcal{Z}}$, $P_{\mathcal{Z}}$ and $I_{\mathcal{Z}}$ for the initial pattern have $2n_1 + 2n_2$ columns. There are $4n_1n_2$ intersection points and, hence, $I_{\mathcal{Z}}$ and $P_{\mathcal{Z}}$ have $4n_1n_2$ rows. We build $I_{\mathcal{Z}}$ and $P_{\mathcal{Z}}$ as follows: the $+$-cells are given by pairs (a, b) in $\{1, \ldots, n_1\} \times \{1, \ldots, n_2\}$. The rows of $I_{\mathcal{Z}}$ which correspond to the four vertices of the $+$-cell (a, b) are have 1 in positions a, $b + n_1$, resp. a, $b + 2n_1 + n_2$, resp. $a + n_1 + n_2$, $b + n_1$, resp. $a + n_1 + n_2$, $b + 2n_1 + n_2$. All other entries in these rows of $I_{\mathcal{Z}}$ are 0. The non-zero entries of the four rows of $P_{\mathcal{Z}}$ which correspond to the vertices of the $+$-cell (a, b) are 1 in position j if $1 \le j \le a$ or $2n_1 + n_2 + 1 \le j \le 2n_1 + n_2 + b$ and -1 in position j if $n_1 + n_2 + 1 \le j \le n_1 + n_2 + a - 1$ or $n_1 + 1 \le j \le n_1 + b - 1$. Figure 10 shows an example with $n_1 = n_2 = 2$.

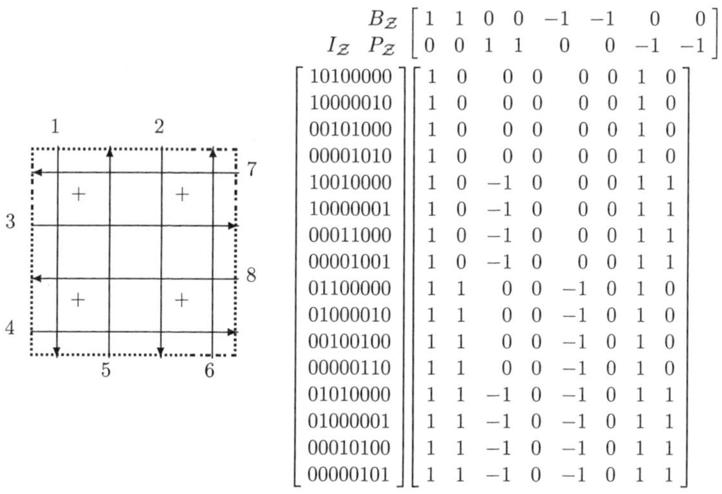

$$\begin{array}{r} B_{\mathcal{Z}} \\ I_{\mathcal{Z}}\ \ P_{\mathcal{Z}} \end{array} \left[\begin{array}{cccccccc} 1 & 1 & 0 & 0 & -1 & -1 & 0 & 0 \\ 0 & 0 & 1 & 1 & 0 & 0 & -1 & -1 \end{array}\right]$$

$$\left[\begin{array}{c} 10100000 \\ 10000010 \\ 00101000 \\ 00001010 \\ 10010000 \\ 10000001 \\ 00011000 \\ 00001001 \\ 01100000 \\ 01000010 \\ 00100100 \\ 00000110 \\ 01010000 \\ 01000001 \\ 00010100 \\ 00000101 \end{array}\right] \left[\begin{array}{cccccccc} 1 & 0 & 0 & 0 & 0 & 0 & 1 & 0 \\ 1 & 0 & 0 & 0 & 0 & 0 & 1 & 0 \\ 1 & 0 & 0 & 0 & 0 & 0 & 1 & 0 \\ 1 & 0 & 0 & 0 & 0 & 0 & 1 & 0 \\ 1 & 0 & -1 & 0 & 0 & 0 & 1 & 1 \\ 1 & 0 & -1 & 0 & 0 & 0 & 1 & 1 \\ 1 & 0 & -1 & 0 & 0 & 0 & 1 & 1 \\ 1 & 0 & -1 & 0 & 0 & 0 & 1 & 1 \\ 1 & 1 & 0 & 0 & -1 & 0 & 1 & 0 \\ 1 & 1 & 0 & 0 & -1 & 0 & 1 & 0 \\ 1 & 1 & 0 & 0 & -1 & 0 & 1 & 0 \\ 1 & 1 & 0 & 0 & -1 & 0 & 1 & 0 \\ 1 & 1 & -1 & 0 & -1 & 0 & 1 & 1 \\ 1 & 1 & -1 & 0 & -1 & 0 & 1 & 1 \\ 1 & 1 & -1 & 0 & -1 & 0 & 1 & 1 \\ 1 & 1 & -1 & 0 & -1 & 0 & 1 & 1 \end{array}\right]$$

Fig. 10. *Pattern to start from*

4.3. At the beginning of an iteration step we have a very good pattern of zigzags $\mathcal{Z} = \{Z_1, \dots, Z_p\}$. The merging moves are determined from the positions of the columns of the matrix $B_{\mathcal{A}}$ with respect to the columns of the matrix $B_{\mathcal{Z}}$. The columns of $B_{\mathcal{Z}}$ are vectors in the plane $\mathbb{R}^2$ and the ordering by increasing index coincides with the counter-clockwise cyclic ordering. In agreement with this cyclic structure we treat the first column of $B_{\mathcal{Z}}$ as consecutive to the last column.

A column v of $B_{\mathcal{A}}$ is either a positive integer multiple of a column of $B_{\mathcal{Z}}$ or there is a unique pair of consecutive columns v_1 and v_2 of $B_{\mathcal{Z}}$ such that $v = c_1 v_1 + c_2 v_2$ with $c_1,\, c_2 \in \mathbb{Q}_{>0}$.

4.4. The **algorithm terminates** automatically when all columns of $B_{\mathcal{A}}$ are multiples of columns of $B_{\mathcal{Z}}$. Since the merging moves decrease the number of zigzags the algorithm will surely terminate.

4.5. Cramer's rule explicates the relation $v = c_1 v_1 + c_2 v_2$:

$$\det(v_1, v_2)\, v \;=\; \det(v, v_2)\, v_1 \;-\; \det(v, v_1)\, v_2\,. \tag{5}$$

In the pattern we start with the determinants of consecutive pairs of non-equal columns of $B_{\mathcal{Z}}$ are 1. The merging of two zigzags in a very good pattern $\mathcal{Z}$ with exactly one intersection point replaces the corresponding columns of $B_{\mathcal{Z}}$ by their sum. Thus in the next iteration step in the algorithm Equation (5) becomes

$$\det(v_1, v_2)\, v \;=\; (\det(v, v_2) - m)\, v_1 \;+\; m(v_1 + v_2) \;-\; (\det(v, v_1) + m)\, v_2$$

with $m = \min(\det(v_1, v), \det(v, v_2))$. This then gives v either as a multiple of $v_1 + v_2$ or as a positive linear combination of v_1 and $v_1 + v_2$ or of $v_1 + v_2$ and v_2. Note that $\det(v_1, v_1 + v_2) = \det(v_1 + v_2, v_2) = \det(v_1, v_2)$.

Conclusion: *In all cases in which Equation (5) is used in the algorithm* $\det(v_1, v_2) = 1$ *and the equation actually reads*

$$v \;=\; \det(v, v_2)\, v_1 \;-\; \det(v, v_1)\, v_2\,. \tag{6}$$

Column v of $B_{\mathcal{A}}$ thus leads to the command that m zigzags of the pattern $\mathcal{Z}$ in the homology class corresponding with the column v_1 of $B_{\mathcal{Z}}$ must merge with m zigzags in the homology class corresponding with the column v_2; here $m = \min(\det(v_1, v), \det(v, v_2))$.

4.6. As Equation (6) indicates we need the determinants of the 2×2-matrices with first column from $B_{\mathcal{A}}$ and second column from $B_{\mathcal{Z}}$. These are simultaneously given as the entries of the $N \times p$-matrix

$$S = B_{\mathcal{A}}^t \, J \, B_{\mathcal{Z}} \qquad \text{with} \quad J = \begin{bmatrix} 0 & 1 \\ -1 & 0 \end{bmatrix}.$$

The columns of $B_{\mathcal{A}}$ correspond with the rows of S. One can implement the discussion in 4.3 and 4.5 for all columns of $B_{\mathcal{A}}$ simultaneously as follows. Let $S^c = S(:, [2:p, 1])$ be the $N \times p$-matrix obtained from S by cyclically permuting the columns so that the first column comes in the last position. Let R be the $N \times p$-matrix with (i,j)-entry

$$\begin{array}{lcll} R_{ij} & = & \frac{1}{2}(|S_{ij}| + |S^c_{ij}| - |S_{ij} + S^c_{ij}|) & \text{if } S_{ij} < 0\,, \\ R_{ij} & = & 0 & \text{if } S_{ij} \geq 0\,. \end{array}$$

We define functions ρ and λ on $\{1, \ldots, p\}$ by

$$\rho(j) = \sum_{i=1}^{N} R_{ij}\,, \qquad \lambda(j) = \rho(j-1) \text{ if } j > 1\,, \quad \lambda(1) = \rho(p)\,.$$

Next we define for a homology class of zigzags $[Z]$ in the pattern $\mathcal{Z} = (Z_1, \ldots, Z_p)$:

$$\begin{array}{lcl} \widetilde{\rho}([Z]) & = & \max\{\rho(j) \mid Z_j \in [Z]\}\,, \\ \widetilde{\lambda}([Z]) & = & \max\{\lambda(j) \mid Z_j \in [Z]\}\,, \\ \widetilde{\mu}([Z]) & = & \sharp([Z]) \;-\; \widetilde{\rho}([Z]) \;-\; \widetilde{\lambda}([Z])\,. \end{array}$$

Then $\widetilde{\rho}([Z])$ (resp. $\widetilde{\lambda}([Z])$) is the number of zigzags in $[Z]$ which must merge with a zigzag in the homology class immediately after (resp. before) $[Z]$ and $\widetilde{\mu}([Z])$ is the number of zigzags in $[Z]$ which must not merge with another zigzag. The merging step defined in 4.8 decreases the number of zigzags by

$$q = \sum_{h=1}^{p} \lambda(h)$$

and uses the map $\varphi : \{1, \ldots, p\} \longrightarrow \{1, \ldots, p-q\}$,

$$\varphi(j) = \begin{cases} j - \sum_{h=1}^{j} \lambda(h) & \text{if} \quad j > \lambda(1) \\ p - q + j - \sum_{h=1}^{j} \lambda(h) & \text{if} \quad j \leq \lambda(1)\,. \end{cases} \tag{7}$$

4.7. In order to eventually satisfy Requirement 2.12.9 and to benefit from Example 3.4 we permute the zigzags in each homology class as follows. First we define for each homology class of zigzags $[Z]$ for which the opposite class $-[Z]$ also occurs in the pattern $\mathcal{Z}$:

$$\widehat{\rho}([Z]) = \min\{\widetilde{\rho}([Z]), \widetilde{\rho}(-[Z]))\,, \qquad \widehat{\lambda}([Z]) = \min\{\widetilde{\lambda}([Z]), \widetilde{\lambda}(-[Z]))\,,$$
$$\widehat{\mu}([Z]) = \min\{\widetilde{\mu}([Z]), \widetilde{\mu}(-[Z]))\,.$$

If $-[Z]$ does not occur in $\mathcal{Z}$ we put $\widehat{\rho}([Z]) = \widehat{\lambda}([Z]) = \widehat{\mu}([Z]) = 0$.
Next we write for every homology class $[Z]$ in $\mathcal{Z}$:

$$\overline{\rho}([Z]) = \widetilde{\rho}([Z]) - \widehat{\rho}([Z]),\ \overline{\lambda}([Z]) = \widetilde{\lambda}([Z]) - \widehat{\lambda}([Z]),\ \overline{\mu}([Z]) = \widetilde{\mu}([Z]) - \widehat{\mu}([Z]).$$

The permutation we apply to the zigzags in homology class $[Z]$ is a so-called shuffle. This means that $[Z]$ is split into disjoint intervals which are permuted, while inside each interval the ordering is unchanged. The shuffle we apply to the zigzags in $[Z]$ is depicted in Figure 11; the numbers $\widehat{\lambda}([Z])$ etc. indicate the length of the interval.

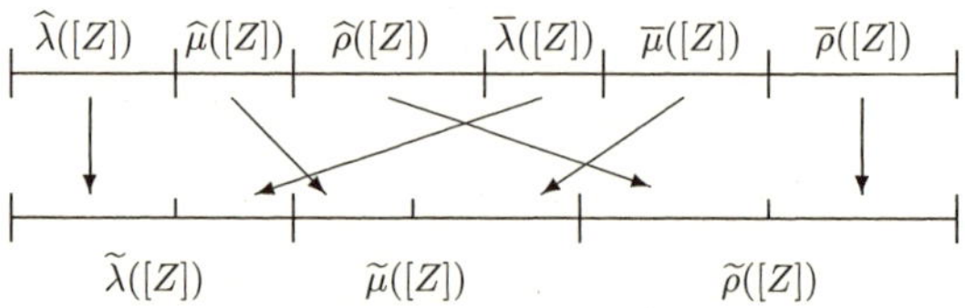

Fig. 11. *Shuffle within one homology class*

Such shuffles must be applied to each homology class in the pattern $\mathcal{Z} = (Z_1, \ldots, Z_p)$. The composite result is the shuffle permutation

$$(8) \qquad \sigma : \{1, \ldots, p\} \longrightarrow \{1, \ldots, p\}\,.$$

4.8. Definition We define the *merging matrix* $M_{\mathcal{A}\mathcal{Z}}$ for the set $\mathcal{A}$ and the pattern of zigzags $\mathcal{Z}$ to be the $p \times (p - q)$-matrix with (i, j)-entry

$$(9) \qquad (M_{\mathcal{A}\mathcal{Z}})_{ij} = 1 \text{ if } j = \varphi(\sigma(i))\,, \quad (M_{\mathcal{A}\mathcal{Z}})_{ij} = 0 \text{ if } j \neq \varphi(\sigma(i))\,,$$

with φ and σ as in (7) and (8).

The *merging step* in the algorithm multiplies the matrices $B_{\mathcal{Z}}$, $I_{\mathcal{Z}}$, $P_{\mathcal{Z}}$ and $Q_{\mathcal{Z}}$ from the right with the matrix $M_{\mathcal{A}\mathcal{Z}}$ and subsequently deletes the rows which correspond to intersection points in the pattern $\mathcal{Z}$ which disappear in the merging process. These are recognized as the rows of $I_{\mathcal{Z}}\, M_{\mathcal{A}\mathcal{Z}}$ with only one non-zero entry (namely 2).

Thus the merging $\mathcal{Z} \rightsquigarrow \mathcal{Z}'$ is realized by

$$(10)\qquad \begin{cases} B_{\mathcal{Z}'} = B_{\mathcal{Z}}\, M_{\mathcal{AZ}} \\ I_{\mathcal{Z}'} = \text{delete rows from } I_{\mathcal{Z}}\, M_{\mathcal{AZ}} \\ P_{\mathcal{Z}'} = \text{delete rows from } P_{\mathcal{Z}}\, M_{\mathcal{AZ}} \\ Q_{\mathcal{Z}'} = \text{delete rows from } Q_{\mathcal{Z}}\, M_{\mathcal{AZ}}\,. \end{cases}$$

4.9. The transformation of patterns of zigzags $\mathcal{Z} \rightsquigarrow \mathcal{Z}'$ in (10) has been organized so that if $\mathcal{Z}$ is a very good pattern, then $\mathcal{Z}'$ satisfies the Conditions 2.2.1–6, 2.3.7 and the property formulated on the first line of Equation (2). However, $\mathcal{Z}'$ need not satisfy 2.3.8, i.e. the equality

$$|Z'_i \wedge Z'_j| = \sharp(Z'_i \cap Z'_j)$$

need not hold for all pairs of zigzags (Z'_i, Z'_j) in $\mathcal{Z}'$. This equality can only be violated if Z'_i results from merging zigzags Z_{i_1} and Z_{i_2} from $\mathcal{Z}$ such that $Z_{i_1} \wedge Z_{i_2} = 1$ and $(Z_{i_1} \wedge Z'_j)(Z_{i_2} \wedge Z'_j) < 0$. Since $[Z'_j]$ can only be the homology class of a zigzag in $\mathcal{Z}$ or the sum of two such, this can only happen if $[Z'_j] = \pm([Z_{i_1}] + [Z_{i_2}]) = \pm[Z'_i]$.

The merging process in (10) is such that if Z'_i results from merging zigzags Z_{i_1} and Z_{i_2} from $\mathcal{Z}$, then every zigzag Z'_j in the homology class $[Z'_i]$ (resp. in $-[Z'_i]$) is the result of merging a zigzag Z_{j_1} from $[Z_{i_1}]$ (resp. $-[Z_{i_1}]$) with a zigzag Z_{j_2} from $[Z_{i_2}]$ (resp. $-[Z_{i_2}]$). Since $\mathcal{Z}$ satisfies Condition 2.3.8. and $Z_{i_1} \wedge Z_{i_2} = 1$ we have in that case

$$Z_{i_1} \cap Z_{j_1} = Z_{i_2} \cap Z_{j_2} = \emptyset\,, \qquad \sharp(Z_{i_1} \cap Z_{j_2}) = \sharp(Z_{i_2} \cap Z_{j_1}) = 1\,.$$

Whence if $Z'_j \in \pm[Z'_i]$ and $Z'_i \neq Z'_j$, then $\sharp(Z'_i \cap Z'_j) = 2$.
Conclusion: *For a pair of zigzags (Z'_i, Z'_j) in $\mathcal{Z}'$ we have:*

$$|Z'_i \wedge Z'_j| \neq \sharp(Z'_i \cap Z'_j) \iff Z'_i \wedge Z'_j = 0 \text{ and } \sharp(Z'_i \cap Z'_j) = 2\,.$$

4.10. We now transform the pattern of zigzags $\mathcal{Z}'$ created in (10) into a very good one. First we apply repairing moves 2 (see 3.3) to those pairs of zigzags (Z'_i, Z'_j) which satisfy $[Z'_i] = [Z'_j]$ and $|Z'_i \wedge Z'_j| \neq \sharp(Z'_i \cap Z'_j)$ and which either both do or both do not belong to a $+$-opposite pair (cf. Definition 2.11). Next we apply repairing moves 1 (see 3.2) wherever possible. See also Example 3.4 and Remark 3.5 for the effect of repairing moves 2 and 1 on alternating sequences of opposite pairs. Thus with repairing moves 2 and 1 and possibly a permutation of columns in $I_{\mathcal{Z}'}$, $P_{\mathcal{Z}'}$, $Q_{\mathcal{Z}'}$ we now have a pattern of zigzags which satisfies also the second half of Equation (2) and in which two zigzags Z_i, Z_j with $[Z_i] = \pm[Z_j]$

do not intersect unless one is member of an opposite pair and the other is not.

Finally we apply repairing moves 3 (see 3.6) as follows for all homology classes $[Z]$ and $-[Z]$ which have resulted from merging and which satisfy $\sharp([Z]) > \sharp(-[Z]) > 0$. In these circumstances the zigzags in $-[Z]$ and the first $\sharp(-[Z])$ zigzags in $[Z]$ form a sequence $(Z_1, \ldots, Z_{2s})$ as in the beginning of 3.6. For the zigzag Z_0 in 3.6 we take the zigzag in $[Z]$ with the highest index. Note that while performing repairing move 3 we have first deleted from the matrices $I_{\mathcal{Z}}$, $P_{\mathcal{Z}}$ and $Q_{\mathcal{Z}}$ all rows which corresponded with intersection points on one of the zigzags in the sequence $(Z_1, \ldots, Z_{2s})$. Subsequently we have inserted rows for intersection points of a zigzag in the new sequence $(Z'_1, \ldots, Z'_{2s})$ with a zigzag which also intersects $Z'_0 = Z_0$. The previously applied repairing moves 2 had already removed all intersection points of Z_0 with the zigzags in $[Z]$ which were not in $(Z_1, \ldots, Z_{2s})$. So after applying repairing moves 3 two zigzags in $[Z] \cup (-[Z])$ do not intersect.

4.11. After the above merging and repairing moves we have produced a very good pattern of zigzags and now **return to 4.6 for the next iteration.**

§5. From $B_{\mathcal{Z}}$, $I_{\mathcal{Z}}$, $P_{\mathcal{Z}}$, $Q_{\mathcal{Z}}$ to $\mathbb{K}_{\mathcal{Z}}$ and back

5.1. The conversion works for every good pattern $\mathcal{Z}$ of, say p, zigzags. Condition 2.12.9 is not needed here. The rows of $P_{\mathcal{Z}}$ and $Q_{\mathcal{Z}}$ must be taken modulo the row space of $B_{\mathcal{Z}}$. This is achieved by multiplying $P_{\mathcal{Z}}$ and $Q_{\mathcal{Z}}$ from the right by a $p \times (p-2)$-matrix $A_{\mathcal{Z}}$ with entries in $\mathbb{Z}$, such that $\mathrm{rank}(A_{\mathcal{Z}}) = p-2$ and $B_{\mathcal{Z}}\, A_{\mathcal{Z}} = \mathbf{0}$.

The rows of $P_{\mathcal{Z}}\, A_{\mathcal{Z}}$ represent points in $\mathbb{Z}^p/\mathbb{Z}^2 B_{\mathcal{Z}}$. We denote the set of these points by $\mathfrak{B}$, because these are in fact the black nodes in the dimer model. In the zigzag pattern these are the $+$-cells. Similarly, we denote the set of rows of $Q_{\mathcal{Z}}\, A_{\mathcal{Z}}$ by $\mathfrak{W}$. These are the white nodes in the dimer model and the $-$-cells in the zigzag pattern.

5.2. Definition(cf. [5] Definition 8.2, [6] Definition 1) The *generalized Kasteleyn matrix* $\mathbb{K}_{\mathcal{Z}}(\mathbf{z}, \mathbf{u})$ of a good pattern of zigzags $\mathcal{Z} = (Z_1, \ldots, Z_p)$ is defined as follows. The rows of $\mathbb{K}_{\mathcal{Z}}(\mathbf{z}, \mathbf{u})$ correspond $1:1$ with the elements of $\mathfrak{B}$ and the columns correspond $1:1$ with the elements of $\mathfrak{W}$. The entries of $\mathbb{K}_{\mathcal{Z}}(\mathbf{z}, \mathbf{u})$ are polynomials in two sets of variables $\mathbf{z}$ and $\mathbf{u}$. The variables in $\mathbf{u} = \{u_1, \ldots, u_p\}$ correspond $1:1$ with the zigzags in $\mathcal{Z}$, and, hence, with the columns of $B_{\mathcal{Z}}$, $I_{\mathcal{Z}}$, $P_{\mathcal{Z}}$ and $Q_{\mathcal{Z}}$. The variables in $\mathbf{z} = \{z_e\}$ correspond $1:1$ with the intersection points e of $\mathcal{Z}$ and, hence, with the rows of $I_{\mathcal{Z}}$, $P_{\mathcal{Z}}$ and $Q_{\mathcal{Z}}$.

For an intersection point e we denote by $\mathbf{b}(e)$ the element of $\mathfrak{B}$ which "is" the e-th row of $P_{\mathcal{Z}}\,A_{\mathcal{Z}}$. Similarly, $\mathbf{w}(e) \in \mathfrak{W}$ "is" the e-th row of $Q_{\mathcal{Z}}\,A_{\mathcal{Z}}$. Finally, $i(e)$ and $j(e)$ are such that $I_{\mathcal{Z}}(e, i(e)) = I_{\mathcal{Z}}(e, j(e)) = 1$.

Finally, for $\mathbf{b} \in \mathfrak{B}$ and $\mathbf{w} \in \mathfrak{W}$ we define:

$$\text{(11)} \qquad \textit{the } (\mathbf{b}, \mathbf{w})\textit{-entry of } \mathbb{K}_{\mathcal{Z}}(\mathbf{z}, \mathbf{u}) \textit{ is } \sum_{e:\, \mathbf{b}(e)=\mathbf{b},\, \mathbf{w}(e)=\mathbf{w}} z_e u_{i(e)} u_{j(e)}\,.$$

5.3. Example For the pattern of zigzags in Figure 2 the generalized Kasteleyn matrix is:

$$\mathbb{K}_{\mathcal{Z}}(\mathbf{z}, \mathbf{u}) = \begin{bmatrix} z_1u_1u_5 & z_2u_1u_2 + z_3u_4u_5 & z_4u_2u_4 \\ z_5u_1u_3 & z_6u_1u_6 + z_7u_3u_4 & z_8u_4u_6 \\ z_9u_3u_5 & z_{10}u_5u_6 + z_{11}u_2u_3 & z_{12}u_2u_6 \end{bmatrix}.$$

5.4. Definition (cf. [6] Definition 3) The *complementary generalized Kasteleyn matrix* $\mathbb{K}^c_{\mathcal{Z}}(\mathbf{z}, \mathbf{u})$ of a good pattern $\mathcal{Z}$ of p zigzags is

$$\mathbb{K}^c_{\mathcal{Z}}(\mathbf{z}, \mathbf{u}) = u_1 \cdot \ldots \cdot u_p \mathbb{K}_{\mathcal{Z}}(\mathbf{z}, u_1^{-1}, \ldots, u_p^{-1})\,.$$

5.5. Remark The information in $\mathbb{K}_{\mathcal{Z}}(\mathbf{z}, \mathbf{u})$ is in fact equivalent with that in $B_{\mathcal{Z}}$, $I_{\mathcal{Z}}$, $P_{\mathcal{Z}}$. By Theorem 9.3 and §3.7 in [5] the columns of $B_{\mathcal{Z}}$ are the primitive vectors along the sides of the Newton polygon of $\det \mathbb{K}_{\mathcal{Z}}(\mathbf{z}, \mathbf{u})$ w.r.t. $u_1, \ldots, u_p$. One recovers $P_{\mathcal{Z}}$ and $I_{\mathcal{Z}}$ as follows. Let $\mathcal{P}$ and $\mathcal{Q}$, respectively, be the sets of exponent vectors of the monomials in $u_1, \ldots, u_p$ which appear in the first columns of the matrices

$$\mathbb{K}_{\mathcal{Z}}(\mathbf{z}, \mathbf{u}) \left(\mathbb{K}^t_{\mathcal{Z}}(\mathbf{z}, \mathbf{u})\, \mathbb{K}_{\mathcal{Z}}(\mathbf{z}, \mathbf{u})\right)^n \quad \text{resp.} \quad \left(\mathbb{K}^t_{\mathcal{Z}}(\mathbf{z}, \mathbf{u})\, \mathbb{K}_{\mathcal{Z}}(\mathbf{z}, \mathbf{u})\right)^{n+1}$$

for $n \in \mathbb{Z}_{\geq 0}$. Translations on $\mathbb{Z}^p$ by vectors in $\mathbb{Z}^2 B_{\mathcal{Z}}$ preserve $\mathcal{P}$ and $\mathcal{Q}$. Now take a 'fundamental domain' $\mathcal{P}^* \subset \mathcal{P}$ for the $\mathbb{Z}^2 B_{\mathcal{Z}}$-action on $\mathcal{P}$. For each vector α in $\mathcal{P}^*$ let $\mathcal{Q}_\alpha$ be the set of vectors β in $\mathcal{Q}$ such that $\alpha - \beta$ has precisely two non-zero entries and these are both 1. The rows of $P_{\mathcal{Z}}$ resp. $I_{\mathcal{Z}}$ are α resp. $\alpha - \beta$ with $\alpha \in \mathcal{P}^*$ and $\beta \in \mathcal{Q}_\alpha$.

§6. The principal $\mathcal{A}$-determinant

6.1. The principal $\mathcal{A}_{\mathcal{Z}}$-determinant for a good pattern of zigzags Let $\mathcal{Z} = (Z_1, \ldots, Z_p)$ be a good pattern of zigzags and let $\mathcal{A}_{\mathcal{Z}}$ denote the set of rows of the matrix $A_{\mathcal{Z}}$ (see 5.1). Let ι be the homomorphism

$$\iota : \mathbb{Z}[z_e \,|\, e \text{ intersection point in } \mathcal{Z}] \longrightarrow \mathbb{Z}\,, \qquad \iota(z_e) = |Z_{i(e)} \wedge Z_{j(e)}|\,.$$

The *principal* $\mathcal{A}_{\mathcal{Z}}$*-determinant* is defined in [2] and Theorem 3 in [6] states that it is equal to

$$\iota\left(\det \mathbb{K}^{c}_{\mathcal{Z}}(\mathbf{z},\mathbf{u})\right) . \tag{12}$$

The relation

$$I_{\mathcal{Z}} = P_{\mathcal{Z}} - Q_{\mathcal{Z}}$$

is precisely the one required in Condition 2 of [6].

6.2. Concluding remark about the principal $\mathcal{A}$-determinant

For the principal $\mathcal{A}$-determinant of the set $\mathcal{A} = \{\mathbf{a}_1, \ldots, \mathbf{a}_N\}$ in the Introduction we may have to make some slight adaptations to formula (12), which reverse in a sense the transformation from $B_{\mathcal{A}}$ to $B_{\mathcal{Z}}$ in 4.1.

In (12) the variables in $\mathbf{u} = (u_1, \ldots, u_p)$ correspond $1:1$ with the columns of $B_{\mathcal{Z}}$. Take a new set of variables $\mathbf{v} = (v_1, \ldots, v_N)$ which correspond $1:1$ with the columns of $B_{\mathcal{A}}$. Recall that $B_{\mathcal{Z}}$ is obtained from $B_{\mathcal{A}}$ by permuting and splitting up columns as in (4). Then, to reverse the transformation from $B_{\mathcal{A}}$ to $B_{\mathcal{Z}}$ one must set $u_i = d_k v_k$ if the i-th column of $B_{\mathcal{Z}}$ comes from the k-th column of $B_{\mathcal{A}}$; here d_k is the g.c.d. of the two entries in the k-th column of $B_{\mathcal{A}}$.

References

[1] A. Dickenstein and B. Sturmfels, Elimination theory in codimension two, J. Symb. Comput., **34** (2002), 119–135; arXiv:math/0102204.

[2] I. M. Gelfand, M. M. Kapranov and A. V. Zelevinsky, Discriminants, Resultants and Multidimensional Determinants, Birkhäuser Boston, 1994.

[3] I. M. Gelfand, A. V. Zelevinskii and M. M. Kapranov, Hypergeometric functions and toral manifolds, Funct. Anal. Appl., **23** (1989), 94–106.

[4] D. Gulotta, Properly ordered dimers, R-charges and an efficient inverse algorithm, J. High Energy Phys., **10** (2008), 014; arXiv:0807.3012.

[5] J. Stienstra, Hypergeometric systems in two variables, quivers, dimers and dessins d'enfants, In: Modular Forms and String Duality, (eds. N. Yui, H. Verrill and C. F. Doran), Fields Inst. Commun., **54**, Amer. Math. Soc., Providence, RI, 2008, pp. 125–161; arXiv:0711.0464.

[6] J. Stienstra, Chow forms, Chow quotients and quivers with superpotential, In: Motives and Algebraic Cycles, a Celebration in Honour of Spencer J. Bloch,(eds. R. de Jeu and J. Lewis), Fields Inst. Commun., **56**, Amer. Math. Soc., Providence, RI, 2009, pp. 327–336; arXiv:0803.3908.

Mathematisch Instituut
Universiteit Utrecht
the Netherlands
E-mail address: J.Stienstra@uu.nl

Advanced Studies in Pure Mathematics 59, 2010
New Developments in Algebraic Geometry,
Integrable Systems and Mirror Symmetry (Kyoto, 2008)
pp. 371–388

Weighted projective lines associated to regular systems of weights of dual type

Atsushi Takahashi

Abstract.

We associate to a regular system of weights a weighted projective line over an algebraically closed field of characteristic zero in two different ways. One is defined as a quotient stack via a hypersurface singularity for a regular system of weights and the other is defined via the signature of the same regular system of weights.

The main result in this paper is that if a regular system of weights is of dual type then these two weighted projective lines have equivalent abelian categories of coherent sheaves. As a corollary, we can show that the triangulated categories of the graded singularity associated to a regular system of weights has a full exceptional collection, which is expected from homological mirror symmetries.

The main theorem of this paper will be generalized to more general one, to the case when a regular system of weights is of genus zero, which will be given in [5]. Since we need more detailed study of regular systems of weights and some knowledge of algebraic geometry of Deligne–Mumford stacks there, the author write a part of the result in this paper to which another simple proof based on the idea by Geigle–Lenzing [2] can be applied.

§1. Introduction

Let $W := (a_1, a_2, a_3; h)$ be a tuple of four positive integers. If it satisfies a certain combinatorial condition, it is called a regular system of weights. Let k be an algebraically closed field of characteristic zero. The condition is equivalent to the condition that for a polynomial in $k[x, y, z]$

$$f_W(x, y, z) = \sum_{a_1 i_1 + a_2 i_2 + a_3 i_3 = h} c_{i_1 i_2 i_3} x^{i_1} y^{i_2} z^{i_3}, \quad c_{i_1 i_2 i_3} \in k,$$

Received September 23, 2008.
2000 *Mathematics Subject Classification.* Primary 14J33; Secondary 53D37, 32S25.

with generic coefficients $c_{i_1i_2i_3}$, $\mathrm{Spec}(k[x,y,z]/(f_W))$ has at most an isolated singularity only at the origin [7]. Since f_W is weighted homogeneous, one naturally associates to W the following quotient stack:

$$\mathcal{C}_{f_W} := [\mathrm{Spec}(k[x,y,z]/(f_W))\backslash\{0\}/k^*],$$

where we set $k^* := \mathrm{Spec}(k\mathbb{Z})$. $\mathcal{C}_{f_W}$ is a Deligne–Mumford stack regarded as a smooth projective curve $\mathrm{Proj}(k[x,y,z]/(f_W))$ with a finite number of isotropic points on it.

On the other hand, to the signature $A_W = (\alpha_1, \dots, \alpha_r)$ of W, a combinatorically defined multi-set which can be identified with the multi-set of orders of isotropy group at isotropic points on $\mathrm{Proj}(k[x,y,z]/(f_W))$ [8], one can associate the algebra following Geigle–Lenzing [1]

$$R_{A_W,\lambda} := k[X_1, \dots, X_r]\,/I_\lambda\,,$$

where I_λ is an ideal generated by $r-2$ homogeneous polynomials

$$X_1^{\alpha_1} + X_2^{\alpha_2} + X_3^{\alpha_3}, X_1^{\alpha_1} + \lambda_i X_2^{\alpha_2} + X_i^{\alpha_i}, \quad \lambda_i \in k\backslash\{0,1\}, \quad i = 4, \dots, r.$$

Since $R_{A_W,\lambda}$ is graded with respect to an abelian group

$$L(A_W) := \bigoplus_{i=1}^{r} \mathbb{Z}\vec{X}_i \Big/ \Big(\alpha_i\vec{X}_i - \alpha_j\vec{X}_j; 1 \le i < j \le r\Big)\,,$$

one can consider the quotient stack

$$\mathcal{C}_{A_W,\lambda} := [\mathrm{Spec}(R_{A_W,\lambda})\backslash\{0\}/\mathrm{Spec}(kL(A_W))]\,.$$

$\mathcal{C}_{A_W,\lambda}$ is a Deligne–Mumford stack whose underlying quotient scheme is a smooth projective line (it is easy to see that $R_{A_W,\lambda} \supset k[X_1^{\alpha_1}, X_2^{\alpha_2}]$ as a sub-ring). Properties of categories $\mathrm{coh}(\mathcal{C}_{A_W,\lambda})$ and $D^b\mathrm{coh}(\mathcal{C}_{A_W,\lambda})$ are extensively studied by Geigle–Lenzing [1].

It is a very natural and interesting problem to compare $\mathcal{C}_{f_W}$ with $\mathcal{C}_{A_W,\lambda}$ as algebraic stacks. We have the following result.

Theorem 1.1 (Main Theorem 5.1). *Let W be a regular system of weights of dual type. Then, there exists a $\mathbb{Z}$-graded sub-ring R_W of $R_{A_W,\lambda}$ which induces equivalences of abelian categories:*

$$\mathrm{coh}(\mathcal{C}_{f_W}) \simeq \mathrm{coh}(\mathcal{C}_W) \simeq \mathrm{coh}(\mathcal{C}_{A_W,\lambda}), \tag{1.1}$$

where $\mathcal{C}_W := [\mathrm{Spec}(R_W)\backslash\{0\}/k^*]$.

If W is a regular system of weights of dual type then we have $r = 3$ and hence the parameter λ does not appear in $R_{A_W,\lambda}$. Note also that as an abelian category $\mathrm{coh}(\mathcal{C}_{f_W})$ is independent of the choice of the polynomial f_W.

One of motivations of this work is to study a qualitative structure of the triangulated category $D^{gr}_{Sg}(R_W) := D^b(\text{gr-}R_W)/D^b(\text{grproj-}R_W)$. In particular, we have been interested in finding a "nice" or "special" full strongly exceptional collection of $D^{gr}_{Sg}(R_W)$ by using graded matrix factorizations. So far, we have succeeded to obtain it in [12] for A_l-type, in [3] for ADE-type and in [4] for the case $\epsilon_W = -1$.

It is in general a very difficult problem, however, by the main theorem in this paper combining with a result by Orlov [6], one has a slightly weaker statement for *any* regular system of weights of dual type:

Corollary 1.2 (Corollary 6.3). *Let W be a regular system of weights of dual type. The triangulated category $D^{gr}_{Sg}(R_W)$ has a full exceptional collection $(E_1, \ldots, E_{\mu_{W^*}})$.*

This is expected from the homological mirror symmetry conjecture that predicts the existence of a triangulated equivalence $D^{gr}_{Sg}(R_W) \simeq D^b\mathrm{Fuk}(f_{W^*}) \simeq D^b\mathrm{Fuk}^{\to}(\{\gamma_\bullet\})$ where $\mathrm{Fuk}^{\to}(\{\gamma_\bullet\})$ is the directed Fukaya category associated to a distinguished basis of vanishing graded Lagrangian sub-manifolds $\{\gamma_\bullet\} = \{\gamma_1, \ldots, \gamma_\mu\}$ in the Milnor fiber of the dual regular system of weights (see Section 7 for details).

We give here an outline of the paper. Section 2 introduces the definition of a regular system of weights and several invariants of it given in [7] and [8]. Section 3 explains the notion of the topological mirror symmetry and the duality of regular systems of weights defined in [11] and [8]. Most of results in this paper rely on the classification of regular systems of weights of dual type and several data given explicitly by them. In Section 4, after preparing some definitions and reviewing a construction of weighted projective line by Geigle–Lenzing [1], we characterize a special sub-ring R_W of their homogeneous coordinate ring R_{A_W} of a weighted projective line, which will play a key role in our story.

Section 5 gives the main theorem of this paper. Its proof uses the data of invariants of regular systems of weights of dual type, which we gave in Appendix. We shall give an application of this result in Section 6. Section 7 explains the homological mirror symmetry conjecture of hypersurface singularities and one of our motivations of the paper. In particular, it is discussed that the bounded derived category of coherent sheaves on a weighted projective line associated to a regular system of weights of dual type is expected to be triangulated equivalent to the derived category of a directed Fukaya category associated to a cusp singularity associated to the dual signature.

The main theorem of this paper will be generalized to more general one, to the case when a regular system of weights W is of genus zero, i.e., when $g(\mathrm{Proj}(R_W)) = 0$, which will be given in [5].

Acknowledgments. The author would like to thank Hiroshige Kajiura and Kyoji Saito for valuable discussion. This work was partly supported by Grant-in Aid for Scientific Research grant numbers 17740036 from the Ministry of Education, Culture, Sports, Science and Technology, Japan.

§2. Regular systems of weights

In this paper, we denote by k an algebraically closed field of characteristic zero.

Definition 2.1 ([7]). Let a_i, $i = 1, 2, 3$ and h be positive integers.

(i) We call a tuple of integers $W := (a_1, a_2, a_3; h)$ a *regular system of weights* if $\gcd(a_1, a_2, a_3; h) = 1$ and a rational function:

$$\chi_W(T) := \prod_{i=1}^{3} \frac{1 - T^{h-a_i}}{1 - T^{a_i}}, \tag{2.1}$$

is a polynomial in T.

(ii) The integers a_i are called *weights* of W and h is called the *Coxeter number* of W.

The next proposition relates combinatorics of regular systems of weights with geometries of hypersurface singularities.

Proposition 2.2 ([7]). *The following conditions are equivalent:*

(i) *$W = (a_1, a_2, a_3; h)$ is a regular system of weights.*

(ii) *There is at least one weighted homogeneous polynomial in $k[x, y, z]$*

$$f(x, y, z) = \sum_{a_1 i_1 + a_2 i_2 + a_3 i_3 = h} c_{i_1 i_2 i_3} x^{i_1} y^{i_2} z^{i_3}, \quad c_{i_1 i_2 i_3} \in k,$$

such that the hypersurface $\{(x, y, z) \in k^3 \mid f(x, y, z) = 0\}$ has at most an isolated singularity only at the origin.

(iii) *There is a non-empty dense subset of*

$$\{g(x, y, z) \in k[x, y, z] \mid g(x, y, z) = \sum_{a_1 i_1 + a_2 i_2 + a_3 i_3 = h} c_{i_1 i_2 i_3} x^{i_1} y^{i_2} z^{i_3}, c_{i_1 i_2 i_3} \in k\}$$

such that any polynomial $f(x,y,z)$ belonging to that set defines the hypersurface $\{(x,y,z) \in k^3 \mid f(x,y,z) = 0\}$ which has at most an isolated singularity only at the origin.

Remark 2.3. We shall see later that one can sometimes choose in *(iii)* a "canonical" polynomial, which will be one of keys to prove the main theorem in this paper.

Definition 2.4 ([7])**.** Let W be a regular system of weights.

(i) The positive integer μ_W defined by

$$\mu_W := \prod_{i=1}^{3} \frac{h - a_i}{a_i} \tag{2.2}$$

is called the *rank* or the *Milnor number* of the regular system of weights W.

(ii) The integer ϵ_W defined by

$$\epsilon_W := \left(\sum_{i=1}^{3} a_i \right) - h \tag{2.3}$$

is called the *minimal exponent* or the *Gorenstein parameter* of W.

(iii) There are finite number of integers $m_1 < m_2 \leq \cdots \leq m_{\mu_W - 1} < m_{\mu_W}$ such that

$$T^{m_1} + \cdots + T^{m_{\mu_W}} = T^{\epsilon_W} \chi_W(T).$$

Each integer m_i is called an *exponent* of W, which satisfies the property

$$m_i + m_{\mu_W - i + 1} = h, \quad i = 1, \ldots, \mu_W. \tag{2.4}$$

Define more "geometric" invariants for regular systems of weights, whose meanings will be clear in later sections.

Definition 2.5 ([7][8])**.** Let $W = (a_1, a_2, a_3; h)$ be a regular system of weights.

(i) The *genus* of W is a non-negative integer g_W defined as the number of exponents of W equal to 0.

(ii) Set $m(a_i, a_j : h) := \#\{(u,v) \in (\mathbb{Z}_{\geq 0})^2 \mid a_i u + a_j v = h\}$. Consider the following multi-set of positive integers

$$A'_W := \{a_i \mid h/a_i \notin \mathbb{Z}, i = 1,2,3\} \coprod \{\gcd(a_i, a_j)^{(m(a_i,a_j:h)-1)} \mid 1 \leq i < j \leq 3\},$$

where we mean by $\gcd(a_i, a_j)^{(m(a_i,a_j:h)-1)}$ that the integer $\gcd(a_i, a_j)$ appears $m(a_i, a_j : h) - 1$ times in A'_W. Let A_W be a subset of A'_W consisting of integers greater than 1 such that $A'_W = A_W \coprod \{1^s\}$ for some $s \in \mathbb{Z}_{\geq 0}$.

(iii) The pair $(g_W; A_W)$ is called the *signature* of W. If g_W is 0, then we often omit g_W and call A_W the signature of W.

Remark 2.6. Usually, we shall write A_W as $(\alpha_1, \dots, \alpha_r)$ so that $2 \leq \alpha_1 \leq \alpha_2 \leq \cdots \leq \alpha_r$ for some $r \in \mathbb{Z}_{\geq 0}$.

Definition 2.7 ([8])**.** Let W be a regular system of weights.

(i) The polynomial

$$\varphi_W(\lambda) := \prod_{i=1}^{\mu_W} (\lambda - \mathbf{e}[\frac{m_i}{h}]), \qquad \mathbf{e}[*] = \exp(2\pi\sqrt{-1}*), \tag{2.5}$$

is called the *characteristic polynomial* of W. $\varphi_W(\lambda)$ is a cyclotomic polynomial.

(ii) $\varphi_W(\lambda)$ has a unique expression

$$\varphi_W(\lambda) = \prod_{i \in \mathbb{Z}_{>0}} (\lambda^i - 1)^{e_W(i)} \tag{2.6}$$

for some integers $e_W(i)$. A set of integers $M(W) := \{i \mid e_W(i) \neq 0\}$ is called the *classifying poset* of W since $M(W)$ is a partially ordered set (poset) with respect to the division relation on integers.

(iii) The *dual characteristic polynomial* $\varphi^*_W(\lambda)$ of W is a polynomial

$$\varphi^*_W(\lambda) := \prod_{i \in M(W)} (\lambda^i - 1)^{-e_W(h/i)}. \tag{2.7}$$

§3. Duality of regular systems of weights

In this subsection, we recall the notion of the duality of regular systems of weights discussed in [8] and [11].

Definition 3.1. Let W be a regular system of weights and let G be a finite abelian subgroup of $GL(3, k)$, whose elements are of the form $\mathrm{diag}(\mathbf{e}[\omega_1\alpha_1], \mathbf{e}[\omega_2\alpha_2], \mathbf{e}[\omega_3\alpha_3])$ where $\omega_i := a_i/h$ and $\alpha_i \in \mathbb{Z}$.

For a pair (W, G), set

$$\chi(W, G)(y, \bar{y}) := \frac{(-1)^3}{|G|} \sum_{\alpha \in G} \chi_\alpha(W, G)(y, \bar{y}), \tag{3.1}$$

(3.2)

$$\chi_\alpha(W,G)(y,\bar{y}) := \sum_{\beta\in G}\prod_{\omega_i\alpha_i\notin\mathbb{Z}}(y\bar{y})^{\frac{1-2\omega_i}{2}}\left(\frac{y}{\bar{y}}\right)^{-\omega_i\alpha_i+[\omega_i\alpha_i]+\frac{1}{2}}$$

$$\times\prod_{\omega_i\alpha_i\in\mathbb{Z}}\mathbf{e}\left[\omega_i\beta_i+\frac{1}{2}\right]\frac{1-\mathbf{e}\left[(1-\omega_i\beta_i)\right](y\bar{y})^{1-\omega_i}}{1-\mathbf{e}\left[\omega_i\beta_i\right](y\bar{y})^{\omega_i}},$$

where $[\omega_i\alpha_i]$ denotes the greatest integer smaller than $\omega_i\alpha_i$. We call $\chi(W,G)(y,\bar{y})$ the *orbifoldized Poincaré polynomial* of (W,G).

Remark 3.2. Note that if we put $T^h = y\bar{y}$, we have

$$\chi_W(T) = \chi(W,\{1\})(y,\bar{y}).$$

By this orbifoldized Poincaré polynomial $\chi(W,G)$, one can define the following notion of the duality among pairs (W,G):

Definition 3.3 (Topological Mirror Symmetry)**.** Let (W,G) and (W^*,G^*) be pairs as in Definition 3.1. A pair (W^*,G^*) is called *topological mirror dual* to a pair (W,G), if

$$\chi(W^*,G^*)(y,\bar{y}) = (-1)^3\bar{y}^{\hat{c}_W}\chi(W,G)(y,\bar{y}^{-1}), \tag{3.3}$$

where $\hat{c}_W := 1-2\frac{\epsilon_W}{h}$.

Let W be a regular system of weights. We shall call the group generated by $\mathrm{diag}(\mathbf{e}[\omega_1],\mathbf{e}[\omega_2],\mathbf{e}[\omega_3])$ the *principal discrete group* for W and denote it by G_W^0. As a special case of topological mirror symmetry, we define the following duality of regular systems of weights:

Definition 3.4 ([11])**.** Let W and W^* be regular systems of weights. W^* is said to be *dual* to W if

$$\chi(W^*,\{1\})(y,\bar{y}) = (-1)^3\bar{y}^{\hat{c}_W}\chi(W,G_W^0)(y,\bar{y}^{-1}). \tag{3.4}$$

Proposition 3.5 ([11])**.** *Let W and W^* be regular systems of weights. W^* is dual to W if and only if W^* is $*$-dual to W in the sense of K. Saito* [8]*, more precisely, if and only if*

(3.5)

$$\varphi_{W^*}(\lambda) = \prod_{i\in M(W^*)}(\lambda^i-1)^{e_{W^*}(i)} = \prod_{i\in M(W)}(\lambda^i-1)^{-e_W(h/i)} = \varphi_W^*(\lambda),$$

and $W^* = (lm-l+1, mk-m+1, kl-k+1; klm+1)$ when $W = (lm-m+1, mk-k+1, kl-l+1; klm+1)$.

Remark 3.6. If W^* is dual to W, then W^* is uniquely determined by W and $(W^*)^* = W$. Therefore, one sees that these two equivalent notions of duality define equivalence relations.

Definition 3.7. A regular system of weights W is called of *dual type* if W has the dual regular system of weights W^*.

Proposition 3.8 ([8]). *Let W be a regular system of weights W of dual type. Then, the signature $(g_W; A_W)$ of W is of the form $(0; \alpha_1, \alpha_2, \alpha_3)$, i.e., $g_W = 0$ and $r = 3$.*

Remark 3.9. Regular systems of weights of dual type are classified into five types with respect to the classifying poset $M(W)$. See Appendix for details of their data, i.e., weights, dual weights, signatures, and so on.

§4. Weighted projective lines associated to A_W

Definition 4.1 ([1]). Let $W = (a_1, a_2, a_3; h)$ be a regular system of weights of genus 0 and let $A_W = (\alpha_1, \ldots, \alpha_r)$ be its signature.

(i) Consider an abelian group generated by r-letters $\vec{X}_i$ $(i = 1, \ldots, r)$,

$$L(A_W) := \bigoplus_{i=1}^{r} \mathbb{Z}\vec{X}_i \Big/ \left(\alpha_i \vec{X}_i - \alpha_j \vec{X}_j ; 1 \le i < j \le r\right). \tag{4.1}$$

It is an ordered group with $L(A_W)_+ := \sum_{i=1}^{r} \mathbb{Z}_{>0}\vec{X}_i$ as its positive elements.

(ii) An element $\omega_{A_W} := (r-2)\cdot\vec{c} - \sum_{i=1}^{r} \vec{X}_i$ is called the *dualizing element* of $L(A_W)$, where $\vec{c} := \alpha_1 \vec{X}_1 = \cdots = \alpha_r \vec{X}_r$.

(iii) Let $\alpha := \mathrm{lcm}(\alpha_1, \ldots, \alpha_r)$. The *degree map* is a group homomorphism $\deg : L(A_W) \longrightarrow \mathbb{Z}$ defined on generators by $\deg(\vec{X}_i) := \alpha/\alpha_i$. The degree map deg is an epimorphism and $\deg(\vec{X}) = 0$ if $\alpha \cdot \vec{X} = 0$. Note that $\deg(\vec{c}) = \alpha$.

(iv) Define an $L(A_W)$-graded k-algebra by

$$R_{A_W,\lambda} := k[X_1, \ldots, X_r] / I_\lambda , \tag{4.2}$$

where I_λ is an ideal generated by $r-2$ homogeneous polynomials

$$X_1^{\alpha_1} + X_2^{\alpha_2} + X_3^{\alpha_3}, X_1^{\alpha_1} + \lambda_i X_2^{\alpha_2} + X_i^{\alpha_i}, \quad \lambda_i \in k\backslash\{0,1\}, \quad i = 4, \ldots, r. \tag{4.3}$$

$R_{A_W,\lambda}$ is of non-negatively graded.

If $r = 3$, then we shall write $R_{A_W,\lambda}$ as R_{A_W} for simplicity.

Denote by $\mathrm{mod}_{L(A_W)}$-$R_{A_W,\lambda}$ the abelian category of finitely generated $L(A_W)$-graded $R_{A_W,\lambda}$-modules and denote by $\mathrm{tor}_{L(A_W)}$-$R_{A_W,\lambda}$ the full subcategory of $\mathrm{mod}_{L(A_W)}$-$R_{A_W,\lambda}$ whose objects are finite dimensional $L(A_W)$-graded $R_{A_W,\lambda}$-modules.

Definition 4.2. Define an algebraic stack $\mathcal{C}_{A_W,\lambda}$ by

$$\mathcal{C}_{A_W,\lambda} := \left[\mathrm{Spec}(R_{A_W,\lambda})\backslash\{0\}/\mathrm{Spec}(k\cdot L(A_W))\right]. \tag{4.4}$$

The abelian category $\mathrm{coh}(\mathcal{C}_{A_W,\lambda})$ of coherent sheaves on the algebraic stack $\mathcal{C}_{A_W,\lambda}$ is equivalent to $\mathrm{mod}_{L(A_W)}$-$R_{A_W,\lambda}/\mathrm{tor}_{L(A_W)}$-$R_{A_W,\lambda}$.

If $r=3$, we shall denote $\mathcal{C}_{A_W,\lambda}$ by $\mathcal{C}_{A_W}$ since λ does not appear. Note that if W is a regular system of weights of dual type then $r=3$.

$L(A_W)$ has a special subgroup isomorphic to $\mathbb{Z}$ as we shall see below, which will be identified with the Lie algebra of k^* acting on the hypersurface singularity defined by W.

Definition 4.3. Let $W=(a_1,a_2,a_3;h)$ be a regular system of weights of dual type. Choose a triple $\vec{l}_W := (\vec{l}_1,\vec{l}_2,\vec{l}_3)$ of elements in $L(A_W)_+$ as follows (see also Appendix):

Type of W	$\vec{l}_1$	$\vec{l}_2$	$\vec{l}_3$
I	$\vec{X}_1$	$\vec{X}_2$	$\vec{X}_3$
II	$\vec{X}_1+\vec{X}_3$	$p_1\vec{X}_3$	$\vec{X}_2$
III	$\vec{X}_1+\vec{X}_2+\vec{X}_3$	$p_1\vec{X}_3$	$p_1\vec{X}_2$
IV	$\frac{p_3}{p_2}\vec{X}_2+\vec{X}_3$	$p_1\vec{X}_3$	$p_1\vec{X}_2$
V	$\vec{X}_2+l\vec{X}_3$	$\vec{X}_1+k\vec{X}_2$	$\vec{X}_3+m\vec{X}_1$

We shall call $\vec{l}_W$ the *principal generators* of W.

Lemma 4.4. *Let $W=(a_1,a_2,a_3;h)$ be a regular system of weights of dual type and let $\vec{l}_W=(\vec{l}_1,\vec{l}_2,\vec{l}_3)$ be its principal generators. There exists a unique element $\vec{\omega}_W\in L(A_W)$ such that*

(i) $-\epsilon_W\cdot\vec{\omega}_W=\vec{\omega}_{A_W}$
(ii) $a_i\cdot\vec{\omega}_W=\vec{l}_i$ *for* $i=1,2,3$.

Proof. By direct calculations, one sees that

$$a_i\cdot\vec{\omega}_{A_W}=-\epsilon_W\cdot\vec{l}_i \quad\text{and}\quad a_i\cdot\vec{l}_j=a_j\cdot\vec{l}_i, \quad \text{for all} \quad i,j=1,2,3. \tag{4.5}$$

Note also that there exists $k_1,k_2,k_3\in\mathbb{Z}$ such that $\sum_{i=1}^{3}k_ia_i=1$ since $\gcd(a_1,a_2,a_3)=1$.

(Existence) Set $\vec{\omega}_W := \sum_{i=1}^3 k_i \cdot \vec{l_i}$. Then

$$-\epsilon_W \cdot \vec{\omega}_W = \sum_{i=1}^3 k_i(-\epsilon_W \cdot \vec{l_i}) = \sum_{i=1}^3 k_i(a_i \cdot \vec{\omega}_{A_W}) = (\sum_{i=1}^3 k_i a_i) \cdot \vec{\omega}_{A_W} = \vec{\omega}_{A_W}.$$

Moreover,

$$\begin{aligned} a_1 \cdot \vec{\omega}_W =& a_1 k_1 \cdot \vec{l_1} + a_1 k_2 \cdot \vec{l_2} + a_1 k_3 \cdot \vec{l_3} \\ =& (1 - k_2 a_2 - k_3 a_3) \cdot \vec{l_1} + a_1 k_2 \cdot \vec{l_2} + a_1 k_3 \cdot \vec{l_3} \\ =& \vec{l_1} + k_2(a_1 \cdot \vec{l_2} - a_2 \cdot \vec{l_1}) + k_3(a_1 \cdot \vec{l_3} - a_3 \cdot \vec{l_1}) \\ =& \vec{l_1}. \end{aligned}$$

In a similar way, it is shown that $a_2 \cdot \vec{\omega}_W = \vec{l_2}$ and $a_3 \cdot \vec{\omega}_W = \vec{l_3}$.

(Uniqueness) Suppose that an element $\vec{\omega}'_W \in L(A_W)$ satisfies the two conditions (i) and (ii) above. Then,

$$\vec{\omega}'_W = (\sum_{i=1}^3 k_i a_i) \cdot \vec{\omega}'_W = \sum_{i=1}^3 k_i(a_i \cdot \vec{\omega}'_W) = \sum_{i=1}^3 k_i \cdot \vec{l_i} = \vec{\omega}_W.$$

Q.E.D.

Remark 4.5. Let $W = (a_1, a_2, a_3; h)$ be a regular system of weights of genus 0 and let $A_W = (\alpha_1, \ldots, \alpha_r)$ be its signature. Then one has

$$\begin{aligned} \deg(\vec{\omega}_W) =& -\frac{\deg(\omega_{A_W})}{\epsilon_W} = -\frac{(r-2)\cdot\deg(\vec{c}) - \sum_{i=1}^r \deg(\vec{X_i})}{\epsilon_W} \\ =& \frac{h}{a_1 a_2 a_3} \cdot \alpha, \end{aligned} \tag{4.6}$$

where $\alpha := \mathrm{lcm}(\alpha_1, \ldots, \alpha_r)$.

This element $\vec{\omega}_W \in L(A_W)$ leads to the following definition:

Definition 4.6. Let W be a regular system of weights of dual type and let $\vec{\omega}_W \in L(A_W)$ be the element given by Lemma 4.4.

(i) The subgroup $L_W := \mathbb{Z}\vec{\omega}_W$ of $L(A_W)$ is called the *principal lattice* of W, which we shall often identify with $\mathbb{Z}$.

(ii) Define an L_W-graded k-algebra R_W by

$$R_W := \bigoplus_{d \in \mathbb{Z}_{\geq 0}} R_{W,d}, \quad R_{W,d} := R_{A_W, d\cdot\vec{\omega}_W}.$$

Denote by gr-R_W the abelian category of finitely generated L_W-graded R_W-modules and by tor-R_W the full abelian subcategory of gr-R_W whose objects are finite dimensional L_W-graded R_W-modules.

Definition 4.7. Define an algebraic stack $\mathcal{C}_W$ by

$$\mathcal{C}_W := [\mathrm{Spec}(R_W)\backslash\{0\}/\mathrm{Spec}(kL_W)]\,. \tag{4.7}$$

The abelian category $\mathrm{coh}(\mathcal{C}_W)$ of coherent sheaves on the algebraic stack $\mathcal{C}_W$ is equivalent to gr-R_W/tor-R_W.

Our main purpose is to compare $\mathcal{C}_{A_W}$ with $\mathcal{C}_W$ as algebraic stacks, or equivalently, $\mathrm{coh}(\mathcal{C}_{A_W}) = \mathrm{mod}_{L(A_W)}\text{-}R_{A_W}/\mathrm{tor}_{L(A_W)}\text{-}R_{A_W}$ with $\mathrm{coh}(\mathcal{C}_W)$ $=$ gr-R_W/tor-R_W as abelian categories, and relate them to the hypersurface singularity with k^*-action defined by W.

§5. Main theorem

Theorem 5.1. *Let W be a regular system of weights of dual type.*

(i) *There exist homogeneous elements x, y, $z \in R_W$ of degree a_1, a_2, a_3 with respect to L_W such that $R_W \simeq k[x,y,z]/(f_W)$ for some weighted homogeneous polynomial $f_W \in k[x,y,z]$ of degree h which defines at most an isolated singularity only at the origin. In other words, R_W is the graded ring of functions on the isolated hypersurface singularity defined by W.*

(ii) *There exists an equivalence of abelian categories:*

$$\mathrm{mod}_{L(A_W)}\text{-}R_{A_W}/\mathrm{tor}_{L(A_W)}\text{-}R_{A_W} \simeq \text{gr-}R_W/\text{tor-}R_W\,. \tag{5.1}$$

Proof. Proofs for both of two statements are generalizations of those given by Geigle–Lenzing [2] for the case $\epsilon_W = 1$, which can be applied to the general cases here.

(i) The statement is shown by direct calculations using the classification of regular systems of weights of dual type. Generators x, y, z and their unique relation $f_W(x,y,z)$ are listed in Appendix below.

(ii) One sees that the functor

$$F : \mathrm{mod}_{L(A_W)}\text{-}R_{A_W} \to \text{gr-}R_W, \quad M \mapsto M|_{L_W}, \tag{5.2}$$

defined by the natural inclusion $R_W \subset R_{A_W}$ is exact and essentially surjective since by Kan extension there exists the right adjoint G of F such that $F \circ G(M') \simeq M'$ for any $M' \in$ gr-R_W. Therefore, what has to be shown is only that any module $N \in$ tor-R_W is an essential image of a module $N' \in$ tor-R_{A_W}, however, this also follows from Proposition 1.3 in [1]. Q.E.D.

§6. Application

Corollary 6.1. *Let W be a regular system of weights of dual type with $\epsilon_W < 0$. The triangulated category $D^{gr}_{Sg}(R_W) := D^b(\text{gr-}R_W)/D^b(\text{grproj-}R_W)$ has a full exceptional collection $(E_1, \dots, E_{\mu_{W^*}})$.*

Proof. Orlov shows (Theorem 2.5 in [6]) that there exists a semi-orthogonal decomposition

(6.1)
$$D^{gr}_{Sg}(R_W) \simeq \langle R_W/\mathfrak{m}_W(a), \dots, R_W/\mathfrak{m}_W(a+\epsilon_W+1), D^b\text{coh}(\mathcal{C}_W)\rangle$$

for any $a \in \mathbb{Z}$ where we denote by $(\cdot)$ the auto-equivalence induced by the grading shift functor on gr-R_W and by $\mathfrak{m}_W$ the unique graded maximal ideal in R_W. Combining our main theorem with the Proposition 4.1 in [1], the statement follows. Q.E.D.

Remark 6.2. There is an isomorphism of functors $(h) \simeq [2]$ on the triangulated category $D^{gr}_{Sg}(R_W)$ where $[1]$ is the translation functor.

Corollary 6.3. *Let W be a regular system of weights of dual type. The triangulated category $D^{gr}_{Sg}(R_W)$ has a full exceptional collection $(E_1, \dots, E_{\mu_{W^*}})$.*

Proof. Since W is of dual type, it is of ADE type or $\epsilon_W < 0$. Since a regular system of weights of type ADE has a full strongly exceptional collection [3], it is the direct consequence of the previous corollary. Q.E.D.

§7. Conjectures

In this section, we shall assume that $k = \mathbb{C}$.

Definition 7.1. Let W be a regular system of weights of dual type.

(i) Set $B_W := A_{W^*}$ and call it the *dual signature of W*.

(ii) *Let $B_W = (\beta_1, \beta_2, \beta_3)$ be the dual signature of W. Denote by f_{B_W} the defining polynomial of the cusp singularity, namely, $f_{B_W} := x_1^{\beta_1} + x_2^{\beta_2} + x_3^{\beta_3} + x_1x_2x_3$.*

Remark 7.2. For singularities associated to regular system of weights with $\epsilon_W = -1$, i.e., Arnold's 14 exceptional singularities, B_W is called the *Gabrielov number.*

Kontsevich observed that the origin of mirror symmetry is a symmetry between complex geometry and symplectic geometry. In particular, he conjectures that the algebraically constructed triangulated category in complex side should be triangulated equivalent to the geometrically constructed one in symplectic side, and that mirror phenomena should be explained from this triangulated equivalence.

Let f be a polynomial which defines an isolated singularity only at the origin $0 \in \mathbb{C}^n$. Consider a polynomial f as a holomorphic map $f : (\mathbb{C}^n, 0) \to (\mathbb{C}, 0)$ and choose a *Morsification* $\{f_t\}_{0 \le t < 1}$ of f, i.e., a smooth family of polynomials with $f_0 = f$ and such that critical points of f_t $(t \neq 0)$ are isolated and non-degenerate. For very small $0 < t \ll \delta \ll \epsilon \ll 1$, put

$$X := \{x \in \mathbb{C}^n \mid ||x|| \le \epsilon, |f_t| \le \delta\}, \tag{7.1}$$

$$D := \{y \in \mathbb{C} \mid |y| \le \delta\}. \tag{7.2}$$

Then, after suitable modifications near the boundary, the holomorphic map $f_t : X \to D$ defines an *exact Lefschetz fibration* and the relative holomorphic $n-1$-form $dx_1 \wedge \cdots \wedge dx_n / df_t$ defines a *relative Maslov map* which enables one to define *graded Lagrangian sub-manifolds.*

Theorem 7.3 (Seidel [9] [10]). *There exists an A_∞-category* $\mathrm{Fuk}(f)$ *called the Fukaya category of the exact Lefschetz fibration such that*

$$D^b\mathrm{Fuk}(f) \simeq D^b\mathrm{Fuk}^{\to}(\{\gamma_\bullet\}), \tag{7.3}$$

as a triangulated category, where $\mathrm{Fuk}^{\to}(\{\gamma_\bullet\})$ *is the directed Fukaya category of a distinguished basis of vanishing graded Lagrangian submanifolds* $\{\gamma_\bullet\} := \{\gamma_1, \ldots, \gamma_\mu\}$ *in the Milnor fiber* $X_y = f_t^{-1}(y)$. *In particular,* $D^b\mathrm{Fuk}(f)$ *is independent of all choices and is an invariant of* f.

We expect that the topological mirror symmetry for isolated hypersurface singularities can also be "categorified". The homological mirror symmetry principle leads the following conjecture:

Conjecture 7.4 (Homological mirror symmetry for weighted projective lines). *Let W be a regular system of weights of dual type. There should exist a triangulated equivalence*

$$D^b\mathrm{coh}(\mathcal{C}_W) \simeq D^b\mathrm{Fuk}(f_{B_W}). \tag{7.4}$$

Remark 7.5. The above conjecture can be verified at the level of the Grothendieck group. More precisely, one has the following isomorphism as lattices

$$(K_0(D^b\mathrm{coh}(\mathcal{C}_W)), \chi + {}^t\chi) \simeq \left(H_2(f_{B_W,t}^{-1}(y), \mathbb{Z}), -I\right). \tag{7.5}$$

We expect that this conjecture follows from the following one since semi-orthogonal decompositions of triangulated categories in algebraic geometry side should correspond to operations "blowing down" which are mirror dual to unfoldings of singularities in symplectic geometry side.

Conjecture 7.6 (HMS for regular systems of weights of dual type). *Let W be a regular system of weights of dual type with $\epsilon_W < 0$.*

(i) *There should exist a triangulated equivalence*

$$D_{Sg}^{gr}(R_W) \simeq D^b\mathrm{Fuk}(f_{W^*}). \tag{7.6}$$

(ii) $D^b\mathrm{Fuk}(f_{W^*})$ *has the following semi-orthogonal decomposition*

$$D^b\mathrm{Fuk}(f_{W^*}) \simeq \langle \mathcal{L}(a), \ldots, \mathcal{L}(a + \epsilon_W + 1), D^b\mathrm{Fuk}(f_{B_W}) \rangle, \tag{7.7}$$

for any $a \in \mathbb{Z}$, where $\mathcal{L}$ is an object of $D^b\mathrm{Fuk}(f_{W^*})$ and (1) *is the auto-equivalence induced by the shift of the gradings of Lagrangian sub-manifolds by $2/h$ such that $(h) = [2]$ where $[1]$ is the translation functor on $D^b\mathrm{Fuk}(f_{W^*})$.*

Remark 7.7. We can define an abelian group L_{f_W} for R_W called the *maximal grading* of R_W in a similar way to Definition 4.1 by introducing the degree vectors for x_i and letting all monomials appearing in f_W have the same degree vectors. As a result, R_W becomes L_{f_W}-graded and hence we can consider the triangulated category $D_{Sg}^{L_{f_W}}(R_W)$. It is expected that if f_W is of the form listed in Appendix (without any conditions for their exponents) then $(W, \mathrm{Spec}(kL_W))$ is topological mirror dual to $(W^*, \{1\})$ with the same W^* in the list. In this case, we can formulate the homological mirror symmetry conjecture as a triangulated equivalence $D_{Sg}^{L_{f_W}}(R_W) \simeq D^b\mathrm{Fuk}(f_{W^*})$. The above conjecture is the special case for regular systems of weights with $L_W = \mathbb{Z}$ (see Definition 3.4).

This homological mirror symmetry conjecture is one of motivations of our study of the triangulated category $D_{Sg}^{gr}(R_W)$, from which we can expect that $D_{Sg}^{gr}(R_W)$ has a full strongly exceptional collection (see [12]). Indeed, we prove this for several important classes of regular systems of weights in [3] (for W corresponding to ADE singularities), [4] (for W

with $\epsilon_W = -1$ including Arnold's 14 exceptional singularities) and [13] (for W of Type I, II and III), where we also checked the homological mirror symmetry conjecture at the level of Grothendieck group.

§8. Appendix

The followings are data of weights, dual weights, signatures, generators and relations of R_W and of R_{W^*}, for regular systems of weights of dual type.

8.1. Type I

$$W = W^* = (p_2p_3, p_3p_1, p_1p_2; p_1p_2p_3), \quad (p_i, p_j) = 1, i = 1, 2, 3. \tag{8.1}$$

$$A_W = A^*_W = (p_1, p_2, p_3). \tag{8.2}$$

$$\vec{l}_W = (\vec{X}_1, \vec{X}_2, \vec{X}_3). \tag{8.3}$$

$$x = X_1, \quad y = X_2, \quad z = X_3. \tag{8.4}$$

$$f_W(x, y, z) = x^{p_1} + y^{p_2} + z^{p_3}. \tag{8.5}$$

8.2. Type II

$$W = (p_3, \frac{p_1p_3}{p_2}, (p_2 - 1)p_1; p_1p_3), \tag{8.6}$$

$$W^* = (p_3, p_1p_2, (\frac{p_3}{p_2} - 1)p_1; p_1p_3), \tag{8.7}$$

where $p_2 \neq p_3$, $p_2|p_3$, $(p_1, p_3) = 1$, $(p_2-1, p_3) = 1$ and $(p_3/p_2-1, p_3) = 1$.

$$A_W = (p_1, \frac{p_3}{p_2}, (p_2 - 1)p_1), \quad A_{W^*} = (p_1, p_2, (\frac{p_3}{p_2} - 1)p_1). \tag{8.8}$$

$$\vec{l}_W = (\vec{X}_1 + \vec{X}_3, p_1\vec{X}_3, \vec{X}_2). \tag{8.9}$$

$$x = X_1X_3, \quad y = X_3^{p_1}, \quad z = X_2. \tag{8.10}$$

$$f_W(x, y, z) = x^{p_1} + y^{p_2} + yz^{\frac{p_3}{p_2}}. \tag{8.11}$$

8.3. Type III

$$W = W^* = (p_2, p_1q_2, p_1q_3; p_1p_2), \tag{8.12}$$

where $(p_1, p_2) = 1$, $p_2 + 1 = (q_2 + 1)(q_3 + 1)$ and $(q_2, q_3) = 1$.

$$A_W = A_{W^*} = (p_1, p_1q_2, p_1q_3). \tag{8.13}$$

$$\vec{l}_W = (\vec{X}_1 + \vec{X}_2 + \vec{X}_3, p_1\vec{X}_3, p_1\vec{X}_2). \tag{8.14}$$

$$x = X_1X_2X_3, \quad y = X_3^{p_1}, \quad z = X_2^{p_1}. \tag{8.15}$$

$$f_W(x, y, z) = x^{p_1} + y^{q_3+1}z + yz^{q_2+1}. \tag{8.16}$$

8.4. Type IV

$$W = (\frac{p_3}{p_1}, (p_1 - 1)\frac{p_3}{p_2}, p_2 - p_1 + 1; p_3), \tag{8.17}$$

$$W^* = (p_2, (\frac{p_3}{p_2} - 1)p_1, \frac{p_3}{p_1} - \frac{p_3}{p_2} + 1; p_3), \tag{8.18}$$

where $p_1 \neq p_2 \neq p_3$, $p_1|p_2$, $p_2|p_3$, $(p_1 - 1, p_2) = 1$, $(p_2 - p_1 + 1, p_3) = 1$, $(p_3/p_2 - 1, p_3/p_1) = 1$ and $(p_3/p_1 - p_3/p_2 + 1, p_3) = 1$.

(8.19)
$$A_W = (\frac{p_3}{p_2}, (p_1 - 1)\frac{p_3}{p_2}, p_2 - p_1 + 1), \quad A_W = (p_1, (\frac{p_3}{p_2} - 1)p_1, \frac{p_3}{p_1} - \frac{p_3}{p_2} + 1).$$

$$\vec{l}_W = \left(\frac{p_3}{p_2}\vec{X}_2 + \vec{X}_3, p_1\vec{X}_3, \vec{X}_1 + \vec{X}_2\right). \tag{8.20}$$

$$x = X_2^{\frac{p_3}{p_2}}X_3, \quad y = X_3^{p_1}, \quad z = X_1X_2. \tag{8.21}$$

$$f_W(x, y, z) = x^{p_1} + xy^{\frac{p_2}{p_1}} + yz^{\frac{p_3}{p_2}}. \tag{8.22}$$

8.5. Type V

$$W = (lm - m + 1, mk - k + 1, kl - l + 1; klm + 1), \tag{8.23}$$

$$W^* = (lm - l + 1, mk - m + 1, kl - k + 1; klm + 1). \tag{8.24}$$

where $(lm - m + 1, klm + 1) = 1$, $(mk - k + 1, klm + 1) = 1$, and $(kl - l + 1, klm + 1) = 1$.

$$\begin{aligned} A_W &= (lm - m + 1, mk - k + 1, kl - l + 1), \\ A_{W^*} &= (lm - l + 1, mk - m + 1, kl - k + 1). \end{aligned} \tag{8.25}$$

$$\vec{l}_W = (\vec{X}_2 + l\vec{X}_3, \vec{X}_1 + k\vec{X}_2, \vec{X}_3 + m\vec{X}_1). \tag{8.26}$$

$$x = X_2 X_3^l, \quad y = X_1 X_2^k, \quad z = X_3 X_1^m. \tag{8.27}$$

$$f_W(x, y, z) = zx^k + xy^m + yz^l. \tag{8.28}$$

References

[1] W. Geigle and H. Lenzing, A class of weighted projective curves arising in representation theory of finite-dimensional algebras, In: Singularities, Representation of Algebras, and Vector Bundles, Lambrecht, 1985, Lecture Notes in Math., **1273**, Springer-Verlag, Berlin, 1987, pp. 9–34.

[2] W. Geigle and H. Lenzing, Perpendicular categories with applications to representations and sheaves, J. Algebra, **144** (1991), 273–343.

[3] H. Kajiura, K. Saito and A. Takahashi, Matrix factorizations and representations of quivers II: type ADE case, Adv. in Math., **211** (2007), 327–362; arXiv:math/0511155[math.AG].

[4] H. Kajiura, K. Saito and A. Takahashi, Triangulated categories of matrix factorizations for regular systems of weights with $\varepsilon = -1$, arXiv: 0708.0210.

[5] H. Kajiura, K. Saito and A. Takahashi, Weighted projective lines associated to regular systems of weights, work in progress.

[6] D. Orlov, Derived categories of coherent sheaves and triangulated categories of singularities, arXiv:math/0503632[math.AG].

[7] K. Saito, Regular systems of weights and associated singularities, In: Complex Analytic Singularities, Adv. Stud. Pure Math., **8**, North-Holland, Amsterdam, 1986, pp. 479–526.

[8] K. Saito, Duality for regular systems of weights, Asian. J. Math., **2** (1998), 983–1048.

[9] P. Seidel, More about vanishing cycles and mutation, symplectic geometry and mirror symmetry, In: Proceedings of the 4th KIAS Annual International Conference, World Scientific, 2001, pp. 429–465.

[10] P. Seidel, Fukaya categories and Picard–Lefschetz theory, to appear in the ETH Lecture Notes series of the European Math. Soc.

[11] A. Takahashi, K. Saito's duality for regular weight systems and duality for orbifoldized Poincaré polynomials, Commun. Math. Phys., **205** (1999), 571–586.

[12] A. Takahashi, Matrix factorizations and representations of quivers I, arXiv: math/0506347[math.AG].

[13] A. Takahashi, Homological mirror symmetry for isolated hypersurface singularities in dimension one, in preparation.

Department of Mathematics
Graduate School of Science
Osaka University
Toyonaka, Osaka 560-0043
Japan
E-mail address: takahashi@math.sci.osaka-u.ac.jp

Advanced Studies in Pure Mathematics 59, 2010
New Developments in Algebraic Geometry,
Integrable Systems and Mirror Symmetry (Kyoto, 2008)
pp. 389–434

Generating functions of stable pair invariants via wall-crossings in derived categories

Yukinobu Toda

Abstract.

The notion of limit stability on Calabi–Yau 3-folds is introduced by the author to construct an approximation of Bridgeland–Douglas stability conditions at the large volume limit. It has also turned out that the wall-crossing phenomena of limit stable objects seem relevant to the rationality conjecture of the generating functions of Pandharipande–Thomas invariants. In this article, we shall make it clear how wall-crossing formula of the counting invariants of limit stable objects solves the above conjecture.

§1. Introduction

A theory of curve counting on Calabi–Yau 3-folds is interesting in both algebraic geometry and string theory. Now there are three such theories, called Gromov–Witten (GW) theory, Donaldson–Thomas (DT) theory, and Pandharipande–Thomas (PT) theory. Conjecturally these theories are equivalent in terms of generating functions. To formulate this equivalence, however, we also need a conjectural rationality property of those functions for DT-theory and PT-theory, to formulate that equivalence. The purpose of this article is to interpret the rationality conjecture for PT-theory from the viewpoint of wall-crossing phenomena in derived categories of coherent sheaves.

1.1. GW-DT-PT correspondences

First of all, let us recall the conjectural GW-DT-PT correspondences on curve counting theories. Suppose that X is a smooth projective

---

Received July 3, 2008.
2000 *Mathematics Subject Classification.* Primary 14D20; Secondary 14J32, 18E30.
Key words and phrases. Stability condition, Donaldson–Thomas invariant.
Partly supported by JSPS Grant-in-Aid for Scientific Research (S-19104002).

Calabi–Yau 3-fold over $\mathbb{C}$, i.e. there is a nowhere vanishing holomorphic 3-form on X. For $g \geq 0$ and $\beta \in H_2(X, \mathbb{Z})$, the *GW-invariant* $N_{g,\beta}$ is defined by the integration of the virtual class,

$$N_{g,\beta} = \int_{[\overline{M}_g(X,\beta)]^{\mathrm{vir}}} 1 \in \mathbb{Q},$$

where $\overline{M}_g(X, \beta)$ is the moduli stack of stable maps $f\colon C \to X$ with $g(C) = g$ and $f_*[C] = \beta$. The *GW-potential* is given by the following generating function,

$$Z_{\mathrm{GW}} = \exp\left(\sum_{g,\beta \neq 0} N_{g,\beta}\lambda^{2g-2}v^{\beta}\right).$$

For $n \in \mathbb{Z}$ and $\beta \in H_2(X, \mathbb{Z})$, let $I_n(X, \beta)$ be the Hilbert scheme of 1-dimensional subschemes $Z \subset X$ satisfying

$$[Z] = \beta, \quad \chi(\mathcal{O}_Z) = n.$$

The obstruction theory on $I_n(X, \beta)$ is obtained by viewing it as a moduli space of ideal sheaves, and the *DT-invariant* $I_{n,\beta}$ is defined by

$$I_{n,\beta} = \int_{[I_n(X,\beta)]^{\mathrm{vir}}} 1 \in \mathbb{Z}.$$

The generating function of the *reduced DT-theory* is

$$Z'_{\mathrm{DT}} = \sum_{n,\beta} I_{n,\beta}q^n v^{\beta} / \sum_{n} I_{n,0}q^n.$$

The theory of stable pairs and their counting invariants are introduced and studied by Pandharipande and Thomas [20], [21], [22] to give a geometric interpretation of the reduced DT-theory. By definition, a *stable pair* is data (F, s),

$$s\colon \mathcal{O}_X \longrightarrow F,$$

where F is a pure one dimensional sheaf on X, and s is a morphism with a zero dimensional cokernel. For $\beta \in H_2(X, \mathbb{Z})$ and $n \in \mathbb{Z}$, the moduli space of stable pairs (F, s) with

$$[F] = \beta, \quad \chi(F) = n,$$

is constructed in [20], denoted by $P_n(X,\beta)$. The obstruction theory on $P_n(X,\beta)$ is obtained by viewing stable pairs (F,s) as two term complexes,

$$\cdots \longrightarrow 0 \longrightarrow \mathcal{O}_X \stackrel{s}{\longrightarrow} F \longrightarrow 0 \longrightarrow \cdots . \tag{1}$$

The *PT-invariant* $P_{n,\beta}$ is defined by

$$P_{n,\beta} = \int_{[P_n(X,\beta)]^{\mathrm{vir}}} 1 \in \mathbb{Z}.$$

The corresponding generating function is

$$Z_{\mathrm{PT}} = \sum_{n,\beta} P_{n,\beta} q^v v^{\beta}.$$

The functions Z_{GW}, Z'_{DT} and Z_{PT} are conjecturally equal after suitable variable change. In order to state this, we need the following conjecture, called *rationality conjecture.*

Conjecture 1.1. [19, Conjecture 2], [20, Conjecture 3.2] *For a fixed β, the generating series*

$$I_{\beta}(q) = \sum_{n\in\mathbb{Z}} I_{n,\beta}q^n / \sum_{n\in\mathbb{Z}} I_{n,0}q^n, \quad P_{\beta}(q) = \sum_{n\in\mathbb{Z}} P_{n,\beta}q^n,$$

are Laurent expansions of rational functions of q, invariant under $q \leftrightarrow 1/q$.

The above conjecture is solved for $I_{\beta}(q)$ when X is a toric local Calabi–Yau 3-fold [19], and for $P_{\beta}(q)$ when β is an irreducible curve class [22]. Now we can state the conjectural GW-DT-PT-correspondences.

Conjecture 1.2. [19, Conjecture 3], [20, Conjecture 3.3] *After the variable change $q = -e^{i\lambda}$, we have*

$$Z_{\mathrm{GW}} = Z'_{\mathrm{DT}} = Z_{\mathrm{PT}}.$$

The variable change $q = -e^{i\lambda}$ is well-defined by Conjecture 1.1.

Note that ideal sheaves $I \subset \mathcal{O}_X$ are objects in $D^b(X)$, where $D^b(X)$ is the bounded derived category of coherent sheaves on X. We can also interpret stable pairs (F,s) as objects in $D^b(X)$ by viewing them as two term complexes (1). As discussed in [20, Secction 3], the equality $Z'_{\mathrm{DT}} = Z_{\mathrm{PT}}$ should be interpreted as a wall-crossing formula for counting invariants in the category $D^b(X)$. The purpose of this article is to show that Conjecture 1.1 is also interpreted as a wall-crossing formula in $D^b(X)$, using the method of limit stability [24] together with Joyce's works [9], [10], [11], [14].

1.2. Limit stability

The notion of limit stability on a Calabi–Yau 3-fold X is introduced in [24] to construct an approximation of Bridgeland–Douglas stability conditions [4], [6], [7] on $D^b(X)$ at the large volume limit. It is a certain stability condition on the category of perverse coherent sheaves

$$\mathcal{A}^p \subset D^b(X),$$

in the sense of Bezrukavnikov [3] and Kashiwara [15]. (See Definition 3.2.) An element $\sigma \in A(X)_{\mathbb{C}}$ determines σ-limit (semi)stable objects in $\mathcal{A}^p$, where $A(X)_{\mathbb{C}}$ is the complexified ample cone,

$$A(X)_{\mathbb{C}} = \{B + i\omega \in H^2(X, \mathbb{C}) \mid \omega \text{ is an ample class}\}.$$

It has also turned out in [24] that the objects (1) appear as σ-limit stable objects for some $\sigma \in A(X)_{\mathbb{C}}$, thus studying stable pairs and limit stable objects are closely related. The objects E given by (1) satisfy

$$(2) \quad (\mathrm{ch}_0(E), \mathrm{ch}_1(E), \mathrm{ch}_2(E), \mathrm{ch}_3(E)) = (-1, 0, \beta, n), \quad \det E = \mathcal{O}_X,$$

for some β and n. Under the above observation, we have constructed in [24] the moduli space of σ-limit stable objects $E \in \mathcal{A}^p$ satisfying (2) as an algebraic space of finite type, denoted by $\mathcal{L}_n^{\sigma}(X, \beta)$. Using that moduli space, the counting invariant of limit stable objects

$$(3) \qquad L_{n,\beta}(\sigma) \in \mathbb{Z}$$

is also defined in [24] as a weighted Euler characteristic with respect to Behrend's constructible function [2], and (3) coincides with the integration of the virtual class if $\mathcal{L}_n^{\sigma}(X, \beta)$ is a projective variety. A particular choice of σ yields an equality $L_{n,\beta}(\sigma) = P_{n,\beta}$, however $L_{n,\beta}(\sigma)$ becomes different from $P_{n,\beta}$ if we deform σ. As discussed in [24, Section 4], a transformation formula of the invariants $L_{n,\beta}(\sigma)$ under change of σ seems relevant to solving Conjecture 1.1 for PT-theory.

1.3. Main result

In this article, we shall proceed the above idea further, using D. Joyce's works [9], [10], [11], [14] on counting invariants of semistable objects on abelian categories and their wall-crossing formulas. We will make it clear how such a formula for counting invariants of objects in $\mathcal{A}^p$ implies Conjecture 1.1 for PT-theory. Unfortunately we are unable to solve Conjecture 1.1 at this moment, as Joyce's theory is applied only for the motivic invariants (e.g. Euler characteristic) of the moduli spaces, so they do not involve virtual classes. On the other hand, the invariant $P_{n,\beta}$ coincides with the Euler characteristic of $P_n(X,\beta)$ (up to sign),

$$P^{eu}_{n,\beta} := e(P_n(X,\beta)) \in \mathbb{Z},$$

if $P_n(X,\beta)$ is non-singular. In general $P_{n,\beta}$ is written as a weighted Euler characteristic with respect to Behrend's constructible function [2], so $P_{n,\beta}$ resembles $P^{eu}_{n,\beta}$ in this sense. So instead of solving Conjecture 1.1, we shall show the motivic version of Conjecture 1.1, i.e. the rationality of the generating series,

$$P^{eu}_{\beta}(q) = \sum_{n\in\mathbb{Z}} P^{eu}_{n,\beta}q^n.$$

The limit stability does not work well to combine Joyce's works, so we will introduce the notion of *μ_σ-limit stability* for $\sigma \in A(X)_{\mathbb{C}}$, which is a coarse version of σ-limit stability. Then we will introduce the Joyce type invariants, (cf. Definition 4.1, Remark 4.2,)

$$L^{eu}_{n,\beta} \in \mathbb{Q}, \quad N^{eu}_{n,\beta} \in \mathbb{Q}.$$

Roughly speaking, $L^{eu}_{n,\beta}$, (resp. $N^{eu}_{n,\beta}$) is the "Euler characteristic" of the moduli stack of $\mu_{i\omega}$-limit semistable objects $E \in \mathcal{A}^p$ with $\det E = \mathcal{O}_X$, (resp. one dimensional ω-Gieseker semistable sheaves F,) satisfying

$$\mathrm{ch}(E) = (-1, 0, \beta, n), \quad (\text{resp. } \mathrm{ch}(F) = (0, 0, \beta, n).)$$

We will consider the generating series,

$$L^{eu}_{\beta}(q) = \sum_{n\in\mathbb{Z}} L^{eu}_{n,\beta}q^n, \quad N^{eu}_{\beta}(q) = \sum_{n\geq 0} nN^{eu}_{n,\beta}q^n.$$

It will turn out that $L^{eu}_{\beta}(q)$ is a polynomial of $q^{\pm 1}$, $N^{eu}_{\beta}(q)$ is the Laurent expansion of a rational function of q, and they are invariant under $q \leftrightarrow 1/q$. (cf. Lemma 4.5, Lemma 4.6.) Somewhat surprisingly, Joyce's wall-crossing formula yields the following equality of those generating functions.

Theorem 1.3. **[Theorem 4.7]** *We have the following equality of the generating series,*

$$\sum_{\beta\geq 0} P^{eu}_{\beta}(q)v^{\beta} = \left(\sum_{\beta\geq 0} L^{eu}_{\beta}(q)v^{\beta}\right) \cdot \exp\left(\sum_{\beta>0} N^{eu}_{\beta}(q)v^{\beta}\right). \tag{4}$$

Here $\beta > 0$ means that β is a numerical class of an effective one cycle on X. As a corollary, we have the following.

Corollary 1.4. **[Corollary 4.8]** *The generating series $P_\beta^{eu}(q)$ is the Laurent expansion of a rational function of q, invariant under $q \leftrightarrow 1/q$.*

The series Z_{PT} also should have a decomposition such as (4). In Problem 4.18 we will address a certain technical problem on the Ringel–Hall Lie algebra of $\mathcal{A}^p$, which enables us to decompose Z_{PT} and solve Conjecture 1.1 for PT-theory. As a conclusion, we have obtained a conceptual understanding of the rationality conjecture and DT-PT correspondences in terms of wall-crossing phenomena in the derived category, and they have been reduced to showing a rather technical problem, namely a compatibility of Ringel–Hall Lie algebra structure of $\mathcal{A}^p$ with taking virtual classes via Behrend's constructible functions.

1.4. Acknowledgement

The author thanks R. Thomas, R. Pandharipande for valuable comments, and D. Joyce for the comment on Problem 4.18. This work is supported by World Premier International Research Center Initiative (WPI Initiative), MEXT, Japan.

1.5. Convention

All the varieties and schemes are defined over $\mathbb{C}$. For a variety X, the category of coherent sheaves on X is denoted by $\mathrm{Coh}(X)$. We say $E \in \mathrm{Coh}(X)$ is d-dimensional if $\dim \mathrm{Supp}(E) = d$.

§2. Review of Joyce's work

This section is devoted to review Joyce's works [9], [10], [11], [14] on counting invariants of semistable objects on abelian categories. We discuss in a general framework rather than working with the category of perverse coherent sheaves $\mathcal{A}^p$, which we will introduce in the next section.

2.1. Setting

We begin with a generality of (weak) stability conditions on abelian categories. Let $\mathcal{A}$ be a $\mathbb{C}$-linear abelian category, and $K(\mathcal{A})$ its Grothendieck group. We put the same assumption as in [14], i.e. $\mathrm{Hom}(E,F)$, $\mathrm{Ext}^1(E,F)$ for any $E, F \in \mathcal{A}$ are finite dimensional $\mathbb{C}$-vector spaces, and compositions $\mathrm{Ext}^i(E,F) \times \mathrm{Ext}^j(F,G) \to \mathrm{Ext}^{i+j}(E,G)$ for $i, j, i+j = 0, 1$ are bilinear. These conditions are satisfied in several good cases, i.e.

$\mathcal{A} = \operatorname{mod} A$ for a finite dimensional algebra A, or $\mathcal{A} = \operatorname{Coh}(X)$ for a projective variety X. In the first case, the group $K(\mathcal{A})$ is finitely generated, but this is not true in the latter case. So instead we fix a quotient space,

$$\mathcal{N}(\mathcal{A}) := K(\mathcal{A})/\equiv,$$

for some equivalence relation $\equiv$ such that a class $[E] \in \mathcal{N}(\mathcal{A})$ is non-zero for any $0 \neq E \in \mathcal{A}$. For instance if $\mathcal{A} = \operatorname{Coh}(X)$, an equivalence relation $\equiv$ can be taken by

$$E_1 \equiv E_2 \overset{\text{def}}{\leftrightarrow} \chi(E_1, F) = \chi(E_2, F) \quad \text{for any } F \in \mathcal{A}, \tag{5}$$

where $\chi(E, F)$ is defined by

$$\chi(E, F) = \sum_{i \in \mathbb{Z}} (-1)^i \dim \operatorname{Ext}^i(E, F). \tag{6}$$

Then $\mathcal{N}(\mathcal{A})$ is embedded into $H^*(X, \mathbb{Q})$, and it is a finitely generated $\mathbb{Z}$-module. The *closed positive cone* and the *positive cone* of $\mathcal{A}$ are defined by

$$\overline{C}(\mathcal{A}) := \operatorname{im}(\mathcal{A} \to K(\mathcal{A}) \to \mathcal{N}(\mathcal{A})),$$
$$C(\mathcal{A}) := \overline{C}(\mathcal{A}) \setminus \{0\},$$

respectively. For a subcategory $\mathcal{B} \subset \mathcal{A}$, we shall use the notation $C(\mathcal{B}) := \operatorname{im}(\mathcal{B} \to C(\mathcal{A})) \subset C(\mathcal{A})$, etc. For an object $E \in \mathcal{A}$, its class is denoted by $[E] \in \overline{C}(\mathcal{A})$, or we omit [ ] if there is no confusion. Let $(T, \succeq)$ be a totally ordered set.

Definition 2.1. A *weak stability function* is a map,

$$Z \colon C(\mathcal{A}) \longrightarrow T,$$

such that if $E, F, G \in C(\mathcal{A})$ satisfies $E = F + G$, we have either

$$Z(F) \preceq Z(E) \preceq Z(G), \quad \text{or}$$
$$Z(F) \succeq Z(E) \succeq Z(G).$$

A weak stability function is a *stability function* if, for E, F, G as above, we have either

$$Z(F) \prec Z(E) \prec Z(G), \quad \text{or}$$
$$Z(F) \succ Z(E) \succ Z(G), \quad \text{or}$$
$$Z(F) = Z(E) = Z(G).$$

Given a weak stability function, we can define the set of (semi)stable objects.

Definition 2.2. Let $Z\colon C(\mathcal{A}) \to T$ be a weak stability function. An object $E \in \mathcal{A}$ is called *Z-(semi)stable* if for any nonzero subobject $F \subset E$, we have

$$Z(F) \prec Z(E/F), \quad (\text{resp. } Z(F) \preceq Z(E/F).)$$

The notion of (weak) stability conditions is defined as follows.

Definition 2.3. A (weak) stability function $Z\colon C(\mathcal{A}) \to T$ is a *(weak) stability condition* if for any object $E \in \mathcal{A}$, there is a filtration

$$0 = E_0 \subset E_1 \subset \cdots \subset E_n = E, \tag{7}$$

such that each subquotient $F_i = E_i/E_{i-1}$ is Z-semistable with

$$Z(F_1) \succ Z(F_2) \succ \cdots \succ Z(F_n).$$

It is easy to see that the filtration (7) is unique up to an isomorphism, if exists. The filtration (7) is called a *Harder–Narasimhan filtration*. Here we give some examples.

Example 2.4. (i) For an abelian category $\mathcal{A}$, let $W\colon \mathcal{N}(\mathcal{A}) \to \mathbb{C}$ be a group homomorphism such that for any $E \in \mathcal{A} \setminus \{0\}$, we have

$$W(E) \in \mathbb{H} := \{r \exp(i\pi\phi) \mid 0 < \phi \leq 1\}.$$

For instance if $\mathcal{A} = \operatorname{mod} A$ for a finite dimensional $\mathbb{C}$-algebra A, the positive cone $C(\mathcal{A})$ is spanned by finite number of simple objects $S_1, \dots, S_n \in \mathcal{A}$, and such W is obtained by choosing the image of $[S_i] \in C(\mathcal{A})$ for $1 \leq i \leq n$ under W. We set $(T, \succeq) = ((0,1], \geq)$, and

$$Z\colon C(\mathcal{A}) \ni E \longmapsto \frac{1}{\pi} \operatorname{Im} \log Z(E) \in T.$$

Then Z is a stability condition on $\mathcal{A}$. This is Bridgeland's approach of stability conditions [4].

(ii) Let X be a smooth projective surface and set $\mathcal{A} = \operatorname{Coh}(X)$. Let ω be an ample divisor on X. For $E \in \operatorname{Coh}(X)$ we set

$$\mu_\omega(E) = \begin{cases} \frac{c_1(E)\cdot\omega}{\operatorname{rk}(E)} & \text{if } E \text{ is not torsion.} \\ \infty & \text{if } E \text{ is torsion.} \end{cases}$$

Then the map $C(\mathcal{A}) \ni E \mapsto \mu_\omega(E) \in \mathbb{Q} \cup \{\infty\}$ is a weak stability condition on $\mathcal{A}$, but not a stability condition on $\mathcal{A}$.

Remark 2.5. Here we mention that a theory of stability conditions on triangulated categories is developed by Bridgeland [4], motivated by M. Douglas's Π-stability [6], [7]. For a triangulated category $\mathcal{D}$, Bridgeland's stability condition consists of $(W, \mathcal{A})$, where $\mathcal{A} \subset \mathcal{D}$ is the heart of a bounded t-structure on $\mathcal{D}$, and W is a group homomorphism $K(\mathcal{A}) \to \mathbb{C}$, as in Example 2.4 (i). Especially W determines a stability condition on the abelian category $\mathcal{A}$. He then shows that the set of "good" stability conditions form a complex manifold $\mathrm{Stab}(\mathcal{D})$. Although Bridgeland's theory is quite powerful, we shall study in this paper more general notion of (weak) stability conditions, which is used in Joyce's works.

2.2. Ringel–Hall algebras

In this subsection, we introduce the algebra $\mathcal{H}(\mathcal{A})$ associated to an abelian category $\mathcal{A}$, whose details are seen in [10]. Let $Z\colon C(\mathcal{A}) \to T$ be a weak stability function. At this moment, we put the following assumption.

Assumption 2.6.
- *$\mathcal{A}$ is noetherian and Z-artinian.*
- *The moduli stack of objects $E \in \mathcal{A}$, denoted by $\mathfrak{Obj}(\mathcal{A})$, is an Artin stack locally of finite type over $\mathbb{C}$.*
- *For $v \in \overline{C}(\mathcal{A})$, let $\mathfrak{M}^v(Z) \subset \mathfrak{Obj}(\mathcal{A})$ be the substack of Z-semistable objects $E \in \mathcal{A}$ with $[E] = v$. Then $\mathfrak{M}^v(Z)$ is an open substack of $\mathfrak{Obj}(\mathcal{A})$, and it is of finite type over $\mathbb{C}$.*

Here we say $\mathcal{A}$ is *Z-artinian* if there is no infinite sequence

$$\cdots \subset E_n \subset E_{n-1} \subset \cdots \subset E_1 \subset E_0,$$

such that $E_{i+1} \neq E_i$ and $Z(E_{i+1}) \succeq Z(E_i/E_{i+1})$ for any i. The first condition of Assumption 2.6 ensures the existence of Harder–Narasimhan filtrations, and hence Z is a weak stability condition, by the same argument as in [23, Theorem 2]. In order to state the second assumption, we need to know about the notion algebraic families of objects and morphisms in $\mathcal{A}$. This notion is obvious if $\mathcal{A} = \mathrm{Coh}(X)$ for a variety X, but in general we need some additional extra data, which is given in [9, Assumptions 7.1, 8.1]. For the introduction of Artin stacks, one can consult [17]. For instance, Assumption 2.6 is satisfied when $\mathcal{A} = \mathrm{mod}\, A$ for a finite dimensional $\mathbb{C}$-algebra A, and Z is given as in Example 2.4 (i).

For a variety Y, recall that the Grothendieck ring of varieties over Y is defined by

$$K_0(\mathrm{Var}/Y) = \bigoplus_{(X,\rho)} \mathbb{Z}[(X,\rho)]/\sim,$$

where X is a variety with a morphism $\rho\colon X \to Y$, and equivalence relations are given by

$$[(X,\rho)] \sim [(X^{\dagger}, \rho|_{X^{\dagger}})] + [(X \setminus X^{\dagger}, \rho|_{X\setminus X^{\dagger}})],$$

where $X^{\dagger}$ is a closed subvariety of X. Taking the fiber products over Y, there is a natural product on $K_0(\mathrm{Var}/Y)$,

$$[(X,\rho)] \cdot [(X',\rho')] = [(X \times_Y X', \rho \circ p)], \tag{8}$$

where p is the projection $X \times_Y X' \to X$.

In order to introduce $\mathcal{H}(\mathcal{A})$, let us introduce the notion of Grothendieck rings of Artin stacks.

Definition 2.7. [13] Let $\mathcal{Y}$ be an Artin stack locally of finite type over $\mathbb{C}$. Define the $\mathbb{Q}$-vector space $K_0(\mathrm{St}/\mathcal{Y})$ to be

$$K_0(\mathrm{St}/\mathcal{Y}) := \bigoplus_{(\mathcal{X},\rho)} \mathbb{Q}[(\mathcal{X},\rho)]/\sim,$$

where $(\mathcal{X},\rho)$ is a pair such that $\mathcal{X}$ is an Artin $\mathbb{C}$-stack of finite type with affine geometric stabilizers, and $\rho\colon \mathcal{X} \to \mathcal{Y}$ is a 1-morphism. The relations $\sim$ are given by

$$[(\mathcal{X},\rho)] \sim [(\mathcal{X}^{\dagger}, \rho|_{\mathcal{X}^{\dagger}})] + [(\mathcal{X} \setminus \mathcal{X}^{\dagger}, \rho|_{\mathcal{X}\setminus \mathcal{X}^{\dagger}})],$$

for closed substacks $\mathcal{X}^{\dagger} \subset \mathcal{X}$.

Again taking the fiber products over $\mathcal{Y}$ gives a product $\cdot$ on $K_0(\mathrm{St}/\mathcal{Y})$,

$$[(\mathcal{X},\rho)] \cdot [(\mathcal{X}',\rho')] = [(\mathcal{X} \times_{\mathcal{Y}} \mathcal{X}', \rho \circ p)], \tag{9}$$

where p is the projection $\mathcal{X} \times_{\mathcal{Y}} \mathcal{X}' \to \mathcal{X}$.

Definition 2.8. [10] Let $\mathcal{A}$ be an abelian category satisfying the second condition of Assumption 2.6. We define the $\mathbb{Q}$-vector space $\mathcal{H}(\mathcal{A})$ to be

$$\mathcal{H}(\mathcal{A}) := K_0(\mathrm{St}/\mathfrak{Obj}(\mathcal{A})).$$

The vector space $\mathcal{H}(\mathcal{A})$ is graded by $v \in \overline{C}(\mathcal{A})$,

$$\mathcal{H}(\mathcal{A}) = \bigoplus_{v \in \overline{C}(\mathcal{A})} \mathcal{H}^{v}(\mathcal{A}), \quad \mathcal{H}^{v}(\mathcal{A}) := K_0(\mathrm{St}/\mathfrak{Obj}^{v}(\mathcal{A})),$$

where $\mathfrak{Obj}^{v}(\mathcal{A})$ is the stack of objects $E \in \mathcal{A}$ with $[E] = v$. There is an associative multiplication $*$ on $\mathcal{H}(\mathcal{A})$, based on *Ringel–Hall algebras*, which differs from the product (9). Let $\mathfrak{Ex}(\mathcal{A})$ be the moduli stack of exact sequences $0 \to E_1 \to E_2 \to E_3 \to 0$ in $\mathcal{A}$. It is shown in [9, Theorem 8.2] that $\mathfrak{Ex}(\mathcal{A})$ is an Artin stack locally of finite type over $\mathbb{C}$. We have the following 1-morphisms,

$$p_i \colon \mathfrak{Ex}(\mathcal{A}) \ni (0 \to E_1 \to E_2 \to E_3 \to 0) \longmapsto E_i \in \mathfrak{Obj}(\mathcal{A}),$$

for $i = 1, 2, 3$. Take $f_i = [(\mathcal{X}_i, \rho_i)] \in \mathcal{H}(\mathcal{A})$ for $i = 1, 2$. We have the following diagram,

$$\begin{array}{ccccc} (p_1, p_3)^*(\mathcal{X}_1 \times \mathcal{X}_2) & \xrightarrow{u} & \mathfrak{Ex}(\mathcal{A}) & \xrightarrow{p_2} & \mathfrak{Obj}(\mathcal{A}) \\ \downarrow & & \downarrow (p_1, p_3) & & \\ \mathcal{X}_1 \times \mathcal{X}_2 & \xrightarrow{(\rho_1, \rho_2)} & \mathfrak{Obj}(\mathcal{A}) \times \mathfrak{Obj}(\mathcal{A}). & & \end{array}$$

Here the left diagram is a Cartesian diagram.

Definition 2.9. We define the $*$-product $f_1 * f_2$ by

$$f_1 * f_2 = [((p_1, p_3)^*(\mathcal{X}_1 \times \mathcal{X}_2),\ p_2 \circ u)].$$

It is shown in [10, Theorem 5.2] that $*$ is associative and $\mathcal{H}(\mathcal{A})$ is a $\mathbb{Q}$-algebra with identity $[0 \hookrightarrow \mathfrak{Obj}(\mathcal{A})]$.

Remark 2.10. Our algebra $\mathcal{H}(\mathcal{A})$ is denoted by $\underline{\mathrm{SF}}(\mathfrak{Obj}(\mathcal{A}))$ in Joyce's paper [10], and an element of $\underline{\mathrm{SF}}(\mathfrak{Obj}(\mathcal{A}))$ is called a *stack function* on $\mathfrak{Obj}(\mathcal{A})$. There is another version of Ringel–Hall type algebra discussed in [10], defined as the set of constructible functions on $\mathfrak{Obj}(\mathcal{A})$, denoted by $\mathrm{CF}(\mathfrak{Obj}(\mathcal{A}))$ in [10]. Although most of the readers might be more familiar with constructible functions than stack functions, we use the latter one since we want to apply [10, Theorem 6.12] which is formulated only for stack functions.

2.3. Elements $\delta^v(Z)$, $\epsilon^v(Z)$

Let $Z \colon C(\mathcal{A}) \to T$ be a weak stability condition, satisfying Assumption 2.6. For an Artin substack $i \colon \mathfrak{M} \hookrightarrow \mathfrak{Obj}(\mathcal{A})$, we write the element $[(\mathfrak{M}, i)] \in \mathcal{H}(\mathcal{A})$ as $[\mathfrak{M} \hookrightarrow \mathfrak{Obj}(\mathcal{A})]$.

Definition 2.11. For $v \in C(\mathcal{A})$, we define $\delta^v(Z), \epsilon^v(Z) \in \mathcal{H}(\mathcal{A})$ to be

$$\delta^v(Z) = [\mathfrak{M}^v(Z) \hookrightarrow \mathfrak{Obj}(\mathcal{A})] \in \mathcal{H}^v(\mathcal{A}),$$

$$\epsilon^v(Z) = \sum_{\substack{l \geq 1,\ v_i \in C(\mathcal{A}), \\ v_1 + \cdots + v_l = v, \\ Z(v_i) = Z(v),\ 1 \leq i \leq l}} \frac{(-1)^{l-1}}{l} \delta^{v_1}(Z) * \cdots * \delta^{v_l}(Z) \in \mathcal{H}^v(\mathcal{A}). \tag{10}$$

Under Assumption 2.6, the sum (10) is a finite sum, (see [11, Proposition 4.9],) and hence $\epsilon^v(Z)$ is an element of $\mathcal{H}(\mathcal{A})$. In [11, Theorem 8.7], Joyce shows that $\epsilon^v(Z)$ is an element of a certain Lie subalgebra of $\mathcal{H}(\mathcal{A})$, called *Ringel–Hall Lie algebra,*

$$\epsilon^v(Z) \in \mathfrak{G}(\mathcal{A}) \subset \mathcal{H}(\mathcal{A}). \tag{11}$$

The Lie algebra $\mathfrak{G}(\mathcal{A})$ is denoted by $\mathrm{SF}^{\mathrm{ind}}_{\mathrm{al}}(\mathfrak{Obj}(\mathcal{A}))$ in Joyce's paper [10]. If we work over the Hall-type algebra $\mathrm{CF}(\mathfrak{Obj}(\mathcal{A}))$, (see Remark 2.10,) the corresponding Lie algebra $\mathrm{CF}^{\mathrm{ind}}(\mathfrak{Obj}(\mathcal{A}))$ is the set of constructible functions on $\mathfrak{Obj}(\mathcal{A})$, supported on indecomposable objects. One might expect, as an analogue for $\mathcal{H}(\mathcal{A})$, that an element $[(\mathcal{X},\rho)]$ is contained in $\mathfrak{G}(\mathcal{A})$ if the image of ρ in $\mathfrak{Obj}(\mathcal{A})$ is supported on indecomposable objects. However Joyce suggests that this definition is not the best analogue, and he introduces the notion of "virtual indecomposable objects", and defines $\mathrm{SF}^{\mathrm{ind}}_{\mathrm{al}}(\mathfrak{Obj}(\mathcal{A}))$ in [10, Definition 5.13] as the set of stack functions supported on virtual indecomposable objects. We omit the precise definition of $\mathfrak{G}(\mathcal{A})$ here, as we will not use this. The Lie algebra $\mathfrak{G}(\mathcal{A})$ also has the decomposition,

$$\mathfrak{G}(\mathcal{A}) = \bigoplus_{v\in\overline{C}(\mathcal{A})} \mathfrak{G}^v(\mathcal{A}), \quad \mathfrak{G}^v(\mathcal{A}) := \mathcal{H}^v(\mathcal{A}) \cap \mathfrak{G}(\mathcal{A}),$$

and $\epsilon^v(Z)$ is an element of $\mathfrak{G}^v(\mathcal{A})$. The conceptual meaning of the definition of $\epsilon^v(Z)$ is that they are "logarithms" of $\delta^v(Z)$, i.e. for $t\in T$, we have formally

$$\sum_{v\in Z^{-1}(t)} \epsilon^v(Z) = \log\left(1 + \sum_{v\in Z^{-1}(t)} \delta^v(Z)\right).$$

Also see [5] for more arguments on the elements $\epsilon^v(Z)$.

2.4. Transformation of the elements $\delta^v(Z)$, $\epsilon^v(Z)$

The descriptions of the variations of the elements $\delta^v(Z)$, $\epsilon^v(Z)$ under change of Z are investigated in [14]. Let us briefly recall the main idea of [14, Theorem 5.2] in this subsection. We first introduce the following definition.

Definition 2.12. Let $Z, Z' \colon C(\mathcal{A}) \to T$ be weak stability conditions and take $v\in C(\mathcal{A})$. We say Z' *dominates* Z with respect to v if for $v_1, v_2 \in C_{\le v}(\mathcal{A})$, $Z(v_1) \succeq Z(v_2)$ implies $Z'(v_1) \succeq Z'(v_2)$. Here $C_{\le v}(\mathcal{A})$ is defined by

$$C_{\le v}(\mathcal{A}) = \{v' \in C(\mathcal{A}) \mid \text{ there is } v'' \in \overline{C}(\mathcal{A}) \text{ with } v'+v''=v\}. \tag{12}$$

The first step is to show the following theorem.

Theorem 2.13. [14, Theorem 5.11] *For weak stability conditions $Z, Z' \colon C(\mathcal{A}) \to T$ satisfying Assumption 2.6, suppose that Z' dominates Z with respect to v. Then we have*

$$\delta^v(Z') = \sum_{\substack{l \geq 1,\ v_i \in C(\mathcal{A}),\ v_1 + \cdots + v_l = v, \\ Z(v_i) \succ Z(v_{i-1}),\ Z'(v_i) = Z'(v),\ 1 \leq i \leq l}} \delta^{v_1}(Z) * \cdots * \delta^{v_l}(Z). \tag{13}$$

The sum (13) may not be a finite sum, but it converges in the sense of [14, Definition 2.16]

Proof. We just explain the idea of the proof. For the full proof, see [14, Theorem 5.11]. For a Z'-semistable object $E \in \mathcal{A}$ with $[E] = v$, there is a Harder–Narasimhan filtration with respect to Z, i.e. there is a unique filtration

$$0 = E_0 \subset E_1 \subset \cdots \subset E_l = E, \tag{14}$$

such that each $F_i = E_i/E_{i-1}$ is Z-semistable with $Z(F_i) \succ Z(F_{i-1})$. Since Z' dominates Z with respect to v, and each class $[F_i] \in C(\mathcal{A})$ is contained in $C_{\leq v}(\mathcal{A})$, we have $Z'(F_i) \succeq Z'(F_{i-1})$. Hence Z'-semistability of E implies $Z'(F_i) = Z'(F_{i-1})$.

Conversely for an object $E \in \mathcal{A}$, suppose that there is a filtration (14) such that $F_i = E_i/E_{i-1}$ is Z-semistable with $Z(F_i) \succ Z(F_{i-1})$ and $Z'(F_i) = Z'(F_{i-1})$ for all i. Since Z' dominates Z with respect to v, the object F_i is also Z'-semistable, and hence E is Z'-semistable.

As a consequence, an object $E \in \mathcal{A}$ with $[E] = v$ is Z'-semistable if and only if there is a unique filtration (14) such that each $F_i = E_i/E_{i-1}$ is Z-semistable and $v_i = [F_i] \in C(\mathcal{A})$ for $i = 1, \ldots, l$ satisfy

$$\begin{aligned} & v_1 + \cdots + v_l = v, \\ & Z(v_1) \succ Z(v_2) \succ \cdots \succ Z(v_l), \\ & Z'(v_1) = Z'(v_2) = \cdots = Z'(v_l). \end{aligned} \tag{15}$$

This observation is expressed as (13) in terms of the algebra $\mathcal{H}(\mathcal{A})$. Q.E.D.

We omit the definition of the convergence [14, Definition 2.16] here, as we will only treat the cases that the relevant sums have only finitely many terms. The next step is to invert (13), and give the formula,

$$(16)\quad \delta^v(Z) = \sum_{\substack{l\geq 1,\ v_i\in C(\mathcal{A}),\\ v_1+\cdots+v_l=v,\\ Z'(v_i)=Z'(v),\ 1\leq i\leq l,\\ Z(v_1+\cdots+v_i)\succ Z(v_{i+1}+\cdots+v_l)}} (-1)^{l-1}\delta^{v_1}(Z')*\cdots*\delta^{v_l}(Z').$$

The proof is provided in [14, Theorem 5.12]. The sum (16) may not converge in the sense of [14, Definition 2.16], but if we impose the assumption that the change from Z to Z' is *locally finite*, (we omit the definition of the local finiteness, see [14, Definition 5.1],) then the sum (16) converges.

Finally for two weak stability conditions Z, Z', consider the following situation (♠).

- There are weak stability conditions $Z=Z_1, Z_2, \dots, Z_m=Z'$, $W_1,\dots,W_{m-1}$ satisfying Assumption 2.6, such that W_i dominates Z_i, Z_{i-1} w.r.t. v, and all changes from Z_i to W_i, W_{i-1} are locally finite. $\cdots$ (♠).

Then in principle one can express $\delta^v(Z')$ in terms of $\delta^v(Z)$ in the algebra $\mathcal{H}(\mathcal{A})$, by applying the formulas (13), (16) successively. The transformation coefficients are determined purely combinatory, and they are given as follows.

Definition 2.14. [14, Definition 4.2] Take $v_1,\dots,v_l\in C(\mathcal{A})$ and weak stability conditions $Z, Z'\colon C(\mathcal{A})\to T$. If for each $i=1,\dots,l-1$, we have either (17) or (18),

$$(17)\quad Z(v_i)\preceq Z(v_{i+1}) \text{ and } Z'(v_1+\cdots+v_i)\succ Z'(v_{i+1}+\cdots+v_l),$$

$$(18)\quad Z(v_i)\succ Z(v_{i+1}) \text{ and } Z'(v_1+\cdots+v_i)\preceq Z'(v_{i+1}+\cdots+v_l).$$

then define $S(\{v_1,\dots,v_l\},Z,Z')$ to be $(-1)^r$, where r is the number of $i=1,\dots,l-1$ satisfying (17). Otherwise we define $S(\{v_1,\dots,v_l\},Z, Z')=0$.

We have the following formula.

Theorem 2.15. [14, Theorem 5.2] *Under the situation* (♠), *we have*

$$(19)\quad \delta^v(Z') = \sum_{\substack{l\geq 1,\ v_i\in C(\mathcal{A}),\\ v_1+\cdots+v_l=v}} S(\{v_1,\dots,v_l\},Z,Z')\delta^{v_1}(Z)*\cdots*\delta^{v_l}(Z).$$

The sum (19) converges in the sense of [14, Definition 2.16].

Remark 2.16. Our condition "$*'$ dominates $*$ w.r.t. v" in Definition 2.12 is weaker than Joyce's condition "$*'$ dominates $*$" given in [14, Definition 3.16], and [14, Theorem 5.2] is formulated using the latter condition. However if we want to know (19) for a fixed $v \in C(\mathcal{A})$, it is enough to assume "$*'$ dominates $*$ w.r.t. v" in (♠), since all the v_i in the sum (19) are contained in $C_{\leq v}(\mathcal{A})$.

The relationship between $\epsilon^v(Z')$ and $\epsilon^v(Z)$ is deduced from (10), (19), and the inverse of (10),

$$\delta^v(Z) = \sum_{\substack{l\geq 1,\ v_i\in C(\mathcal{A}),\ v_1+\cdots+v_l=v,\\ Z(v_i)=Z(v),\ 1\leq i\leq l}} \frac{1}{l!}\epsilon^{v_1}(Z) * \cdots * \epsilon^{v_l}(Z). \tag{20}$$

The proof of (20) is provided in [11, Theorem 8.2]. The transformation coefficients are given as follows.

Definition 2.17. [14, Definition 4.4] For $v_1, \ldots, v_l \in C(\mathcal{A})$, we define $U(\{v_1, \ldots, v_l\}, Z, Z') \in \mathbb{Q}$ to be

$$U(\{v_1, \ldots, v_l\}, Z, Z') = \sum_{1\leq m'\leq m\leq l} \sum_{\substack{\text{surjective } \psi\colon \{1,\ldots,l\}\to\{1,\ldots,m\},\\ \text{surjective } \xi\colon \{1,\ldots,m\}\to\{1,\ldots,m'\},\\ i\leq j \text{ implies } \psi(i)\leq\psi(j),\ \xi(i)\leq\xi(j),\\ \psi \text{ and } \xi \text{ satisfy } (\diamondsuit)}}$$

$$\prod_{a=1}^{m'} S(\{w_i\}_{i\in\xi^{-1}(a)}, Z, Z') \cdot \frac{(-1)^{m'}}{m'} \cdot \prod_{b=1}^{m} \frac{1}{|\psi^{-1}(b)|!}. \tag{21}$$

Here the condition ($\diamondsuit$) is as follows.

- For $1 \leq i, j \leq l$ with $\psi(i) = \psi(j)$, we have $Z(v_i) = Z(v_j)$, and for $1 \leq i, j \leq m'$, we have

$$Z'\Big(\sum_{k\in\psi^{-1}\xi^{-1}(i)} v_k\Big) = Z'\Big(\sum_{k\in\psi^{-1}\xi^{-1}(j)} v_k\Big). \quad \cdots(\diamondsuit)$$

Also w_i for $1 \leq i \leq m$ is defined as

$$w_i = \sum_{j\in\psi^{-1}(i)} v_j \in C(\mathcal{A}).$$

Theorem 2.18. [14, Theorem 5.2] *In the situation (♠), the following holds.*

$$\epsilon^v(Z') = \sum_{\substack{l\geq 1,\ v_i\in C(\mathcal{A}),\\ v_1+\cdots+v_l=v}} U(\{v_1, \ldots, v_l\}, Z, Z')\epsilon^{v_1}(Z) * \cdots * \epsilon^{v_l}(Z). \tag{22}$$

The sum (22) converges in the sense of [14, Definition 2.16].

Remark 2.19. It is possible to rewrite (22) by a $\mathbb{Q}$-linear combination of multiple commutators of $\epsilon^{v_i}(Z)$ such as

$$[[\cdots[[\epsilon^{v_1}(Z),\epsilon^{v_2}(Z)],\epsilon^{v_3}(Z)],\cdots],\epsilon^{v_l}(Z)],$$

so (22) is an equality in $\mathfrak{G}(\mathcal{A})$, rather than in $\mathcal{H}(\mathcal{A})$. The proof of this fact is given in [14, Theorem 5.4].

2.5. Motivic invariants of stacks

As a final step, we integrate the elements $\epsilon^v(Z) \in \mathfrak{G}^v(\mathcal{A})$ to give $\mathbb{Q}$-valued invariants, and establish the transformation formula of these invariants. Let us recall that a motivic invariant is a ring homomorphism

$$\Upsilon\colon K_0(\mathrm{Var}/\operatorname{Spec}\mathbb{C}) \longrightarrow \Lambda,$$

where Λ is a $\mathbb{Q}$-algebra and a ring structure on $K_0(\mathrm{Var}/\operatorname{Spec}\mathbb{C})$ is given by (8). In order to simplify the arguments, we only consider the special case that $\Lambda = \mathbb{Q}(t)$ and

$$\Upsilon([Y]) = \sum_i (-1)^i \dim H^i(Y,\mathbb{C})t^i,$$

where Y is a smooth projective variety. Since $K_0(\mathrm{Var}/\operatorname{Spec}\mathbb{C})$ is generated by $[Y]$ for smooth projective varieties Y, the above data uniquely determines Υ. In this situation, there is a unique extension of Υ,

$$\Upsilon'\colon K_0(\mathrm{St}/\operatorname{Spec}\mathbb{C}) \longrightarrow \mathbb{Q}(t),$$

such that if G is a special algebraic group acting on Y, we have (cf. [13, Theorem 4.9])

$$\Upsilon'([Y/G]) = \Upsilon([Y])/\Upsilon([G]).$$

Here an algebraic group is *special* if every principal G-bundle is locally trivial in Zariski topology. In what follows, we assume that $\mathcal{A}$ satisfies the following condition.

$(\star)$: there is an anti-symmetric biadditive-paring $\chi\colon \mathcal{N}(\mathcal{A})\times\mathcal{N}(\mathcal{A})\to\mathbb{Z}$ such that for any $E,F\in\mathcal{A}$, we have

$$\chi(E,F) = \dim \operatorname{Hom}(E,F) - \dim \operatorname{Ext}^1(E,F) + \dim \operatorname{Ext}^1(F,E) - \dim \operatorname{Hom}(F,E).$$

For instance if $\mathcal{A} = \operatorname{Coh}(X)$ for a smooth projective Calabi–Yau 3-fold X, the usual Euler pairing (6) descends to the pairing on $\mathcal{N}(\mathcal{A})$, which satisfies $(\star)$ by Serre duality. Using the pairing χ, we can define the following Lie algebra.

Definition 2.20. For an abelian category $\mathcal{A}$ satisfying $(\star)$, we define the Lie algebra $\mathfrak{g}(\mathcal{A})$ to be the $\mathbb{Q}$-vector space,

$$\mathfrak{g}(\mathcal{A}) := \bigoplus_{v \in \mathcal{N}(\mathcal{A})} \mathbb{Q} c_v,$$

with its Lie-brackets given by $[c_v, c_{v'}] = \chi(v, v') c_{v+v'}$.

Let Π_v be the composition,

$$\Pi_v \colon \mathfrak{G}^v(\mathcal{A}) \subset \mathcal{H}(\mathcal{A}) \xrightarrow{\pi_*} K_0(\mathrm{St}/\operatorname{Spec}\mathbb{C}) \xrightarrow{\Upsilon'} \mathbb{Q}(t),$$

where the map π_* sends $[(\mathcal{X}, \rho)]$ to $[(\mathcal{X}, \pi \circ \rho)]$ and $\pi \colon \mathfrak{Obj}(\mathcal{A}) \to \operatorname{Spec}\mathbb{C}$ is the structure morphism. It is shown in [14, Section 6.2] that for $\epsilon \in \mathfrak{G}^v(\mathcal{A})$, the rational function $\Pi_v(\epsilon) \in \mathbb{Q}(t)$ has a pole of order at most one at $t = 1$. Hence the following definition makes sense,

$$\Theta_v(\epsilon) = (t^2 - 1)\Pi_v(\epsilon)|_{t=1} \in \mathbb{Q}. \tag{23}$$

Definition 2.21. We define the invariant $J^v(Z) \in \mathbb{Q}$ by

$$J^v(Z) = \Theta_v(\epsilon^v(Z)) \in \mathbb{Q}.$$

Remark 2.22. If all the Z-semistable objects $E \in \mathcal{A}$ with $[E] = v$ are in fact Z-stable, and their moduli problem is represented by a scheme, then $\epsilon^v(Z)$ is written as $[M^v(Z)/\mathbb{G}_m]$ for a scheme $M^v(Z)$. Here $\mathbb{G}_m$ is acting on $M^v(Z)$ trivially. In this case $J^v(Z)$ equals to the Euler characteristic of $M^v(Z)$. Note that the factor $\Upsilon([\mathbb{G}_m]) = t^2 - 1$ in (23) is required to cancel out the contribution of the stabilizer group $\operatorname{Aut}(E) \cong \mathbb{G}_m$.

We have the following theorem.

Theorem 2.23. [10, Theorem 6.12] *The map,*

$$\Theta \colon \mathfrak{G}(\mathcal{A}) = \bigoplus_{v \in \overline{C}(\mathcal{A})} \mathfrak{G}^v(\mathcal{A}) \ni \{\epsilon^v\}_v \longmapsto \sum_{v \in \overline{C}(\mathcal{A})} \Theta_v(\epsilon^v) c_v \in \mathfrak{g}(\mathcal{A}), \tag{24}$$

is a Lie algebra homomorphism.

Since (22) is a relationship in the Lie algebra $\mathfrak{G}(\mathcal{A})$, we can obtain the relationship between $J^v(Z')$ and $J^v(Z)$ by applying Θ. The result is as follows.

Theorem 2.24. [14, Theorem 6.28, Equation (130)] *In the situation of (♠), assume that there are only finitely many terms in (22). Applying* Θ *to (22) yields the formula,*

$$J^v(Z') = \sum_{\substack{l\ge 1,\ v_i\in C(\mathcal{A}),\\ v_1+\cdots+v_l=v}} \sum_{\substack{\Gamma \text{ is a connected, simply connected oriented} \\ \text{graph with vertex } \{1,\dots,l\},\ \overset{i}{\bullet}\to\overset{j}{\bullet} \text{ implies } i<j}}$$

$$\frac{1}{2^{l-1}}U(\{v_1,\dots,v_l\},Z,Z') \prod_{\overset{i}{\bullet}\to\overset{j}{\bullet} \text{ in } \Gamma} \chi(v_i,v_j)\prod_{i=1}^{l} J^{v_i}(Z). \tag{25}$$

2.6. Generalization to quasi-abelian categories

For our purpose it is useful to give a slight generalization of Theorem 2.24, especially we want to relax Assumption 2.6, as our abelian category $\mathcal{A}^p$ which will be introduced in the next section does not satisfy that assumption. Let $\mathcal{A}$ be an abelian category and $Z\colon C(\mathcal{A})\to T$ a weak stability condition. Here we do not assume Assumption 2.6. For $t\in T$, we set

$$\mathcal{A}_{Z\succeq t} = \langle E \mid E \text{ is } Z\text{-semistable with } Z(E)\succeq t\rangle,$$
$$\mathcal{A}_{Z\prec t} = \langle E \mid E \text{ is } Z\text{-semistable with } Z(E)\prec t\rangle.$$

Here for a set of objects S in $\mathcal{A}$, we denote by $\langle S\rangle\subset\mathcal{A}$ the smallest extension closed subcategory of $\mathcal{A}$ which contains S. Equivalently, an object $E\in\mathcal{A}$ is contained in $\mathcal{A}_{Z\succeq t}$ (resp. $\mathcal{A}_{Z\prec t}$) if and only if any Z-semistable factor F of E satisfies $Z(F)\succeq t$, (resp. $Z(F)\prec t$.) It can be shown that $\mathcal{A}_{Z\succeq t}$, $\mathcal{A}_{Z\prec t}$ are quasi-abelian categories. See [4, Section 4] for the detail on quasi-abelian categories.

Definition 2.25. For objects $E,F\in\mathcal{A}_{Z\prec t}$ and a morphism $f\colon E\to F$, it is called a *strict monomorphism* if f is injective in $\mathcal{A}$ and $\mathrm{Coker}(f)\in\mathcal{A}_{Z\prec t}$. Similarly f is called a *strict epimorphism* if f is surjective in $\mathcal{A}$ and $\mathrm{Ker}(f)\in\mathcal{A}_{Z\prec t}$.

For $v\in C(\mathcal{A}_{Z\prec t})$, we set

$$C_{\le v}(\mathcal{A}_{Z\prec t}) = \{v'\in C(\mathcal{A}_{Z\prec t}) \mid v-v'\in\overline{C}(\mathcal{A}_{Z\prec t})\}. \tag{26}$$

For Z,t,v as above, we put the following assumption instead of Assumption 2.6.

Assumption 2.26.

- *The category $\mathcal{A}_{Z\prec t}$ is noetherian and artinian with respect to strict monomorphisms.*
- *There is an Artin stack locally of finite type $\mathfrak{Obj}(\mathcal{A})$, which parametrizes objects $E\in\mathcal{A}$.*

- *For any $v' \in C_{\leq v}(\mathcal{A}_{Z\prec t})$, the substack $\mathfrak{M}^{v'}(Z) \subset \mathfrak{Obj}(\mathcal{A})$ is an open substack, and it is of finite type.*

We also modify the dominant conditions.

Definition 2.27. Let $Z, Z' \colon C(\mathcal{A}) \to T$ be weak stability conditions. For $t \in T$ and $v \in C(\mathcal{A}_{Z\prec t})$, we say Z' *dominates Z with respect to (v,t)* if the following holds.

- We have $\mathcal{A}_{Z\succeq t} = \mathcal{A}_{Z'\succeq t}$ and $\mathcal{A}_{Z\prec t} = \mathcal{A}_{Z'\prec t}$.
- For $v_1, v_2 \in C_{\leq v}(\mathcal{A}_{Z\prec t})$, if $Z(v_1) \succeq Z(v_2)$ then $Z'(v_1) \succeq Z'(v_2)$.

For two weak stability conditions Z, Z', we consider the following situation.

- There are weak stability conditions $Z = Z_1, Z_2, \ldots, Z_m = Z'$, $W_1, \ldots, W_{m-1}$ such that W_i dominates Z_i, Z_{i+1} w.r.t. (v,t), all changes from Z_i to W_i, W_{i-1} are locally finite, and Z_i, W_i satisfy Assumption 2.26 for (v,t). $\cdots$ ($\spadesuit'$)

We have the following generalization of Theorem 2.24.

Theorem 2.28. *For weak stability conditions Z, Z' on $\mathcal{A}$, $t \in T$ and $v \in C(\mathcal{A}_{Z\prec t})$, suppose that the condition ($\spadesuit'$) holds. Then the equation (22) with each $v_i \in C(\mathcal{A}_{Z\prec t})$ holds. If there are only finitely many terms in (22) with $v_i \in C(\mathcal{A}_{Z\prec t})$, then (25) holds, with each $v_i \in C(\mathcal{A}_{Z\prec t})$.*

Proof. First suppose that Z' dominates Z with respect to (v,t), and check that (13) holds with each $v_i \in C(\mathcal{A}_{Z\prec t})$. Let $E \in \mathcal{A}$ be Z'-semistable with $[E] = v$. As in the proof of Theorem 2.13, we have a unique filtration (14). Let $F_i = E_i/E_{i-1}$, $v_i = [F_i] \in C(\mathcal{A})$. Since $v \in C(\mathcal{A}_{Z\prec t}) = C(\mathcal{A}_{Z'\prec t})$, we have $E \in \mathcal{A}_{Z'\prec t} = \mathcal{A}_{Z\prec t}$. Hence we have $Z(v_i) \prec t$, and $v_i \in C(\mathcal{A}_{Z\prec t})$ follows.

Conversely given a filtration (14), suppose that each F_i is Z-semistable with $v_i \in C(\mathcal{A}_{Z\prec t})$. Then $F_i \in \mathcal{A}_{Z\prec t} = \mathcal{A}_{Z'\prec t}$, and hence F_i is also Z'-semistable as Z' dominates Z w.r.t. (v,t).

As a summary, an object $E \in \mathcal{A}_{Z\prec t}$ is Z-semistable if and only if there is a filtration (14), satisfying (15) with each $v_i \in C(\mathcal{A}_{Z\prec t})$. Then the same proof as in [14, Theorem 5.11] works and gives the formula (13) with each $v_i \in C(\mathcal{A}_{Z\prec t})$. Note that to state the formula (13), it is enough to assume that $\mathfrak{M}^{v'}(Z) \subset \mathfrak{Obj}(\mathcal{A})$ is open and of finite type for any $v' \in C_{\leq v}(\mathcal{A}_{Z\prec t})$.

By the same idea, we can also show the formulas (16), (19), (20), (22) hold with each $v_i \in C(\mathcal{A}_{Z\prec t})$. We leave the readers to follow Joyce's work and that the same proofs are applied in this case. Q.E.D.

§3. Limit stability and μ-limit stability

In this section, we recall the notion of limit stability on a Calabi–Yau 3-fold X introduced in [24], and also introduce the notion of μ-limit stability. Below we always assume that X is a projective complex 3-fold with a trivial canonical class, i.e. X is a Calabi–Yau 3-fold. We denote by $D^b(X)$ the bounded derived category of $\mathrm{Coh}(X)$, and $K(X)$ the Grothendieck group of $\mathrm{Coh}(X)$. The *numerical Grothendieck group* of $\mathrm{Coh}(X)$ is given by

$$\mathcal{N}(X) = K(X)/\equiv,$$

where the numerical equivalence relation $\equiv$ is given by (5). Note that if $\mathcal{A} \subset D^b(X)$ is the heart of a bounded t-structure on $D^b(X)$, then the group $\mathcal{N}(\mathcal{A}) = K(\mathcal{A})/\equiv$, where $\equiv$ is (5) as above, coincides with $\mathcal{N}(X)$. So we always regard $C(\mathcal{A})$ as a subset of $\mathcal{N}(X)$.

Let us fix notation of the numerical classes of curves on X. An element $\beta \in H^4(X, \mathbb{Z})$ is called an *effective class* if there is a one dimensional subscheme $C \subset X$ such that β is the Poincaré dual of the fundamental cycle of C. We set $C(X)$, $\overline{C}(X)$ as

$$C(X) := \{\beta \in H^4(X,\mathbb{Z}) \mid \beta \text{ is an effective class }\},$$
$$\overline{C}(X) := C(X) \cup \{0\}.$$

3.1. Definition of limit stability

The limit stability introduced in [24] is a stability condition on the category of perverse coherent sheaves $\mathcal{A}^p$ in the sense of Bezrukavnikov [3] and Kashiwara [15], and it is also one of the polynomial stability conditions introduced by Bayer [1] independently. In order to introduce $\mathcal{A}^p$, let us define the subcategories $(\mathrm{Coh}_{\le 1}(X), \mathrm{Coh}_{\ge 2}(X))$ of $\mathrm{Coh}(X)$, as follows.

Definition 3.1. We define the pair of subcategories, $(\mathrm{Coh}_{\le 1}(X), \mathrm{Coh}_{\ge 2}(X))$, to be

$$\mathrm{Coh}_{\le 1}(X) := \{E \in \mathrm{Coh}(X) \mid \dim \mathrm{Supp}(E) \le 1\},$$
$$\mathrm{Coh}_{\ge 2}(X) := \{E \in \mathrm{Coh}(X) \mid \mathrm{Hom}(\mathrm{Coh}_{\le 1}(X), E) = 0\}.$$

The category $\mathcal{A}^p$ is defined as follows.

Definition 3.2. We define the subcategory $\mathcal{A}^p \subset D^b(X)$ to be

$$\mathcal{A}^p := \left\{ E \in D^b(X) : \begin{array}{l} \mathcal{H}^{-1}(E) \in \mathrm{Coh}_{\ge 2}(X), \mathcal{H}^0(E) \in \mathrm{Coh}_{\le 1}(X), \\ \text{and } \mathcal{H}^i(E) = 0 \text{ for } i \ne -1, 0. \end{array} \right\}.$$

It is easy to see that $(\mathrm{Coh}_{\le 1}(X), \mathrm{Coh}_{\ge 2}(X))$ determines a torsion theory on $\mathrm{Coh}(X)$, and $\mathcal{A}^p$ is the corresponding tilting. (cf. [24, Definition 2.9, Lemma 2.10].) Therefore $\mathcal{A}^p$ is the heart of a bounded t-structure on $D^b(X)$, in particular $\mathcal{A}^p$ is an abelian category.

Next recall that the *complexified ample cone* is defined by

$$A(X)_{\mathbb{C}} = \{B + i\omega \in H^2(X, \mathbb{C}) \mid \omega \text{ is an ample class}\}.$$

Given $\sigma = B + i\omega \in A(X)_{\mathbb{C}}$, one can define the map $Z_\sigma \colon K(X) \to \mathbb{C}$,

$$Z_\sigma \colon K(X) \ni E \longmapsto -\int e^{-(B+i\omega)} \operatorname{ch}(E)\sqrt{\operatorname{td}_X} \in \mathbb{C}.$$

Explicitly we have

$$Z_\sigma(E) = \left(-v_3^B(E) + \frac{1}{2}\omega^2 v_1^B(E)\right) + \left(\omega v_2^B(E) - \frac{1}{6}\omega^3 v_0^B(E)\right) i, \tag{27}$$

where $v_i^B \in H^{2i}(X, \mathbb{R})$ for $0 \le i \le 3$ are given by

$$e^{-B} \operatorname{ch}(E)\sqrt{\operatorname{td}_X} = (v_0^B(E), v_1^B(E), v_2^B(E), v_3^B(E)) \in H^{\mathrm{even}}(X, \mathbb{R}).$$

For $\sigma_m = B + im\omega$ with $m \in \mathbb{R}$, one can show the following: for each non-zero object $E \in \mathcal{A}^p$, one has

$$Z_{\sigma_m}(E) \in \left\{ r\exp(i\pi\phi) : r > 0,\ \frac{1}{4} < \phi < \frac{5}{4} \right\}, \tag{28}$$

for $m \gg 0$. (See [24, Lemma 2.20].) Hence the *phase* of E is well-defined for $m \gg 0$ as follows,

$$\phi_{\sigma_m}(E) = \frac{1}{\pi} \operatorname{Im} \log Z_{\sigma_m}(E) \in \left(\frac{1}{4}, \frac{5}{4}\right).$$

Definition 3.3. [14, Definition 2.21] An object $E \in \mathcal{A}^p$ is *σ-limit (semi)stable* if for any non-zero subobject $F \subset E$ in $\mathcal{A}^p$, we have

$$\phi_{\sigma_m}(F) < \phi_{\sigma_m}(E), \quad (\text{resp. } \phi_{\sigma_m}(F) \le \phi_{\sigma_m}(E),) \tag{29}$$

for $m \gg 0$.

Let us interpret the above stability in terms of Definition 2.1. Let T be the one variable function field $\mathbb{R}(m)$. We define the total order on $\mathbb{R}(m)$ to be

$$f(m) \succeq g(m) \quad \stackrel{\text{def}}{\leftrightarrow} \quad f(m) \ge g(m) \text{ for } m \gg 0.$$

Note that we have

$$\operatorname{Im} e^{-\pi i/4} Z_{\sigma_m}(E) > 0, \tag{30}$$

for $m \gg 0$ since (28) holds. In particular (30) is non-zero as a polynomial of m, thus the following map is well-defined.

$$Z_{\sigma}^{T} : C(\mathcal{A}^p) \ni E \longmapsto -\frac{\operatorname{Re} e^{-\pi i/4} Z_{\sigma_m}(E)}{\operatorname{Im} e^{-\pi i/4} Z_{\sigma_m}(E)} \in T.$$

Lemma 3.4. *The map Z_{σ}^{T} is a stability condition on $\mathcal{A}^p$, and an object $E \in \mathcal{A}^p$ is Z_{σ}^{T}-(semi)stable if and only if E is σ-limit (semi)stable.*

Proof. Since (30) holds, it is obvious that for $v_1, v_2 \in C(\mathcal{A}_p)$, the inequality $\phi_{\sigma_m}(v_1) \le \phi_{\sigma_m}(v_2)$ holds for $m \gg 0$ if and only if $Z_{\sigma}^{T}(v_1) \preceq Z_{\sigma}^{T}(v_2)$ holds in T. Hence Z_{σ}^{T} is a stability function, and the latter statement also follows. The existence of Harder–Narasimhan filtrations for limit stability is proved in [24, Theorem 2.28], and hence Z_{σ}^{T} is a stability condition. Q.E.D.

3.2. μ-limit stability

In this subsection, we introduce a weak stability condition on $\mathcal{A}^p$, which we call *μ-limit stability.* Let us introduce the following notation.

Definition 3.5. Let $f = \sum_{i=0}^{d} a_i(\sigma) m^i$ be a polynomial such that each coefficient $a_i(\sigma)$ is a $\mathbb{R}$-valued function on $A(X)_{\mathbb{C}}$, and $a_d(\sigma) \not\equiv 0$. We define $f^{\dagger} = a_d(\sigma) m^d$.

By the formula (27), $\operatorname{Re} Z_{\sigma_m}(E)$ and $\operatorname{Im} Z_{\sigma_m}(E)$ are written as polynomials of m whose coefficients are $\mathbb{R}$-valued functions on $A(X)_{\mathbb{C}}$. Thus the following makes sense,

$$Z_{\sigma_m}^{\dagger}(E) := (\operatorname{Re} Z_{\sigma_m}(E))^{\dagger} + (\operatorname{Im} Z_{\sigma_m}(E))^{\dagger} i.$$

The same argument as in [24, Lemma 2.20] shows that

$$Z_{\sigma_m}^{\dagger}(E) \in \left\{ r \exp(i\pi\phi) : r > 0,\ \frac{1}{4} < \phi < \frac{5}{4} \right\},$$

for $m \gg 0$. Hence as before we can define the following map,

$$Z_{\mu_\sigma} : C(\mathcal{A}^p) \ni E \longmapsto -\frac{\operatorname{Re} e^{-\pi i/4} Z_{\sigma_m}^{\dagger}(E)}{\operatorname{Im} e^{-\pi i/4} Z_{\sigma_m}^{\dagger}(E)} \in T.$$

We have the following lemma.

Lemma 3.6. *The map Z_{μ_σ} is a weak stability function on $\mathcal{A}^p$.*

Proof. Since the operation $f \mapsto f^{\dagger}$ is just taking the initial term of the polynomials, we have the implication

$$Z_{\sigma}^{T}(E) \succ Z_{\sigma}^{T}(F) \quad \Rightarrow \quad Z_{\mu_\sigma}(E) \succeq Z_{\mu_\sigma}(F), \tag{31}$$

for $E, F \in C(\mathcal{A}^p)$. Hence Z_{μ_σ} is a weak stability function. Q.E.D.

It is easy to see that for $0 \neq E \in \mathcal{A}^p$, one has $Z_{\mu_\sigma}(E) \to 1$ or -1 as $m \to \infty$. To see that Z_{μ_σ} is a weak stability condition, we introduce a pair of subcategories in $\mathcal{A}^p$. (cf. [24, subsection 2.3].)

Definition 3.7. We define $(\mathcal{A}_1^p, \mathcal{A}_{1/2}^p)$ to be

$$\mathcal{A}_1^p = \{E \in \mathcal{A}^p \mid \dim \operatorname{Supp} \mathcal{H}^0(E) = 0 \text{ and } \mathcal{H}^{-1}(E) \text{ is a torsion sheaf}\},$$
$$\mathcal{A}_{1/2}^p = \{E \in \mathcal{A}^p \mid \operatorname{Hom}(F, E) = 0 \text{ for any } F \in \mathcal{A}_1^p\}.$$

It is shown in [24, Lemma 2.16] that $(\mathcal{A}_1^p, \mathcal{A}_{1/2}^p)$ determines a torsion theory on $\mathcal{A}^p$. We shall use the same notation of "strict monomorphisms", "strict epimorphisms" in $\mathcal{A}_i^p$ as in Definition 2.25, i.e. we replace $\mathcal{A}_{Z \prec t}$ by $\mathcal{A}_i^p$ to define them. We have the following.

Lemma 3.8. *An object $E \in \mathcal{A}^p$ is Z_{μ_σ}-semistable with $Z_{\mu_\sigma}(E) \to -1$ (resp. 1) as $m \to \infty$ if and only if $E \in \mathcal{A}_{1/2}^p$, (resp. $\mathcal{A}_1^p$,) and for any strict monomorphism $0 \neq F \hookrightarrow E$ in $\mathcal{A}_{1/2}^p$, (resp. $\mathcal{A}_1^p$,) one has $Z_{\mu_\sigma}(F) \preceq Z_{\mu_\sigma}(E/F)$.*

Proof. The proof is same as in [24, Lemma 2.26] for the limit stability, by noting that

$$Z_{\mu_\sigma}(E) \to -1, \quad (\text{resp. } Z_{\mu_\sigma}(E) \to 1,)$$

for $E \in \mathcal{A}_{1/2}^p$, (resp. $E \in \mathcal{A}_1^p$,) and $m \to \infty$. Q.E.D.

We also have the following.

Lemma 3.9. *The weak stability function Z_{μ_σ} is a weak stability condition.*

Proof. The existence of Harder–Narasimhan filtrations follows from the same argument for the limit stability. The proof of [24, Theorem 2.28] also works in this case, by noting Lemma 3.8. Q.E.D.

We say $E \in \mathcal{A}^p$ is μ_σ-limit (semi)stable if it is (semi)stable in Z_{μ_σ}-weak stability. To explain this notation, let us recall that the (usual) μ-stability is defined by cutting off the lower degree terms of the reduced Hilbert polynomials. In this sense, our μ_σ-limit stability resembles to

μ-stability, as we also cut off the lower degree terms of the polynomial $Z_{\sigma_m}(*)$. By (31), we have the following implications,

$$\mu_\sigma\text{-limit stable} \Rightarrow \sigma\text{-limit stable} \Rightarrow \sigma\text{-limit semistable} \Rightarrow \mu_\sigma\text{-limit semistable}.$$

Remark 3.10. For $0 \neq E \in \mathcal{A}^p$, let $\phi^{\dagger}_{\sigma_m}(E)$ be

$$\phi^{\dagger}_{\sigma_m}(E) = \frac{1}{\pi}\operatorname{Im}\log Z^{\dagger}_{\sigma_m}(E) \in \left(\frac{1}{4}, \frac{5}{4}\right).$$

Then obviously an object $E \in \mathcal{A}^p$ is μ_σ-limit (semi)stable if and only if for any non-zero subobject $F \subset E$, one has

$$\phi^{\dagger}_{\sigma_m}(F) < \phi^{\dagger}_{\sigma_m}(E/F), \quad (\text{resp. } \phi^{\dagger}_{\sigma_m}(F) \le \phi^{\dagger}_{\sigma_m}(E/F),)$$

for $m \gg 0$.

Remark 3.11. Let us take $F \in \operatorname{Coh}_{\le 1}(X)$. In this case we have $Z^{\dagger}_{\sigma_m}(F) = Z_{\sigma_m}(F)$, and $\operatorname{Coh}_{\le 1}(X) \subset \mathcal{A}^p$ is closed under taking subobjects and quotients. So F is μ_σ-limit (semi)stable if and only if it is σ-limit (semi)stable. On the other hand for $\sigma = B + i\omega \in A(X)_{\mathbb{C}}$, let $\mu_\sigma(F) \in \mathbb{Q}$ be

$$\mu_\sigma(F) = \frac{\operatorname{ch}_3(F) - B\operatorname{ch}_2(F)}{\omega\operatorname{ch}_2(F)} \in \mathbb{Q}. \tag{32}$$

As in [24, Example 2.24 (ii)], the object F is σ-limit (semi)stable if and only if for any subsheaf $0 \neq F' \subset F$ we have $\mu_\sigma(F') < \mu_\sigma(F)$, (resp. $\mu_\sigma(F') \le \mu_\sigma(F)$,) i.e. F is a (B, ω)-twisted (semi)stable sheaf. If $B = k\omega$ for $k \in \mathbb{R}$, then F is σ-limit (semi)stable if and only if F is a ω-Gieseker (semi)stable sheaf, and this notion does not depend on k.

Remark 3.12. By Lemma 3.8, it is obvious that

$$\mathcal{A}^p_{Z_{\mu_\sigma} \prec 0} = \mathcal{A}^p_{1/2}, \quad \mathcal{A}^p_{Z_{\mu_\sigma} \succeq 0} = \mathcal{A}^p_1, \tag{33}$$

in the notation of subsection 2.6. Since $\mathcal{A}^p_{1/2}$ and $\mathcal{A}^p_1$ are of finite length with respect to strict monomorphisms and strict epimorphisms, (cf. [24, Lemma 2.19],) the categories $\mathcal{A}^p_{Z_{\mu_\sigma} \prec 0}$ and $\mathcal{A}^p_{Z_{\mu_\sigma} \succeq 0}$ also have such properties.

3.3. Characterization of μ-limit semistable objects

Take $v \in C(\mathcal{A}^p)$ satisfying

$$(\mathrm{ch}_0(v), \mathrm{ch}_1(v), \mathrm{ch}_2(v), \mathrm{ch}_3(v)) = (-1, 0, \beta, n), \tag{34}$$

for some $\beta \in H^4(X, \mathbb{Q})$ and $n \in H^6(X, \mathbb{Q}) \cong \mathbb{Q}$. In this subsection we give a characterization of μ-limit semistable objects $E \in \mathcal{A}^p$ of numerical type v. i.e. $[E] = v \in C(\mathcal{A}^p)$. Note that such objects satisfy $Z_{\mu_\sigma}(E) \to -1$ as $m \to \infty$, and hence we have

$$E \in \mathcal{A}^p_{1/2}, \quad v \in C(\mathcal{A}^p_{1/2}).$$

Also if such E exists, the classes β, n are contained in $\overline{C}(X)$, $H^6(X, \mathbb{Z}) \cong \mathbb{Z}$ respectively, by [24, Remark 3.3]. We have the following proposition, whose corresponding result for limit stability is seen in [24, Section 3].

Proposition 3.13. *Take $\sigma = B + i\omega \in A(X)_{\mathbb{C}}$. For an object $E \in \mathcal{A}^p_{1/2}$ of numerical type v, it is μ_σ-limit (semi)stable if and only if the following conditions hold.*

(a) For any pure one dimensional sheaf $G \neq 0$ which admits a strict epimorphism $E \twoheadrightarrow G$ in $\mathcal{A}^p_{1/2}$, one has

$$\mu_\sigma(G) > -\frac{3B\omega^2}{\omega^3}. \quad \left(\textit{resp. } \mu_\sigma(G) \geq -\frac{3B\omega^2}{\omega^3}.\right) \tag{35}$$

(b) For any pure one dimensional sheaf $F \neq 0$ which admits a strict monomorphism $F \hookrightarrow E$ in $\mathcal{A}^p_{1/2}$, one has

$$\mu_\sigma(F) < -\frac{3B\omega^2}{\omega^3}. \quad \left(\textit{resp. } \mu_\sigma(F) \leq -\frac{3B\omega^2}{\omega^3}.\right) \tag{36}$$

Proof. By Lemma 3.8 and applying the same argument as in [24, Lemma 3.4], an object $E \in \mathcal{A}^p_{1/2}$ of numerical type v is Z_{μ_σ}-limit semistable if and only if

(a') For any pure one dimensional sheaf $G \neq 0$ which admits an exact sequence $0 \to F \to E \to G \to 0$ in $\mathcal{A}^p_{1/2}$, one has $Z_{\mu_\sigma}(F) \preceq Z_{\mu_\sigma}(G)$.

(b') For any pure one dimensional sheaf $F \neq 0$ which admits an exact sequence $0 \to F \to E \to G \to 0$ in $\mathcal{A}^p_{1/2}$, one has $Z_{\mu_\sigma}(F) \preceq Z_{\mu_\sigma}(G)$.

By Lemma 3.14 below, the conditions (a'), (b') are equivalent to (a), (b) respectively. Q.E.D.

Lemma 3.14. *Take $v_1, v_2 \in C(\mathcal{A}^p_{1/2})$ with $\mathrm{ch}(v_1) = (-1, 0, \beta_1, n_1)$, $\mathrm{ch}(v_2) = (0, 0, \beta_2, n_2)$, and $\beta_2 \neq 0$. Then*

$$Z_{\mu_\sigma}(v_1) \preceq Z_{\mu_\sigma}(v_2), \quad (\textit{resp. } Z_{\mu_\sigma}(v_1) \succeq Z_{\mu_\sigma}(v_2),)$$

if and only if

$$\mu_\sigma(v_2) \geq -\frac{3B\omega^2}{\omega^3}, \quad (\textit{resp. } \mu_\sigma(v_2) \leq -\frac{3B\omega^2}{\omega^3}). \tag{37}$$

If $\sigma = k\omega + i\omega$ for $k \in \mathbb{R}$, (37) is equivalent to

$$k \geq -\frac{1}{2}\mu_{i\omega}(v_2), \quad (\textit{resp. } k \leq -\frac{1}{2}\mu_{i\omega}(v_2)). \tag{38}$$

Proof. An easy computation shows,

$$\begin{aligned} Z^{\dagger}_{\sigma_m}(v_1) &= \frac{1}{2}m^2\omega^2 B + \frac{1}{6}m^3\omega^3 i, \\ Z^{\dagger}_{\sigma_m}(v_2) &= -n_2 + B\beta_2 + m\omega\beta_2 i. \end{aligned} \tag{39}$$

Since $\omega\beta_2 > 0$, we have

$$\begin{aligned} Z_{\mu_\sigma}(v_1) \preceq Z_{\mu_\sigma}(v_2) \quad &\Leftrightarrow \quad -\frac{m^2\omega^2 B/2}{m^3\omega^3/6} \preceq \frac{\mu_\sigma(v_2)}{m} \\ &\Leftrightarrow \quad \mu_\sigma(v_2) \geq -\frac{3\omega^2 B}{\omega^3}. \end{aligned}$$

If $\sigma = k\omega + i\omega$ with $k \in \mathbb{R}$, then $\mu_\sigma(v_2) = \mu_{i\omega}(v_2) - k$ and $-3B\omega^2/\omega^3 = -3k$. Hence (37) is equivalent to (38). Q.E.D.

3.4. Moduli theory of μ-limit semistable objects

In this subsection, we establish a moduli theory of μ-limit semistable objects. In [24, Theorem 1.1], a moduli theory of σ-limit stable objects is studied, and the resulting moduli space is an algebraic subspace of Inaba's algebraic space [8]. Since we need a moduli theory not only for stable objects but also for semistable objects, the resulting space should not be an algebraic space in general, but an Artin stack. So instead of working with Inaba's algebraic space, we use Lieblich's algebraic stack of objects $E \in D^b(X)$, satisfying the condition,

$$\operatorname{Ext}^i_X(E, E) = 0, \qquad \text{for all } i < 0, \tag{40}$$

which we denote by $\mathfrak{M}$. More precisely, the stack $\mathfrak{M}$ is defined by the 2-functor,

$$\mathfrak{M} \colon (\mathrm{Sch}/\mathbb{C}) \longrightarrow (\mathrm{groupoid}),$$

which takes a $\mathbb{C}$-scheme S to the groupoid $\mathfrak{M}(S)$, whose objects consist of relatively perfect object $\mathcal{E} \in D^b(X \times S)$ such that $\mathcal{E}_s$ satisfies (40) for any closed point $s \in S$. Lieblich [18] shows the following.

Theorem 3.15. ([18]) *The 2-functor $\mathfrak{M}$ is an Artin stack locally of finite type.*

For $v \in C(\mathcal{A}^p_{1/2})$ as in (34), we consider a moduli problem of μ-limit semistable objects of numerical type $v' \in C_{\le v}(\mathcal{A}^p_{1/2})$, where $C_{\le v}(\mathcal{A}^p_{1/2})$ is given in (26). First we show the following.

Lemma 3.16. *For any $v' \in C_{\le v}(\mathcal{A}^p_{1/2})$, we have one of the following.*

- *There is $\beta' \in C(X)$ and $n' \in \mathbb{Z}$ such that* $\mathrm{ch}(v') = (-1, 0, \beta', n')$.
- *We have* $\mathrm{ch}(v') = (-1, 0, 0, 0)$.
- *There is $\beta' \in C(X)$ and $n' \in \mathbb{Z}$ such that* $\mathrm{ch}(v') = (0, 0, \beta', n')$.

Proof. For $v' \in C_{\le v}(\mathcal{A}^p_{1/2})$, let $v'' = v - v' \in C(\mathcal{A}^p_{1/2})$. Since $\mathrm{ch}_0(v) = -1$, we have $\mathrm{ch}_0(v') = 0$ or $\mathrm{ch}_0(v') = -1$. Suppose that $\mathrm{ch}_0(v') = 0$ and take $E \in \mathcal{A}^p_{1/2}$ with $[E] = v'$. Then $\mathcal{H}^{-1}(E)$ must be torsion, and hence $\mathcal{H}^{-1}(E) = 0$ since $E \in \mathcal{A}^p_{1/2}$. Therefore E is a non-zero one dimensional sheaf, thus $\mathrm{ch}(v') = (0, 0, \beta', n')$ for some $\beta' \in C(X)$ and $n' \in \mathbb{Z}$.

In the latter case, we have $\mathrm{ch}_0(v'') = 0$ thus $\mathrm{ch}_0(v'') = (0, 0, \beta'', n'')$ for some $\beta'' \in C(X)$ and $n'' \in \mathbb{Z}$. Therefore $\mathrm{ch}_0(v') = (-1, 0, \beta', n')$ for some $\beta' \in \overline{C}(X)$ and $n' \in \mathbb{Z}$. If $\beta' = 0$, then $\mathcal{H}^{-1}(E)$ is a line bundle and $\mathcal{H}^0(E)$ is a zero dimensional sheaf, by [24, Lemma 3.2]. Thus E is isomorphic to a direct sum of $\mathcal{H}^{-1}(E)[1]$ and $\mathcal{H}^0(E)$, which contradicts to $E \in \mathcal{A}^p_{1/2}$ unless $n' = \dim \mathcal{H}^0(E) = 0$. Q.E.D.

For $\sigma = B + i\omega \in A(X)_{\mathbb{C}}$ and $v' \in C_{\le v}(\mathcal{A}^p_{1/2})$, let us consider the following (abstract) stacks,

$$\mathfrak{M}^{v'}(Z_{\mu_\sigma}) \subset \mathfrak{Obj}(\mathcal{A}^p) \subset \mathfrak{M},$$

where $\mathfrak{Obj}(\mathcal{A}^p)$ is the stack of objects $E \in \mathcal{A}^p$, and $\mathfrak{M}^{v'}(Z_{\mu_\sigma})$ is the stack of μ_σ-limit semistable objects $E \in \mathcal{A}^p_{1/2}$ of numerical type v'. We have the following.

Proposition 3.17. *The substacks $\mathfrak{Obj}(\mathcal{A}^p)$ and $\mathfrak{M}^{v'}(Z_{\mu_\sigma})$ are open substacks of $\mathfrak{M}$, and hence they are Artin stacks locally of finite type. Moreover $\mathfrak{M}^{v'}(Z_{\mu_\sigma})$ is of finite type.*

Proof. The openness of $\mathfrak{Obj}(\mathcal{A}^p) \subset \mathfrak{M}$ follows from [24, Lemma 3.14]. Let us take $v' \in C_{\le v}(\mathcal{A}^p_{1/2})$. Suppose first that $\mathrm{ch}(v') = (0, 0, \beta', n')$ for $\beta' \in C(X)$ and $n' \in \mathbb{Z}$. Then any μ_σ-limit semistable object of numerical type v' is a (B, ω)-twisted semistable sheaf. (cf. Remark 3.11.) Then it is well-known that $\mathfrak{M}^{v'}(Z_{\mu_\sigma})$ is open in $\mathfrak{M}$, and it is of finite type. (cf. [25, Proposition 3.9].)

Next suppose that $\mathrm{ch}(v') = (-1, 0, \beta', n')$. In this case the claim for $\mathfrak{M}^{v'}(Z_{\mu_\sigma})$ follows from the straightforward adaptation of the argument of [24, Section 3]. In fact using Lemma 3.13, we can show the boundedness of μ_σ-limit semistable objects of numerical type v', and destabilizing objects in a family of objects in $\mathcal{A}^p_{1/2}$, along with the same arguments of [24, Proposition 3.13, Lemma 3.15]. Then the same proof as in [24, Theorem 3.20] works to show the openness of $\mathfrak{M}^{v'}(Z_{\mu_\sigma}) \subset \mathfrak{M}$. The boundedness of relevant μ_σ-limit semistable objects implies that $\mathfrak{M}^{v'}(Z_{\mu_\sigma})$ is of finite type. Q.E.D.

3.5. Stable pairs and μ-limit semistable objects

The notion of stable pairs and their counting invariants are introduced by Pandharipande and Thomas [20] to interpret the reduced Donaldson–Thomas theory geometrically. In [24, Section 4], the relationship between PT-invariants and counting invariants of limit stable objects are discussed. In this subsection we state the similar result for μ-limit semistable objects. Since the proofs are straightforward adaptation of the arguments in [24, Section 4], we again leave the readers to check the detail. First let us recall the definition of stable pairs.

Definition 3.18. A *stable pair* on a Calabi–Yau 3-fold X is data (F, s), where F is a pure one dimensional sheaf on X, and s is a morphism

$$s\colon \mathcal{O}_X \longrightarrow F,$$

whose cokernel is a zero dimensional sheaf.

To simplify the notation, we also include the pair $(F = 0, s = 0)$ in the definition of stable pairs. For a stable pair (F, s), we have the associated two term complex,

$$I^\bullet = (\mathcal{O}_X \stackrel{s}{\longrightarrow} F) \in D^b(X), \tag{41}$$

where F is located in degree zero. Note that the object $I^\bullet$ satisfies

$$I^\bullet \in \mathcal{A}^p_{1/2}, \quad \det I^\bullet = \mathcal{O}_X,$$
$$\mathrm{ch}(I^\bullet) = (-1, 0, \beta, n),$$

for $\beta = \mathrm{ch}_2(F)$, $n = \mathrm{ch}_3(F)$. By abuse of notation, we also call an object (41) as a stable pair. In [20], the moduli space of stable pairs is constructed as a projective variety, and denoted by $P_n(X, \beta)$,

$$P_n(X, \beta) := \{(F, s) \mid (F, s) \text{ is a stable pair, } (\mathrm{ch}_2(F), \mathrm{ch}_3(F)) = (\beta, n)\}.$$

Let $\mathfrak{Obj}_0(\mathcal{A}^p)$ be the closed fiber at the point $[\mathcal{O}_X] \in \mathrm{Pic}(X)$ of the following morphism,

$$\det\colon \mathfrak{Obj}(\mathcal{A}^p) \ni E \longmapsto \det E \in \mathrm{Pic}(X).$$

Definition 3.19. For $\beta \in \overline{C}(X)$, $n \in \mathbb{Z}$, and $\sigma \in A(X)_{\mathbb{C}}$, define $\mathfrak{L}_n^{\mu_\sigma}(X,\beta)$ to be

$$\mathfrak{L}_n^{\mu_\sigma}(X,\beta) := \mathfrak{M}^v(Z_{\mu_\sigma}) \cap \mathfrak{Obj}_0(\mathcal{A}^p), \tag{42}$$

where $v \in C(\mathcal{A}^p_{1/2})$ satisfies $\mathrm{ch}(E) = (-1, 0, \beta, n)$.

Note that $\mathfrak{L}_n^{\mu_\sigma}(X,\beta)$ is the moduli stack of μ_σ-limit semistable objects $E \in \mathcal{A}^p$ with $\det E = \mathcal{O}_X$ and $[E] = v$. We shall compare $\mathfrak{L}_n^{\mu_\sigma}(X,\beta)$ and $P_n(X,\beta)$, when σ is written as $\sigma = k\omega + i\omega$ with $k \in \mathbb{R}$. For $\beta \in \overline{C}(X)$, we set

$$\begin{aligned} \overline{C}_{\le\beta}(X) &:= \{\beta' \in \overline{C}(X) \mid \beta - \beta' \in \overline{C}(X)\}, \\ C_{\le\beta}(X) &:= \overline{C}_{\le\beta}(X) \setminus \{0\}. \end{aligned} \tag{43}$$

Definition 3.20. For $\beta \in \overline{C}(X)$, we define $m(\beta)$ as follows. If $\beta = 0$, we set $m(\beta) = 0$. Otherwise $m(\beta)$ is

$$m(\beta) := \min\{\mathrm{ch}_3(\mathcal{O}_C) \mid C \subset X \text{ satisfies } \dim C = 1, [C] \in \overline{C}_{\le\beta}(X)\}.$$

It is well-known that $\overline{C}_{\le\beta}(X)$ is a finite set and $m(\beta) > -\infty$, whose proofs are seen in [24, Lemma 3.9, Lemma 3.10]. Thus Definition 3.20 makes sense. For $\beta \in C(X)$ and $n \in \mathbb{Z}$, we define $\mu_{n,\beta} \in \mathbb{Q}$ to be

$$\mu_{n,\beta} := \max\left\{\frac{n - m(\beta - \beta')}{\omega\beta'} : \beta' \in C_{\le\beta}(X)\right\}. \tag{44}$$

The following is μ-stability version of [24, Theorem 4.7].

Theorem 3.21. *Let $\sigma = k\omega + i\omega$ for $k \in \mathbb{R}$. We have*

$$\mathfrak{L}_n^{\mu_\sigma}(X,\beta) \cong [P_n(X,\beta)/\mathbb{G}_m], \quad \text{if } k < -\mu_{n,\beta}/2, \tag{45}$$

$$\mathfrak{L}_n^{\mu_\sigma}(X,\beta) \cong [P_{-n}(X,\beta)/\mathbb{G}_m], \quad \text{if } k > \mu_{-n,\beta}/2. \tag{46}$$

Here $\mathbb{G}_m$ is acting on $P_{\pm n}(X,\beta)$ trivially.

Proof. The same proof as in [24, Theorem 4.7] shows that, if $k < -\mu_{n,\beta}/2$, then $E \in \mathcal{A}^p_{1/2}$ is μ_σ-limit semistable if and only if E is isomorphic to a stable pair (41). Note that [24, Lemma 4.6] is crucial in [24, Theorem 4.7], and in our case Proposition 3.13 and (38) are applied instead of [24, Lemma 4.6]. Thus the $\mathbb{C}$-valued points of $\mathfrak{L}_n^{\mu_\sigma}(X,\beta)$ and $P_n(X,\beta)$ are identified.

The existence of a universal stable pair on $X \times P_n(X,\beta)$ (cf. [20, Section 2]) yields a 1-morphism $P_n(X,\beta) \to \mathcal{L}_n^{\mu_\sigma}(X,\beta)$, which descends to

$$[P_n(X,\beta)/\mathbb{G}_m] \longrightarrow \mathcal{L}_n^{\mu_\sigma}(X,\beta). \tag{47}$$

Since any $E \in P_n(X,\beta)$ is τ-limit stable for some τ by [24, Theorem 4.7], we have $\mathrm{Hom}(E,E) = \mathbb{C}$ and $\mathrm{Aut}(E) = \mathbb{G}_m$. Therefore (47) is an equivalence of groupoids on $\mathbb{C}$-valued points. As proved in [20, Theorem 2.7], the stack $[P_n(X,\beta)/\mathbb{G}_m]$ is considered as an open substack of $\mathfrak{Obj}_0(\mathcal{A}^p)$. By Proposition 3.17, $\mathcal{L}_n^{\mu_\sigma}(X,\beta)$ is also open in $\mathfrak{Obj}_0(\mathcal{A}^p)$, and hence (47) gives an isomorphism of Artin stacks. The isomorphism (46) is also similarly proved. Q.E.D.

§4. Generating functions of stable pair invariants

In this section, we combine the arguments in the previous sections to show the rationality of the generating functions of stable pair invariants. As in the previous section, X is a projective Calabi–Yau 3-fold, $\mathcal{A}^p \subset D^b(X)$ is the heart of a perverse t-structure on $D^b(X)$.

4.1. Counting invariants of μ-limit stable objects

In this subsection, we construct counting invariants of μ-limit semistable objects. Take $v \in C(\mathcal{A}^p_{1/2})$ which satisfies (34) with $\beta \in \overline{C}(X)$ and $n \in \mathbb{Z}$. As in subsection 2.3, there is a Ringel–Hall Lie-algebra $\mathfrak{G}(\mathcal{A}^p) \subset \mathcal{H}(\mathcal{A}^p)$ and the elements,

$$\epsilon^{v'}(Z_{\mu_\sigma}) \in \mathfrak{G}^{v'}(\mathcal{A}^p) \subset \mathcal{H}(\mathcal{A}^p),$$

for any $v' \in C_{\le v}(\mathcal{A}^p_{1/2})$ and $\sigma \in A(X)_{\mathbb{C}}$. Here we have used Proposition 3.17 which ensures the existence of $\mathcal{H}(\mathcal{A}^p)$ and $\epsilon^{v'}(Z_{\mu_\sigma})$. We also use the following map on $\mathcal{H}(\mathcal{A}^p)$ to construct the counting invariants,

$$\Xi\colon \mathfrak{G}(\mathcal{A}^p) \ni f \longmapsto f \cdot [\mathfrak{Obj}_0(\mathcal{A}^p) \hookrightarrow \mathfrak{Obj}(\mathcal{A}^p)] \in \mathfrak{G}(\mathcal{A}^p). \tag{48}$$

The product $\cdot$ is given by (9). In the following, we use the notation of subsection 2.5.

Definition 4.1. For $\beta \in \overline{C}(X)$ and $n \in \mathbb{Z}$, we define [1] $P^{eu}_{n,\beta} \in \mathbb{Z}$, $L^{eu}_{n,\beta}(\sigma) \in \mathbb{Q}$ and $N^{eu}_{n,\beta}(\sigma) \in \mathbb{Q}$ to be

$$\begin{aligned} P^{eu}_{n,\beta} &:= e(P_n(X,\beta)), \\ L^{eu}_{n,\beta}(\sigma) &:= \Theta_v \Xi \epsilon^v(Z_{\mu_\sigma}), \quad \text{where } \mathrm{ch}(v) = (-1,0,\beta,n), \\ N^{eu}_{n,\beta}(\sigma) &:= \Theta_{v'} \epsilon^{v'}(Z_{\mu_\sigma}) = J^{v'}(Z_{\mu_\sigma}), \quad \text{where } \mathrm{ch}(v') = (0,0,\beta,n). \end{aligned}$$

Here $e(*)$ is the topological Euler characteristic.

For simplicity we set

$$L^{eu}_{n,\beta} := L^{eu}_{n,\beta}(i\omega), \quad N^{eu}_{n,\beta} := N^{eu}_{n,\beta}(i\omega).$$

Remark 4.2. Suppose that $\sigma = k\omega + i\omega$ for $k \in \mathbb{R}$. Then we have

$$N^{eu}_{n,\beta}(\sigma) = N^{eu}_{n,\beta},$$

by noting Remark 3.11. Also if $k < -\mu_{n,\beta}/2$, then Theorem 3.21 and Remark 2.22 imply

$$L^{eu}_{n,\beta}(\sigma) = P^{eu}_{n,\beta}.$$

Let us recall that for a fixed β, the moduli space $P_n(X,\beta)$ is empty for a sufficiently negative n. (See [20].) So we can take $N(\beta) \in \mathbb{Z}$ such that

$$P^{eu}_{n,\beta} = 0 \quad \text{for } n < N(\beta). \tag{49}$$

In particular the series $P^{eu}_{\beta}(q)$ is a Laurent series of q.

4.2. Generating functions of counting invariants of μ-limit semistable objects

In this subsection, we study the generating functions of the invariants given in Definition 4.1. Below we fix an ample divisor ω on X and only consider the case $\sigma = k\omega + i\omega$ for $k \in \mathbb{R}$. For $\sigma = k\omega + i\omega$, we set $\sigma^{\vee} = -k\omega + i\omega$. We have the following symmetry for the invariants $L^{eu}_{n,\beta}(\sigma)$ and $N^{eu}_{n,\beta}$.

Lemma 4.3. *(i) We have the equalities,*

$$L^{eu}_{n,\beta}(\sigma) = L^{eu}_{-n,\beta}(\sigma^{\vee}), \quad N^{eu}_{n,\beta} = N^{eu}_{-n,\beta}. \tag{50}$$

(ii) For $d := \omega \cdot \beta$, we have $N^{eu}_{n+d,\beta} = N^{eu}_{n,\beta}$ for any $n \in \mathbb{Z}$.

---

[1]The subscript $*^{eu}$ means "Euler characteristic " of the moduli spaces.

Proof. (i) Let $\mathbb{D}\colon D^b(X) \to D^b(X)^{\mathrm{op}}$ be the dualizing functor,

$$\mathbb{D}(E) = \mathbf{R}\mathcal{H}om(E, \mathcal{O}_X[2]).$$

The functor $\mathbb{D}$ induces an isomorphism of rings,

$$\mathbb{D}\colon \mathcal{H}(\mathcal{A}^p) \xrightarrow{\sim} \mathcal{H}(\mathbb{D}(\mathcal{A}^p))^{\mathrm{op}}. \tag{51}$$

On the other hand, the functor $\mathbb{D}$ preserves the subcategory $\mathcal{A}^p_{1/2} \subset D^b(X)$ by [24, Lemma 2.18]. Moreover the same argument as in [24, Lemma 2.27] shows that an object $E \in \mathcal{A}^p_{1/2}$ is μ_σ-limit semistable if and only if $\mathbb{D}(E) \in \mathcal{A}^p_{1/2}$ is $\mu_{\sigma^\vee}$-limit semistable. Hence the map (51) takes $\delta^{v'}(Z_{\mu_\sigma})$ to $\delta^{v'^\vee}(Z_{\mu_{\sigma^\vee}})$ for any $v' \in C_{\le v}(\mathcal{A}^p_{1/2})$. Here if v' is given by $\mathrm{ch}(v') = (r, 0, \beta', n')$ for $r = 0$ or -1, then $v'^\vee$ is given by $\mathrm{ch}(v'^\vee) = (r, 0, \beta', -n')$. Hence (50) follows by the definitions of $L^{eu}_{n,\beta}(\sigma)$ and $N^{eu}_{n,\beta}$.

(ii) Let us take an ample line bundle $\mathcal{L} \in \mathrm{Pic}(X)$ with $c_1(\mathcal{L}) = \omega$. The equivalence $\otimes\mathcal{L}\colon \mathcal{A}^p \to \mathcal{A}^p$ induces an isomorphism of algebras,

$$\otimes\mathcal{L}\colon \mathcal{H}(\mathcal{A}^p) \longrightarrow \mathcal{H}(\mathcal{A}^p). \tag{52}$$

On the other hand, it is easy to see that an object $E \in \mathcal{A}^p_{1/2}$ is μ_σ-limit semistable if and only if $E \otimes \mathcal{L}$ is μ_σ-limit semistable. Thus the map (52) takes $\delta^{v'}(Z_{\mu_\sigma})$ to $\delta^{v''}(Z_{\mu_\sigma})$, where $\mathrm{ch}(v') = (0, 0, \beta, n)$ and $\mathrm{ch}(v'') = (0, 0, \beta, n + d)$. Hence we can conclude $N^{eu}_{n+d,\beta} = N^{eu}_{n,\beta}$. Q.E.D.

Next we show the following finiteness result.

Lemma 4.4. *For a fixed $\beta \in \overline{C}(X)$ and $\sigma = k\omega + i\omega$, the set*

$$\{n \in \mathbb{Z} \mid L^{eu}_{n,\beta}(\sigma) \neq 0\} \tag{53}$$

is a finite set.

Proof. If $\beta = 0$, then $L^{eu}_{n,\beta}(\sigma) = 0$ unless $n = 0$ by Lemma 3.16. Suppose that $\beta \neq 0$, and take an integer $N(\beta)$ as in (49). Assume that $L^{eu}_{n,\beta}(\sigma) \neq 0$ for $n < N(\beta)$. Then at least the moduli stack $\mathfrak{L}^\sigma_n(X, \beta)$ is non-empty, and hence we must have $k \ge -\frac{1}{2}\mu_{n,\beta}$ by Theorem 3.21. By the definition of $\mu_{n,\beta}$, there is $\beta' \in C_{\le\beta}(X)$ such that

$$k \ge -\frac{n - m(\beta - \beta')}{2\omega\beta'}.$$

Hence we have either

$$n \ge -2k(\omega\beta') + m(\beta - \beta'), \quad \text{or} \quad n \ge N(\beta).$$

Thus the set (53) is bounded below. Since $L^{eu}_{n,\beta}(\sigma) = L^{eu}_{-n,\beta}(\sigma^\vee)$ by Lemma 4.3, the set (53) is also bounded above. Q.E.D.

Lemma 4.5. *For a fixed $\beta \in \overline{C}(X)$, the generating series*

$$L^{eu}_\beta(q) = \sum_{n\in\mathbb{Z}} L^{eu}_{n,\beta} q^n \tag{54}$$

is a polynomial of $q^{\pm 1}$, and hence a rational function of q, invariant under $q \leftrightarrow 1/q$.

Proof. By Lemma 4.4, the series $L^{eu}_\beta(q)$ is a polynomial of $q^{\pm 1}$. Since

$$L^{eu}_{n,\beta} = L^{eu}_{n,\beta}(i\omega) = L^{eu}_{-n,\beta}(i\omega^\vee) = L^{eu}_{-n,\beta}(i\omega) = L^{eu}_{-n,\beta},$$

by Lemma 4.3, the polynomial $L^{eu}_\beta(q)$ is invariant under $q \leftrightarrow 1/q$. Q.E.D.

Lemma 4.6. *The generating series*

$$N^{eu}_\beta(q) = \sum_{n\geq 0} n N^{eu}_{n,\beta} q^n,$$

is the Laurent expansion of a rational function of q, invariant under $q \leftrightarrow 1/q$.

Proof. Let $d \in \mathbb{Z}$ be as in Lemma 4.3. Applying Lemma 4.3, we have

$$\begin{aligned}
&N^{eu}_\beta(q)\\
&= \sum_{j=0}^{d-1} \sum_{m\geq 0} (dm+j) N^{eu}_{j,\beta} q^{dm+j}\\
&= \frac{1}{2} \sum_{j=0}^{d-1} \sum_{m\geq 0} \left\{ (dm+j) N^{eu}_{j,\beta} q^{dm+j} + (dm+d-j) N^{eu}_{d-j,\beta} q^{dm+d-j} \right\}\\
&= \sum_{j=0}^{d-1} \frac{N^{eu}_{j,\beta}}{2} \sum_{m\geq 0} \left\{ (dm+j) q^{dm+j} + (dm+d-j) q^{dm+d-j} \right\}.
\end{aligned}$$

We can calculate as

$$\begin{aligned}
&\sum_{m\geq 0} \left\{ (dm+j) q^{dm+j} + (dm+d-j) q^{dm+d-j} \right\}\\
&= \frac{(d-j)(q^{j+d} + q^{d-j}) + j(q^j + q^{2d-j})}{(1-q^d)^2}.
\end{aligned} \tag{55}$$

Then the assertion follows since (55) is a rational function of q, invariant under $q \leftrightarrow 1/q$. Q.E.D.

For a fixed $\beta \in \overline{C}(X)$, we consider the following generating series,

$$P_{\beta}^{eu}(q) = \sum_{n \in \mathbb{Z}} P_{n,\beta}^{eu} q^n.$$

Now we state our main theorem in this paper.

Theorem 4.7. *We have the following equality of the generating series,*

$$\sum_{\beta \geq 0} P_{\beta}^{eu}(q) v^{\beta} = \left(\sum_{\beta \geq 0} L_{\beta}^{eu}(q) v^{\beta} \right) \cdot \exp \left(\sum_{\beta > 0} N_{\beta}^{eu}(q) v^{\beta} \right). \tag{56}$$

Here $\beta \geq 0$, $\beta > 0$ mean $\beta \in \overline{C}(X)$, $\beta \in C(X)$ respectively. Combining Theorem 4.7 with Lemma 4.5 and Lemma 4.6, we obtain the following.

Corollary 4.8. *The generating series $P_{\beta}^{eu}(q)$ is the Laurent expansion of a rational function of q, invariant under $q \leftrightarrow 1/q$.*

The proof of Theorem 4.7 will be given in subsection 4.5 below.

4.3. Transformation of the invariants $L_{n,\beta}^{eu}(\sigma)$

In this subsection, we investigate the transformation formula of our invariants $L_{n,\beta}^{eu}(\sigma)$ under change of $\sigma = k\omega + i\omega$. For $\beta \in C(X)$, we set $S(\beta) \subset \mathbb{R}$ as

$$S(\beta) := \left\{ \frac{m}{2\omega\beta'} \mid \beta' \in C_{\leq \beta}(X), m \in \mathbb{Z} \right\} \subset \mathbb{R}.$$

Note that $S(\beta)$ is a discrete subset in $\mathbb{R}$ because $C_{\leq \beta}(X)$ is a finite set. For $k_0 \in S(\beta)$, let $\mathcal{C}_{\pm} \subset \mathbb{R} \setminus S(\beta)$ be the connected components such that $\mathcal{C}_{-} \subset \mathbb{R}_{<k_0}$ and $\mathcal{C}_{+} \subset \mathbb{R}_{>k_0}$. We take $k_{\pm} \in \mathcal{C}_{\pm}$ and set $\sigma_{*} = k_{*}\omega + i\omega$ for $* = \pm, 0$. We have the following.

Lemma 4.9. *Take $v \in C(\mathcal{A}_{1/2}^{p})$ as in (34) and $0 \in T = \mathbb{R}(m)$. The weak stability condition $Z_{\mu_{\sigma_0}}$ dominates $Z_{\mu_{\sigma_{\pm}}}$ with respect to $(v, 0)$.*

Proof. Take $v_1, v_2 \in C_{\leq v}(\mathcal{A}_{1/2}^{p})$, and suppose that $Z_{\mu_{\sigma_{\pm}}}(v_1) \preceq Z_{\mu_{\sigma_{\pm}}}(v_2)$. We want to show that

$$Z_{\mu_{\sigma_0}}(v_1) \preceq Z_{\mu_{\sigma_0}}(v_2). \tag{57}$$

By Lemma 3.16, we have $\mathrm{ch}(v_i) = (r_i, 0, \beta_i, n_i)$ with $r_i = 0$ or -1. If $r_1 = r_2 = -1$, it is easy to see that $Z_{\mu_\sigma}(v_1) = Z_{\mu_\sigma}(v_2)$ for any σ. If $r_1 = r_2 = 0$, (57) follows easily from (39). If $r_1 \neq r_2$, (57) follows from Lemma 3.14. Then the assertion follows by noting Remark 3.12. Q.E.D.

Remark 4.10. Lemma 4.9 is not true for the limit stability. This is the reason why we use μ-limit stability rather than the limit stability.

For simplicity, we fix $k \in \mathbb{R}_{<0}$ and write

$$Z = Z_{\mu_\sigma} \text{ for } \sigma = k\omega + i\omega \text{ with } k < 0, \quad Z' = Z_{\mu_{i\omega}}. \tag{58}$$

Applying the results in the previous sections, we obtain the following proposition.

Proposition 4.11. *For $v \in C(\mathcal{A}^p_{1/2})$ as in (34), and Z, Z' as in (58) w.r.t. $(v, 0)$, we have the following.*

$$\epsilon^v(Z') = \sum_{\substack{l \geq 1, \\ v_i \in C(\mathcal{A}^p_{1/2}), \\ v_1 + \cdots + v_l = v}} U(\{v_1, \ldots, v_l\}, Z, Z')\epsilon^{v_1}(Z) * \cdots * \epsilon^{v_l}(Z). \tag{59}$$

The sum (59) has only finitely many non-zero terms.

Proof. By Proposition 3.17 and Remark 3.12, Z, Z' satisfy Assumption 2.26 w.r.t. $(v, 0)$. Furthermore by Lemma 4.9, the condition ($\spadesuit'$) is also satisfied with respect to $(v, 0)$, except the local finiteness condition which is satisfied if we knew that (59) has only finitely many non-zero terms. The finiteness of (59) will be shown in Proposition 4.14 (iii) below. Therefore (59) follows from Theorem 2.28. Q.E.D.

In the formula (59), $\{v_i\}_{i=1}^l \subset C(\mathcal{A}^p_{1/2})$ satisfy the following by Lemma 3.16.

(60) There is a unique $1 \leq e \leq l$ such that $\mathrm{ch}(v_i) = (0, 0, \beta_i, n_i)$ for $i \neq e$ with $\beta_i \in C(X), n_i \in \mathbb{Z}$, and $\mathrm{ch}(v_e) = (-1, 0, \beta_e, n_e)$ with $\beta_e \in \overline{C}(X), n_e \in \mathbb{Z}$.

Let us see that (59) is a finite sum. For simplicity we write

$$\mu_i := \mu_{i\omega}(v_i) = \frac{n_i}{\beta_i \omega} \in \mathbb{Q},$$

if $v_i \in C(\mathcal{A}^p_{1/2})$ satisfies $\mathrm{ch}(v_i) = (0, 0, \beta_i, n_i)$.

Lemma 4.12. *Take $v_1, \dots, v_l \in C(\mathcal{A}^p_{1/2})$ such that*

$$\mathrm{ch}(v_i) = (0, 0, \beta_i, n_i),$$

for all i with $\beta_i \in C(X)$ and $n_i \in \mathbb{Z}$. For Z, Z' as in (58), we have

$$S(\{v_1, \dots, v_l\}, Z, Z') = \begin{cases} 1, & l = 1, \\ 0, & l \geq 2. \end{cases}$$

Proof. Note that we have

$$Z(v_i) \preceq Z(v_j) \quad \Leftrightarrow \quad \mu_i \leq \mu_j \quad \Leftrightarrow \quad Z'(v_i) \preceq Z'(v_j). \tag{61}$$

Then one can check that the same proof as in [9, Theorem 4.5] is applied. Q.E.D.

Lemma 4.13. *Take $v_1, \dots, v_l \in C(\mathcal{A}^p_{1/2})$ as in (60). For Z, Z' as in (58), $S(\{v_1, \dots, v_l\}, Z, Z')$ is non-zero only if*

$$0 < \mu_1 \leq \mu_2 \leq \cdots \leq \mu_{e-1} \leq -2k > \mu_{e+1} > \mu_{e+2} > \cdots > \mu_l \geq 0. \tag{62}$$

Moreover in this case, we have $S(\{v_1, \dots, v_l\}, Z, Z') = (-1)^{e-1}$.

Proof. Note that for v_i, v_j with $i, j \neq e$, the same implications (61) hold. Also for $i \neq e$, we have

$$\begin{aligned} Z(v_i) \preceq Z(v_e) \quad &\Leftrightarrow \quad \mu_i \leq -2k, \\ Z'(v_i) \preceq Z'(v_e) \quad &\Leftrightarrow \quad \mu_i \leq 0. \end{aligned}$$

Suppose that $S(\{v_1, \dots, v_l\}, Z, Z') \neq 0$. We say $1 < i < l$ is of type A (resp. B) if the following holds,

$$Z(v_{i-1}) \succ Z(v_i) \preceq Z(v_{i+1}), \quad (\text{resp. } Z(v_{i-1}) \preceq Z(v_i) \succ Z(v_{i+1})).$$

If $1 < i \leq e - 1$ is of type A, we have

$$\begin{aligned} Z'(v_1 + \cdots + v_{i-1}) &\preceq Z'(v_i + \cdots + v_l), \\ Z'(v_1 + \cdots + v_i) &\succ Z'(v_{i+1} + \cdots + v_l). \end{aligned}$$

Hence we have

$$\begin{aligned} \mu_{i\omega}(v_1 + \cdots + v_{i-1}) &\leq 0, \\ \mu_{i\omega}(v_1 + \cdots + v_{i-1} + v_i) &> 0, \end{aligned}$$

which implies $\mu_i > 0$. Similarly if i is of type B, we have $\mu_i < 0$.

Suppose that there is $1 \le i \le e-1$ of type A or B, and take the smallest such i. We assume i is of type A, and hence $\mu_i > 0$. We have

$$Z(v_1) \succ \cdots \succ Z(v_{i-1}) \succ Z(v_i) \preceq Z(v_{i+1}) \cdots,$$

thus $\mu_1 > \cdots > \mu_i > 0$ holds. On the other hand, we have $Z'(v_1) \preceq Z'(v_2 + \cdots + v_l)$, thus $\mu_1 \le 0$. This is a contradiction, so there is no $1 \le i \le e-1$ of type A. Similarly there is no $1 \le i \le e-1$ of type B.

By the above argument, one of (63) or (64) holds.

$$Z(v_1) \succ Z(v_2) \succ \cdots \succ Z(v_{e-1}) \succ Z(v_e), \tag{63}$$

$$Z(v_1) \preceq Z(v_2) \preceq \cdots \preceq Z(v_{e-1}) \preceq Z(v_e). \tag{64}$$

Assume by a contradiction that (63) holds. Then $Z'(v_1) \preceq Z'(v_2 + \cdots + v_l)$, thus (63) implies

$$0 \ge \mu_1 > \mu_2 > \cdots > \mu_{e-1} > -2k. \tag{65}$$

The inequality (65) does not occur since we took $k < 0$. Therefore we must have (64).

A similar argument for $v_{e+1}, \ldots, v_l$ shows

$$Z(v_1) \preceq \cdots \preceq Z(v_e) \succ Z(v_{e+1}) \succ \cdots \succ Z(v_l). \tag{66}$$

Obviously (66) together with $S(\{v_1, \ldots, v_l\}, Z, Z') \ne 0$ imply (62).

Q.E.D.

Proposition 4.14. *(i) Take $v_1, \ldots, v_l \in C(\mathcal{A}^p_{1/2})$ as in (60), and let Z, Z' be as in (58). Then $U(\{v_1, \ldots, v_l\}, Z, Z')$ is non-zero only if*

$$0 \le \mu_1 \le \mu_2 \le \cdots \le \mu_{e-1} \le -2k \ge \mu_{e+1} \ge \cdots \ge \mu_l \ge 0. \tag{67}$$

(ii) In the same situation of (i), suppose that the following holds,

$$U(\{v_1, \ldots, v_l\}, Z, Z') \prod_{i \ne e} n_i \ne 0.$$

Then we have

$$U(\{v_1, \ldots, v_l\}, Z, Z') = \sum_{1 \le m \le l} \sum_{\substack{\text{surjective } \psi \colon \{1,\ldots,l\} \to \{1,\ldots,m\}, \\ i \le j \text{ implies } \psi(i) \le \psi(j), \text{ and satisfies (69)}}} (-1)^{\psi(e)-1} \prod_{b=1}^{m} \frac{1}{|\psi^{-1}(b)|!}. \tag{68}$$

Here ψ satisfies the following.

$$\begin{aligned}&\text{For } i,j<e \text{ with } \psi(i)=\psi(j), \text{ we have } \mu_i=\mu_j,\\ &\text{if } \psi(i)=\psi(e) \text{ then } \mu_i=-2k, \text{ and for } e<i,j,\\ &\text{we have } \psi(i)=\psi(j) \text{ if and only if } \mu_i=\mu_j.\end{aligned} \tag{69}$$

(iii) The formula (59) is a finite sum.

Proof. (i) Suppose that $U(\{v_1,\dots,v_l\},Z,Z')\neq 0$ and take

$$\psi\colon\{1,\dots,l\}\to\{1,\dots,m\},\quad \xi\colon\{1,\dots,m\}\to\{1,\dots,m'\}, \tag{70}$$

as in (21). Let us set $c=\xi\psi(e)\in\{1,\dots,m'\}$ and take $a\in\{1,\dots,m'\}$ with $a\neq c$. By Lemma 4.12, the set $\xi^{-1}(a)$ consists of one element, say $b\in\{1,\dots,m\}$. Then by the definition of $U(\{v_1,\dots,v_l\},Z,Z')$, we have $\mu_i=\mu_j$ for $i,j\in\psi^{-1}(b)$ and

$$Z'(\sum_{i\in\psi^{-1}(b)}v_i)=Z'(\sum_{i\in\psi^{-1}\xi^{-1}(c)}v_i). \tag{71}$$

The condition (71) implies $\mu_{i\omega}(\sum_{i\in\psi^{-1}(b)}v_i)=0$, and hence $\mu_i=0$ for any $i\in\psi^{-1}(b)$, i.e. $\mu_i=0$ for any $i\notin\psi^{-1}\xi^{-1}(c)$. By Lemma 4.13, we must have (67).

(ii) Suppose that

$$U(\{v_1,\dots,v_l\},Z,Z')\prod_{i\neq e}n_i\neq 0, \tag{72}$$

and take $\psi\colon\{1,\dots,l\}\to\{1,\dots,m\}$, $\xi\colon\{1,\dots,m\}\to\{1,\dots,m'\}$ as in (i). Then the proof of (i) shows that (72) is non-zero only if $m'=1$. Then (68) follows from the definition of $U(\{v_1,\dots,v_l\},Z,Z')$ and Lemma 4.13.

(iii) Since $v_1,\dots,v_l$ satisfy (60), the number l is bounded, and there is only finite number of possibilities for $\beta_i=\mathrm{ch}_2(v_i)$. Hence we may fix l and $\beta_1,\dots,\beta_l$. Then the values $n_i=\mathrm{ch}_3(v_i)$ have only finite number of possibilities by (67). Q.E.D.

Now we have the wall-crossing formula of the invariants $L^{eu}_{n,\beta}(\sigma)$.

Proposition 4.15. *For $\sigma = k\omega + i\omega$ with $k < 0$, $\beta \in \overline{C}(X)$ and $n \in \mathbb{Z}$, we have the following formula,*

$$L^{eu}_{n,\beta} = \sum_{\substack{l\ge 1,\ 1\le e\le l,\ \beta_i\in C(X) \text{ for } i\neq e,\ \beta_e\in\overline{C}(X),\ n_i\in\mathbb{Z},\\ \beta_1+\cdots+\beta_l=\beta,\ n_1+\cdots+n_l=n,\ \mu_i=n_i/\beta_i\omega \text{ satisfy}\\ 0<\mu_1\le\mu_2\le\cdots\le\mu_{e-1}\le -2k\ge\mu_{e+1}\ge\cdots\ge\mu_l>0}} \sum_{\substack{1\le m\le l,\ \text{surjective}\\ \psi\colon\{1,\ldots,l\}\to\{1,\ldots,m\},\\ i\le j \text{ implies } \psi(i)\le\psi(j),\\ \text{and satisfies (69)}}}$$

$$\left(-\frac{1}{2}\right)^{l-1}\prod_{b=1}^{m}\frac{(-1)^{\psi(e)-e}}{|\psi^{-1}(b)|!}\prod_{i\neq e} n_i N^{eu}_{n_i,\beta_i} L^{eu}_{n_e,\beta_e}(\sigma). \tag{73}$$

Proof. Let $v \in C(\mathcal{A}^p_{1/2})$ be as in (34), and Z, Z' be as in (58). Applying Ξ given in (48) to (59), we obtain

$$\Xi\epsilon^v(Z') = \sum_{l\ge 1,\ 1\le e\le l} \sum_{\substack{v_i\in C(\mathcal{A}^p_{1/2}),\ v_1+\cdots+v_l=v,\\ \mathrm{ch}_0(v_i)=0 \text{ for } i\neq e,\ \mathrm{ch}_0(v_e)=-1}} U(\{v_1,\ldots,v_l\},Z,Z')$$

$$\epsilon^{v_1}(Z) * \cdots * \Xi\epsilon^{v_e}(Z) * \cdots * \epsilon^{v_l}(Z). \tag{74}$$

Note that for $i \neq e$, the element $\epsilon^{v_i}(Z) \in \mathcal{H}(\mathcal{A}^p)$ is supported on $\mathfrak{Obj}_0(\mathcal{A}^p) \subset \mathfrak{Obj}(\mathcal{A}^p)$, and hence $\Xi(\epsilon^{v_i}(Z) * \epsilon) = \epsilon^{v_i}(Z) * \Xi(\epsilon)$ follows for any $\epsilon \in \mathcal{H}(\mathcal{A}^p)$. Thus (74) follows from (59). Also we note that (74) is a finite sum by Proposition 4.14 (iii). Hence applying Θ given in (24) and using the same argument as in Theorem 2.24, we obtain

$$L^{eu}_{n,\beta} = \sum_{\substack{l\ge 1,\ v_i\in C(\mathcal{A}^p_{1/2}),\\ v_1+\cdots+v_l=v}} \sum_{\substack{\Gamma \text{ is a connected, simply connected oriented}\\ \text{graph with vertex } \{1,\ldots,l\},\ \overset{i}{\bullet}\to\overset{j}{\bullet} \text{ implies } i<j}}$$

$$U(\{v_1,\ldots,v_l\},Z,Z')\frac{1}{2^{l-1}}\prod_{\overset{i}{\bullet}\to\overset{j}{\bullet} \text{ in } \Gamma}\chi(v_i,v_j)\prod_{i\neq e} N^{eu}_{n_i,\beta_i} L^{eu}_{n_e,\beta_e}(\sigma), \tag{75}$$

where $v_1,\ldots,v_l$ satisfy (60) and we have used the notation of (60). By Riemann–Roch theorem, we have $\chi(v_i,v_j)=0$ for $i,j\neq e$, $\chi(v_i,v_e)=n_i$ and $\chi(v_e,v_i)=-n_i$. Hence a term in the sum (75) is non-zero only if the oriented graph Γ is of the following form,

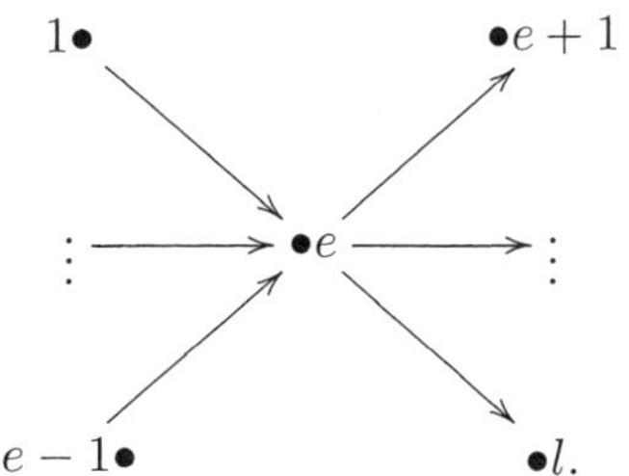

Hence applying Proposition 4.14 (ii), we obtain the formula (73).

Q.E.D.

4.4. Relationship between $L^{eu}_{n,\beta}$ and $P^{eu}_{n,\beta}$

We next establish a relationship between $L^{eu}_{n,\beta}$ and $P^{eu}_{n,\beta}$. Let us take $N(\beta) \in \mathbb{Z}$ as in (49). We choose $k < 0$ so that

$$k < -\frac{1}{2}(n - N(\beta')), \quad k < -\frac{1}{2}\mu_{n,\beta'} \quad \text{for any } \beta' \in C_{\le \beta}(X). \tag{76}$$

In this particular choice of k, we have the following formula.

Proposition 4.16. *If k satisfies (76), then (73) implies the following.*

$$L^{eu}_{n,\beta} = \sum_{\substack{l\ge 1,\ 1\le e\le l,\ 0\le t\le e-1,\ 0\le s\le l-e, \\ 0=m_0<m_1<\cdots<m_t=e-1, \\ e=m'_0<m'_1<\cdots<m'_s=l}} \sum_{\substack{\beta_i\in C(X) \text{ for } i\neq e,\ \beta_e\in\overline{C}(X),\ n_i\in\mathbb{Z}, \\ \beta_1+\cdots+\beta_l=\beta,\ n_1+\cdots+n_l=n,\ \mu_i=n_i/\omega\beta_i \\ \text{satisfy } 0<\mu_1=\cdots=\mu_{m_1}<\mu_{m_1+1}=\cdots, \\ 0<\mu_l=\cdots=\mu_{m'_{s-1}+1}<\mu_{m'_{s-1}}=\cdots.}}$$

$$\left(-\frac{1}{2}\right)^{l-1} \prod_{i=1}^{t} \frac{1}{(m_i - m_{i-1})!} \prod_{i=1}^{s} \frac{1}{(m'_i - m'_{i-1})!} \prod_{i\neq e} n_i N^{eu}_{n_i,\beta_i} P^{eu}_{n_e,\beta_e}. \tag{77}$$

Proof. First note that all the n_i in the formula (73) are positive except $i = e$. Thus we have $n_e \le n$, and hence

$$k < -\mu_{n,\beta_e}/2 \le -\mu_{n_e,\beta_e}/2.$$

Therefore in the formula (73), we have

$$L^{eu}_{n_e,\beta_e}(\sigma) = P^{eu}_{n_e,\beta_e},$$

by Remark 4.2. Thus we may assume $n_e \ge N(\beta_e)$ in the formula (73). Then the condition $\mu_i \le -2k$ in (73) is automatically satisfied, since

$$0 < \mu_i \le n_i \le n - n_e \le n - N(\beta_e) < -2k,$$

by our choice of k. Hence we can eliminate the condition $\mu_i \le -2k$ in (73), and obtain the formula,

$$L^{eu}_{n,\beta} = \sum_{\substack{l\ge 1,\ 1\le e\le l,\ \beta_i\in C(X) \text{ for } i\neq e,\ \beta_e\in\overline{C}(X),\ n_i\in\mathbb{Z}, \\ \beta_1+\cdots+\beta_l=\beta,\ n_1+\cdots+n_l=n,\ \mu_i=n_i/\beta_i\omega \text{ satisfy} \\ 0<\mu_1\le\mu_2\le\cdots\le\mu_{e-1},\ 0<\mu_l\le\mu_{l-1}\le\cdots\le\mu_{e+1}}} \sum_{\substack{1\le m\le l, \text{ surjective} \\ \psi\colon \{1,\dots,l\}\to\{1,\dots,m\}, \\ i\le j \text{ implies } \psi(i)\le\psi(j), \\ \text{and satisfies (69)}}}$$

$$\left(-\frac{1}{2}\right)^{l-1} \prod_{b=1}^{m} \frac{(-1)^{\psi(e)-e}}{|\psi^{-1}(b)|!} \prod_{i\neq e} n_i N^{eu}_{n_i,\beta_i} P^{eu}_{n_e,\beta_e}. \tag{78}$$

We rearrange the sum (78) by first choosing partitions $0 = m_0 < m_1 < \cdots < m_l = e-1$, $e = m'_0 < m'_1 < \cdots < m'_s = l$ and then choosing β_i, n_i so that $0 < \mu_1 = \cdots = \mu_{m_1} < \mu_{m_1+1} = \cdots$ and $0 < \mu_l = \cdots = \mu_{m'_{s-1}+1} < \mu_{m'_{s-1}} = \cdots$ are satisfied. Noting that ψ satisfies (69), we obtain

$$L^{eu}_{n,\beta} = \sum_{\substack{l\geq 1,\ 1\leq e\leq l,\\ 0\leq t\leq e-1,\ 0\leq s\leq l-e,\\ 0=m_0<m_1<\cdots<m_t=e-1,\\ e=m'_0<m'_1<\cdots<m'_s=l}} \sum_{\substack{\beta_i\in C(X)\text{ for } i\neq e,\ \beta_e\in\overline{C}(X),\ n_i\in\mathbb{Z},\\ \beta_1+\cdots+\beta_l=\beta,\ n_1+\cdots+n_l=n,\ \mu_i=n_i/\omega\beta_i\\ \text{satisfy } 0<\mu_1=\cdots=\mu_{m_1}<\mu_{m_1+1}=\cdots,\\ 0<\mu_l=\cdots=\mu_{m'_{s-1}+1}<\mu_{m'_{s-1}}=\cdots.}}$$

$$\left(-\frac{1}{2}\right)^{l-1}\prod_{i=1}^{t}\left(\sum_{\substack{l_i=m_i-m_{i-1},\ 1\leq m\leq l_i,\\ \text{surjective } \psi:\ \{1,\ldots,l_i\}\to\{1,\ldots,m\},\\ i\leq j \text{ implies } \psi(i)\leq\psi(j)}}\prod_{b=1}^{m}\frac{(-1)^{l_i-m}}{|\psi^{-1}(b)|!}\right)$$

$$\prod_{i=1}^{s}\frac{1}{(m'_i-m'_{i-1})!}\prod_{i\neq e} n_i N^{eu}_{n_i,\beta_i} P^{eu}_{n_e,\beta_e}. \tag{79}$$

Then (77) follows from Lemma 4.17 below. Q.E.D.

Lemma 4.17. *For a fixed l, we have*

$$\sum_{\substack{1\leq m\leq l,\ \text{surjective } \psi:\ \{1,\ldots,l\}\to\{1,\ldots,m\},\\ i\leq j \text{ implies } \psi(i)\leq\psi(j)}}\prod_{b=1}^{m}\frac{(-1)^{l-m}}{|\psi^{-1}(b)|!} = \frac{1}{l!}. \tag{80}$$

Proof. The proof is elementary, and this is a special case of [14, Proposition 4.9]. Q.E.D.

4.5. Proof of Theorem 4.7

We finally give a proof of Theorem 4.7.

Proof. For a fixed data $l \geq 1$, $1 \leq e \leq l$, $\beta_i \in C(X)$ $(i \neq e)$ and $\beta_e \in \overline{C}(X)$, we set

$$F_e(\beta_1,\ldots,\beta_l) = \sum_{n\in\mathbb{Z}} \sum_{\substack{0\leq t\leq e-1,\ 0\leq s\leq l-e,\\ 0=m_0<m_1<\cdots<m_t=e-1,\\ e=m'_0<m'_1<\cdots<m'_s=l.}} \sum_{\substack{n_i\in\mathbb{Z},\ n_1+\cdots+n_l=n,\\ \mu_i=n_i/\omega\beta_i \text{ satisfy}\\ 0<\mu_1=\cdots=\mu_{m_1}<\mu_{m_1+1}=\cdots,\\ 0<\mu_l=\cdots=\mu_{m'_{s-1}+1}<\mu_{m'_{s-1}}=\cdots.}}$$

$$\prod_{i=1}^{t}\frac{1}{(m_i-m_{i-1})!}\prod_{i=1}^{s}\frac{1}{(m'_i-m'_{i-1})!}\prod_{i\neq e} n_i N^{eu}_{n_i,\beta_i} P^{eu}_{n_e,\beta_e} q^n. \tag{81}$$

Let us fix $\beta \in \overline{C}(X)$. By the formula (77), we have

$$\begin{aligned}
&\sum_{n\in\mathbb{Z}} L^{eu}_{n,\beta} q^n \\
&= \sum_{l\ge 1, 1\le e\le l} \left(-\frac{1}{2}\right)^{l-1} \sum_{\substack{\beta_i \in C(X) \text{ for } i\neq e, \\ \beta_e \in \overline{C}(X), \\ \beta_1+\cdots+\beta_l=\beta}} F_e(\beta_1,\dots,\beta_e,\dots,\beta_l) \\
&= \sum_{l\ge 1, 1\le e\le l} \sum_{\substack{\kappa_1 \colon I_1 \to C(X),\ \kappa_2 \colon I_2\to C(X), \\ \beta_e\in\overline{C}(X), \\ \sum_{i\in I_1}\kappa_1(i)+\sum_{i\in I_2}\kappa_2(i)+\beta_e=\beta}} \sum_{\substack{\lambda_1 \colon \{1,\dots,e-1\}\xrightarrow{\sim} I_1, \\ \lambda_2\colon\{e+1,\dots,l\}\xrightarrow{\sim} I_2}} \\
&\qquad \frac{1}{(e-1)!(l-e)!}\left(-\frac{1}{2}\right)^{l-1} \\
&\qquad F_e(\kappa_1\lambda_1(1),\dots,\kappa_1\lambda_1(e-1),\beta_e,\kappa_2\lambda_2(e+1),\dots,\kappa_2\lambda_2(l)).
\end{aligned} \tag{82}$$

Here I_1 and I_2 are finite sets with $|I_1| = e-1$, $|I_2| = l-e$. Let us fix data $l \ge 1$, $1 \le e \le l$, $\kappa_1 \colon I_1 \to C(X)$, $\kappa_2 \colon I_2 \to C(X)$ and $\beta_e \in \overline{C}(X)$, and consider the last sum of (82). If we also fix bijections $\lambda_1' \colon \{1,\dots,e-1\} \to I_1$ and $\lambda_2' \colon \{e+1,\dots,l\} \to I_2$, then the choices of λ_1, λ_2 in (82) correspond to the elements of the symmetric groups $\gamma \in \mathfrak{S}_{e-1}$, $\gamma' \in \mathfrak{S}_{l-e}$ respectively. Here an element $\gamma' \in \mathfrak{S}_{l-e}$ is regarded as a permutation on $\{e+1,\dots,l\}$. Let us rewrite $\beta_i = \kappa_1\lambda_1'(i)$ for $1 \le i \le e-1$ and $\beta_i = \kappa_2\lambda_2'(i)$ for $e+1 \le i \le l$. Then we have

$$\begin{aligned}
&\sum_{\substack{\lambda_1 \colon \{1,\dots,e-1\}\xrightarrow{\sim} I_1, \\ \lambda_2\colon\{e+1,\dots,l\}\xrightarrow{\sim} I_2}} F_e(\kappa_1\lambda_1(1),\dots,\beta_e,\dots,\kappa_2\lambda_2(l)) \\
&= \sum_{\substack{\gamma\in\mathfrak{S}_{e-1} \\ \gamma'\in\mathfrak{S}_{l-e}}} F_e(\beta_{\gamma(1)},\dots,\beta_{\gamma(e-1)},\beta_e,\beta_{\gamma'(e+1)},\dots,\beta_{\gamma'(l)})
\end{aligned}$$

$$= \sum_{\substack{0\le t\le e-1,\ 0\le s\le l-e, \\ 0=m_0<m_1<\cdots<m_t=e-1 \\ e=m_0'<m_1'\cdots<m_s'=l, \\ \gamma\in\mathfrak{S}_{e-1}\ \gamma'\in\mathfrak{S}_{l-e}}} \prod_{i=1}^{t}\frac{1}{(m_i-m_{i-1})!}\prod_{i=1}^{s}\frac{1}{(m_i'-m_{i-1}')!} G_{\gamma,\gamma'}, \tag{83}$$

where $G_{\gamma,\gamma'}$ is given by

$$G_{\gamma,\gamma'} = \sum_{n\in\mathbb{Z}} \sum_{\substack{n_i\in\mathbb{Z},\ n_1+\cdots+n_l=n,\ \mu_i=n_i/\omega\beta_i \text{ satisfy} \\ 0<\mu_{\gamma(1)}=\cdots=\mu_{\gamma(m_1)}<\mu_{\gamma(m_1+1)}=\cdots, \\ 0<\mu_{\gamma'(l)}=\cdots=\mu_{\gamma'(m'_{s-1}+1)}<\mu_{\gamma'(m'_{s-1})}=\cdots}} \prod_{i\neq e} n_i N^{eu}_{n_i,\beta_i} P^{eu}_{n_e,\beta_e} q^n.$$

Note that for $\gamma^\sharp \in \prod_{i=1}^t \mathfrak{S}_{m_i-m_{i-1}} \subset \mathfrak{S}_{e-1}$ and $\gamma^\flat \in \prod_{i=1}^s \mathfrak{S}_{m'_i-m'_{i-1}} \subset \mathfrak{S}_{l-e}$, we have

$$G_{\gamma,\gamma'} = G_{\gamma\gamma^\sharp,\gamma'\gamma^\flat}.$$

Since we have

$$\prod_{i=1}^t |\mathfrak{S}_{m_i-m_{i-1}}| = \prod_{i=1}^t (m_i - m_{i-1})!, \quad \prod_{i=1}^s |\mathfrak{S}_{m'_i-m'_{i-1}}| = \prod_{i=1}^s (m'_i - m'_{i-1})!,$$

(83) is written as

$$(84) \qquad (83) = \sum_{\substack{0\le t\le e-1,\ 0\le s\le l-e, \\ 0=m_0<m_1<\cdots<m_t=e-1, \\ e=m'_0<m'_1<\cdots<m'_s=l}} \sum_{\substack{\gamma\in\mathfrak{S}_{e-1}, \gamma(i)<\gamma(i') \text{ if} \\ i,i'\in[m_j+1,m_{j+1}] \text{ for some } j \\ \gamma'\in\mathfrak{S}_{l-e}, \gamma'(i)<\gamma'(i') \text{ if} \\ i,i'\in[m'_j+1,m_{j'+1}] \text{ for some } j}} G_{\gamma,\gamma'}.$$

On the other hand, if we are given $n_i \in \mathbb{Z}_{>0}$ for $i \neq e$ and $n_e \in \mathbb{Z}$ with $n_1 + \cdots + n_l = n$, there are unique $\gamma \in \mathfrak{S}_{e-1}$, $\gamma' \in \mathfrak{S}_{l-e}$ and partitions $0 = m_0 < m_1 < \cdots < m_t = e-1$, $e = m'_0 < m'_1 < \cdots < m'_s = l$ such that $\gamma(i) < \gamma(i')$ for $i, i' \in [m_j + 1, m_{j+1}]$, $\gamma'(i) < \gamma(i')$ for $i, i' \in [m'_j + 1, m'_{j+1}]$, and $\mu_i = n_i/\omega\beta_i$ satisfy

$$0 < \mu_{\gamma(1)} = \cdots = \mu_{\gamma(m_1)} < \mu_{\gamma(m_1+1)} = \cdots,$$
$$0 < \mu_{\gamma'(l)} = \cdots = \mu_{\gamma'(m'_{s-1}+1)} < \mu_{\gamma'(m'_{s-1})} = \cdots.$$

Therefore (84) is written as

$$(84) = \sum_{n\in\mathbb{Z}} \sum_{\substack{n_1+\cdots+n_l=n, \\ n_i\in\mathbb{Z}_{>0} \text{ for } i\neq e}} \prod_{i\neq e} n_i N^{eu}_{n_i,\beta_i} P^{eu}_{n_e,\beta_e} q^n$$

$$(85) \qquad = \prod_{i\neq e} N^{eu}_{\beta_i}(q) \cdot P^{eu}_{\beta_e}(q).$$

Noting

$$\sum_{1\le e\le l} \frac{1}{(e-1)!(l-e)!2^{l-1}} = \frac{1}{(l-1)!},$$

we obtain

$$(82) = \sum_{l \geq 1} \sum_{\substack{\beta_i \in C(X) \text{ for } i \neq l, \ \beta_l \in \overline{C}(X) \\ \beta_1 + \cdots + \beta_l = \beta}} \frac{(-1)^{l-1}}{(l-1)!} \prod_{i \neq l} N^{eu}_{\beta_i}(q) \cdot P^{eu}_{\beta_l}(q). \tag{86}$$

The formula (86) implies (56) as desired. Q.E.D.

4.6. Problem of incorporating virtual classes to Joyce's work

Since invariants defined in Definition 4.1 are interpreted as Euler characteristics of moduli stacks, they are unlikely to be unchanged under deformations of X. In order to construct invariants which are unchanged under deformations, we need to construct virtual moduli cycles on the moduli spaces and integrate them. The resulting invariants are Euler characteristics of the moduli spaces (up to sign) if the moduli spaces are non-singular, but in general they differ from Euler characteristics. Thus in order to solve Conjecture 1.1, we have to construct invariants involving virtual classes and establish the formulas like (25). At this moment we are unable to overcome this problem. However if we could involve virtual classes with Joyce's theory, then Conjecture 1.1 for PT-theory follows along with the same argument in this paper. To state this, let us recall that the integrations of virtual classes are also realized as weighted sums of certain constructible functions introduced by Behrend [2]. He shows that, for any scheme M, there is a canonical constructible function $\chi_M \colon M \to \mathbb{Z}$ such that $\chi_M = (-1)^{\dim M}$ if M is non-singular, and if M carries a symmetric perfect obstruction theory, we have

$$\sharp^{\mathrm{vir}} M = \sum_{n \in \mathbb{Z}} ne(\chi_M^{-1}(n)).$$

Under the situation in this section, we shall address the following question. [2]

Problem 4.18. *Does there exist a map*

$$\Theta' \colon \mathfrak{G}(\mathcal{A}^p) \longrightarrow \mathfrak{g}(\mathcal{A}^p),$$

such that the following conditions hold?

- *For $v \in C(\mathcal{A}^p)$, suppose that $\mathfrak{M}^v(Z_{\mu_\sigma})$ is written as $[M/\mathbb{G}_m]$ for a scheme M. Then*

$$\Theta'(\epsilon^v(Z_{\mu_\sigma})) = \sum_{n \in \mathbb{Z}} -n\Theta([[\chi_M^{-1}(n)/\mathbb{G}_m] \hookrightarrow \mathfrak{Obj}(\mathcal{A}^p)]).$$

[2]The formulation of Problem 4.18 is taught to the author by D. Joyce.

- *For $v_1, v_2 \in C(\mathcal{A}^p)$, we have*

$$[\Theta'(\epsilon^{v_1}(Z)), \Theta'(\epsilon^{v_2}(Z))] = (-1)^{\chi(v_1,v_2)}\Theta'[\epsilon^{v_1}(Z), \epsilon^{v_2}(Z)]. \tag{87}$$

There should be sign change in (87), because $\chi_M = (-1)^{\dim M}$ on a smooth variety M. We are unable to solve Problem 4.18 at this moment, but the techniques given in this paper yield the following.

Theorem 4.19. *Suppose that Problem 4.18 is true. Then Conjecture 1.1 is true for PT-theory.*

Proof. It is enough to work over the invariants, defined by Θ'. As a modification of Definition 4.1, let us define $L_{n,\beta}(\sigma)$, $N_{n,\beta}(\sigma)$ to be

$$L_{n,\beta}(\sigma) := \Theta'\Xi\epsilon^v(Z_{\mu_\sigma}), \quad \text{where } \mathrm{ch}(v) = (-1, 0, \beta, n),$$
$$N_{n,\beta}(\sigma) := \Theta'\epsilon^v(Z_{\mu_\sigma}), \quad \text{where } \mathrm{ch}(v) = (0, 0, \beta, n).$$

Then (87) yields a similar wall-crossing formula for $L_{n,\beta}(\sigma)$, and

$$L_{n,\beta}(\sigma) = (-1)^{\dim \mathrm{Pic}(X) - 1} P_{n,\beta},$$

for $\sigma = k\omega + i\omega$ with $k < -\mu_{n,\beta}/2$. Therefore the same proof as in Theorem 4.7 works, and we have the similar expansion of the generating series Z_{PT} as in (56). Then Conjecture 1.1 for $P_\beta(q)$ follows as a corollary. Q.E.D.

At this moment, the author does not know how to solve Problem 4.18. It seems that the recent work of Kontsevich and Soibelman [16] gives a crucial idea of this problem.

References

[1] A. Bayer, Polynomial Bridgeland stability conditions and the large volume limit, preprint, arXiv:math/0712.1083[math.AG].

[2] K. Behrend, Donaldson–Thomas invariants via microlocal geometry, Ann. of Math. (2), to appear; arXiv:math/0507523[math.AG].

[3] R. Bezrukavnikov, Perverse coherent sheaves (after Deligne), preprint, arXiv:math/0005152[math.AG].

[4] T. Bridgeland, Stability conditions on triangulated categories, Ann. of Math. (2), **166** (2007), 317–345.

[5] T. Bridgeland and V. Toledano Laredo, Stability conditions and Stokes factors, preprint, arXiv:math/0801.3974[math.AG].

[6] M. Douglas, D-branes, categories and $N = 1$ supersymmetry, J. Math. Phys., **42** (2001), 2818–2843.

[7] M. Douglas, Dirichlet branes, homological mirror symmetry, and stability, In: Proceedings of the 1998 ICM, 2002, pp. 395–408, arXiv:math/0207021[math.AG].

[8] M. Inaba, Toward a definition of moduli of complexes of coherent sheaves on a projective scheme, J. Math. Kyoto Univ., **42** (2002), 317–329.

[9] D. Joyce, Configurations in abelian categories I. Basic properties and moduli stack, Adv. Math., **203** (2006), 194–255.

[10] D. Joyce, Configurations in abelian categories II. Ringel–Hall algebras, Adv. Math., **210** (2007), 635–706.

[11] D. Joyce, Configurations in abelian categories III. Stability conditions and identities, Adv. Math., **215** (2007), 153–219.

[12] D. Joyce, Holomorphic generating functions for invariants counting coherent sheaves on Calabi–Yau 3-folds, Geom. Topol., **11** (2007), 667–725.

[13] D. Joyce, Motivic invariants of Artin stacks and 'stack functions', Q. J. Math., **58** (2007), 345–392.

[14] D. Joyce, Configurations in abelian categories IV. Invariants and changing stability conditions, Adv. Math., **217** (2008), 125–204.

[15] M. Kashiwara, t-structures on the derived categories of holonomic $\mathcal{D}$-modules and coherent $\mathcal{O}$-modules, Mosc. Math. J., **981** (2004), 847–868.

[16] M. Kontsevich and Y. Soibelman, Stability structures, motivic Donaldson–Thomas invariants and cluster transformations, in preparation.

[17] G. Laumon and L. Moret-Bailly, Champs Algébriques, Ergeb. Math. Grenzgeb. (3), **39**, Springer-Verlag, Berlin, 2000.

[18] M. Lieblich, Moduli of complexes on a proper morphism, J. Algebraic Geom., **15** (2006), 175–206.

[19] D. Maulik, N. Nekrasov, A. Okounkov and R. Pandharipande, Gromov–Witten theory and Donaldson–Thomas theory. I, Compositio. Math., **142** (2006), 1263–1285.

[20] R. Pandharipande and R. P. Thomas, Curve counting via stable pairs in the derived category, preprint, arXiv:math/0707.2348[math.AG].

[21] R. Pandharipande and R. P. Thomas, The 3-fold vertex via stable pairs, preprint, arXiv:math/0709.3823[math.AG].

[22] R. Pandharipande and R. P. Thomas, Stable pairs and BPS invariants, preprint, arXiv:math/0711.3899[math.AG].

[23] A. Rudakov, Stability for an abelian category, J. Algebra, **197** (1997), 231–235.

[24] Y. Toda, Limit stable objects on Calabi–Yau 3-folds, preprint, arXiv:math/0803.2356[math.AG].

[25] Y. Toda, Birational Calabi–Yau 3-folds and BPS state counting, Commun. Number Theory Phys., **2** (2008), 63–112.

Institute for the Physics and Mathematics of the Universe (IPMU)
University of Tokyo
Kashiwano-ha 5-1-5, Kashiwa City, Chiba 277-8582
Japan
E-mail address: `toda@ms.u-tokyo.ac.jp`

Advanced Studies in Pure Mathematics

A SERIES OF UP-TO-DATE GUIDES OF LASTING INTEREST TO ADVANCED MATHEMATICS

Volume 1 Algebraic Varieties and Analytic Varieties.
Edited by S. Iitaka. February, 1983

Volume 2 Galois Groups and their Representations.
Edited by Y. Ihara. December, 1983

Volume 3 Geometry of Geodesics and Related Topics.
Edited by K. Shiohama. June, 1984

Volume 4 Group Representations and Systems of Differential Equations.
Edited by K. Okamoto. March, 1985

Volume 5 Foliations.
Edited by I. Tamura. February, 1986

Volume 6 Algebraic Groups and Related Topics.
Edited by R. Hotta. March, 1985

Volume 7 Automorphic Forms and Number Theory.
Edited by I. Satake. February, 1986

Volume 8 Complex Analytic Singularities.
Edited by T. Suwa and P. Wagreich. February, 1987

Volume 9 Homotopy Theory and Related Topics.
Edited by H. Toda. February, 1987

Volume 10 Algebraic Geometry, Sendai, 1985.
Edited by T. Oda. July, 1987

Volume 11 Commutative Algebra and Combinatorics.
Edited by M. Nagata and H. Matsumura.
October, 1987

Volume 12 Galois Representations and Arithmetic Algebraic Geometry.
Edited by Y. Ihara. November, 1987

Volume 13 Investigations in Number Theory.
Edited by T. Kubota. March, 1988

Volume 14 Representations of Lie Groups, Kyoto, Hiroshima, 1986.
Edited by K. Okamoto and T. Oshima. February, 1989

Volume 15 Automorphic Forms and Geometry of Arithmetic Varieties.
Edited by K. Hashimoto and Y. Namikawa. July, 1989

Volume 16 Conformal Field Theory and Solvable Lattice Models.
Edited by M. Jimbo, T. Miwa and A. Tsuchiya.
June, 1988

Volume 17 Algebraic Number Theory—in honor of K. Iwasawa.
Edited by J. Coates, R. Greenberg, B. Mazur and
I. Satake. August, 1989

Volume 18 I—Recent Topics in Differential and Analytic Geometry.
Edited by T. Ochiai. November, 1990
II—Kähler Metric and Moduli Spaces.
Edited by Edited by T. Ochiai. November, 1990

Volume 19 Integrable Systems in Quantum Field Theory and
Statistical Mechanics.
Edited by M. Jimbo, T. Miwa and A. Tsuchiya.
November, 1989

Volume 20 Aspects of Low Dimensional Manifolds.
Edited by Y. Matsumoto and S. Morita.
December, 1992

Volume 21 Zeta Functions in Geometry.
Edited by N. Kurokawa and T. Sunada.
December, 1992

Volume 22 Progress in Differential Geometry.
Edited by K. Shiohama. September, 1993

Volume 23 Spectral and Scattering Theory and Applications.
Edited by K. Yajima. February, 1994

Volume 24 Progress in Algebraic Combinatorics.
Edited by E. Bannai and A. Munemasa. May, 1996

Volume 25 CR-Geometry and Overdetermined Systems.
Edited by T. Akahori, G. Komatsu, K. Miyajima,
M. Namba and K. Yamaguchi. July, 1997

Volume 26 Analysis on Homogeneous Spaces and Representation
Theory of Lie Groups, Okayama-Kyoto.
Edited by M. Kashiwara, T. Kobayashi, T. Matsuki,
K. Nishiyama and T. Oshima. April, 2000

Volume 27 Arrangements – Tokyo 1998.
Edited by M. Falk and H. Terao. August, 2000

Volume 28 Combinatorial Methods in Representation Theory.
Edited by M. Kashiwara, K. Koike, S. Okada, I. Terada,
and H.-F. Yamada. December, 2000

Volume 29	Singularities – Sapporo 1998. Edited by J.-P. Brasselet and T. Suwa. December, 2000
Volume 30	Class Field Theory – Its Centenary and Prospect. Edited by K. Miyake. April, 2001
Volume 31	Taniguchi Conference on Mathematics Nara '98. Edited by M. Maruyama and T. Sunada. May, 2001
Volume 32	Groups and Combinatorics — in memory of Michio Suzuki. Edited by E. Bannai, H. Suzuki, H. Yamaki and T. Yoshida. October, 2001
Volume 33	Computational Commutative Algebra and Combinatorics. Edited by T. Hibi. February, 2002
Volume 34	Minimal Surfaces, Geometric Analysis and Symplectic Geometry. Edited by K. Fukaya, S. Nishikawa and J. Spruck. June, 2002
Volume 35	Higher Dimensional Birational Geometry. Edited by S. Mori and Y. Miyaoka. July, 2002
Volume 36	Algebraic Geometry 2000, Azumino. Edited by S. Usui, M. Green, L. Illusie, K. Kato, E. Looijenga, S. Mukai and S. Saito. November, 2002
Volume 37	Lie Groups, Geometric Structures and Differential Equations — One Hundred Years after Sophus Lie —. Edited by T. Morimoto, H. Sato and K. Yamaguchi. December, 2002
Volume 38	Operator Algebras and Applications. Edited by H. Kosaki. January, 2004
Volume 39	Stochastic Analysis on Large Scale Interacting Systems. Edited by T. Funaki and H. Osada. March, 2004
Volume 40	Representation Theory of Algebraic Groups and Quantum Groups. Edited by T. Shoji, M. Kashiwara, N. Kawanaka, G. Lusztig and K. Shinoda. March, 2004
Volume 41	Stochastic Analysis and Related Topics in Kyoto *In honour of Kiyosi Itô.* Edited by H. Kunita, S. Watanabe and Y. Takahashi. March, 2004

Volume 42 Complex Analysis in Several Variables
– Memorial Conference of Kiyoshi Oka's Centennial Birthday, Kyoto/Nara 2001.
Edited by K. Miyajima, M. Furushima, H. Kazama, A. Kodama, J. Noguchi, T. Ohsawa, H. Tsuji and T. Ueda. June, 2004

Volume 43 Singularity Theory and Its Applications.
Edited by S. Izumiya, G. Ishikawa, H. Tokunaga, I. Shimada and T. Sano. January, 2007

Volume 44 Potential Theory in Matsue.
Edited by H. Aikawa, T. Kumagai, Y. Mizuta and N. Suzuki. August, 2006

Volume 45 Moduli Spaces and Arithmetic Geometry (Kyoto, 2004).
Edited by S. Mukai, Y. Miyaoka, S. Mori, A. Moriwaki and I. Nakamura. January, 2007

Volume 46 Singularities in Geometry and Topology 2004.
Edited by J.-P. Brassele and T. Suwa. February, 2007

Volume 47 Asymptotic Analysis and Singularities.
1—Hyperbolic and dispersive PDEs and fluid mechanics.
2—Elliptic and parabolic PDEs and related problems.
Edited by H. Kozono, T. Ogawa, K. Tanaka, Y. Tsutsumi and E. Yanagida. October, 2007

Volume 48 Finsler Geometry, Sapporo 2005
— In Memory of Makoto Matsumoto.
Edited by S. V. Sabau and H. Shimada.
November, 2007

Volume 49 Probability and Number Theory — Kanazawa 2005.
Edited by S. Akiyama, K. Matsumoto, L. Murata and H. Sugita. December, 2007

Volume 50 Algebraic Geometry in East Asia — Hanoi 2005.
Edited by K. Konno and V. Nguyen-Khac.
February, 2008

Volume 51 Surveys on Geometry and Integrable Systems.
Edited by M. Guest, R. Miyaoka and Y. Ohnita.
November, 2008

Volume 52 Groups of Diffeomorphisms
in honor of Shigeyuki Morita
on the occasion of his 60th birthday.
Edited by R. Penner, D. Kotschick, T. Tsuboi, N. Kawazumi, T. Kitano and Y. Mitsumatsu.
November, 2008

Volume 53 Advances in Discrete Dynamical Systems.
Edited by S. Elaydi, K. Nishimura, M. Shishikura and N. Tose. June, 2009

Volume 54 Algebraic Analysis and Around
in honor of Professor Masaki Kashiwara's 60th birthday.
Edited by T. Miwa, A. Matsuo, T. Nakashima and Y. Saito. February, 2009

Volume 55 Noncommutativity and Singularities
Proceedings of French–Japanese symposia held at IHÉS in 2006.
Edited by J.-P. Bourguignon, M. Kotani, Y. Maeda and N. Tose. July, 2009

Volume 56 Singularities — Niigata–Toyama 2007.
Edited by J.-P. Brasselet, S. Ishii, T. Suwa and M. Vaquie. November, 2009

Volume 57 Probabilistic Approach to Geometry.
Edited by M. Kotani, M. Hino and T. Kumagai.
March, 2010

Volume 58 Algebraic and Arithmetic Structures of Moduli Spaces (Sapporo 2007).
Edited by I. Nakamura and L. Weng. June, 2010

Volume 59 New Developments in Algebraic Geometry, Integrable Systems and Mirror Symmetry (RIMS, Kyoto, 2008).
Edited by M.-H. Saito, S. Hosono and K. Yoshioka.
August, 2010

To be continued